Abstract Algebra

David M. Burton
University of New Hampshire

wcb
WM. C. BROWN PUBLISHERS
Dubuque, Iowa

Library of Congress Catalog Card Number: 87-27096

ISBN 0-697-06761-0

Printed in the United States of America

10 9 8 7 6 5 4 3 2 1

Contents

Preface

A mathematician, like a painter or a poet, is a maker of patterns. If his patterns are more permanent than theirs, it is because they are made with ideas.

G. H. HARDY

The purpose of this book is to give a simple account of the basic ideas, methods, and results of abstract (which some prefer to call "modern") algebra. While chiefly intended for use as a textbook by undergraduate mathematics majors, it may be profitably read by those in related disciplines who seek an introduction to some of the fundamental algebraic systems. In a beginning text in mathematics like this one, terseness is not a primary virtue. With an eye to the student, I have aimed for clarity rather than conciseness and for a presentation that begins at a leisurely pace, yet is carefully developed in accordance with current standards of rigor. Proofs are nearly always developed in full detail, and concrete examples of the concepts discussed are given as much as possible. It is hoped that the reader will gain en route an appreciation of the power and elegance of the abstract approach in mathematics.

A steady diet of theorems prepared in a crust of dry technical matter soon becomes unpalatable. Therefore I have taken pains to weave into the presentation interesting historical remarks and biographical sketches of the great masters of the past. In addition to enlivening the theoretical side of the exposition, these help to bring out the point that abstract algebra in particular as well as mathematics generally is a living discipline, fed by the efforts of many practitioners. Algebra has developed bit by bit, with the work of each individual contributor built upon the research of many others; decades of endeavor have often been required before significant steps were made.

Although the text was written with the mathematics major in mind, very little is demanded in the way of prerequisites. A degree of mathematical maturity is certainly helpful, since there is no doubt that the chief difficulty for the reader is the abstractness of the subject matter. To smooth the way, I have devoted the first chapter to a fairly thorough review of the essential facts concerning sets, functions, and number theory. Chapter 1 also serves as a suitable vehicle for introducing some of the notation and terminology to be used later. Those who have prior experience in these matters may prefer to skip this opening chapter and refer to it only as the need arises.

A word about the problems is in order. Most sections close with a substantial number of problems ranging in difficulty from purely mechanical to challenging theoretical questions. These are an integral part of the book and require the reader's active participation, since nobody can learn mathematics without solving problems. The computational exercises develop basic techniques and test understanding of concepts, while the more theoretical problems give practice in constructing proofs. Besides conveying additional information about the material covered earlier, the problems introduce a variety of ideas that are not treated in the body of the text. I have resisted the temptation to use the problems to introduce results that will be needed thereafter. As a consequence, the reader need not work all the exercises to digest the rest of the book. Problems whose solutions do not seem straightforward are frequently accompanied by hints.

The text treatment includes more material than can be covered satisfactorily during a one-semester course. This should provide the flexibility that is desirable for a diverse audience; it permits instructors to choose topics in accordance with personal tastes, and it presents students with the opportunity for further reading in the subject. Experience indicates that a standard course can be built up from Chapters 1, 2, and 4, along with the first three sections of Chapter 5; if an instructor is pressed for time, it would be feasible to omit Sections 4.6 and 4.7 without destroying the continuity. Chapter 3 and Sections 5.4 and 5.5, in which the results on groups and rings are really put to work, might be described as specialized topics. They are included for those whose curricula afford the luxury of a year-long course in algebra.

Abstract algebra is such a vast, rich area of mathematics that an introductory text cannot do justice to its entirety. My view is that a deeper coverage of a lesser number of topics is preferable to a superficial discussion of many. In selecting textual material I was governed to a certain extent by personal interest, condensing or omitting altogether some aspects of the theory that other authors might have pursued more vigorously. Despite these omissions the coverage should meet the needs of most readers. Those who are stimulated to broaden their algebraic horizons will find themselves prepared to specialize in various directions.

I would like to thank the reviewers of this text: Neal C. Raber, University of Akron; Robert A. Nowlan, Southern Connecticut University; and Douglas Cenzer, University of Florida. A special word of thanks goes to my wife, Martha, whose generous help with the book at all stages of development was indispensable.

1

Preliminary Notions

Mathematics in general is fundamentally the science of self-evident truths.

FELIX KLEIN

1.1 THE ALGEBRA OF SETS

This chapter summarizes some of the basic ideas concerning sets, functions, and number theory. It also serves as a vehicle for establishing certain conventions in notation and terminology that are used throughout the text. Inasmuch as this material is intended primarily for background purposes, the reader who is already acquainted with the ideas in this chapter may prefer to embark directly on the next.

The theory of sets as a mathematical discipline was developed by Georg Cantor (1845–1918) in a series of remarkable papers published between 1873 and 1897. Cantor introduced the notion of set to distinguish between different kinds of infinite numbers and to extend the operations of ordinary arithmetic to them. The first sentence of his major work *Contributions to the Founding of the Theory of Transfinite Numbers* (1895) contains his classic definition: "By an aggregate [set] we are to understand any collection into a whole M of definite and distinct objects m of our intuition or thought." That is, a collection of objects, even an infinite collection, is conceived of as a single entity. Cantor's insistence on treating infinite sets as mathematically permissible objects on a par with finite sets and counting numbers departed radically from standard interpretations and brought forth a storm of protest from his contemporaries. One of Cantor's former professors at Berlin, Leopold Kronecker, considered Cantor a scientific charlatan, a renegade, and a "corrupter of youth." (As a result of continuing animosity such as this, Cantor suffered periods of mental breakdown and spent the last five years of his life in a sanatorium.) Fortunately, Cantor's bold step was welcomed by others; David Hilbert, the leading mathematician of his day, declared stoutly, "No one shall expel us from the paradise which Cantor created for us."

About the turn of the century, just when Cantor's ideas were beginning to gain acceptance, a number of entirely unexpected logical contradictions, termed "paradoxes," were discovered on the fringes of set theory. In his desire to produce a concept as general as possible, Cantor had allowed any collection to be a set. He intuitively assumed that if one could describe a sensible property of objects, then one could also speak of the set of all objects that possess the property. This naive and unsystematic approach led to upsetting consequences. Certain "large" collections, such as the "set of all sets," had to be discarded to circumvent the troublesome paradoxes. The remedy was to construct a sufficiently restrictive system of postulates for deciding which objects were to be regarded as sets and what properties sets should have. The leader of the movement of rehabilitation was Ernst Zermelo (1871–1953), who published the first axiomatic treatment of set theory in 1908. Zermelo's system of axioms later (1922) required strengthening by Abraham Fraenkel, giving rise to a list of codifying assumptions, widely used up to the present day, that are known as the Zermelo-Fraenkel axioms.

Rather than attempting to list the undefined notions of set theory and the various axioms relating them, we shall take an informal approach to the subject. To this end, the term *set* will be understood to mean any identifiable collection of objects of any sort. The objects that make up a particular set are called its *elements* or *members*. (Almost every attempt to give an intuitive explanation of the word "set" results in replacing it by a synonym such as "collection," "aggregate," or "class"; the least painful attitude is to accept the terms "set" and "element" as primitive, undefined concepts.)

Sets will generally be designated by capital italic letters and their elements by lowercase letters. A few sets occur so frequently that it is a good idea to establish a fixed notation for them:

Z	the set of integers
Q	the set of rational numbers
$R^{\#}$	the set of real numbers

The symbols Z_+, Q_+, and $R^{\#}_+$ will stand for the positive elements of these sets.

If x is an element of the set A, it is customary to employ the notation $x \in A$ and to read the symbol $\in$ as "belongs to." On the other hand, when x fails to be an element of the set A, we shall indicate this situation by writing $x \notin A$. The use of the symbol $\in$ (a stylized form of the Greek epsilon) to denote the membership relation was initiated by the Italian logician Giuseppe Peano in 1889; it abbreviates the Greek word $\varepsilon\sigma\tau\iota$, which means "is."

There are two common methods for specifying a particular set. First, we may list all its elements within braces, as with the set $\{-1, 0, 1, 2\}$, or we may merely list some of its elements and use three dots to indicate the fact that certain obvious elements have been omitted, as with the set $\{1, 2, 3, 4, \ldots\}$. When such a listing is not practical, we may indicate instead a characteristic property whereby we can determine whether or not a given object is an element of the set. More specifically, if $P(x)$ is a statement depending on the variable x, then the set of all elements x for which the statement $P(x)$ is true is denoted by $\{x \mid P(x)\}$. For example, we might have $\{x \mid x \text{ is an odd integer greater than } 21\}$. Certain sets may be described both ways:

$$\{0, 1\}$$

$$\{x \mid x \in Z \text{ and } -1 < x < 2\}$$

$$\{x \mid x \in Z \text{ and } x^2 = x\}$$

all define the same set of integers. In what follows, we shall depart slightly from the basic notation and write $\{x \in A | P(x)\}$ in place of the awkward $\{x | x \in A \text{ and } P(x)\}$.

DEFINITION 1.1 Two sets A and B are *equal*, symbolized by $A = B$, if every element of A is an element of B and every element of B is an element of A; in short, $A = B$ if they are made up of exactly the same elements. The negation of the statement $A = B$ is written $A \neq B$.

Thus a set is completely determined once its elements are known. For instance,

$$\{1, 2, 3\} = \{3, 1, 2, 2\},$$

since each set consists of only the integers 1, 2, and 3. The order in which the elements are listed in a set is immaterial, and repetition conveys no additional information about the set.

It is certainly possible to state properties that cannot be satisfied by any object. If we agree that each property will describe a set, then we are led to admit the existence of a set having no elements at all. (In a strict sense, it must be agreed that to each set A and each condition $P(x)$ concerning x there is a set whose elements are exactly those elements x of A for which $P(x)$ holds.) This set, known as the *empty* or *null set*, is represented by the symbol $\varnothing$. Typical examples are

$$\varnothing = \{x \in R^{\#} | x^2 < 0\} \qquad \text{and} \qquad \varnothing = \{x \in Z | x \neq x\}.$$

Under our definition of equality of sets the empty set is uniquely determined; it is logically sound to conclude that any two empty sets possess the same elements—there are none!. In essence, there is just one empty set, so we are free to speak of *the* empty set $\varnothing$.

A set with just one element, say x, is referred to as a *singleton* or *the singleton of* x and is denoted by $\{x\}$:

$$\{x\} = \{y | y = x\}.$$

We observe that $x \in \{x\}$; in particular, $\{\varnothing\} \neq \varnothing$, since $\varnothing \in \{\varnothing\}$. A clear distinction is made between the element x and the set $\{x\}$ whose only member is this element.

A basic relationship between sets is that of inclusion, defined next.

DEFINITION 1.2 We say that the set A is a *subset* of the set B (A is *contained* in B, A is *included* in B), and we write $A \subseteq B$ provided that every element of A is also an element of B.

Thus $A \subseteq B$ is a convenient abbreviation for the statement "if $x \in A$, then $x \in B$." Definition 1.2 does not exclude the possibility that A and B might be equal. If $A \subseteq B$, but $A \neq B$, then A is termed a *proper subset* of B; notationally, this fact is recognized by writing $A \subset B$.

The reader should take care not to confuse the symbols $\in$ and $\subseteq$. We place the symbol $\in$ between an element and a set to which it belongs, while $\subseteq$ indicates a certain relation between two sets. One more point to notice: $\in$ represents an undefined term of our theory, but $\subseteq$ is defined in terms of $\in$.

It is helpful to regard all sets under consideration as being subsets of some constant set U, called the *universe* (*universal set, universe of discourse*). The particular choice of universe may be different in different contexts but is always held fixed throughout any given discussion. For instance, in elementary number theory the universe is generally taken to be the set Z, whereas in analysis one might have $R^{\#}$ for the universe.

With these remarks out of the way, we turn to some of the consequences of the definition of set inclusion.

THEOREM 1.1 *If A, B, and C are subsets of the universe U, then*

a) $A \subseteq A$, $\varnothing \subseteq A$, $A \subseteq U$;

b) $A \subseteq \varnothing$ *if and only if* $A = \varnothing$;

c) $\{x\} \subseteq A$ *if and only if* $x \in A$;

d) *if* $A \subseteq B$ *and* $B \subseteq C$, *then* $A \subseteq C$;

e) $A = B$ *if and only if* $A \subseteq B$ *and* $B \subseteq A$.

Observe that the assertion $\varnothing \subseteq A$ follows from the logical principle that a false hypothesis implies any conclusion whatsoever; thus "if $x \in \varnothing$, then $x \in A$" is a true statement, since $x \in \varnothing$ is always false.

The last part of Theorem 1.1 is the basis of one of the most frequently employed methods for establishing the equality of two sets, the so-called *double-inclusion argument*. An illustration of this method will be presented shortly.

We now consider several important ways in which sets may be combined with one another. If A and B are subsets of some universe U, the operations of union, intersection, and difference are defined as follows:

DEFINITION 1.3 The *union* of A and B, denoted by $A \cup B$, is the subset of U defined by

$$A \cup B = \{x | x \in A \text{ or } x \in B\}.$$

The *intersection* of A and B, denoted by $A \cap B$, is the subset of U defined by

$$A \cap B = \{x | x \in A \text{ and } x \in B\}.$$

The *difference* of A and B (or the *relative complement* of B in A), denoted by $A - B$, is the subset of U defined by

$$A - B = \{x | x \in A \text{ but } x \notin B\}.$$

In the definition of union the word "or" is employed in the "and/or" sense. Thus the statement "$x \in A$ or $x \in B$" includes the case in which x lies in both A and B.

The particular difference $U - A$ is called the (absolute) *complement* of A and is designated simply by $-A$:

$$-A = \{x | x \notin A\}.$$

This abbreviation is convenient when a set U is given and the subsequent reasoning involves only subsets of U. It might also be noted parenthetically that, utilizing the notion of difference, we could define A to be a proper subset of B provided that $A \subseteq B$ and $B - A \neq \varnothing$.

A necessary and sufficient condition for two sets A and B to have elements in common is that $A \cap B \neq \varnothing$; if it happens that $A \cap B = \varnothing$, then the sets are said to be *disjoint*.

After this string of definitions, let us consider an actual example.

Example 1.1

Take $U = \{0, 1, 2, 3, 4, 5, 6\}$ for the universe and let

$$A = \{1, 2, 4\}, \qquad B = \{2, 3, 5\}.$$

Then $A \cup B = \{1, 2, 3, 4, 5\}$ and $A \cap B = \{2\}$, while the relative complements are

$$A - B = \{1, 4\}, \qquad B - A = \{3, 5\}.$$

Also

$$-A = \{0, 3, 5, 6\}, \qquad -B = \{0, 1, 4, 6\}.$$

Observe that $A - B$ and $B - A$ are disjoint and unequal; in fact, $A - B = B - A$ if and only if $A = B$.

If A, B, and C are sets for which $C \subseteq A$ and $C \subseteq B$, then it is clear that $C \subseteq A \cap B$. Thus with due allowance for the hazards of generalization, one can think of $A \cap B$ as the largest set that is a subset of both A and B. Analogously, $A \cup B$ can be interpreted as the smallest set that contains both A and B.

In the following theorem, some simple consequences of the definitions of union, intersection, and complementation are listed.

THEOREM 1.2 *If A, B, and C are subsets of some universe U, then*

a) $A \cup A = A$, $A \cap A = A$;

b) $A \cup B = B \cup A$, $A \cap B = B \cap A$;

c) $A \cup (B \cup C) = (A \cup B) \cup C$, $A \cap (B \cap C) = (A \cap B) \cap C$;
d) $A \cup (B \cap C) = (A \cup B) \cap (A \cup C)$, $A \cap (B \cup C) = (A \cap B) \cup (A \cap C)$;
e) $A \cup \varnothing = A$, $A \cap \varnothing = \varnothing$;
f) $A \cup U = U$, $A \cap U = A$.

We shall verify the first equality of part (d), since its proof illustrates a technique mentioned previously. Suppose that $x \in A \cup (B \cap C)$. Then either $x \in A$ or $x \in B \cap C$. Now, if $x \in A$, then clearly both $x \in A \cup B$ and $x \in A \cup C$, so $x \in (A \cup B) \cap (A \cup C)$. On the other hand, if $x \in B \cap C$, then $x \in B$ and therefore $x \in A \cup B$; also $x \in C$ and therefore $x \in A \cup C$. The two conditions together imply that

$$x \in (A \cup B) \cap (A \cup C).$$

This establishes the inclusion

$$A \cup (B \cap C) \subseteq (A \cup B) \cap (A \cup C).$$

Conversely, suppose that $x \in (A \cup B) \cap (A \cup C)$. Then both $x \in A \cup B$ and $x \in A \cup C$. Since $x \in A \cup B$, either $x \in A$ or $x \in B$; at the same time, since $x \in A \cup C$, either $x \in A$ or $x \in C$. Together, these conditions mean that $x \in A$ or $x \in B \cap C$; that is, $x \in A \cup (B \cap C)$. This proves the opposite inclusion

$$(A \cup B) \cap (A \cup C) \subseteq A \cup (B \cap C).$$

By part (e) of Theorem 1.1 the two inclusions are sufficient to establish the equality

$$A \cup (B \cap C) = (A \cup B) \cap (A \cup C).$$

We should caution the reader that the "cancellation laws" do not hold for union and intersection. That is, neither one of the equations

$$A \cup B = A \cup C \qquad \text{or} \qquad A \cap B = A \cap C$$

by itself implies, in general, that $B = C$. To see this, simply take $A = U$ in the first equation and $A = \varnothing$ in the second.

The following properties of complementation are immediate:

LEMMA *If A and B are subsets of the universe U, then*

a) $-(-A) = A$;
b) $-\varnothing = U$, $-U = \varnothing$;
c) $A \cup (-A) = U$, $A \cap (-A) = \varnothing$;
d) $A \subseteq B$ *if and only if* $-B \subseteq -A$.

There are various identities that relate the operation of complementation to those of union and intersection. The most important of these are known as De Morgan's rules:

$$-(A \cup B) = (-A) \cap (-B), \qquad -(A \cap B) = (-A) \cup (-B).$$

The name is that of Augustus De Morgan (1806–1871), a distinguished mathematical logician and contributor to algebra. De Morgan was born in India, the son of a British Army colonel, but was educated at Trinity College, Cambridge. At the early age of 22 he was appointed professor of mathematics at the newly formed University of London and continued to teach there until 1867, except for the short period 1831–1835. Through his works and pupils (most notably J. J. Sylvester), De Morgan exercised considerable influence on the development of British mathematics in the nineteenth century. A voluminous writer, he contributed no fewer than 850 features on mathematics and its history to the *Penny Cyclopaedia* alone; in one of these, written in 1838, he invented the term "mathematical induction" to describe a process that had previously been used, but without much clarity, by mathematicians. De Morgan's rules as we now know them appeared in a paper on symbolic logic published in 1858.

Let us proceed to derive

THEOREM 1.3 (De Morgan's Rules) *If A and B are subsets of the universe U, then*

a) $-(A \cup B) = (-A) \cap (-B)$,
b) $-(A \cap B) = (-A) \cup (-B)$.

Proof. To establish the first of the above assertions, let x be an arbitrary element of $-(A \cup B)$. Then $x \notin A \cup B$. Hence, $x \notin A$ and $x \notin B$. This implies that $x \in -A$ and $x \in -B$, from which it follows that $x \in (-A) \cap (-B)$. Thus $-(A \cup B) \subseteq (-A) \cap (-B)$.

Conversely, if $x \in (-A) \cap (-B)$, then x belongs to both $-A$ and $-B$. In other words, $x \notin A$ and $x \notin B$. This guarantees that $x \notin A \cup B$; equivalently, $x \in -(A \cup B)$. We consequently have the inclusion $(-A) \cap (-B) \subseteq -(A \cup B)$, and the desired equality holds.

The proof of (b) is similar. However, it can just as well be deduced from the first part of the theorem and the following complementation:

$$-[(-A) \cup (-B)] = -(-A) \cap -(-B) = A \cap B.$$

Now apply part (a) of the last lemma again to obtain the result in question.

At times we wish to consider sets whose elements themselves are sets; to avoid the awkward repetition "set of sets," we refer to these as *families* of sets. One family that will prove to be of considerable importance is the so-called power set of a given set.

DEFINITION 1.4 If A is an arbitrary set, then the set whose elements are all the subsets of A is known as the *power set* of A and denoted by $P(A)$:

$$P(A) = \{B \mid B \subseteq A\}.$$

A few remarks are in order before considering a specific example. First, since $\varnothing \subseteq A$ and $A \subseteq A$, we always have $\{\varnothing, A\} \subseteq P(A)$ no matter what the nature of the set A. If $A = \varnothing$, then of course $P(A) = \{\varnothing\}$. The next thing to observe is that if $x \in A$, then $\{x\} \subseteq A$; hence $\{x\} \in P(A)$. From this we infer that the power set of A has, at the very least, as many elements as the set A. Indeed, it can be shown that whenever A is a finite set with n elements, then $P(A)$ is itself a finite set having 2^n elements. For this reason the power set of A is often represented by the symbol 2^A.

Example 1.2

Suppose the set $A = \{a, b, c\}$. The power set of A, which has as its elements all the subsets of $\{a, b, c\}$, is then

$$P(A) = \{\varnothing, \{a\}, \{b\}, \{c\}, \{a, b\}, \{a, c\}, \{b, c\}, A\}.$$

It is possible to extend our definitions of union and intersection from two sets to any finite number, or even an infinite number, of sets. If $A_1, A_2, \ldots, A_n$ are subsets of the universe U, we write

$$\bigcup_{i=1}^{n} A_i$$

to indicate their union, that is, to indicate the set of all elements that belong to at least one of the sets $A_1, A_2, \ldots, A_n$. Similarly, the intersection of the given sets $A_1, A_2, \ldots, A_n$ is denoted by the symbol

$$\bigcap_{i=1}^{n} A_i,$$

by which we mean the set of all elements that are common to $A_1, A_2, \ldots, A_n$.

More generally, the union and intersection of an arbitrary (non-empty) family $\mathscr{A}$ of subsets of the universe U are defined by

$$\cup\mathscr{A} = \{x \mid x \in A \text{ for some set } A \in \mathscr{A}\}$$

and

$$\cap\mathscr{A} = \{x \mid x \in A \text{ for every set } A \in \mathscr{A}\}.$$

At times we shall resort to an indexing set to define these notions. To be more precise, let I be a set, finite or infinite, and with each $i \in I$ associate a set A_i. The resulting family of sets,

$$\mathscr{A} = \{A_i \mid i \in I\},$$

is then said to be *indexed* by the elements of I, and the set I is called an *index set* for $\mathscr{A}$. When this notation is employed, it is customary to denote the union and intersection of the family $\mathscr{A}$ by

$$\cup\{A_i|i\in I\} \qquad \text{and} \qquad \cap\{A_i|i\in I\}.$$

If the nature of the index set I is clearly understood or if the emphasizing of it is inessential for some reason, we simply write

$$\cup A_i \qquad \text{and} \qquad \cap A_i.$$

Example 1.3

If $A_n = \{x\in R^{\#}| -1/n \leqslant x \leqslant 1/n\}$ for $n\in Z_+$, then

$$\cup\{A_n|n\in Z_+\} = \{x|x\in A_n \text{ for some } n\in Z_+\} = A_1,$$

$$\cap\{A_n|n\in Z_+\} = \{x|x\in A_n \text{ for every } n\in Z_+\} = \{0\}.$$

In passing, we should remark that by a *chain of sets* is meant a nonempty family $\mathscr{A}$ of subsets of some universe U such that if $A, B\in\mathscr{A}$, then either $A\subseteq B$ or $B\subseteq A$. For instance, the family in Example 1.3 constitutes a chain of sets in $R^{\#}$.

From our definition of set equality we have $\{a, b\} = \{b, a\}$, since both sets contain the same two elements a and b. The set concept carries no notion of order among its members, so no preference is given to one element over the other. When we wish to distinguish one of these elements as being the first, say a, we write (a, b) and call this an ordered pair.

The characteristic property of ordered pairs can be expressed unofficially by saying that (a, b) and (c, d) are equal pairs if and only if $a = c$ and $b = d$. One can give a purely set-theoretic definition of the notion of ordered pairs that fulfills this criterion:

DEFINITION 1.5 The *ordered pair* of elements a and b, with its *first component* a and *second component* b, denoted by (a, b) is the set

$$(a, b) = \{\{a, b\}, \{a\}\}.$$

Note that according to this definition, a and b are not elements of (a, b) but rather components. The actual elements of the set (a, b) are $\{a, b\}$, the unordered pair involved, and $\{a\}$, that member of the unordered pair that has been selected to be first. This agrees with our intuition that an ordered pair should be an entity representing two elements in a given order.

For $a \neq b$ the sets $\{\{a, b\}, \{a\}\}$ and $\{\{b, a\}, \{b\}\}$ are unequal, having different elements, so that $(a, b) \neq (b, a)$. Hence if a and b are distinct, there are two distinct ordered pairs whose components are a and b, namely, the pairs (a, b) and (b, a). Ordered pairs thus provide a way of handling two things as one while losing track of neither. A point to notice is that if $a = b$, then the ordered pair $(a, a) = \{\{a\}\}$ is not the same as the set $\{a\}$.

The fundamental theorem on the equality of ordered pairs is indicated next; the proof, which we omit, relies mainly on Definitions 1.1 and 1.5.

THEOREM 1.4 *Two ordered pairs (a, b) and (c, d) are equal if and only if $a = c$ and $b = d$.*

Having faced the problem of defining ordered pairs, we naturally consider ordered triples, ordered quadruples, and, for that matter, ordered n-tuples. What simplifies affairs is that these notions can be formulated in terms of ordered pairs. For instance, the *ordered triple* of a, b, and c is just an ordered pair whose first component is itself an ordered pair:

$$(a, b, c) = ((a, b), c).$$

Assuming that ordered $(n-1)$-tuples have been defined, we shall take the *ordered n-tuple* of $a_1, a_2, \ldots, a_n$ to mean $((a_1, a_2, \ldots, a_{n-1}), a_n)$, abbreviated by $(a_1, a_2, \ldots, a_n)$. It should come as no surprise that two ordered n-tuples are equal whenever their corresponding components are equal; in other words, $(a_1, a_2, \ldots, a_n) = (b_1, b_2, \ldots, b_n)$ if and only if $a_k = b_k$ for $k = 1, 2, \ldots, n$.

One of the most important constructions in set theory is that of the Cartesian product, defined below.

DEFINITION 1.6 The *Cartesian product* of two sets A and B, distinct or not, is the set

$$A \times B = \{(a, b) | a \in A, b \in B\}.$$

This definition implies that $A \times B = \varnothing$ if and only if $A = \varnothing$ or $B = \varnothing$. Observe also that if the set A contains n elements and B contains m elements, then $A \times B$ has nm elements, which accounts for the use of the word "product" in Cartesian product.

Example 1.4

Let $A = \{-1, 0, 1\}$ and $B = \{0, 2\}$. Then

$$A \times B = \{(-1, 0), (-1, 2), (0, 0), (0, 2), (1, 0), (1, 2)\},$$

while

$$B \times A = \{(0, -1), (0, 0), (0, 1), (2, -1), (2, 0), (2, 1)\}.$$

Clearly, the sets $A \times B$ and $B \times A$ are not identical. In fact, $A \times B = B \times A$ if and only if $A = B$, $A = \varnothing$, or $B = \varnothing$.

In general, if $A_1, A_2, \ldots, A_n$ are arbitrary sets, one may define their Cartesian product $A_1 \times A_2 \times \cdots \times A_n$, or more compactly $\times A_i$, as the set of all ordered n-tuples $(a_1, a_2, \ldots, a_n)$ where $a_1 \in A_1, \ldots, a_n \in A_n$. If all the sets coincide, say $A_1 = \cdots = A_n = A$, let us agree to write A^n for $\times A_i$ and call this the nth *power* of the set A.

PROBLEMS 1.1

In the following set of problems A, B, C, and D denote subsets of some universe U.

1. Given the inclusion $A \subseteq B$, show that

a) $A \cup C \subseteq B \cup C$,

b) $A \cap C \subseteq B \cap C$,

c) $A \cup (B - A) = B$,

d) $A = B - (B - A)$.

2. Prove that $A \cap B = \varnothing$ is equivalent to the two conditions $A - B = A$ and $B - A = B$.

3. To see that the notion of set inclusion may be expressed in terms of either union or intersection, verify that

a) $A \subseteq B$ if and only if $A \cup B = B$,

b) $A \subseteq B$ if and only if $A \cap B = A$.

4. For any two sets A and B, prove that $A = (A \cap B) \cup (A - B)$ is a representation of A as a disjoint union.

5. Prove that $A - B = A \cap (-B)$ and use this fact to establish each of the following:

a) $A - \varnothing = A$, $\varnothing - A = \varnothing$, $A - A = \varnothing$;

b) $A - B = A - (A \cap B) = (A \cup B) - B$;

c) $(A - B) \cap (B - A) = \varnothing$;

d) $A \cup B = (A \cap B) \cup (B - A) \cup (A - B)$;

e) $A \cup B = (A - B) \cup B$; $A \cap B = A - (A - B)$.

6. Derive the identity $A \cap (B \cup C) = (A \cap B) \cup (A \cap C)$.

7. Prove that $(A \cap B \cap C) \cup ((-A) \cup (-B) \cup (-C)) = U$.

8. Establish the following results on differences:

a) $(A - B) - C = A - (B \cup C)$,

b) $A - (B - C) = (A - B) \cup (A \cap C)$,

c) $A \cup (B - C) = (A \cup B) - (C - A)$,

d) $A \cap (B - C) = (A \cap B) - (A \cap C)$.

9. Verify the assertions below:

a) If $A \subseteq B$ and $A \subseteq -B$, then $A = \varnothing$.

b) If $A \subseteq B$ and $-A \subseteq B$, then $B = U$.

10. Obtain the so-called *absorption laws*:

$$A \cup (A \cap B) = A, \qquad A \cap (A \cup B) = A.$$

11. Assuming that A, B, and C are sets for which

$$A \cup B = A \cup C \quad \text{and} \quad A \cap B = A \cap C,$$

prove that $B = C$. [*Hint*: $B = B \cap (B \cup A)$.]

12. Show that $P(\varnothing) = \{\varnothing\}$, while $P(\{\varnothing\}) = \{\varnothing, \{\varnothing\}\}$.

13. For any set A, establish that

a) $\cap\{B | B \in P(A)\} = \varnothing$,

b) $\cup\{B | B \in P(A)\} = A$.

14. If $\{A_i\}$ is an indexed family of subsets of U, prove that

a) $\cup P(A_i) \subseteq P(\cup A_i)$,

b) $\cap P(A_i) = P(\cap A_i)$.

15. Given that $A_n = \{x \in R^{\#} | x \leqslant -n\}$ for $n = 1, 2, \ldots$, what is $\bigcup_{n \in Z_+} A_n$?

16. Let $\{A_n\}$ be a family of subsets indexed by the positive integers. Define a new family $\{B_n\}$ as follows:

$$B_1 = A_1; \qquad B_n = A_n - \bigcup_{k=1}^{n-1} A_k \qquad \text{for } n > 1.$$

Prove that the members of $\{B_n\}$ are disjoint sets and that

$$\bigcup_{n \in Z_+} A_n = \bigcup_{n \in Z_+} B_n.$$

17. Establish the following results on Cartesian products:

a) $A \times (B \cup C) = (A \times B) \cup (A \times C)$,

b) $A \times (B \cap C) = (A \times B) \cap (A \times C)$,

c) $A \times (B - C) = (A \times B) - (A \times C)$,

d) if $C \times D \neq \varnothing$, then $C \times D \subseteq A \times B$ if and only if $C \subseteq A$ and $D \subseteq B$.

18. Prove that if A and B are sets for which $A \times A = B \times B$, then $A = B$.

19. Suppose that A, B, C are sets such that $A \neq \varnothing$, $B \neq \varnothing$, and

$$C \times C = (A \times B) \cup (B \times A).$$

Verify that $A = B = C$.

20. If $\{a, b\} = \{a, c\}$, prove that $b = c$.

21. Using Theorem 1.4, show that the ordered triples (a, b, c) and (a', b', c') are equal if and only if $a = a'$, $b = b'$, and $c = c'$.

1.2 FUNCTIONS AND RELATIONS

The purpose of this section is to introduce two types of relations that are basic to all branches of mathematics—functions and equivalence relations. We shall avoid the traditional view of a function as a "rule of correspondence" and give instead a definition phrased in terms of ordered pairs. What this latter approach lacks in naturalness is more than compensated for by its clarity and precision.

Intuitively, a relation between two sets X and Y provides a criterion by which, for each ordered pair (x, y) of elements of $X \times Y$, one can determine whether the statement "x is related to y" is meaningful (in the sense of being true or false according to the choice of elements x and y). The relation is completely described once we know those pairs for which the first component stands in relation to the second. This vague idea can best be formulated in set-theoretic language as

DEFINITION 1.7 For sets X and Y, a (binary) *relation* R between X and Y is any subset of the Cartesian product $X \times Y$. In the case where $X = Y$ we speak of R as being a relation in X.

As an example of a relation in, say $R^{\#}$, we might cite the familiar less-than relation

$$R = \{(x, y) | x, y \in R^{\#}; x < y\}.$$

It should not go unnoticed that, since $X \times Y$ and $Y \times X$ are not generally the same set, a relation between X and Y may be quite different mathematical object from a relation between Y and X.

Let $R \subseteq X \times Y$ be a relation. The *domain* of R (in symbols, dom R) is the set of all first components occurring in the ordered pairs of R, while the *range* of R (in symbols, rng R) is the set of all second components of elements of R; formally,

$$\text{dom } R = \{x \in X | (x, y) \in R \text{ for some } y \in Y\},$$

$$\text{rng } R = \{y \in Y | (x, y) \in R \text{ for some } x \in X\}.$$

From these definitions we see that dom $R \subseteq X$, rng $R \subseteq Y$ and $R \subseteq \text{dom } R \times \text{rng } R$. It may happen that dom $R = X$, a situation that is described by saying that *R is a relation from X into Y.*

Certain kinds of relations play a prominent role in mathematics and perhaps the most important are the functions. These are relations f (a more familiar letter than R in this context) involving a uniqueness property, described below:

DEFINITION 1.8 A *function* or *mapping* from X into Y is a relation f from X into Y such that whenever $(x, y_1) \in f$ and $(x, y_2) \in f$, then

$y_1 = y_2$; in other words, no two distinct ordered pairs of f have the same first component.

In brief, a function f is a relation from X into Y in which each $x \in X$ is related to exactly one $y \in Y$. The element y is said to be the *functional value* or *image* of x under f and is denoted by $f(x)$; that is, the symbol $f(x)$ represents the unique second component of that ordered pair of f having x as its first component. With this notation the function f may be given by

$$f = \{(x, f(x)) | x \in X\}.$$

We shall on a few occasions omit the parentheses and write fx in place of $f(x)$.

Example 1.5
If the function f is the finite set of ordered pairs

$$f = \{(-1, 0), (0, 0), (1, 2), (2, 1)\},$$

then

$$\text{dom } f = \{-1, 0, 1, 2\}, \qquad \text{rng } f = \{0, 1, 2\},$$

and we write $f(-1) = 0$, $f(0) = 0$, $f(1) = 2$ and $f(2) = 1$.

Example 1.6
The set of ordered pairs

$$R = \{(x, y) | x, y \in R^{\#}; x^2 + y^2 = 25\}$$

constitutes a relation in $R^{\#}$, but not a function; both (3, 4) and (3, -4) are elements of this set.

It is often convenient to describe a function by giving a formula for its ordered pairs. By way of example we might have

$$f = \{(x, x^2 + 2) | x \in R^{\#}\}.$$

Using the functional value notation, one would then write $f(x) = x^2 + 2$ for each $x \in R^{\#}$. Needless to say, there are functions whose ordered pairs would be difficult—if not impossible—to express in terms of a formula. To make things more specific, there is no general formula whose value at the integer $n \in Z_+$ is the nth prime number. The discerning reader is advised to keep in mind the distinction between a function and its values or, as the case may be, its formula; although the notation sometimes leads to confusion, these concepts are obviously quite different.

To convey the information that f is a function from X into Y, we shall write $f: X \to Y$; in particular, this notation signifies that $\text{dom} f = X$. For most functions $f: X \to Y$ it is not the case that $\text{rng} f = Y$, which is suggested by saying that f *maps X into Y.* This leads us directly to the notion of an *onto* function: the function f *maps X onto Y* if for each $y \in Y$ there exists at least one $x \in X$ such that $f(x) = y$.

Before proceeding further we should reach an understanding on the equality of functions. Since functions are sets, Definition 1.1 furnishes a ready-made criterion for equality; namely, two functions f and g are identical if and only if they are composed of the same ordered pairs. Equivalently, $f = g$ if and only if $\operatorname{dom} f = \operatorname{dom} g$ and $f(x) = g(x)$ for each element x in their common domain.

Let us turn to the problem of combining functions in one way or another. Suppose that f and g are two functions whose ranges are subsets of a system in which addition and multiplication are permissible (for definiteness, one might think of functions with range in Z or $R^{\#}$). The following formulas define functions $f + g$ and $f \cdot g$ by specifying the values of these functions at each point of their domain:

$$\begin{aligned} (f + g)(x) &= f(x) + g(x) \\ (f \cdot g)(x) &= f(x)g(x) \end{aligned} \qquad \text{where } x \in \operatorname{dom} f \cap \operatorname{dom} g.$$

One calls $f + g$ and $f \cdot g$ the *pointwise sum* and *product* of f and g.

Example 1.7
Suppose that

$$f = \{(x, \sqrt{4 - x^2}) \mid -2 \leqslant x \leqslant 2\} \quad \text{and} \quad g = \{(x, 2/x) \mid x \in R^{\#} - \{0\}\},$$

so that

$$f(x) = \sqrt{4 - x^2} \qquad \text{for } -2 \leqslant x \leqslant 2,$$

$$g(x) = 2/x \qquad \text{for } x \in R^{\#} - \{0\}.$$

If $x \in \operatorname{dom} f \cap \operatorname{dom} g = \operatorname{dom} f - \{0\}$, then the functions $f + g$ and $f \cdot g$ are defined by

$$(f + g)(x) = \sqrt{4 - x^2} + 2/x,$$

$$(f \cdot g)(x) = (\sqrt{4 - x^2})2/x.$$

The pointwise operations just considered depend on the algebraic properties of the range; the domain merely furnishes the points for these operations. The most important way of producing new functions from given ones, functional composition, is independent of such structure and depends only on the underlying sets.

DEFINITION 1.9 The *composition* of two functions f and g, denoted by $f \circ g$, is the function

$$f \circ g = \{(x, y) \mid \text{for some } z, (x, z) \in g \text{ and } (z, y) \in f\}.$$

Written in terms of functional values, this gives the formula

$$(f \circ g)(x) = f(g(x)), \qquad \text{where } x \in \operatorname{dom} g \text{ and } g(x) \in \operatorname{dom} f.$$

The restrictions on the domains are understandable: Before there can be an $f(g(x))$ there must be a $g(x)$, and it must lie in the domain of f. The fact that $f \circ g$ is actually a function, and not just a relation, is a consequence of the fact that f and g are both functions. We omit the proof of this assertion, leaving its verification as an exercise. As a general comment, observe that $\operatorname{dom} f \circ g \subseteq \operatorname{dom} g$ and $\operatorname{rng} f \circ g \subseteq \operatorname{rng} f$.

The order in which the symbols appear in a composition is quite important. The notation $f \circ g$ is chosen so that g will stand directly beside x when we write $(f \circ g)(x)$; this is in keeping with the fact that g is applied to x, whereas f operates on the image $g(x)$.

It is possible that the composition of f and g will be empty (indeed, $f \circ g \neq \varnothing$ if and only if $\operatorname{rng} g \cap \operatorname{dom} f \neq \varnothing$). In practice, however, we seldom form the composition unless a substantial part of the range of g lies in the domain of f. One of the most important cases arises when $\operatorname{rng} g \subseteq \operatorname{dom} f$; in this setting,

$$f \circ g = \{(x, f(g(x))) | x \in \operatorname{dom} g\}.$$

Example 1.8

Let f, g be the functions

$$f = \{(x, \sqrt{x}) | x \in R^{\#};\ x \geqslant 0\} \qquad \text{and} \qquad g = \{(x, 2x + 3) | x \in R^{\#}\},$$

so that

$$f(x) = \sqrt{x} \qquad \text{for } x \geqslant 0,$$
$$g(x) = 2x + 3 \qquad \text{for } x \in R^{\#}.$$

Forming the composition $f \circ g$, we get

$$(f \circ g)(x) = f(g(x)) = f(2x + 3) = \sqrt{2x + 3},$$

where

$$\begin{aligned} \operatorname{dom} f \circ g &= \{x \in \operatorname{dom} g | g(x) \in \operatorname{dom} f\} \\ &= \{x \in R^{\#} | 2x + 3 \in \operatorname{dom} f\} = \{x \in R^{\#} | 2x + 3 \geqslant 0\}. \end{aligned}$$

On the other hand,

$$(g \circ f)(x) = g(f(x)) = g(\sqrt{x}) = 2\sqrt{x} + 3,$$

where

$$\begin{aligned} \operatorname{dom} g \circ f &= \{x \in \operatorname{dom} f | f(x) \in \operatorname{dom} g\} \\ &= \{x \geqslant 0 | \sqrt{x} \in R^{\#}\} = \{x \in R^{\#} | x \geqslant 0\}. \end{aligned}$$

We observe that $f \circ g$ is different from $g \circ f$; rarely does $f \circ g = g \circ f$, even when both compositions are defined.

The next theorem concerns some of the basic properties of the operation of functional composition. Its proof is an exercise in the use of the definitions of this section.

THEOREM 1.5 *If f, g, and h are functions for which the following operations are defined, then*

a) $(f \circ g) \circ h = f \circ (g \circ h)$,
b) $(f + g) \circ h = (f \circ h) + (g \circ h)$,
c) $(f \cdot g) \circ h = (f \circ h) \cdot (g \circ h)$.

Proof. We establish here only property (a). The other parts of the theorem are obtained in a similar fashion and so are left as an exercise. Observe that for $x \in \operatorname{dom}(f \cdot g) \circ h$, we have

$$\begin{aligned}[(f \cdot g) \circ h](x) &= (f \cdot g)(h(x)) = f(h(x))g(h(x)) \\ &= (f \circ h)(x) \cdot (g \circ h(x) = [(f \circ h) \cdot (g \circ h)](x),\end{aligned}$$

which, according to the definition of equality of functions, shows that

$$(f \cdot g) \circ h = (f \circ h) \cdot (g \circ h).$$

Once again, consider an arbitrary function $f\colon X \to Y$. While no element of X can be mapped under f onto more than one element of Y, it is clearly possible that several (perhaps even all) elements may map onto the same element of Y. To rule out this behavior, the notion of a one-to-one function is useful. The formal definition follows.

DEFINITION 1.10 A function f is called *one-to-one* if and only if x_1, $x_2 \in \operatorname{dom} f$, with $x_1 \neq x_2$, implies $f(x_1) \neq f(x_2)$; that is, distinct elements in the domain have distinct functional values.

When establishing one-to-oneness, it will often prove to be more convenient to use the contrapositive of Definition 1.10:

$$f(x_1) = f(x_2) \text{ implies that } x_1 = x_2.$$

In terms of ordered pairs, a function f is one-to-one if and only if no two distinct ordered pairs of f have the same second component. Thus the collection of ordered pairs obtained by interchanging the components of the pairs of f is itself a function. More specifically, the *inverse* of a one-to-one function f, symbolized by f^{-1}, is the set of ordered pairs,

$$f^{-1} = \{(y, x) | (x, y) \in f\}.$$

It should be underscored that the assumption that f is one-to-one is precisely the condition that will guarantee that this new set is a function. Notice, incidentally, that $\operatorname{dom} f^{-1} = \operatorname{rng} f$ and $\operatorname{rng} f^{-1} = \operatorname{dom} f$.

For the combined action of f and f^{-1} it is easy to verify that

$$(f^{-1} \circ f)(x) = x \qquad \text{for} \qquad x \in \operatorname{dom} f,$$

$$(f \circ f^{-1})(y) = y \qquad \text{for} \qquad y \in \operatorname{dom} f^{-1} = \operatorname{rng} f.$$

To state this result a little more concisely, let us introduce some special terminology.

DEFINITION 1.11 Given a nonempty set X, the function $i_X: X \to X$ defined by $i_X(x) = x$ for each $x \in X$ is called the *identity function* on X; that is to say, i_X merely sends each element of X onto itself.

Expressed in terms of the identity function, what was just seen is that for any function $f: X \to Y$ that is both one-to-one and onto Y,

$$f^{-1} \circ f = i_X \qquad \text{and} \qquad f \circ f^{-1} = i_Y.$$

In the event that f is a one-to-one mapping of the set X onto itself, we obtain

$$f^{-1} \circ f = i_X = f \circ f^{-1}.$$

It might also be mentioned at this point that the identity function i_X is itself a one-to-one mapping onto the set X such that $i_X^{-1} = i_X$.

Example 1.9
The function $f = \{(x, 3x - 2) | x \in R^{\#}\}$ is one-to-one, for $3x_1 - 2 = 3x_2 - 2$ implies that $x_1 = x_2$. Consequently, the inverse of f exists and is the set of ordered pairs $f^{-1} = \{(3x - 2, x) | x \in R^{\#}\}$. It is preferable, however, to have f^{-1} defined in terms of its domain and the value at each point of the domain. Observing that

$$\{(3x - 2, x) | x \in R^{\#}\} = \{(x, \tfrac{1}{3}(x + 2)) | x \in R^{\#}\},$$

we choose to write $f^{-1} = \{(x, \frac{1}{3}(x + 2)) | x \in R^{\#}\}$. In terms of functional values,

$$f^{-1}(x) = \tfrac{1}{3}(x + 2)$$

for each $x \in R^{\#}$.

An important situation arises when we wish to consider the behavior of a function on a subset of its domain. For example, it is frequently advantageous to limit the domain to where the function is one-to-one. Suppose, in general, that $f: X \to Y$ is an arbitrary function and the subset $A \subseteq X$. The composition $f \circ i_A: A \to Y$ is known as the *restriction* of f to the set A and is, by established custom, denoted by $f|A$. For the reader who prefers the ordered-pair approach,

$$f|A = \{(x, y)|(x, y) \in f \text{ and } x \in A\}.$$

In any event, if the element $x \in A$, then $(f|A)(x) = f(x)$ so that both f and $f|A$ agree on the set A. It is well worth noting that while there is only one restriction of the given function f to the subset A, f is not necessarily uniquely determined by $f|A$. The particular restriction $i_X|A = i_A$, when viewed as a function from A into X, is termed the *inclusion* or *injection* map from A to X.

Dually, suppose that we are given two functions $f: X \to Y$ and $g: W \to Y$, where $X \subseteq W$. If $f = g|X$, then g is said to be an *extension* of f. That is, g is an extension of f if and only if f is the restriction of g to some subset of the domain of g.

The next definition indicates that any function $f: X \to Y$ gives rise, in a rather natural way, to functions from $P(X)$ to $P(Y)$ and from $P(Y)$ to $P(X)$.

DEFINITION 1.12 Consider a function $f: X \to Y$. If $A \subseteq X$, then the *direct image of A under f*, denoted by $f(A)$, is the subset of Y defined by

$$f(A) = \{f(x)|x \in A\}.$$

On the other hand, if $B \subseteq Y$, then the *inverse image of B under f*, denoted by $f^{-1}(B)$, is the subset of X defined by

$$f^{-1}(B) = \{x|f(x) \in B\}.$$

Simply put, $f(A)$ is the set of all images under f of elements from A, while $f^{-1}(B)$ is the set of all elements of X whose images under f lie in B. Note in particular that $f(X) = \operatorname{rng} f$. These notions are meaningful when f is neither a one-to-one nor an onto function. If f is not one-to-one, then $f^{-1}(B)$ may consist of more than one element, even when B is a singleton; if f fails to map onto Y, then $f^{-1}(B)$ may be the empty set even though B itself is nonempty.

A commonly employed notation for the direct and inverse image sets is $f^{\rightarrow}(A)$ and $f^{\leftarrow}(B)$, but we shall avoid this symbolism.

It will be our convention to omit unnecessary parentheses whenever possible. In regard to singletons, for instance, we shall write direct and inverse images as $f(x)$ and $f^{-1}(x)$, rather than $f(\{x\})$ and $f^{-1}(\{x\})$. The reader who worries about notation may feel somewhat uneasy about this double use of the symbol f^{-1}. The abuse of notation should not cause any confusion, since in any given context it should be perfectly clear how f^{-1} is to be interpreted.

Certain properties of a function $f: X \to Y$ may be conveniently characterized in terms of inverse images. To be more explicit, f is a one-to-one function if and only if $f^{-1}(y)$ contains at most one element for each $y \in Y$, whereas f maps onto Y if and only if $f^{-1}(y)$ is nonempty for each $y \in Y$. Taking these together, we conclude that a necessary and sufficient condition for f to be both one-to-one and onto is that $f^{-1}(y)$ consist of exactly one element for each y in Y.

This just about completes our study of functions. Before leaving the subject we present two short theorems that will establish the relationship between direct and inverse images; these results are rather trivial but will be essential for our later work.

THEOREM 1.6 *If $f: X \to Y$, then for each subset $B \subseteq Y$, $f(f^{-1}(B)) \subseteq B$.*

Proof. If $b \in f(f^{-1}(B))$, then $b = f(a)$ for some element a in $f^{-1}(B)$. From this it follows that $f(a) \in B$, and consequently $b \in B$.

COROLLARY *If, in addition, f maps onto the set Y, then $f(f^{-1}(B)) = B$.*

Proof. In view of the inclusion proved in the theorem, we need only establish that, under the existing hypothesis, $B \subseteq f(f^{-1}(B))$. For this, let $b \in B$; then, as f is by supposition an onto function, $b = f(a)$ for some choice of a in X. Since $a \in f^{-1}(B)$,

$$f(a) \in f(f^{-1}(B)),$$

which says that

$$b \in f(f^{-1}(B)).$$

THEOREM 1.7 *If $f: X \to Y$, then for each subset $A \subseteq X$, $A \subseteq f^{-1}(f(A))$.*

Proof. The proof is almost obvious; if $a \in A$, then $f(a) \in f(A)$; hence

$$a \in f^{-1}(f(A)).$$

COROLLARY *If, in addition, f is a one-to-one function, then $A = f^{-1}(f(A))$.*

Proof. It plainly suffices to establish the one inclusion $f^{-1}(f(A)) \subseteq A$. To start with, let $a \in f^{-1}(f(A))$. We then have $f(a) \in f(A)$ and thus $f(a) = f(a')$ for some a' in A. Since f is assumed to be one-to-one, $a = a'$ and so $a \in A$.

Example 1.10

Consider the function $f: R^{\#} \to R^{\#}$ given by

$$f = \{(x, x^2 + 2) | x \in R^{\#}\};$$

that is, $f(x) = x^2 + 2$. If $[a, b]$ denotes the set $\{x \in R^{\#} | a \leqslant x \leqslant b\}$, then we have

$$f^{-1}(f([0, 1]) = f^{-1}([2, 3]) = [-1, 1],$$

$$f(f^{-1}([0, 3]) = f([-1, 1]) = [2, 3].$$

Our next concern is with equivalence relations. In practice, these arise whenever it is desirable to identify, as a single entity, all those elements of a set that have some preassigned characteristic.

DEFINITION 1.13 A relation R in a set X is an *equivalence relation* in X provided that it satisfies the following three properties:

1. $(a, a) \in R$ for all $a \in X$ (*reflexive property*);
2. if $(a, b) \in R$, then $(b, a) \in R$ (*symmetric property*);
3. if $(a, b) \in R$ and $(b, c) \in R$, then $(a, c) \in R$ (*transitive property*).

Equivalence relations are usually denoted by the symbol $\sim$ (pronounced "wiggle") rather than by R as heretofore. We shall also indicate the fact that $(a, b) \in \sim$ by writing $a \sim b$. With these changes in notation the above definition may be recast in a more familiar form:

DEFINITION 1.14 A relation $\sim$ in a set X is an equivalence relation in X if for all a, b, c in X,

1. $a \sim a$,
2. $a \sim b$ implies that $b \sim a$,
3. $a \sim b$ and $b \sim c$ together imply that $a \sim c$.

In the following set of miscellaneous examples we leave to the reader the task of verifying that each relation described actually is an equivalence relation.

Example 1.11
Let X be a nonempty set. For $a, b \in X$, define $a \sim b$ if and only if $a = b$, in other words, if a and b are the same element. Then $\sim$ satisfies the requirements of Definition 1.14, showing that an equivalence relation generalizes the concept of equality. It is evident that $\sim$ is just the subset $\{(a, a) | a \in X\}$ of $X \times X$.

Example 1.12
In the Cartesian product $R^{\#} \times R^{\#}$, let $(a, b) \sim (c, d)$ signify that $a - c$ and $b - d$ are both integers. A simple calculation reveals that $\sim$, so defined, is an equivalence relation in $R^{\#} \times R^{\#}$.

Example 1.13
An equivalence relation $\sim$ can be defined in the set Z by requiring that $a \sim b$ if and only if $a - b \in Z_e$, the set of even integers.

Example 1.14
As another illustration, consider the set L of all lines in a plane. Then the phrase "is parallel to" is meaningful when applied to the elements of L and may be used to define a relation in L. If we agree that any line is parallel to itself, this yields an equivalence relation.

Example 1.15
Let $f: X \to Y$ be a given mapping. Take for $\sim$ the relation $a \sim b$ if and only if $f(a) = f(b)$; then $\sim$ is an equivalence relation in X, called the *equivalence relation associated with the mapping f.*

One is often led to conclude, incorrectly, that the reflexive property is redundant in Definition 1.14. The argument proceeds like this: If $a \sim b$, then the symmetric property implies that $b \sim a$; from $a \sim b$ and $b \sim a$, together with the transitivity of $\sim$, it follows that $a \sim a$. Thus there appears to be no necessity for the reflexive condition at all. The flaw in this reasoning lies, of course, in the fact that, for some element $a \in X$, there may not exist any b in X such that $a \sim b$. As a result, we would not have $a \sim a$ for every member of X, as the reflexive property requires.

Any equivalence relation $\sim$ determines a separation of the set X into a collection of subsets of a kind which we now describe. For each $a \in X$, let $[a]$ denote the subset of X consisting of all elements that are equivalent to a:

$$[a] = \{b \in X | b \sim a\}.$$

This set $[a]$ is referred to as the *equivalence class determined by a.* The reader should realize that, in general, the equivalence class $[a]$ is the same as the class $[a']$ for many elements $a' \in X$. As a notational device, let us henceforth use the symbol $X/\sim$ to represent the set of all equivalence classes of the relation $\sim$; that is,

$$X/\sim \; = \{[a] | a \in X\}.$$

Some of the basic properties of equivalence classes are listed in the theorem below.

THEOREM 1.8 *Let $\sim$ be an equivalence relation in the set X. Then*

1. *for each $a \in X$, the class $[a] \neq \varnothing$;*
2. *if $b \in [a]$, then $[a] = [b]$; in other words, any element of an equivalence class determines that class;*
3. *for all a, $b \in X$, $[a] = [b]$ if and only if $a \sim b$;*
4. *for all a, $b \in X$, either $[a] \cap [b] = \varnothing$ or $[a] = [b]$;*
5. *$\cup X/\sim \; = X$.*

Proof. Clearly, the element $a \in [a]$, for $a \sim a$. To prove the second assertion, let $b \in [a]$, so that $b \sim a$. Now suppose that $x \in [a]$, which means $x \sim a$. Using the symmetric and transitive properties of $\sim$, we thus obtain $x \sim b$, whence $x \in [b]$. Since x is an arbitrary member of $[a]$, this establishes the inclusion $[a] \subseteq [b]$. A similar argument yields the reverse inclusion and equality follows. As regards (3), first assume that $[a] = [b]$; then $a \in [a] = [b]$ and so $a \sim b$. Conversely, if we let $a \sim b$, then the element $a \in [b]$; hence $[a] = [b]$ from (2).

To derive (4), suppose that $[a]$ and $[b]$ have an element in common, say, $c \in [a] \cap [b]$. Statement (2) then informs us that $[a] = [c] = [b]$. In brief, if $[a] \cap [b] \neq \emptyset$, then we must have $[a] = [b]$. Finally, since each class $[a] \subseteq X$, the inclusion $\cup\{[a] | a \in X\} \subseteq X$ certainly holds. For the opposite inclusion it is enough to show that each element $a \in X$ belongs to some equivalence class; but this is no problem, for $a \in [a]$.

As evidenced by Example 1.15, any mapping determines an equivalence relation in its domain. The following corollary indicates that every equivalence relation arises in this manner; that is to say, each equivalence relation is the associated equivalence relation of some function.

COROLLARY *Let $\sim$ be an equivalence relation in the set X. Then there exists a set Y and a mapping $g: X \to Y$ such that $a \sim b$ if and only if $g(a) = g(b)$.*

Proof. Simply take $Y = X/\sim$ and $g: X \to Y$ to be the mapping defined by $g(a) = [a]$; in other words, send each element of X onto the (necessarily unique) equivalence class to which it belongs. By the foregoing theorem, $a \sim b$ if and only if $[a] = [b]$, or equivalently, $g(a) = g(b)$.

We next connect the notion of an equivalence relation in X with that of a partition of X.

DEFINITION 1.15 By a *partition* of the set X is meant a family $\mathscr{P}$ of subsets of X with the properties

1. $\emptyset \notin \mathscr{P}$;
2. for any $A, B \in \mathscr{P}$, either $A = B$ or $A \cap B = \emptyset$ (pairwise disjoint);
3. $\cup \mathscr{P} = X$.

Expressed otherwise, a partition of X is a collection $\mathscr{P}$ of nonempty subsets of X such that every element of X belongs to one and only one member of $\mathscr{P}$. The set Z of integers, for instance, can be partitioned into subsets of odd and even integers; another partition of Z might consist of the sets of positive integers, negative integers, and $\{0\}$.

Theorem 1.8 may be viewed as asserting that each equivalence relation $\sim$ on a set X yields a partition of X, namely, the partition $X/\sim$ into the equivalence classes for $\sim$. In this connection, note that for the equivalence relation of equality the corresponding classes contain only one element each; hence the resulting partition is the finest possible. We now reverse the situation and show that a given partition of X induces an equivalence relation in X.

THEOREM 1.9 *If $\mathscr{P}$ is a partition of the set S, then there is a unique equivalence relation in S whose equivalence classes are precisely the members of $\mathscr{P}$.*

Proof. The fact that $\mathscr{P}$ partitions S guarantees that each element of S lies in exactly one member of $\mathscr{P}$. Given $a, b \in S$, we shall write $a \sim b$ if and only if a and b belong to the same subset in $\mathscr{P}$, that is, if there exists some $P \in \mathscr{P}$ with $a, b \in P$. The reader may easily check that $\sim$ is an equivalence relation in S.

Let us prove that the partition $\mathscr{P}$ has the form $S/\sim$. If the subset $P \in \mathscr{P}$, then $a \in P$ for some a in S. Now the element $b \in P$ if and only if $b \sim a$, or, what amounts to the same thing, if and only if $b \in [a]$. This demonstrates the equality $P = [a] \in S/\sim$. On the other hand, let $[a]$ be an arbitrary equivalence class and P be the partition set in $\mathscr{P}$ to which the element a belongs. By similar reasoning we conclude that $[a] = P$; hence $S/\sim \subseteq \mathscr{P}$. Thus the set of equivalence classes for $\sim$ coincides with the partition $\mathscr{P}$.

To summarize, there is a natural one-to-one correspondence between the equivalence relations in a set and the partitions of that set; every equivalence relation gives rise to a partition, and vice versa. We have a single idea, which has been considered from two different points of view.

Example 1.16
Let $X = R^{\#} \times R^{\#}$ and define the equivalence relation $\sim$ in X by taking $(a, b) \sim (c, d)$ if and only if $a - c = b - d$. The equivalence class determined by a typical element of X, say (a, b), is just

$$[(a, b)] = \{(x, y) | x - a = y - b\}.$$

This set may be represented geometrically as a straight line in the plane with slope 1 and passing through the point (a, b). Thus the relation $\sim$ can be thought of as partitioning X into a family of parallel lines.

PROBLEMS 1.2

1. For arbitrary sets X and Y, prove that
   a) R is a relation between X and Y if and only if $R \in P(X \times Y)$.
   b) $\varnothing$ is a relation between X and Y, with dom $\varnothing$ = rng $\varnothing = \varnothing$.
   c) $X \times Y$ is a relation between X and Y, with $\text{dom}(X \times Y) = X$ and $\text{rng}(X \times Y) = Y$.
2. Given a nonempty set X, define the relation R by

$$R = \{(x, A) | x \in A;\ A \subseteq X\}.$$

   Find the domain and range of R.
3. Give an example of two functions f and g from $R^{\#}$ into $R^{\#}$ with $f \neq g$ for which $f \circ g = g \circ f$.
4. Determine $f \circ g$, $g \circ f$, and their respective domains, given that
   a) $f = \{(x, x^2 + x) | x \in R^{\#}\}$, $g = \{(x, 3x + 4) | x \in R^{\#}\}$;
   b) $f = \{(x, x/(x^2 + 1)) | x \in R^{\#}\}$, $g = \{(x, 1/x) | x \in R^{\#} - \{0\}\}$.
5. Let f, g and h be functions such that $\text{rng}\, h \subseteq \text{dom}\, g$ and $\text{rng}\, g \subseteq \text{dom}\, f$. Establish the identity $f \circ (g \circ h) = (f \circ g) \circ h$.
6. For functions $f, g, h : R^{\#} \to R^{\#}$, verify that

$$(f/g) \circ h = (f \circ h)/(g \circ h)$$

   where f/g is the function defined by

$$(f/g)(x) = f(x)/g(x) \qquad x \in \text{dom}\, f \cap \text{dom}\, g \cap \{x | g(x) \neq 0\}.$$

7. Show that any one-to-one function from a finite set into itself is necessarily onto.
8. Using functions $f : R^{\#} \to R^{\#}$, give an example of a function that is
   a) one-to-one but not onto $R^{\#}$,
   b) onto $R^{\#}$ but not one-to-one.
9. For every pair of real numbers a and b, define a function $f_{ab} : R^{\#} \to R^{\#}$ by the formula $f_{ab}(x) = ax + b$ for each $x \in R^{\#}$.
   a) Show that $f_{1b} \circ f_{a0} = f_{ab}$.
   b) For $a \neq 0$, prove that f_{ab} is both one-to-one and onto $R^{\#}$.
   c) For $a \neq 0$, obtain f_{ab}^{-1}.
10. For functions $g : X \to Y$ and $f : Y \to W$, show that the following statements are true:
    a) If $f \circ g$ is an onto function, then f is also.
    b) If $f \circ g$ is a one-to-one function, then g is also.
    c) If f and g are both one-to-one functions, then $f \circ g$ is also one-to-one and $(f \circ g)^{-1} = g^{-1} \circ f^{-1}$.

11. Assuming that $f: X \to Y$, verify that
$$f \circ i_X = f = i_Y \circ f.$$

12. If $f: X \to Y$ is one-to-one and $B \subseteq Y$, then $f^{-1}(B)$ has two interpretations: as an inverse image under f and as a direct image under f^{-1}. Show that these interpretations agree.

13. If $f: X \to Y$, establish that the following conditions are equivalent:
   a) f maps X onto Y,
   b) for all functions $g, h: Y \to W$, $g \circ f = h \circ f$ implies that $g = h$,
   c) for all subsets $A \subseteq X$, $f(X - A) \supseteq Y - f(A)$.

14. If $f: X \to Y$, establish that the following conditions are equivalent:
   a) f is one-to-one,
   b) for all functions $g, h: W \to X$, $f \circ g = f \circ h$ implies that $g = h$,
   c) for all subsets $A \subseteq X$, $f(X - A) \subseteq Y - f(A)$.

15. Show by example that $f|A = g|A$ does not imply that $f = g$.

16. If $f: X \to Y$, $g; Y \to X$ and $A \subseteq X$, $B \subseteq Y$, prove that
   a) $(g \circ f)|A = g \circ (f|A)$,
   b) $(f|A)^{-1}(B) = A \cap f^{-1}(B)$.

17. Given $f: X \to Y$ and $A, B \subseteq X$, show that
   a) $f(A \cup B) = f(A) \cup f(B)$,
   b) $f(A \cap B) \subseteq f(A) \cap f(B)$,
   c) $f(A) - f(B) \subseteq f(A - B)$,
   d) if $A \subseteq B$, then $f(A) \subseteq f(B)$.

18. Given $f: X \to Y$ and $A, B \subseteq Y$, show that
   a) $f^{-1}(A \cup B) = f^{-1}(A) \cup f^{-1}(B)$,
   b) $f^{-1}(A \cap B) = f^{-1}(A) \cap f^{-1}(B)$,
   c) $f^{-1}(A) - f^{-1}(B) = f^{-1}(A - B)$,
   d) if $A \subseteq B$, then $f^{-1}(A) \subseteq f^{-1}(B)$.

   Comparing these results with those of Problem 17, one can see that inverse images tend to be better behaved than direct images.

19. If $f: X \to Y$, prove that f is a one-to-one function if and only if $f(A \cap B) = f(A) \cap f(B)$ for all sets $A, B \subseteq X$.

20. Classify each of the following relations R in the set Z of integers as to whether they do or do not have the properties of being reflexive, symmetric, and transitive:
   a) $(a, b) \in R$ if and only if $a < b$,
   b) $(a, b) \in R$ if and only if $a - b$ is an odd integer,
   c) $(a, b) \in R$ if and only if $ab \geqslant 0$,
   d) $(a, b) \in R$ if and only if $a^2 = b^2$,

e) $(a, b) \in R$ if and only if $|a - b| \leqslant 1$,

f) $(a, b) \in R$ if and only if $|a| = |b|$.

21. Let Y be a fixed subset of the set X. Show that the relation defined by $P(X)$ by taking $A \sim B$ if and only if $A \cap Y = B \cap Y$ is an equivalence relation.

22. Prove that the following relations $\sim$ are equivalence relations in the Cartesian product $R^{\#} \times R^{\#}_{+}$:

a) $(a, b) \sim (c, d)$ if and only if $ad = bc$,

b) $(a, b) \sim (c, d)$ if and only if $a + d = b + c$,

c) $(a, b) \sim (c, d)$ if and only if $a - c \in Z$, $b = d$,

d) $(a, b) \sim (c, d)$ if and only if $ab = cd$,

e) $(a, b) \sim (c, d)$ if and only if $a^2 + b^2 = c^2 + d^2$.

23. If X is a finite set, define the relation $\sim$ in $P(X)$ by taking $A \sim B$ if and only if A and B contain the same number of elements. Prove that $\sim$ is an equivalence relation in $P(X)$ and determine its equivalence classes.

24. Let $\{A_i\}$ and $\{B_j\}$ be partitions of the sets X and Y, respectively. Show that the family of sets $\{A_i \times B_j\}$ forms a partition of the Cartesian product $X \times Y$.

25. Assume that the function $f: X \to Y$ maps onto Y and that $\mathscr{P}'$ is a partition of Y. Establish that the family of sets.

$$\mathscr{P} = \{f^{-1}(P) | P \in \mathscr{P}'\}$$

is a partition of X.

1.3 ELEMENTARY NUMBER THEORY

This section deals with some of the more frequently used factorization properties of the integers. There are few prerequisites other than some familiarity with the axiomatic foundations of the integers. Our development may be regarded as a brief introduction to the theory of numbers, one of the oldest branches of mathematics. Although the primary aim is to assemble a few of the technical facts needed in the sequel, it is hoped that the results encountered along the way will prove interesting and rewarding in themselves. Indeed, the elegance of the subject led Gauss to characterize number theory as the "queen of mathematics."

An axiom of the integers, which will be used implicitly time and time again, is the so-called Well-Ordering Principle:

WELL-ORDERING PRINCIPLE Every nonempty subset S of nonnegative integers contains a least element; that is, there exists some (necessarily unique) integer $a \in S$ such that $a \leqslant b$ for all $b \in S$.

To illustrate the force of this principle, we first prove the following little result concerning the integer 1.

THEOREM 1.10 *If $a \in Z_+$, then $a \geqslant 1$; in other words, 1 is the least positive integer.*

Proof. Let the subset S of Z_+ be defined by $S = \{a \in Z_+ | 0 < a < 1\}$ and assume that $S \neq \varnothing$. From the Well-Ordering Principle, there is a smallest element b in this set; as an element of S, $0 < b < 1$. If we multiply these inequalities by b, it follows that $0 < b^2 < b$. But then b^2 is a member of S smaller than b, contradicting the choice of b as the smallest integer in S. The original assumption that $S \neq \varnothing$ must be false, and so $a \geqslant 1$ for every a in Z_+.

COROLLARY *If a and b are integers with $a < b$, then $a + 1 \leqslant b$.*

Proof. If $a < b$, then $b - a > 0$. Hence by the theorem, $b - a \geqslant 1$, which is to say that $a + 1 \leqslant b$.

With the Well-Ordering Principle available, it is an easy matter to derive the Principle of Finite Induction. The latter principle provides a basis for a method of proof called "mathematical induction." Loosely speaking, the Principle of Finite Induction asserts that if a set of positive integers has two specific properties, then it is the set of all positive integers. To be less cryptic:

THEOREM 1.11 (Principle of Finite Induction) *Let S be a set of positive integers with the properties*

a) *1 belongs to S, and*
b) *whenever the integer k is in S, then the next integer $k + 1$ must also be in S.*

Then S is the set of all positive integers.

Proof. Let T be the set of all positive integers not in S, and assume that T is nonempty. The Well-Ordering Principle tells us that T possesses a least element, which we denote by a. Since 1 is in S, certainly $a > 1$ and so $0 < a - 1 < a$. The choice of a as the smallest positive integer in T implies that $a - 1$ is not a member of T, or equivalently, that $a - 1$ belongs to S. By hypothesis, S must also contain $(a - 1) + 1 = a$, which contradicts the fact that a lies in T. We conclude that the set T is empty and in consequence that S contains all the positive integers.

Here is a typical formula that can be established by mathematical induction:

$$(1) \qquad 1^2 + 2^2 + 3^2 + \cdots + n^2 = \frac{n(2n+1)(n+1)}{6}$$

for $n = 1, 2, 3, \ldots$. In anticipation of using Theorem 1.11, let S denote the set of all positive integers n for which (1) is true. We observe that when $n = 1$, the formula becomes

$$1^2 = \frac{1(2+1)(1+1)}{6} = 1;$$

this means that 1 is in S. Next assume that k belongs to S (where k is a fixed but unspecified integer) so that

$$(2) \qquad 1^2 + 2^2 + 3^2 + \cdots + k^2 = \frac{k(2k+1)(k+1)}{6}.$$

To obtain the sum of the first $k + 1$ squares, we merely add the next one, $(k+1)^2$, to both sides of equation (2). This gives

$$1^2 + 2^2 + \cdots + k^2 + (k+1)^2 = \frac{k(2k+1)(k+1)}{6} + (k+1)^2.$$

After some algebraic manipulation, the right-hand side becomes

$$(k+1)\left[\frac{k(2k+1) + 6(k+1)}{6}\right] = (k+1)\left[\frac{2k^2 + 7k + 6}{6}\right]$$
$$= \frac{(k+1)(2k+3)(k+2)}{6},$$

which is precisely the right-hand member of (1) when $n = k + 1$. Our reasoning shows that the set S contains the integer $k + 1$ whenever it contains the integer k. By Theorem 1.11, S must be all the positive integers; that is, the given formula is true for $n = 1, 2, 3, \ldots$.

In the development of number theory a single theorem acts as a cornerstone for all that follows. This result, generally referred to as the *Division Algorithm*, asserts that a given integer can be "divided" by a positive integer to yield a quotient and positive remainder. Again, our argument relies on the Well-Ordering Principle.

THEOREM 1.12 *If $a, b \in Z$, with $b > 0$, then there exist unique integers q and r such that*

$$a = qb + r, \qquad 0 \leqslant r < b.$$

Proof. We begin by proving that the set

$$S = \{a - xb \mid x \in Z;\ a - xb \geqslant 0\}$$

is not empty. Since $b \geqslant 1$, $|a|b \geqslant |a|$, and

$$a - (-|a|)b = a + |a|b \geqslant a + |a| \geqslant 0.$$

Hence for $x = -|a|$, $a - xb \in S$. By the Well-Ordering Principle, S contains a smallest integer, say r. In other words, there is some $q \in Z$ for which

$$r = a - qb, \qquad r \geqslant 0.$$

We now show that $r < b$. In the contrary case, $r \geqslant b$ and

$$a - (q + 1)b = (a - qb) - b = r - b \geqslant 0,$$

which implies that $r - b \in S$. Since $r - b < r$, this contradicts the choice of r as the smallest element in S; therefore having reached a contradiction, we conclude that $r < b$.

To prove the uniqueness of the integers q and r, suppose that

$$a = qb + r = q'b + r',$$

where $0 \leqslant r < b$, $0 \leqslant r' < b$. Then $r' - r = b(q - q')$; consequently,

$$|r' - r| = b|q - q'|.$$

Adding the inequalities $-b < -r \leqslant 0$ and $0 \leqslant r' < b$, we obtain

$$-b < r' - r < b, \qquad \text{or} \qquad |r' - r| < b.$$

Hence $b|q - q'| < b$, so that

$$0 \leqslant |q - q'| < 1.$$

Since $|q - q'|$ is a nonnegative integer, it follows that $q = q'$, which in turn gives $r = r'$.

COROLLARY (Division Algorithm) *If $a, b \in Z$, with $b \neq 0$, then there exist unique integers q and r such that*

$$a = qb + r, \qquad 0 \leqslant r < |b|.$$

Proof. It suffices to consider the case in which b is negative. Then $|b| > 0$, and the theorem yields unique integers q' and r for which

$$a = q'|b| + r, \qquad 0 \leqslant r < |b|.$$

Since $|b| = -b$, we may take $q = -q'$ to get

$$a = qb + r, \qquad 0 \leqslant r < |b|.$$

Among the basic ideas of number theory is that of the divisibility of integers.

DEFINITION 1.16 Let $a, b \in Z$, with $a \neq 0$. The integer a is said to *divide* b, in symbols $a|b$, provided that there exists some $c \in Z$ such that $b = ac$. If a does not divide b, then we write $a \nmid b$.

Other language for the divisibility property $a|b$ is that a is a factor of b, that b is divisible by a, and that b is a multiple of a. Whenever the notation $a|b$ is employed, it is to be understood that $a \neq 0$.

Some immediate consequences of this definition are noted below; the reader is asked to verify each of them.

THEOREM 1.13 *Let $a, b, c \in Z$. Then*

a) $a|0, \; \pm 1|a, \; \pm a|a$;
b) $a| \pm 1$ *if and only if* $a = \pm 1$;
c) If $a|b$, *then* $ac|bc$;
d) If $a|b$ *and* $b|c$, *then* $a|c$;
e) $a|b$ *and* $b|a$ *if and only if* $a = \pm b$;
f) If $c|a$ *and* $c|b$, *then* $c|(ax + by)$ *for every* $x, y \in Z$.

From part (a) above, we see that each integer $a \neq 0$ is divisible by ± 1 and $\pm a$, divisors that are frequently referred to as *improper divisors*. An integer $p > 1$ having no divisors other than the improper ones is said to be a *prime number*, or simply a *prime*. Another way of putting this is to say that a prime is an integer $p > 1$ whose only positive divisors are itself and 1. (Throughout the rest of the book the letter p will be reserved for primes.) Any integer $a > 1$ that is not prime is termed *composite*. Such an integer is representable as a product $a = bc$, with $1 < b < a$, $1 < c < a$. Interchanging the roles of b and c, if necessary, we may assume that $b \leq c$; hence $b^2 \leq bc = a$, and so $b \leq \sqrt{a}$. These considerations show that if an integer $a > 1$ is composite, then it has a divisor b such that $1 < b \leq \sqrt{a}$.

Among the first ten positive integers 2, 3, 5, and 7 are primes, while 4, 6, 8, 9, and 10 are composite numbers.

If $a, b \in Z$, we say that an integer d is a *common divisor* of a and b if $d|a$ and $d|b$. Since 1 is a divisor of every integer, 1 provides an example of a common divisor of a and b. It follows that the set of positive common divisors of two integers is nonempty. Let us make the following definition.

DEFINITION 1.17 Let a and b be integers, not both of which are zero. The *greatest common divisor* of a and b, denoted by gcd (a, b), is a positive integer d such that

1. $d|a$ and $d|b$,
2. if $c|a$ and $c|b$, then $c|d$.

It is to be observed that if $a = b = 0$, then every integer is a common divisor of a and b. The notion of greatest common divisor of a and b makes sense only when we require that at least one of a and b is different from zero. As the terminology suggests, gcd (a, b) is the largest integer in the set of common divisors of a and b.

When we hereafter write gcd (a, b), it is implicitly assumed that the integers a and b are not both zero.

The foregoing definition does not assert that such a divisor exists or that it is unique when it does exist. These facts remain to be established. As regards uniqueness, suppose that there are two positive integers d and d' that satisfy the conditions of Definition 1.17. Then by condition (2) we must have $d \mid d'$ as well as $d' \mid d$, whence $d = \pm d'$ (Theorem 1.13(e) applies). Since d and d' are both positive integers, it follows that $d = d'$. As a result, the greatest common divisor of a and b is unique when it exists.

The next theorem will establish that any two integers that are not both zero actually do possess a greatest common divisor.

THEOREM 1.14 *If a, b are integers, not both of which are zero, then $\gcd(a, b)$ exists; in fact, there exist integers x and y such that*

$$\gcd(a, b) = ax + by.$$

Proof. First define the set S by

$$S = \{au + bv \mid u, v \in Z;\ au + bv > 0\}.$$

This set S is not empty. For example, if $a \neq 0$, the integer $|a| = au + b0$ will lie in S, where we choose $u = 1$ or -1 depending on whether a is positive or negative. By the Well-Ordering Principle, S must contain a smallest element $d > 0$; that is to say, there exist $x, y \in Z$ for which $d = ax + by$. We assert that $d = \gcd(a, b)$.

From the Division Algorithm, one can obtain integers q and r such that $a = qd + r$, $0 \leqslant r < d$. But then r will be of the form

$$\begin{aligned} r &= a - qd = a - q(ax + by) \\ &= a(1 - qx) + b(-qy). \end{aligned}$$

Were $r > 0$, this representation would imply $r \in S$ and contradict the fact that d is the least integer in S. Thus $r = 0$, so that $d \mid a$. A similar argument establishes that $d \mid b$, making d a common divisor of a and b. On the other hand, if $c \mid a$ and $c \mid d$, then by Theorem 1.13(f), $c \mid (ax + by)$, or rather, $c \mid d$. From these two statements we conclude that d is the greatest common divisor of a and b.

There is a corollary to the foregoing theorem that will play an important role in the future. But first, a definition: Two integers a and b, not both of which are zero, are said to be relatively prime if and only if gcd $(a, b) = 1$. Viewed otherwise, a and b are *relatively prime* if and only if 1 and -1 are their only common divisors. The integers 8 and 15 are relatively prime, for instance, although neither is itself a prime.

COROLLARY *Let $a, b \in Z$, not both zero. Then a and b are relatively prime if and only if there exist integers x and y such that*

$$1 = ax + by.$$

Proof. If a and b are relatively prime, so that $\gcd(a, b) = 1$, Theorem 1.14 guarantees the existence of x and y satisfying $1 = ax + by$. Conversely, suppose that $1 = ax + by$ for suitable $x, y \in Z$ and that $d = \gcd(a, b)$. Since $d|a$, $d|b$, Theorem 1.13(f) implies that $d|(ax + by)$ or rather $d|1$. Because d is positive, this forces $d = 1$ (Theorem 1.13(b)), as desired.

Our next result, while it seems quite mild, is of fundamental importance.

THEOREM 1.15 (Euclid's Lemma) *If $a|bc$, with a and b relatively prime, then $a|c$.*

Proof. Since $\gcd(a, b) = 1$, there exist integers x and y for which $1 = ax + by$. Multiplying this equation by c, we obtain

$$c = (ax + by)c = a(cx) + (bc)y.$$

Now $a|a$ trivially and $a|bc$ by hypothesis, so a must divide the sum $acx + bcy$ (part (f) of Theorem 1.13 plays a role again); hence $a|c$ as asserted.

We shall often have occasion to refer to the following corollary.

COROLLARY *If p is a prime and $p|(a_1 a_2 \cdots a_n)$, then $p|a_k$ for some k, $1 \leqslant k \leqslant n$.*

Proof. Our proof is by induction on n. For $n = 1$ the result obviously holds. Suppose, as the induction hypothesis, that $n > 1$ and that whenever p divides a product of less than n factors, then it divides at least one of the factors of this product. Now let $p|(a_1 a_2 \cdots a_n)$. If p divides a_1, there is nothing to prove. In the contrary case, p and a_1 are relatively prime; hence by the theorem, $p|(a_2 \cdots a_n)$. Since the product $a_2 \cdots a_n$ contains $n - 1$ factors, the induction hypothesis implies that $p|a_k$ for some k with $2 \leqslant k \leqslant n$.

Although Theorem 1.14 asserts the existence of gcd (a, b), the proof of the theorem does not give a convenient method for finding its value. We now present an orderly and systematic process, known as the *Euclidean Algorithm*, for determining the greatest common divisor of two positive integers a and b. Without loss of generality, let $a \geqslant b$. We may apply the Division Algorithm to obtain

$$a = q_1 b + r_1, \qquad 0 \leqslant r_1 < b.$$

Were $r_1 = 0$, then b would divide a and $\gcd(a, b) = b$. If, however, $r_1 \neq 0$, then there exist integers q_2 and r_2 such that

$$b = q_2 r_1 + r_2, \qquad 0 \leqslant r_2 < r_1.$$

If $r_2 = 0$, we stop; otherwise we proceed as before, obtaining

$$r_1 = q_3 r_2 + r_3, \qquad 0 \leqslant r_3 < r_2.$$

We continue this division process until some zero remainder appears, say at the $n + 1$st stage where we divide r_{n-1} by r_n (this must occur sooner or later, since the integers r_i are decreasing yet nonnegative). We can tabulate the results as

$$\begin{aligned}
a &= q_1 b + r_1 & 0 &< r_1 < b \\
b &= q_2 r_1 + r_2 & 0 &< r_2 < r_1 \\
r_1 &= q_3 r_2 + r_3 & 0 &< r_3 < r_2 \\
&\vdots \\
r_{n-2} &= q_n r_{n-1} + r_n & 0 &< r_n < r_{n-1} \\
r_{n-1} &= q_{n+1} r_n
\end{aligned}$$

The greatest common divisor of a and b will turn out to be r_n, the last nonzero remainder obtained in this manner. Now $r_n | r_{n-1}$ by the last equation of the above system. From the equation immediately preceding, it follows that $r_n | r_{n-2}$, since r_{n-2} is a linear combination of r_n and r_{n-1}. Working backward through the equation, we find that r_n divides each of the preceding remainders r_i. Finally, $r_n | b$, and from the first equation $a = q_1 b + r_1$, $r_n | a$. Therefore r_n is a common divisor of a and b. On the other hand, going down the list of equations, it is clear that any common divisor of a and b divides $r_1, r_2, \ldots,$ and ultimately r_n. In consequence, r_n satisfies the second property of Definition 1.17, and so $r_n = \gcd(a, b)$, as asserted.

The Euclidean Algorithm can also be used to represent $\gcd(a, b)$ as a linear combination of the integers a and b. Starting with the next to last equation in the algorithm,

$$r_n = r_{n-2} - q_n r_{n-1},$$

we now substitute the remainder r_{n-1} from the previous equation. We obtain

$$\begin{aligned} r_n &= r_{n-2} - q_n(r_{n-3} - q_{n-1}r_{n-2}) \\ &= (-q_n)r_{n-3} + (1 + q_nq_{n-1})r_{n-2}. \end{aligned}$$

Continuing up the set of equations, we can eliminate the successive remainders $r_{n-1}, r_{n-2}, \ldots, r_2, r_1$ and eventually express r_n in terms of a and b; that is, we can find suitable integers x and y satisfying

$$\gcd(a, b) = r_n = ax + by.$$

To illustrate this method, let us find gcd(657, 306). The successive equations in the Euclidean Algorithm are listed below:

$$\begin{aligned} 657 &= 2\cdot 306 + 45, \\ 306 &= 6\cdot 45 + 36, \\ 45 &= 1\cdot 36 + 9, \\ 36 &= 4\cdot 9. \end{aligned}$$

From the preceding discussion it follows that $9 = \gcd(657, 306)$. To express 9 as a linear combination of 657 and 306, we start with the next-to-last equation and eliminate the remainders 36 and 45; accordingly,

$$\begin{aligned} 9 &= 45 - 36 \\ &= 45 - (306 - 6\cdot 45) \\ &= -306 + 7\cdot 45 \\ &= -306 + 7(657 - 2\cdot 306) \\ &= 7\cdot 657 + (-15)306. \end{aligned}$$

This last-written equation represents gcd(657, 306) in the desired form.

It may be well to record that the integers x and y appearing in the representation $\gcd(a, b) = ax + by$ need not be unique. As was just shown, $\gcd(657, 306) = 9 = 7\cdot 657 + (-15)306$. Among other possibilities, we also have

$$\begin{aligned} 9 &= (7 - 306)657 + (-15 + 657)306 \\ &= 313\cdot 657 + (-672)306. \end{aligned}$$

Our next theorem justly deserves to be described as the fundamental theorem of arithmetic, so many and varied are its applications.

THEOREM 1.16 (Fundamental Theorem of Arithmetic) *Every positive integer $a > 1$ can be expressed as a product of primes; this representation is unique, apart from the order in which the factors occur.*

Proof. The first part of the proof—the existence of a prime factorization—is proved by induction on the values of a. The statement of the theorem is trivially true for the integer 2, since 2 is itself a prime. Assume that the result holds for all positive integers $2 \leqslant x \leqslant a$. If a is already a prime, we are through; otherwise, $a = bc$ for suitable integers b, c with $1 < b < a$, $1 < c < a$. By the induction hypothesis,

$$b = p_1 p_2 \cdots p_r, \qquad c = p'_1 p'_2 \cdots p'_s,$$

with p_i, p'_i all primes. But then,

$$a = bc = p_1 \cdots p_r p'_1 \cdots p'_s$$

is a product of primes.

To establish uniqueness, let us suppose that the integer a can be represented as a product of primes in two ways, say

$$a = p_1 p_2 \cdots p_n = q_1 q_2 \cdots q_m,$$

where the p_i and q_i are primes. The argument proceeds by induction on the integer n. In the case $n = 1$ we have $a = p_1 = q_1(q_2 \cdots q_m)$. Since p_1 is prime, it possesses no proper factorization, so $m = 1$ and $p_1 = q_1$.

Next, assume that $n > 1$ and that whenever a can be expressed as a product of less than n factors, this representation is unique, except for the order of the factors. From the equality $p_1 p_2 \cdots p_n = q_1 q_2 \cdots q_m$, it follows that $p_1 | (q_1 q_2 \cdots q_m)$. Thus by the preceding corollary there is some prime q_k, $1 \leqslant k \leqslant m$, for which $p_1 | q_k$; renumbering if necessary, we may suppose $p_1 | q_1$. But then $p_1 = q_1$, for q_1 has no divisors other than 1 and itself. Canceling this common factor, we conclude that $p_2 \cdots p_n = q_2 \cdots q_m$. According to the induction hypothesis, a product of $n - 1$ primes can be factored in essentially one way. Therefore the primes $q_2, \ldots, q_m$ are simply a rearrangement of the primes $p_2, \ldots, p_n$. The two prime factorizations of a are thus identical, completing the induction step.

Of course, the primes that appear in the factorization of a given integer may be repeated (as is the case with $360 = 2 \cdot 2 \cdot 2 \cdot 3 \cdot 3 \cdot 5$). By collecting together the like primes and changing the indices, if necessary, we can rephrase Theorem 1.16 as follows:

COROLLARY *Any positive integer $a > 1$ can be written uniquely in a standard form*

$$a = p_1^{k_1} p_2^{k_2} \cdots p_r^{k_r}$$

where, for $i = 1, 2, \ldots, r$, each k_i is a positive integer and each p_i a prime with $1 < p_1 < \cdots < p_r$.

Historically, the last few theorems come from the text by the Greek mathematician Euclid, the *Elements*, written around 300 B.C. We know remarkably little of Euclid's life, but he most probably received his training at Plato's Academy in Athens and thereafter taught at the Museum, or "Temple of the Muses," in Alexandria. His fame rests almost exclusively on the *Elements*, the most important and successful mathematical textbook ever composed; with the single exception of the Bible, no work has been more widely studied. The *Elements* consists of thirteen Books, as the parts are called, into which Euclid incorporated nearly all of the mathematical knowledge of his time. Euclid's great achievement lies in his presentation of the material; he organized a collection of isolated results into a logical sequence of theorems with proper proof following inevitably from an initial set of definitions and axioms. Books VII, VIII, and IX deal with elementary number theory, though in terminology unfamiliar to the modern reader. Proposition 14 of Book IX is almost equivalent to what later became known as the Fundamental Theorem of Arithmetic. To quote the proposition itself, "If a number be the least that is measured by prime numbers, it will not be measured by any other prime except those originally measuring it." The first clear statement and proof of the Fundamental Theorem seems to have been given by Gauss in his *Disquisitiones Arithmeticae* of 1801. One of the most interesting propositions of Book IX is Proposition 20, which asserts that there is no largest prime: "Prime numbers are more than any assigned multitude of prime numbers."

As applications of the Fundamental Theorem of Arithmetic, we derive two famous results, namely the irrationality of $\sqrt{2}$ and Euclid's theorem on the infinitude of primes. These are classical examples of proofs by contradiction.

THEOREM 1.17 (Pythagoras) *$\sqrt{2}$ is irrational.*

Proof. Suppose to the contrary that $\sqrt{2}$ is rational; say, $\sqrt{2} = a/b$, where $\gcd(a, b) = 1$. Then $a^2 = 2b^2$, so that $b \mid a^2$. If $b > 1$, then by the Fundamental Theorem of Arithmetic there exists a prime p such that $p \mid b$. It follows that $p \mid a^2$ and, by the corollary to Euclid's Lemma, that $p \mid a$. This contradicts $\gcd(a, b) = 1$, and so it must be the case that $b = 1$. As a result, $a^2 = 2$, which, granting that there is no integer whose square is 2, is absurd.

The foregoing argument lends itself readily to generalization. We leave the reader to prove that $\sqrt[m]{N}$ is irrational unless N is the mth power of an integer.

Our second application answers the question of whether the primes terminate somewhere in the sequence of all positive integers.

THEOREM 1.18 (Euclid) *There are an infinite number of primes.*

Proof. Assume that the statement is false; that is, assume that there are only a finite number of primes $p_1, p_2, \ldots, p_n$. Consider the positive integer

$$a = (p_1 p_2 \cdots p_n) + 1.$$

None of the primes p_i divides a. If a were divisible by p_i, for instance, we would then have $p_i | (a - p_1 p_2 \cdots p_n)$, by Theorem 1.13(f), or $p_i | 1$; this is impossible by part (b) of the same theorem. But, since $a > 1$, the Fundamental Theorem asserts that it must have a prime factor. Accordingly, a is divisible by a prime that is not among our list of all primes. Our argument thus shows that there is no finite listing of the prime integers.

Not only are there infinitely many primes, but the distribution of these primes within the positive integers is most irregular. On the one hand, the difference between consecutive primes can be small, as with 5 and 7, or 17 and 19. It is a famous conjecture that there are infinitely many of these so-called *twin primes*; that is, pairs p and $p + 2$ that are both primes. Going in the other direction, arbitrarily large gaps occur between consecutive primes. Stated otherwise: Given any positive integer n, there exists n consecutive composite integers. Simply consider the integers

$$(n + 1)! + 2, (n + 1)! + 3, \ldots, (n + 1)! + (n + 1),$$

where $(n + 1)! = (n + 1) \cdot n \cdots 3 \cdot 2 \cdot 1$. Clearly, there are n elements listed, and they are consecutive. Each is composite, for $(n + 1)! + 2$ is divisible by 2, $(n + 1)! + 3$ is divisible by 3, and so on.

We can get a bit more information out of Euclid's argument by showing that there are infinitely many primes among the integers of the form $4n + 3$.

THEOREM 1.19 *There are infinitely many primes of the form* $4n + 3$, *where* $n \in Z_+$.

Proof. Let us start by looking at the odd primes, that is, those primes different from 2. When divided by 4, these leave a remainder of 1 or 3. It follows that every odd prime has one of the forms $4n + 1$ or $4n + 3$.

In anticipation of a contradiction, assume that there are only finitely many primes of the form $4n + 3$; call them $p_1, p_2, \ldots, p_r$. This time, consider the integer

$$a = 4(p_1 p_2 \cdots p_r) - 1 = 4(p_1 p_2 \cdots p_r - 1) + 3$$

and let $a = q_1 q_2 \cdots q_s$ be its prime factorization. Since a is an odd integer, $q_i \neq 2$ for all i, so that each q_i is of the form $4n + 1$ or $4n + 3$. However, the product of any number of primes of the form $4n + 1$ is again an integer of this form. In order that a have the form $4n + 3$, as it does, a must possess at least one prime factor q_i of the form $4n + 3$. But no q_i is found among $p_1, p_2, \ldots, p_r$, since otherwise it would follow that $q_i | 1$. Hence for any finite set of primes $4n + 3$, we have shown that there is a prime of this form not in the set; this contradicts our assumption that there are finitely many such primes.

The previous theorem is a special case of a remarkable result proved by Dirichlet. The proof is much too difficult for inclusion here, and so we must be content with just the statement of the theorem.

THEOREM 1.20 (Dirichlet) *If a and b are two positive integers such that $\gcd(a, b) = 1$, then there are infinitely many primes of the form $an + b$, where $n \in Z_+$.*

Any two integers a and b possess *common multiples* as well as common divisors; that is, integers c such that $a|c$ and $b|c$. For instance, the products ab and $-(ab)$ are multiples of both a and b. As might be expected, among the common multiples of a and b, there is one of special significance, called their *least common multiple*. Here is a formal definition.

DEFINITION 1.18 The *least common multiple* of two nonzero integers a and b, denoted by $\mathrm{lcm}(a, b)$, is a positive integer e such that

1. $a|e$ and $b|e$
2. if $a|c$ and $b|c$, then $e|c$.

We remark that there cannot be two different positive integers having the properties stated in Definition 1.18. Indeed, if e and e' are both least common multiples of a and b, it follows from (2) that $e|e'$ and $e'|e$ and hence that $e = e'$.

There remains the problem of showing that any two nonzero integers have a least common multiple. This is revealed by the following theorem, a result that, at the same time, indicates the close relationship of least common multiples to greatest common divisors.

THEOREM 1.21 *If a and b are nonzero integers, then*

$$\gcd(a, b)\,\mathrm{lcm}(a, b) = ab.$$

Proof. It will suffice to prove the theorem for positive integers, since $\gcd(a, -b) = \gcd(a, b)$. Let $d = \gcd(a, b)$ and write $a = da'$,

$b = db'$. If $e = (ab)/d$, then $e = a'b = ab'$, making e a common multiple of a and b. Now suppose that c is another common multiple of a and b; say, $c = ar = bs$. Since there exist integers x and y such that $d = ax + by$, this leads to

$$\frac{c}{e} = \frac{cd}{ab} = \frac{c(ax + by)}{ab} = sx + ry.$$

Therefore $e|c$, and we conclude that $e = \operatorname{lcm}(a, b)$; that is,

$$\operatorname{lcm}(a, b) = \frac{ab}{d} = \frac{ab}{\gcd(a, b)}.$$

COROLLARY *Given two nonzero integers a and b,* $\operatorname{lcm}(a, b) = ab$ *if and only if* $\gcd(a, b) = 1$.

By virtue of this theorem the computation of the least common multiple of two nonzero integers can be made to depend on the computation of their greatest common divisor which, in its turn, can be determined by the Euclidean Algorithm. For instance, we found earlier that $\gcd(657, 306) = 9$; hence

$$\operatorname{lcm}(657, 306) = \frac{657 \cdot 306}{9} = 22{,}338.$$

This completes our survey of some of the fundamental notions concerning sets, functions, and arithmetic in the integers. Although the treatment was purposely sketchy, it is hoped that the reader did not find it too superficial. In the subsequent chapters of the text we shall utilize the foregoing concepts by applying these ideas to certain specific situations.

PROBLEMS 1.3

1. Prove by mathematical induction that the following identities are true for every positive integer n:

a) $1 + 2 + 3 + \cdots + n = \dfrac{n(n + 1)}{2}$;

b) $1 + 3 + 5 + \cdots + (2n - 1) = n^2$;

c) $1 \cdot 2 + 2 \cdot 3 + 3 \cdot 4 + \cdots + n(n + 1) = \dfrac{n(n + 1)(n + 2)}{3}$;

d) $1^2 + 3^2 + 5^2 + \cdots + (2n - 1)^2 = \dfrac{n(4n^2 - 1)}{3}$;

e) $1 + 2 \cdot 2 + 3 \cdot 2^2 + \cdots + n2^{n-1} = (n-1)2^n + 1$;

f) $\dfrac{1}{1 \cdot 3} + \dfrac{1}{3 \cdot 5} + \dfrac{1}{5 \cdot 7} + \cdots + \dfrac{1}{(2n-1)(2n+1)} = \dfrac{n}{2n+1}$.

2. Given nonzero integers a, b, c, verify that

a) if $a|b$, then $a|bc$;

b) if $a|b$ and $a|c$, then $a^2|bc$;

c) $a|b$ if and only if $ac|bc$.

3. For any integer a, show that at least one of the integers a, $a + 2$, $a + 4$ is divisible by 3. [*Hint*: By the Division Algorithm, a must take one of the forms $3k$, $3k + 1$, or $3k + 2$.]

4. Prove each of the divisibility statements below by induction on n:

a) $3|2^n + (-1)^{n+1}$ for all $n \geqslant 1$;

b) $8|5^{2n} + 7$ for all $n \geqslant 1$;
[*Hint*: $5^{2(k+1)} + 7 = 5^2(5^{2k} + 7) + (7 - 5^2 \cdot 7)$];

c) $21|4^{n+1} + 5^{2n-1}$ for all $n \geqslant 1$.

5. Given integers a and b, which are not both zero, establish the following facts concerning $\gcd(a, b)$:

a) $\gcd(a, -b) = \gcd(-a, b) = \gcd(-a, -b) = \gcd(a, b)$;

b) $\gcd(a, b) = \gcd(a, b + ca)$ for any $c \in Z$;

c) $\gcd(ca, cb) = |c| \gcd(a, b)$, provided $c \neq 0$;

d) if p is a prime, then $\gcd(a, p) = 1$ if and only if $p \nmid a$.

6. Assuming that a, b, c are integers, no two of which are zero, prove that

$$\gcd(\gcd(a, b), c) = \gcd(a, \gcd(b, c)) = \gcd(\gcd(a, c), b).$$

7. Let $a, b, c \in Z$, with a and b not both zero, and let $d = \gcd(a, b)$. Show that there exist integers x and y such that $c = ax + by$ if and only if $d|c$.

8. Verify each of the statements below:

a) If $\gcd(a, b) = \gcd(a, c) = 1$, then $\gcd(a, bc) = 1$.

b) If $\gcd(a, b) = 1$, $a|c$ and $b|c$, then $ab|c$.

c) If $d = \gcd(a, b)$ and $a = rd$, $b = sd$, then $\gcd(r, s) = 1$.

d) If $\gcd(a, b) = 1$ and $c|(a + b)$, then $\gcd(a, c) = \gcd(b, c) = 1$. [*Hint*: Set $d = \gcd(a, c)$. Then $d|a$, $d|c$, so that $d|((a + b) - a)$, or $d|b$.]

9. Prove that any two consecutive integers are relatively prime.

10. Given that $\gcd(a, b) = 1$, prove that $\gcd(a^n, b^n) = 1$ for all $n \geqslant 1$. [*Hint*: see Problem 8(a).]

11. Given integers $a, b \in Z_+$, and $n > 1$, prove the following:

a) The relation $a^n|b^n$ implies that $a|b$. [*Hint*: Set $d = \gcd(a, b)$ and write $a = rd$, $b = sd$, where $\gcd(r, s) = 1$. By Problem 10, $\gcd(r^n, s^n) = 1$. Now show that $r = 1$, so that $a = d$.]

b) The relation $a^n|2b^n$ implies that $a|b$.

12. a) Prove that $\sqrt{p}$ is irrational for any prime p.
    b) Suppose that a and n are positive integers and $\sqrt[n]{a}$ is rational. Deduce that $\sqrt[n]{a}$ is an integer.

13. If p is a prime and $p|a^n$, show that $p^n|a^n$ for all $n \geqslant 1$.

14. If p_n denotes the nth prime (that is, number the primes so that $p_1 = 2$, $p_2 = 3$, $p_3 = 5, \ldots$), derive the inequality $p_{n+1} \leqslant p_n^n + 1$. [*Hint*: The integer $q = (2 \cdot 3 \cdot 5 \cdots p_n) + 1 \leqslant p_n^n + 1$.]

15. Modify Euclid's proof that there are infinitely many primes by assuming that there is a largest prime p and using $a = p! + 1$ to arrive at a contradiction.

16. By a slight variation of the proof of Theorem 1.19, deduce that there are infinitely many primes of the form $6n + 5$.

17. Prove that any positive integer of the form $4n + 1$ must have a prime factor of the same form.

18. Let a and b be nonzero integers. Show that the following conditions are equivalent:

    a) $a|b$, b) $\gcd(a, b) = |a|$, c) $\text{lcm}(a, b) = |b|$.

19. Prove that $\gcd(a, b) = \text{lcm}(a, b)$ if and only if $a = b$.

20. Assume that the integer $a > 1$ has the prime factorization $a = p_1^{n_1} p_2^{n_2} \cdots p_r^{n_r}$, where the p_i are distinct primes and $n_i > 0$ for all i. Prove the following statements:
    a) Any divisor b of a is of the form $b = p_1^{k_1} p_2^{k_2} \cdots p_r^{k_r}$, where $0 \leqslant k_i \leqslant n_i$ for each i.
    b) If $v(a)$ denotes the number of distinct divisors of a, then

    $$v(a) = (n_1 + 1)(n_2 + 1) \cdots (n_r + 1).$$

    c) If $\sigma(a)$ denotes the sum of the distinct divisors of a, then

    $$\sigma(a) = \left(\frac{p_1^{n_1+1} - 1}{p_1 - 1}\right)\left(\frac{p_2^{n_2+1} - 1}{p_2 - 1}\right) \cdots \left(\frac{p_r^{n_r+1} - 1}{p_r - 1}\right).$$

    [*Hint*: Utilize the formula $1 + p + \cdots + p^n = (p^{n+1} - 1)/(p - 1)$.]
    d) If $\gcd(a, b) = 1$, then $v(a, b) = v(a)v(b)$ and $\sigma(ab) = \sigma(a)\sigma(b)$.

21. Let $a = p_1^{n_1} p_2^{n_2} \cdots p_r^{n_r}$ and $b = p_1^{m_1} p_2^{m_2} \cdots p_r^{m_r}$ be factorizations of the positive integers a and b into distinct primes, with $n_i \geqslant 0$, $m_i \geqslant 0$ for all i. Establish that

    $$\gcd(a, b) = p_1^{k_1} p_2^{k_2} \cdots p_r^{k_r}$$
    $$\text{lcm}(a, b) = p_1^{j_1} p_2^{j_2} \cdots p_r^{j_r}$$

    where $k_i = \min\{n_i, m_i\}$, the smaller of n_i and m_i; and $j_i = \max\{n_i, m_i\}$, the larger of n_i and m_i. Hence conclude that $\gcd(a, b) = 1$ if and only if $n_i m_i = 0$ for all i.

2

Group Theory

Mathematics, once firmly established on a foundation of a few axioms and definitions, as upon a rock, has grown from age to age, so as to become the most solid fabric that human reason can boast.

THOMAS REID

2.1 ARTHUR CAYLEY

The theory of groups is the oldest branch of abstract algebra, with numerous applications both within mathematics and beyond its boundaries. Groups can provide totally unexpected answers to problems in physics, chemistry, and computer science. In fact, with the emergence of quantum mechanics and general relativity, group theory has played a central role in revolutionizing ideas in physics. The historical roots of the subject extend back to the eighteenth and nineteenth centuries and the work of such mathematicians as Joseph Louis Lagrange (1736–1813), Paulo Ruffini (1765–1822), and Evariste Galois (1811–1832) on the algebraic solvability of equations.

The word "group" made its appearance in a mathematical context when Galois spoke (1830) of a "group of substitutions" without expressly defining the notion; but he used the term inconsistently, employing it also in the everyday sense to refer to an arbitrary collection of objects. Arthur Cayley (1821–1895), the most prominent British mathematician of the nineteenth century, was the first to realize that the group-theoretic ideas existing up to then in various areas of mathematics could be formulated in a more abstract setting. In 1854, Cayley published a paper entitled "On the Theory of Groups as Depending on the Symbolic Equation $\theta^n = 1$," in which a comprehensive group concept is explicitly defined. Since this article is acknowledged as being the earliest work on abstract groups, Cayley is frequently considered the founder of group theory.

Arthur Cayley was born in 1821 at Richmond, in Surrey, during a visit by his parents to England. His earliest years were spent in St. Petersburg, Russia, where his father had settled as a partner in a firm of Russian merchants. In 1829 the family took up permanent residence in England.

At the age of 14, Arthur was sent to Kings College in London. There the young Cayley gave such indications of mathematical genius that school officials persuaded his father to abandon his intentions of bringing the boy up in the family business and to send him instead to Cambridge. Accordingly, Cayley began his university career at Trinity College at the rather early age of 17. In 1841, while still an undergraduate of 20, he published a work in the *Cambridge Mathematical Journal*, thus beginning an astounding series of papers that were eventually to enrich various scientific periodicals. Cayley finished his student days in the following year by winning Cambridge's two highest honors, the Mathematical Tripos and the more difficult Smith's Prize.

(The name "tripos" is of medieval origin; in those days the parties to the dispute sat on three-legged stools.)

At that time there was no honor for a student in any university that was deemed comparable to that of earning the top place in the Tripos, "the first of the firsts." Stories were current in Cambridge of the equanimity with which Cayley treated his success. A concluding feature of the examination was a three-hour paper consisting of problems representing the utmost range of the examiner's fancy. On the evening after the grand competition a friend rushed into Cayley's room with disconcerting news of Cayley's chief rival. "I've just seen Smith," the friend announced, "and he told me that he did all the questions within two hours." Cayley responded, "Likely enough he did: I cleaned up that paper in 45 minutes." Upon completing his degree, Cayley was immediately elected a Fellow of Trinity College, at the youngest age for any student in that century.

In the mid-nineteenth century the government of England's two ancient universities, Oxford and Cambridge, was in the hands of the clergy. A Fellow was required to take Holy Orders or to vacate the position after a reasonable time. So it was with Cayley, who left the academic world at the end of three years to enter the legal profession. He practiced in London for 14 years as a conveyancer, drawing up deeds for the transference of property. But his real occupation was pure mathematics. He wrote something approaching 300 mathematical papers, and these incorporated some of his most significant discoveries. Determined to preserve a portion of his time for research, Cayley rejected much of the legal work that came his way, making a comfortable living but turning has back upon any chance of accumulating wealth. It was during this phase of his life that he met his lifelong friend and collaborator, James Joseph Sylvester (1814–1897), who was likewise dividing his attention between the law and mathematics.

Between 1851 and 1854, Cayley and Sylvester collaborated on building up the algebraic theory of invariants, a joint work through which the two men became known as "the invariant twins." Often they walked about the Courts of Lincoln's Inn, with Cayley, in Sylvester's words, "habitually discoursing pearls and rubies." Their frequent discussions stimulated each other's ideas to the extent that neither would find it easy to say which idea properly belonged to whom.

In 1863, Cambridge created a new Sadlerian professorship of pure mathematics expressly for Cayley. Abandoning any future in the legal profession for a modest provision in his true vocation, Cayley held this chair until his death in 1895. It is said that Cayley brought mathematical glory to Cambridge second only to that of Sir Isaac Newton. Though he wrote but one extensive work, the *Treatise on Elliptic Functions* (1876), his output of papers and memoirs was prodigious. His most productive period, the years between 1863 and 1883, saw the publication of 430 papers. Many of these appeared in the *Quarterly Journal of Mathematics*, a periodical which Cayley and Sylvester helped to found in 1855. As Cayley continued in creative activity until the end of his life, his grand total of 966 articles places him third, behind Euler and Cauchy, as the most prolific writer of mathematics. His *Collected Mathematical Papers* (1889–1898) fills 13 large volumes.

It is customary to view Cayley as the creator of an algebra of matrices that did not require repeated reference to the equations from which their elements were taken. Cayley had a sufficiently clear idea of matrices in the mid-1840s, but it was not until 1858 that he published his first major paper on the subject. In this paper, "A Memoir on the Theory of Matrices," he defined a matrix as "a set of quantities arranged in the form of a square" and observed that this notion arose "from an abbreviated notation for a set of linear equations." Accordingly, Cayley laid down the operations of matrix addition and matrix multiplication, or "compounding" as he called it. He based the operations on the simple idea of transferring the rules for computing with linear transformations to the matrices that represent them. This article lists many of the basic properties of matrices, such as the associative and distributive laws. Sylvester subsequently carried the investigations much further, and to him we owe much of the development of the theory of matrices. The introduction into the mathematical literature of the very term "matrix" appears to be due to Sylvester (1850). Cayley, on the other hand, came up with the notation of a square matrix between a pair of vertical lines to denote a determinant and used a pair of double vertical lines to indicate a matrix (1843).

Given the extent of Cayley's mathematical interests, it was almost inevitable that he should become involved in the movement toward greater generality in the group concept. His pioneering attempt to define an abstract group appears in a paper entitled "On the Theory of Groups as Depending on the Symbolic Equation $\theta^n = 1$," which was published in the *Philosophical Magazine* in 1854. Since the terminology of set theory was not current at that time, Cayley started by making it clear that he was working with abstract "symbols" θ, ϕ, and so on, rather than with concrete objects or numbers; in his words, "It is not necessary (even if this could be done) to attach any meaning to a symbol such as $\theta \pm \phi$ . . . ." With regard to the definition of a group he offered the following:

> A set of symbols $1, \alpha, \beta$ . . . all of them different, and such that the product of any two of them, or the product of any one of them into itself, belongs to the set, is said to be a group.

Cayley went on in the article to introduce the now-accepted procedure of defining a group by listing its elements in the form of a multiplication table, or a *Cayley table* as it is known today. By an analysis of the possible multiplication tables he showed that there are two "essentially distinct" abstract groups with four elements and two with six elements.

Arthur Cayley (1821–1895)

The mathematical climate did not favor an axiomatic approach to a concept having the theory of equations as its only significant application, and so Cayley's development of group theory remained virtually unnoticed for 25 years. An immensely versatile mathematician, Cayley made enormous progress in other areas also; this probably best explains his failure to pursue the abstract route. While Cayley was happily ensconced at Cambridge, his co-worker Sylvester was unemployed for several years. Sylvester found a new intellectual home in the United States when he was invited to join Johns Hopkins University upon its opening in 1876; he was one of its original six faculty members and head of the mathematics department. The two old friends worked together again in 1881 when Cayley gave an extended series of lectures at Hopkins. The presence at the same time of two of the world's most prominent mathematicians was said to have made Baltimore "the stronghold of mathematics in America."

Among Sylvester's contributions to the mathematical awakening in this country was the founding of the *American Journal of Mathematics* (1878), the first periodical in the United States dedicated solely to "the publication of original investigations." To maintain the highest standards of scholarship right from the start, Sylvester secured a short paper by Cayley under the title "The Theory of Groups." A more streamlined version of the 1854 article, Cayley's return to group theory met with an entirely different reception. An increasing acceptance of the axiomatic method and a growing body of concrete examples of groups finally led to the recognition of the importance of Cayley's abstract conception.

2.2 DEFINITION AND EXAMPLES OF GROUPS

In this chapter, and throughout the remainder of the text, we shall deal with mathematical systems that are defined by a prescribed list of properties. Emphasis will be on deriving theorems that follow logically from the postulates and that at the same time help to describe the algebraic structure of the particular system under consideration. This axiomatic approach not only permits us to concentrate on essential ideas, but also unifies the presentation by showing the basic similarities of many diverse and apparently unrelated examples.

We first confine our attention to systems involving just one operation, since they are amenable to the simplest formal description. Despite this simplicity, the axioms permit the construction of a profuse and elegant theory in which one encounters many of the fundamental notions common to all algebraic systems.

Before beginning, however, it is necessary to arrive at some understanding concerning the use of the equivalence relation $=$. We will henceforth take the equality sign to mean, intuitively, "is the same as." In other words, the symbol $=$ asserts that the two particular expressions involved are merely different names for, or descriptions of, one and the same object; just one object is being considered, and it is named twice. To indicate that a and b are not the same object, we shall, naturally enough, write $a \neq b$.

As a first step in our program, we introduce the concept of a binary operation. This idea is the cornerstone of all that follows.

DEFINITION 2.1 Given a nonempty set S, any function from the Cartesian product $S \times S$ into S is called a *binary operation* on S.

A binary operation on S thus assigns to each ordered pair of elements of S a uniquely determined third element of the same set S. For instance, if $P(A)$ denotes the power set of a fixed set A, then both $\cup$ and $\cap$ are binary operations on $P(A)$. In practice, we shall generally use the symbol $*$ to represent a binary operation and write $a * b$, instead of $*((a, b))$, for its value at the ordered pair $(a, b) \in S \times S$. While this convention is at variance with the functional notation developed in the previous chapter, its use in the present situation is dictated by long-standing mathematical tradition. At the very least, it has the advantage of avoiding some rather clumsy notation.

From time to time we shall permit ourselves to make such informal statements as "combine a with b" or "form $a * b$." In a precise sense, what is really meant of course is to apply the function $*$ to the ordered pair (a, b). The most useful aspect of a binary operation is that, having once formed the element $a * b$, we may in turn combine it with other members of S; the result of all such calculations again lies in S.

Needless to say, the particular notation used for the abstract product of two elements is of no great importance. On occasions, some other symbol, as noncommittal as $*$, will be employed. Specifically, we will frequently choose to write $a \circ b$ in place of $a * b$; in this context the symbol $\circ$ is not intended to have any special connection with functional composition. In general, a and b will have no numerical value but will simply be arbitrary elements in our underlying set S, whatever this set may be, while $*$ may well be some law of composition that bears no resemblance to the usual operations of elementary algebra.

Closely allied to the notion of a binary operation is the so-called *closure condition.* For a formal statement of this property, suppose that $*$ is a binary operation on the set S and $A \subseteq S$; the subset A is said to be *closed* under the operation $*$ provided that $a * b \in A$ whenever a and b are in A. The desirable feature here is that when A is closed under the operation $*$, the restriction of $*$ to the subset A is a binary operation on A as well as S.

Example 2.1

Ordinary subtraction is clearly a binary operation on the set Z of integers; the subset Z_+ of positive integers, however, is not closed under subtraction.

When the set S being considered has a relatively small number of elements, the results of applying the operation $*$ to its members may be conveniently represented in what might be called an *operation* or *multiplication table.* We construct this table by first listing the members of S in the same order both vertically and horizontally. The result $a * b$ then appears in the body of the table at the intersection of the row headed by a and the column headed by b. Conversely, such a table could equally well serve to define a binary operation on S, since the result of combining any pair of elements of S would be displayed somewhere in the table.

Example 2.2

A binary operation $*$ may be defined on the three-element set

$S = \{1, 2, 3\}$ by means of the operation table below:

$*$	1	2	3
1	1	2	3
2	3	1	2
3	2	3	1

According to the table, for instance, the product $2 * 3$ is equal to the element 2, located at the intersection of the row marked 2 and the column marked 3.

Given an arbitrary binary operation $*$, there is certainly no reason to expect that $a * b$ will be the same as $b * a$ for all a and b. In fact, it can be seen in the above example that $1 * 2 = 2$, whereas $2 * 1 = 3$. One must consequently take care to refer to $a * b$ as the product of a and b and to $b * a$ as the product of b and a; the distinction is quite important. We should also point out that it is obviously possible to combine an element with itself. That is to say, $a * a$ can be defined.

DEFINITION 2.2 By a *mathematical system* or *algebraic system* we shall mean a nonempty set of elements together with one or more binary operations defined on this set.

A mathematical system consisting of the set S and a single binary operation $*$ will be denoted by the ordered pair $(S, *)$; analogously, a system consisting of the set S and two operations $*$ and $\circ$ will be represented by the ordered triple $(S, *, \circ)$.

Example 2.3
The pair $(S, \cdot)$, where the set $S = \{1, -1, i, -i\}$ and the operation is that of ordinary multiplication, is a mathematical system provided that one defines $i^2 = -1$.

Example 2.4
If Z_e and Z_o denote the even and odd integers, respectively, then $(Z_e, +, \cdot)$ constitutes a mathematical system, while $(Z_o, +, \cdot)$ does not. In the latter case the set Z_o is not closed under addition, since the sum of two odd integers is necessarily even.

The systems we are going to study are classified according to the properties they possess or, to put it another way, according to the axioms they satisfy. Our object will be to present a sequential development of the principal mathematical systems of modern algebra, beginning with those involving relatively few axioms and progressing to systems satisfying more detailed hypotheses.

The axioms that form the starting point of the abstract theory can be, by nature, rather varied. The growing tendency of modern mathematics is to isolate almost any convenient set of properties from its original context, to define a particular system, and to develop the corresponding abstract theory through logical deduction. Some of these formal axiomatic theories, such as the notion of a group, have a fundamental importance to the whole of mathematics and have been instrumental in unifying various apparently unrelated branches; other theories, while satisfying the esthetic and inquisitive needs of the mathematician, are limited in the extent of their applicability. We do not mean to create the impression that it is the usual practice for one to define a new system by arbitrarily (apart from logical considerations) writing down axioms. Although there is no particular necessity for the model to precede the theoretical development, in most cases the axioms are the abstract realization of the properties common to a variety of specific examples. With these general remarks out of the way, let us get down to work.

A set on which a single unrestricted binary operation is defined does not by itself yield a structure that is rich enough for our purposes; the concept, being too general, is poor in content. Certain reasonable limitations must be imposed on the operation if one is to obtain useful results. In the following paragraphs, some of the more basic requirements are named and briefly examined. For conciseness we shall hereafter omit the word "binary," inasmuch as every operation to be considered will necessarily be binary.

Given a mathematical system $(S, *)$, the symbol $a * b * c$ is at the moment completely meaningless, since the operation $*$ has been defined only for pairs of elements of S. However, if we make the stipulation that whenever quantities are enclosed in parentheses, these are to be evaluated first, then both the expressions $a * (b * c)$ and $(a * b) * c$ make sense. Namely, $a * (b * c)$ is to be interpreted as: combine a with what results from combining b with c; while $(a * b) * c$ is to be interpreted as: first combine a with b and then combine the result with c. Of course, the resulting elements $a * (b * c)$ and $(a * b) * c$ will not necessarily be the same.

DEFINITION 2.3 The operation $*$ defined on the set S is said to be *associative* if

$$a * (b * c) = (a * b) * c \qquad \text{(associative law)},$$

for every triple, distinct or not, of elements a, b, and c of S.

Example 2.5

The operation of subtraction on the set $R^{\#}$ of real numbers is not associative, since in general

$$a - (b - c) \neq (a - b) - c.$$

Example 2.6
An associative operation $*$ may be defined on Z, the set of integers, by taking $a * b = a + b + ab$. (We shall frequently delete the dot and write the product of a and b under ordinary multiplication simply as ab.) Then

$$\begin{aligned} a * (b * c) &= a * (b + c + bc) \\ &= a + (b + c + bc) + a(b + c + bc), \end{aligned}$$

while

$$\begin{aligned} (a * b) * c &= (a + b + ab) * c \\ &= (a + b + ab) + c + (a + b + ab)c. \end{aligned}$$

The equality of these two expressions follows in part from the fact that addition and multiplication are themselves associative in Z.

When dealing with a system whose operation is defined by a multiplication table rather than a formula, it is generally quite tedious to establish the associativity of the operation, since one must compute all possible threefold products. On the other hand, it may be far easier to show that the operation is not associative, since all we need do in this case is find three particular elements from the underlying set for which the associative law fails.

Example 2.7
Consider the operation $*$ defined on the set $S = \{1, 2, 3\}$ by the following operation table:

$*$	1	2	3
1	1	2	3
2	3	1	2
3	2	3	1

From this table we see that $2 * (1 * 3) = 2 * 3 = 2$, whereas $(2 * 1) * 3 = 3 * 3 = 1$; that is,

$$2 * (1 * 3) \neq (2 * 1) * 3.$$

The associative law thus fails to hold in the system $(S, *)$.

The mathematical system that we shall use to build up more complicated algebraic structure is known as a semigroup.

DEFINITION 2.4 A *semigroup* is a pair $(S, *)$ consisting of a nonempty set S together with an associative (binary) operation $*$ defined on S.

Let us stress that it is an abuse of language to say a certain set alone is a semigroup without also specifying the operation involved, since it may be quite possible to equip the same set with several associative operations. For this reason we have utilized the ordered pair notation to indicate both the operation and the underlying set of elements.

Observe that since any three elements from the set of a semigroup always associate, there is no particular reason for parentheses. Consequently, in dealing with such a system the symbol $a * b * c$ has meaning in the sense that we are free to interpret it either as $a * (b * c)$ or as $(a * b) * c$. More generally, the notation $a_1 * a_2 * \cdots * a_m$ is unambiguous, for it can be shown that all ways of inserting parentheses so as to give this expression a value yield the same result (Theorem 2.8). An operation that is not associative has the decided disadvantage that the notation for multiple-factored products can become quite unwieldy as a result of the constant need for parentheses.

To solidify the notation of a semigroup, we present several examples.

Example 2.8
There are several semigroups with which the reader is already familiar. If, for instance, Z_+ denotes the set of all positive integers, then both the pairs $(Z_+, +)$ and $(Z_+, \cdot)$ form semigroups. Similar statements hold for the sets Z, Q, and $R^\#$.

Example 2.9
Define the operation $*$ on the real numbers by the rule

$$a * b = \max\{a, b\}, \qquad a, b \in R^\#.$$

That is, $a * b$ is the larger of the two numbers a and b, or either one if $a = b$. Here, we have

$$a * (b * c) = \max\{a, b, c\} = (a * b) * c,$$

so that $(R^\#, *)$ satisfies the requirements of a semigroup.

Example 2.10
For any set X, each of the systems $(P(X), \cup)$ and $(P(X), \cap)$ constitutes a semigroup (Theorem 1.2).

Example 2.11
Let X be a nonempty set and S be the collection of all functions $f: X \to X$. If $\circ$ denotes functional composition, then the pair $(S, \circ)$ provides another illustration of a semigroup (Problem 5, Section 1.2).

As we shall subsequently see, the relevance of the semigroup concept lies in the fact that many important systems contain the semigroup structure as a subsystem.

We have already indicated that the order in which elements occur in a product is quite essential. If it is possible to interchange the order of combining any two elements from our set without affecting the result, then the operation is termed commutative.

DEFINITION 2.5 The operation $*$ defined on the set S is called *commutative* if

$$a * b = b * a \qquad \text{(commutative law)},$$

for every pair of elements $a, b \in S$.

Examples 2.8, 2.9, and 2.10 are of *commutative semigroups*—semigroups whose operations are commutative—while in Example 2.11, functional composition is not, in general, a commutative operation. Although the commutative law may fail to hold throughout an entire system, it may still be valid for particular pairs of elements; accordingly, it will be convenient to make the following definition.

DEFINITION 2.6 Two elements a and b are said to *commute* or *permute* with each other provided $a * b = b * a$.

Employing this terminology, we observe that the operation of the system $(S, *)$ is commutative if and only if every pair of elements of S commute.

Once an operation has been defined on a set, one finds that certain elements play special roles; there may exist identity elements and inverse elements.

DEFINITION 2.7 The system $(S, *)$ is said to have a (two-sided) identity element for the operation $*$ if there exists an element e in S such that

$$a * e = e * a = a$$

for every $a \in S$. An element e having this property is called an *identity element* (unit element, neutral element) for $(S, *)$.

An identity element thus causes each element of the set S to remain stationary under the operation. In particular, note that $e * e = e$. Of course, for a given system an identity element may or may not exist; in case an identity does exist, it must be unique, as the theorem below shows.

THEOREM 2.1 *A mathematical system $(S, *)$ has at most one identity element.*

Proof. For the proof, let us suppose that $(S, *)$ has two identity elements e and e'. Since $e * a = a$ for each $a \in S$, then in particular $e * e' = e'$. But on the other hand, e' is also an identity element, so we must have $e * e' = e$. We thus obtain $e = e * e' = e'$, and consequently, $e = e'$; that is, if the system actually has an identity, then there is precisely one element with this property.

It follows from Theorem 2.1 that, whenever $(S, *)$ has an identity, we are justified in using the expression "the identity element of $(S, *)$"; the symbol e will be reserved exclusively for this identity.

DEFINITION 2.8 A semigroup $(S, *)$ is said to be a *semigroup with identity* if there exists a (unique) identity element for $(S, *)$.

Example 2.12
The semigroup $(Z_+, \cdot)$ possesses an identity element, namely, the positive integer 1. On the other hand, the semigroup $(Z_+, +)$ has none, since $0 \notin Z_+$.

Example 2.13
Both the semigroups $(P(X), \cup)$ and $(P(X), \cap)$ have identities. Here, the empty set $\varnothing$ is the identity element for the union operation, since

$$A \cup \varnothing = \varnothing \cup A = A \qquad \text{for each set} \quad A \subseteq X.$$

As is easy to see, the universal set X acts as the identity element for the operation of intersection, since

$$A \cap X = X \cap A = A \qquad \text{for each set} \quad A \subseteq X.$$

Example 2.14
To record one more example of a semigroup with identity, consider the set of numbers

$$S = \{a + b\sqrt{2} \mid a, b \in Z\},$$

and the operation of ordinary multiplication. First, one is obliged to check that S is actually closed under multiplication; this is fairly clear, for if $a + b\sqrt{2}$ and $c + d\sqrt{2}$ are arbitrary members of S, then

$$(a + b\sqrt{2})(c + d\sqrt{2}) = (ac + 2bd) + (ad + bc)\sqrt{2} \in S.$$

It is not particularly difficult to establish that the pair $(S, \cdot)$ is a commutative semigroup with identity element $1 = 1 + 0\sqrt{2}$; we omit the argument.

When working with an operation that has an identity element, it is natural to inquire which elements of the underlying set, if any, have inverses.

DEFINITION 2.9 Let $(S, *)$ be a mathematical system with identity element e. An element $a \in S$ is said to have a (two-sided) inverse under the operation $*$ if there exists some member a' of S such that

$$a * a' = a' * a = e.$$

An element a' having this property is called an *inverse* of a and is customarily denoted by a^{-1}.

An inverse has the effect of reducing a given element, under the operation, to the identity element. In particular, since $e * e = e$, we may infer that $e^{-1} = e$.

The next theorem establishes that in a semigroup with identity, each element has at most one inverse relative to the (necessarily unique) identity element. In this setting, there is no danger of ambiguity in the symbol a^{-1}, and we are free to use the definite article in speaking of "the inverse of an element."

THEOREM 2.2 *In a semigroup $(S, *)$ with identity, each element has at most one inverse.*

Proof. We argue as in Theorem 2.1, supposing that two inverses exist and then showing that they are actually equal. To this end, let the element $a \in S$ have two inverses, say a' and a''. By the definition of inverse,

$$a * a' = a' * a = e \qquad \text{and} \qquad a * a'' = a'' * a = e$$

so that $a * a' = a * a''$. Multiplying both sides of the last-written equation on the left by a', we obtain

$$a' * (a * a') = a' * (a * a'').$$

The associativity of $*$ now gives $(a' * a) * a' = (a' * a) * a''$. But then $e * a' = e * a''$, or in equivalent terms, $a' = a''$, which ensures that the element a cannot possess more than one inverse.

Example 2.15

On the set Q of rational numbers, define the binary operation $*$ by

$$a * b = a + b + ab.$$

Then the system $(Q, *)$ forms a commutative semigroup with identity element 0. Each $a \in Q$, other than -1, has an inverse, namely, the element

$$a^{-1} = \frac{-a}{1 + a},$$

since

$$a * \frac{-a}{1+a} = a + \frac{-a}{1+a} + a\frac{-a}{1+a}$$
$$= \frac{a(1+a) - a - a^2}{1+a} = 0.$$

Example 2.16
Let X be a nonempty set and S be the collection of all functions $f: X \to X$. It is easy to see that the system $(S, \circ)$ is a semigroup with identity, having as its identity the identity map i_X. A function $f \in S$ will possess an inverse relative to the operation of composition if and only if f is a one-to-one mapping from X onto itself; in this event the inverse of f under $\circ$ is the usual inverse function f^{-1}:

$$f \circ f^{-1} = i_X = f^{-1} \circ f.$$

Example 2.17
As a further illustration of these ideas, let us return to the semigroup $(P(X), \cup)$ of Example 2.13. In this case, just the empty set $\varnothing$ possesses an inverse; if $A \in P(X)$, with $A \neq \varnothing$, there is no subset $A^{-1} \in X$ such that $A \cup A^{-1} = \varnothing$. Likewise, in regard to the semigroup $(P(X), \cap)$, the only member of $P(X)$ that has an inverse is the universal set X.

There is a mathematical system, known as a group, that displays most of the properties we have so far discussed.

DEFINITION 2.10 The pair $(G, *)$ is a *group* if and only if $(G, *)$ is a semigroup with identity, in which each element of G has an inverse.

While the above definition is perfectly acceptable, we prefer to rephrase it in the following more detailed form, merely as a matter of convenience.

DEFINITION 2.11 A *group* is a pair $(G, *)$ consisting of a nonempty set G and a binary operation $*$ defined on G, satisfying the four requirements:

1. G is closed under the operation $*$,
2. the operation $*$ is associative,
3. G contains an identity element e for the operation $*$, and
4. each element a of G has an inverse $a^{-1} \in G$, relative to $*$.

This definition calls for several remarks. For one thing, the first of the requirements cited above could easily have been omitted, since any set is closed with respect to a binary operation defined on it. We merely wish to emphasize that one must always check the closure condition. Observe particularly that commutativity is not required in the definition. If it happens that the group operation satisfies this additional hypothesis, then $(G, *)$ is referred to as a *commutative* or *abelian group*. Let us also point out that it is possible to give a less redundant version of the group axioms from which the present axioms follow as logical consequences; for this, we refer the reader to Problem 12 at the end of the section.

So that the reader may appreciate the generality of the group concept and at the same time gain some familiarity with this idea, we offer the following selection of examples.

Example 2.18

Let a be any nonzero real number and consider the set G of integral multiples of a:

$$G = \{na | n \in Z\}.$$

The system $(G, +)$, where ordinary addition is the operation, forms a commutative group. In this case the identity is $0 = 0a \in G$, while the inverse of a typical element na of G is simply $-(na) = (-n)a \in G$.

Example 2.19

For an example of a noncommutative group, take the set G to consist of all ordered pairs of real numbers with nonzero first component:

$$G = \{(a, b) | a, b \in R^{\#}; a \neq 0\}.$$

If we define the operation $*$ on G by the rule

$$(a, b) * (c, d) = (ac, bc + d),$$

then it is easily checked that $(G, *)$ meets all the requirements of a group.

The associativity of the operation $*$ is a consequence of the familiar properties of the real numbers; to spell out the details,

$$\begin{aligned}
[(a, b) * (c, d)] * (e, f) &= (ac, bc + d) * (e, f) \\
&= ((ac)e, (bc + d)e + f) \\
&= (a(ce), b(ce) + (de + f)) \\
&= (a, b) * (ce, de + f) \\
&= (a, b) * [(c, d) * (e, f)].
\end{aligned}$$

There is no problem in verifying that the ordered pair $(1, 0)$ plays the role of the identity element, while the inverse of $(a, b) \in G$ is $(1/a, -b/a)$. To see that our operation is a noncommutative one, note that

$$(1, 2) * (3, 4) = (3, 10) \neq (3, 6) = (3, 4) * (1, 2).$$

Example 2.20

Consider the power set $P(X)$ of some nonempty set X. As we know, the systems $(P(X), \cup)$ and $(P(X), \cap)$ possess identity elements $\varnothing$ and X, respectively, but neither system admits inverses for elements other than the identity. In short, $P(X)$ does not constitute a group with respect to the formation of either unions or intersections.

It is possible to remedy this drawback by introducing another operation in $P(X)$ that does lead to a group structure. Specifically, let us define the operation Δ by

$$A \,\Delta\, B = (A - B) \cup (B - A), \qquad A, B \in P(X).$$

This operation, known as the *symmetric difference of A and B*, produces the set that is represented by the shaded area in Figure 2.1.

The verification that the symmetric difference operation is both commutative and associative is left as an exercise. For any subset $A \subseteq X$ we have

$$A \,\Delta\, \varnothing = (A - \varnothing) \cup (\varnothing - A) = A \cup \varnothing = A,$$

which guarantees that the empty set $\varnothing$ serves as an identity element for Δ. Furthermore,

$$A \,\Delta\, A = (A - A) \cup (A - A) = \varnothing \cup \varnothing = \varnothing,$$

so each member of $P(X)$ happens to be its own inverse. All this supports the conclusion that $(P(X), \Delta)$ forms a commutative group.

Example 2.21

Here is another illustration of a noncommutative group: Let the set G consist of the six functions f_i ($i = 1, 2, \ldots, 6$) from $R^{\#} - \{0, 1\}$ into

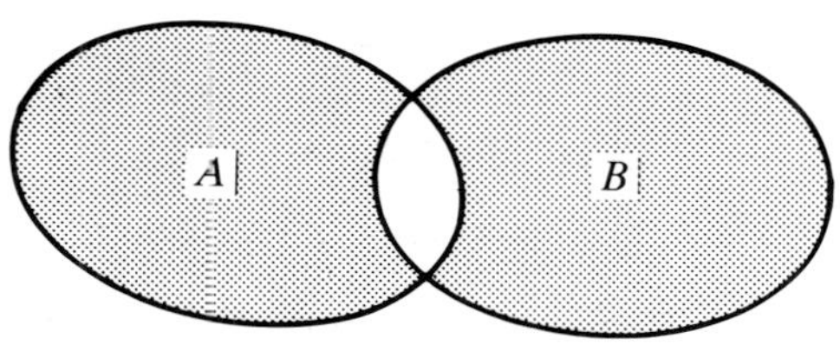

Figure 2.1

itself, where

$$f_1(x) = x, \qquad f_2(x) = 1/x, \qquad f_3(x) = 1 - x,$$

$$f_4(x) = \frac{1}{1-x}, \quad f_5(x) = \frac{x-1}{x}, \quad f_6(x) = \frac{x}{x-1}.$$

The set G together with the operation of functional composition is then a group (since composition is associative, $(G, \circ)$ certainly forms a semigroup).

As regards commutativity, we find that

$$(f_2 \circ f_4)(x) = f_2(f_4(x)) = f_2\left(\frac{1}{1-x}\right)$$

$$= \frac{1}{1/(1-x)} = 1 - x = f_3(x)$$

for all $x \in R^{\#} - \{0, 1\}$, whence the equality $f_2 \circ f_4 = f_3$. By contrast,

$$(f_4 \circ f_2)(x) = f_4(f_2(x)) = f_4(1/x)$$

$$= \frac{1}{1-(1/x)} = \frac{x}{x-1} = f_6(x)$$

so that $f_4 \circ f_2 = f_6$. Thus $(G, \circ)$ is not a commutative group.

We leave it to the reader to check that the following operation table has been correctly filled in.

$\circ$	f_1	f_2	f_3	f_4	f_5	f_6
f_1	f_1	f_2	f_3	f_4	f_5	f_6
f_2	f_2	f_1	f_4	f_3	f_6	f_5
f_3	f_3	f_5	f_1	f_6	f_2	f_4
f_4	f_4	f_6	f_2	f_5	f_1	f_3
f_5	f_5	f_3	f_6	f_1	f_4	f_2
f_6	f_6	f_4	f_5	f_2	f_3	f_1

Examining the table, one sees that f_1 is the identity element and that the various inverses are

$$f_1^{-1} = f_1, \quad f_2^{-1} = f_2, \quad f_3^{-1} = f_3, \quad f_4^{-1} = f_5, \quad f_5^{-1} = f_4,$$
$$f_6^{-1} = f_6.$$

Example 2.22
The group to be introduced next is known as the *group of symmetries of the square*. To visualize this geometrically, imagine a cardboard square having its sides parallel to the axes of a coordinate system and its center at the origin; numbers are painted at the corners, as indicated in Figure 2.2.

The elements of the underlying set G_s are taken to be certain rigid motions of the square, that is, displacements throughout which the shape of the square is unaltered. Permitted motions are the following: the identity motion e in which the square remains fixed; clockwise rotations r_1, r_2, and r_3 about the center through angles of 90, 180, and 270 degrees, respectively; reflections (flips out of the plane and back into it) h and v about horizontal and vertical lines through the center; reflections d_1 and d_2 about the indicated diagonal lines.

Two such motions can be "multiplied" by performing them in succession. If x and y are motions, then $x * y$ is to be interpreted as the motion that achieves the same result as y followed by x. For example, $h * r_1$ is that element of G_s that has the same net effect as r_1 (a clockwise rotation through 90°) followed by h (a flip, exactly once, over the horizontal). By keeping track of where each numbered corner goes we

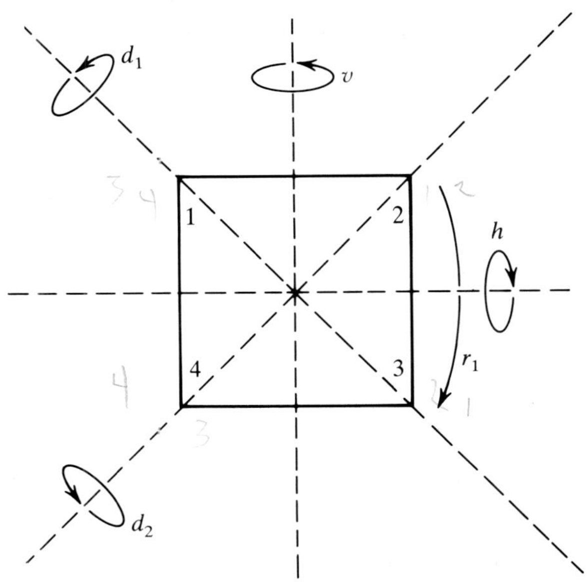

Figure 2.2

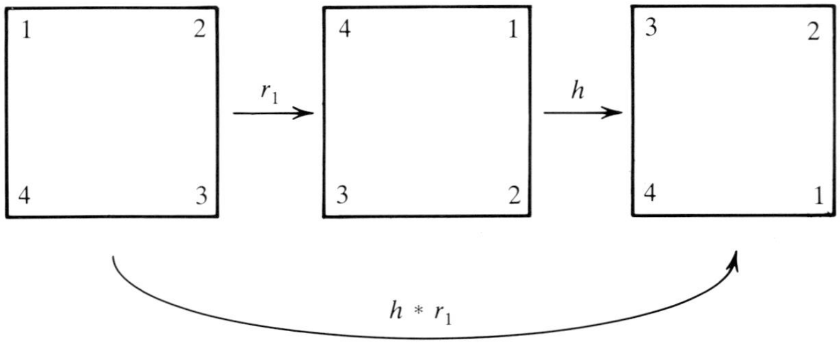

Figure 2.3

see that $h * r_1$ produces the same result as the single motion d_2; hence $h * r_1 = d_2$. This situation is pictured in Figure 2.3.
A similar analysis shows that $r_1 * h = d_1$, from which it may be inferred that our notion of multiplication is not commutative.

The complete multiplication table for $(G_s, *)$ appears below.

$*$	e	r_1	r_2	r_3	h	v	d_1	d_2
e	e	r_1	r_2	r_3	h	v	d_1	d_2
r_1	r_1	r_2	r_3	e	d_1	d_2	v	h
r_2	r_2	r_3	e	r_1	v	h	d_2	d_1
r_3	r_3	e	r_1	r_2	d_2	d_1	h	v
h	h	d_2	v	d_1	e	r_2	r_3	r_1
v	v	d_1	h	d_2	r_2	e	r_1	r_3
d_1	d_1	h	d_2	v	r_1	r_3	e	r_2
d_2	d_2	v	d_1	h	r_3	r_1	r_2	e

A quick inspection of the table tells us that e acts as the identity element and each of e, r_2, h, v, d_1 and d_2 is its own inverse, whereas r_1 and r_3 are inverses of one another. The associative law also holds, but this is not immediately obvious. We shall see later that the symmetries of the square are equivalent to a group of permutations of the set $\{1, 2, 3, 4\}$—observe that a symmetry is completely described by the changes that occur in the vertices—and associativity therefore follows from the associativity of functional composition. Granting this for the moment, the proof that $(G_s, *)$ constitutes a group is complete.

Similar groups may be defined for other geometric figures, in fact, for any regular n-sided polygon. Problem 13 at the end of this section deals with the group of symmetries of the equilateral triangle.

Example 2.23

A 2×2 matrix with entries from $R^{\#}$ is an ordered quadruple (a, b, c, d) of real numbers arranged as a square array enclosed by brackets:

$$A = \begin{bmatrix} a & b \\ c & d \end{bmatrix}.$$

We designate the set of all such matrices by the symbol $M_2(R^{\#})$, that is,

$$M_2(R^{\#}) = \left\{ \begin{bmatrix} a & b \\ c & d \end{bmatrix} \middle| a, b, c, d \in R^{\#} \right\}.$$

Being ordered quadruples, two elements of $M_2(R^{\#})$ are equal whenever their corresponding entries are the same. For instance,

$$\begin{bmatrix} 1 & 2 \\ 3 & 4 \end{bmatrix} = \begin{bmatrix} 2/2 & 2 \\ 3 & 4 \end{bmatrix}, \quad \text{but} \quad \begin{bmatrix} 1 & 2 \\ 3 & 4 \end{bmatrix} \neq \begin{bmatrix} 1 & 3 \\ 2 & 4 \end{bmatrix}.$$

On the set $M_2(R^{\#})$ an operation of addition can be defined by simply adding the corresponding entries of the matrices to obtain the entries of the sum

$$\begin{bmatrix} a & b \\ c & d \end{bmatrix} + \begin{bmatrix} a' & b' \\ c' & d' \end{bmatrix} = \begin{bmatrix} a + a' & b + b' \\ c + c' & d + d' \end{bmatrix}.$$

To consider a specific instance,

$$\begin{bmatrix} 3 & 2 \\ 1 & 1 \end{bmatrix} + \begin{bmatrix} 2 & 4 \\ -1 & 6 \end{bmatrix} = \begin{bmatrix} 5 & 6 \\ 0 & 7 \end{bmatrix}.$$

Since matrices are added entry by entry, the properties of addition that hold for the real numbers are also satisfied by matrix addition. In particular, the associative and commutative laws hold. The matrix

$$O = \begin{bmatrix} 0 & 0 \\ 0 & 0 \end{bmatrix},$$

all of whose entries are zero, is called the zero matrix. There is no difficulty in checking that $A + O = O + A = A$ for all matrices A in $M_2(R^{\#})$, whence the zero matrix acts as the identity element for our operation. The matrix A has an inverse under addition, namely,

$$\begin{bmatrix} -a & -b \\ -c & -d \end{bmatrix}.$$

In short, the system $(M_2(R^{\#}), +)$ forms a commutative group.

A second binary operation, that of multiplication of matrices, can be defined so that an appropriate subset of $M_2(R^{\#})$ yields a noncommutative group. This operation is a little more complicated than the previous one. The definition is

$$\begin{bmatrix} a & b \\ c & d \end{bmatrix} \cdot \begin{bmatrix} a' & b' \\ c' & d' \end{bmatrix} = \begin{bmatrix} aa' + bc' & ab' + bd' \\ ca' + dc' & cb' + dd' \end{bmatrix}.$$

A brief inspection shows that each of the four entries on the right-hand side is obtained by combining a certain row of the first matrix with a certain column of the second matrix in the indicated manner. The scheme for carrying it out is often called row-by-column multiplication. An illustration will make the procedure clearer:

$$\begin{bmatrix} 2 & 1 \\ 0 & 2 \end{bmatrix} \cdot \begin{bmatrix} 1 & 2 \\ 6 & -1 \end{bmatrix} = \begin{bmatrix} 2\cdot 1 + 1\cdot 6 & 2\cdot 2 + 1(-1) \\ 0\cdot 1 + 2\cdot 6 & 0\cdot 2 + 2(-1) \end{bmatrix} = \begin{bmatrix} 8 & 3 \\ 12 & -2 \end{bmatrix}.$$

There is good reason for defining matrix multiplication in this way, but at the moment we are in no position to justify our curious definition other than to say that it arises from consideration of the change of variables in linear equations. (It was the search for a convenient notation for describing systems of linear equations that inspired Cayley to introduce matrices in 1850.)

The burden of performing the arithmetical work required for verifying that matrix multiplication is an associative operation of $M_2(R^{\#})$ is left to the reader; it is a straightforward calculation but somewhat tedious. Of special significance is the matrix

$$I = \begin{bmatrix} 1 & 0 \\ 0 & 1 \end{bmatrix}.$$

It will be found that $A \cdot I = I \cdot A = A$ is satisfied for all matrices A in $M_2(R^{\#})$, making I the multiplicative identity element for this operation.

The matter of inverses is not so simple. It is evident that some matrices, such as the zero matrix, will not possess an inverse under multiplication. To ascertain the subset of matrices in $M_2(R^{\#})$ that do have inverses, let us start by attempting to solve the matrix equation

$$\begin{bmatrix} a & b \\ c & d \end{bmatrix} \cdot \begin{bmatrix} x & y \\ z & w \end{bmatrix} = \begin{bmatrix} 1 & 0 \\ 0 & 1 \end{bmatrix}.$$

When the multiplication is carried out and the corresponding entries in the resulting matrices are equated, we obtain a system of four equations in x, y, z, w that must be satisfied:

$$ax + bz = 1, \qquad ay + bw = 0,$$

$$cx + dz = 0, \qquad cy + dw = 1.$$

A little algebraic manipulation produces

$$x(ad - bc) = d, \qquad y(ad - bc) = -b,$$

$$z(ad - bc) = -c, \qquad w(ad - bc) = a.$$

If the number $\Delta = ad - bc \neq 0$, then the system has a solution

$$x = d/\Delta, \qquad y = -b/\Delta, \qquad z = -c/\Delta, \qquad w = a/\Delta.$$

(The number Δ is called the *determinant* of the matrix A and is frequently denoted by $\det A$.) Putting

$$A^{-1} = \begin{bmatrix} d/\Delta & -b/\Delta \\ -c/\Delta & a/\Delta \end{bmatrix},$$

it is easily checked that $A \cdot A^{-1} = A^{-1} \cdot A = I$, which makes A^{-1} the inverse of A. The point to bear in mind is that a matrix A in $M_2(R^{\#})$ has an inverse under matrix multiplication, provided that $\det A = ad - bc \neq 0$. The set of all such matrices may be conveniently denoted by $M_2^*(R^{\#})$; in other words,

$$M_2^*(R^{\#}) = \left\{ \begin{bmatrix} a & b \\ c & d \end{bmatrix} \middle| a, b, c, d \in R^{\#};\ ad - bc \neq 0 \right\}.$$

Collecting these results, we conclude that the system $(M_2^*(R^{\#}), \cdot)$ forms a group, called the multiplicative group of 2×2 invertible matrices with real entries. To see that this group is not commutative, it will be sufficient to exhibit two matrices A and B in $M_2(R^{\#})$ for which $A \cdot B \neq B \cdot A$. This is easy enough:

$$\begin{bmatrix} 1 & 1 \\ 0 & 1 \end{bmatrix} \cdot \begin{bmatrix} 0 & 1 \\ 1 & 0 \end{bmatrix} = \begin{bmatrix} 1 & 1 \\ 1 & 0 \end{bmatrix},$$

while on the other hand,

$$\begin{bmatrix} 0 & 1 \\ 1 & 0 \end{bmatrix} \cdot \begin{bmatrix} 1 & 1 \\ 0 & 1 \end{bmatrix} = \begin{bmatrix} 0 & 1 \\ 1 & 1 \end{bmatrix}.$$

Consequently, matrix multiplication does not satisfy the commutative property, even though multiplication in $R^{\#}$ is commutative.

It is important to realize that a solid foundation of any abstract theory requires clear and precise definitions of the basic terms used. From an historical perspective the first decisive step in the direction of an abstract group concept was Arthur Cayley's 1854 paper. Cayley observed:

> A set of symbols $1, \alpha, \beta, \ldots$ all of them different, and such that the product of any two of them, or the product of any one of them into itself, belongs to the set, is said to be a group.

We infer that the symbol 1 denotes the identity element or, in his words, "an operation that leaves the operand unaltered." Cayley was aware that the associative law must be valid (the term "associative" seems to have been introduced by W. R. Hamilton in 1835); he stated the axiom in its usual form:

> The symbols θ, ϕ, ... are in general such that $\theta.\phi\chi = \theta\phi.\chi$, etc., so that $\theta\phi\chi$, $\theta\phi\chi\omega$, etc. have a definite signification independent of the particular mode of compounding the symbols.

Allowing for his failure to postulate expressly the existence of inverse elements, the modern conception of a group is clearly recognizable in Cayley's writing.

During the next few years there was no general agreement as to what exactly constituted a group. Most practicing mathematicians were oblivious to the advantages of a definition based on axioms, occasionally using the word "group" as a technical term but with entirely different meanings. It is customary to credit the publication of the earliest explicit set of postulates for an abstract group to the German mathematicians Leopold Kronecker (1823–1891) and Heinrich Weber (1842–1913). Although Weber's definition has the drawback of being limited to finite groups, it does specify an abstract operation on a set of undefined "elements" that satisfies the closure condition, the associative law, and a property that will ensure the existence of an identity and of inverse elements (see Problem 19). As formulated in 1882, it reads:

> A system G of h elements $\theta_1, \theta_2, \ldots, \theta_h$ is called a group if
>
> I. By means of some rule, we obtain from any two elements of the system a new element of the system; in symbols, $\theta_r\theta_s = \theta_t$.
>
> II. $(\theta_r\theta_s)\theta_t = \theta_r(\theta_s\theta_t)$
>
> III. $\theta\theta_r = \theta\theta_s$ and $\theta_r\theta = \theta_s\theta$ always imply $\theta_r = \theta_s$.

A few years later (1893), Weber modified this definition to include the case of infinite groups, starting with the words, "A system G of finitely or infinitely many arbitrary things (elements) forms a group if. . . ."

By 1893, Weber had defined the concept of an abstract group in the accepted technical sense, so that nobody could misunderstand what the term meant. In the years immediately following, various definitions (some being minor variations of each other) began to appear at a much more rapid rate. The set of axioms that is in favor today makes use of an identity element and the notion of inverses; that is to say, condition III in Weber's definition is replaced by

> III′. G has an identity element e such that $ae = ea = a$ for all a in G.
>
> IV′. For each element a in G, there is some element a^{-1} in G with the property that $aa^{-1} = a^{-1}a = e$.

This definition was obtained independently, and almost simultaneously, by the American mathematicians E. H. Moore (1862–1932) and James Pierpont (1866–1938). Moore used the above definition of a group in his lectures at the University of Chicago during the winter of 1896; Pierpont presented it at Buffalo in September 1896, at the first colloquium sponsored by the newly formed American Mathematical Society. Thus the 1890's may be seen as marking the close of the attempt, begun by Cayley in 1854, to build an abstract group concept.

PROBLEMS 2.2

1. Assume that the system $(S, *)$ has an identity element. If the equation

$$(a * b) * (c * d) = (a * c) * (b * d)$$

holds for all choices of $a, b, c, d \in S$, prove that the operation $*$ is both associative and commutative.

2. Verify that the system $(Q_+, *)$ forms a commutative semigroup if the operation $*$ is defined by

$$a * b = \frac{ab}{a + b}.$$

3. Let S be a set containing more than one element. If the operation $*$ is defined by $a * b = b$, verify that $(S, *)$ is a noncommutative semigroup that admits no identity element.

4. Let $(S, *)$ be a semigroup and e be any element that is not in S. Extend the binary operation to $S' = S \cup \{e\}$ by means of the following definitions:

$$e * e = e, \qquad e * a = a * e = a \qquad \text{for } a \in S.$$

Show that the system $(S', *)$ constitutes a semigroup with identity.

5. Define the binary operations $*$ and $*'$ on the set Z_+ of positive integers by taking

$$a * b = \gcd(a, b) \qquad \text{and} \qquad a *' b = \operatorname{lcm}(a, b).$$

Prove that $(Z_+, *)$ and $(Z_+, *')$ are both commutative semigroups and that the latter system has an identity element and the former does not. [*Hint*: See Problem 6, Section 1.3.]

6. If $(S, *)$ is a semigroup with identity and G denotes the set of elements of S that have inverses with respect to the operation $*$, show that $(G, *)$ forms a group.

7. Let $S = \{a, b, c\}$ be a set having three elements. If the operation $*$ is defined by means of the table below, establish that $(S, *)$ is a semigroup with identity, but not a group.

$*$	a	b	c
a	a	b	c
b	b	a	c
c	c	c	c

8. In the following cases, determine whether the systems $(G, *)$ are commutative groups; for those systems failing to be so, indicate which axioms are not satisfied.

a) $G = Z$, $a * b = 0$

b) $G = Z$, $a * b = \min\{a, b\}$ ($\min\{a, b\}$ stands for "the smaller of a and b")

c) $G = Q - \{1\}$, $a * b = a + b - ab$

d) $G = Z_+$, $a * b = \max\{a, b\} - \min\{a, b\}$

e) $G = Z \times Z$, $(a, b) * (c, d) = (a + c, b + d)$

f) $G = R^\# \times R^\#$, $(a, b) * (c, d) = (ac - bd, ad + bc)$

9. Suppose that the element $a \in R^\# - \{0\}$ and consider the set G of integral powers of a; that is, $G = \{a^n | n \in Z\}$. If $\cdot$ denotes ordinary multiplication, prove that $(G, \cdot)$ is a commutative group.

10. Let G be the set of mappings $f_i (i = 1, 2, 3, 4)$ from $R^\# - \{0\}$ into itself, where

$$f_1(x) = x, \qquad f_2(x) = 1/x, \qquad f_3(x) = -x, \qquad f_4(x) = -1/x$$

for each $x \in R^\# - \{0\}$. Verify that the system $(G, \circ)$ forms a commutative group.

11. Prove that the symmetric difference operation Δ defined by

$$A \Delta B = (A - B) \cup (B - A)$$

may also be given by the rule

$$A \Delta B = (A \cup B) - (B \cap A).$$

12. Prove that the following weakened set of axioms for a group are equivalent to those given in Definition 2.11: A group is a mathematical system $(G, *)$ for which

a) $*$ is an associative operation,

b) G contains an element e such that $a * e = a$ for all $a \in G$ (existence of a right identity),

c) for each $a \in G$ there exists an element $a^{-1} \in G$ such that $a * a^{-1} = e$ (existence of right inverses).

[*Hint*: Put $b = a^{-1} * a$ and show that $b^2 = b$; then $b = b * b^{-1} = e$.]

13. Let the set G_t consist of certain rigid motions of an equilateral triangle. Permitted motions are as follows: the identity motion e which leaves the triangle alone; two clockwise rotations r_1 and r_2 about the center through angles of 120 and 240 degrees, respectively; three reflections l_1, l_2, and l_3 about the lines L_1, L_2, and L_3 as indicated (see Figure 2.4). If $x * y$ is understood to mean "first perform the motion y and then x," it follows that $*$ is a binary operation on G_t. Show that the system $(G_t, *)$ forms a group, known as the *group of symmetries of the equilateral triangle.*

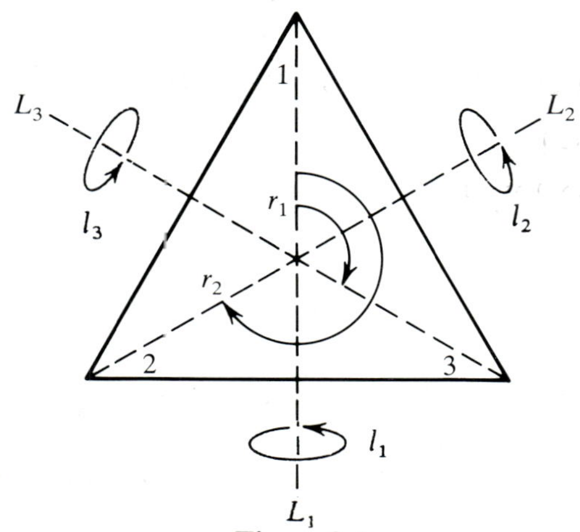

Figure 2.4

14. Let the set G be the subset of $M_2(R^\#)$ consisting of the matrices

$$\begin{bmatrix}1&0\\0&1\end{bmatrix},\ \begin{bmatrix}-1&0\\0&1\end{bmatrix},\ \begin{bmatrix}1&0\\0&-1\end{bmatrix},\ \begin{bmatrix}-1&0\\0&-1\end{bmatrix}.$$

If $\cdot$ denotes matrix multiplication, show that $(G, \cdot)$ is a commutative group. Write the multiplication table for this group.

15. Using the sets G below and taking $\cdot$ to be matrix multiplication, prove that each of the systems $(G, \cdot)$ is a group:

a) $G = \left\{\begin{bmatrix}a&b\\0&d\end{bmatrix}\middle| a, b, d \in R^\#;\ ad \neq 0\right\}$;

b) $G = \left\{\begin{bmatrix}a&0\\c&1\end{bmatrix}\middle| a, c \in R^\#;\ a \neq 0\right\}$;

c) $G = \left\{\begin{bmatrix}a&-b\\b&a\end{bmatrix}\middle| a, b \in R^\#;\ a \text{ and } b \text{ not both zero}\right\}$;

d) $G = \left\{\begin{bmatrix}a&a\\0&0\end{bmatrix}\middle| a \in R^\# - \{0\}\right\}$;

e) $G = \left\{\begin{bmatrix}1&b\\0&1\end{bmatrix}\middle| b \in Z\right\}$

16. Let the set G consist of the following six matrices:

$$\begin{bmatrix}1&0\\0&1\end{bmatrix},\ \begin{bmatrix}1&0\\0&-1\end{bmatrix},\ \begin{bmatrix}-1/2&\sqrt{3}/2\\\sqrt{3}/2&1/2\end{bmatrix},$$

$$\begin{bmatrix}-1/2&-\sqrt{3}/2\\-\sqrt{3}/2&1/2\end{bmatrix},\ \begin{bmatrix}-1/2&\sqrt{3}/2\\-\sqrt{3}/2&-1/2\end{bmatrix},\ \begin{bmatrix}-1/2&-\sqrt{3}/2\\\sqrt{3}/2&-1/2\end{bmatrix}.$$

Establish that $(G, \cdot)$ is a group, where $\cdot$ is matrix multiplication. Construct a multiplication table.

17. Using the indicated sets G and taking $\cdot$ to be ordinary multiplication, show that each of the systems $(G, \cdot)$ forms a group:

a) $G = \{a + b\sqrt{2} \mid a, b \in Q;\ a \text{ and } b \text{ not both zero}\}$;

b) $G = \{1, r, r^2\}$, where r is a root of the equation $x^2 + x + 1 = 0$;

c) $G = \left\{1,\ -1,\ \dfrac{1+\sqrt{3}i}{2},\ \dfrac{1-\sqrt{3}i}{2},\ \dfrac{-1+\sqrt{3}i}{2},\ \dfrac{-1-\sqrt{3}i}{2}\right\}$, where $i^2 = -1$.

18. Let $(S, *)$ be a semigroup with the property that for any two elements a, $b \in S$ the equations

$$a * x = b \quad \text{and} \quad y * a = b$$

have solutions x and y in S. Verify that $(S, *)$ forms a group. [*Hint*: Appeal to Problem 12.]

19. Assume that $(S, *)$ is a semigroup in which the operation $*$ satisfies the following condition: If $a, b, c \in S$, then $a * b = a * c$ or $b * a = c * a$ implies that $b = c$.

a) If the set S is finite, prove that $(S, *)$ is a group. [*Hint*: Let $a_1, a_2, \ldots, a_n$ be the elements of S and a be a particular element. The products $aa_1, aa_2, \ldots, aa_n$ are all distinct and hence account for the elements of S. Thus if $b \in S$, then $b = aa_i$ for some i. Now refer to Problem 18.]

b) Show that the conclusion of part (a) may be false if S is an infinite set.

2.3 CERTAIN ELEMENTARY THEOREMS ON GROUPS

As remarked earlier, our approach to the subject matter in this text emphasizes the use of the abstract axiomatic methods. Thus each system to be investigated consists of a set of elements and one or more operations that are undefined except that they are assumed to obey certain rules known as the axioms or postulates of the system. Once the axioms are granted, all else is obtained by rigorous, logical argument. Each new theorem is deduced from the definitions, the axioms, and the theorems proved previously. The end result of studying algebraic systems by considering only their abstract characterizations is the development of a single mathematical theory that is applicable to a great many diverse examples; any particular example that we encounter that satisfies the axioms of a given system will also satisfy all the theorems that are true for that system (this relieves us from the task of going over what amounts to the same line of reasoning again and again). Although the uninitiated reader may find the axiomatic approach uncomfortably abstract at first, the feeling should diminish as one grows more accustomed to it. Along the way, we hope that the reader will gain an appreciation of mathematics as an exacting, logical discipline.

A few words about notational convention are in order: If there is no danger of confusion, it is customary to designate a group $(G, *)$ simply by the set symbol G. This allows us to speak of "the group G," when logically the emphasis should not be on the set alone; the context will make it clear whether G is to be interpreted as a set or as a set together with an associated binary operation. Since the symbol used to represent the group operation is generally not important, another common practice is to omit this symbol from all calculations. In consequence, multiplication of elements is hereafter denoted by juxtaposition, ab taking the place of $a * b$. The letter e will continue to stand for the group identity, while the inverse of an arbitrary element $a \in G$ is indicated by a^{-1}. All this being agreed, let us turn now to our formal development of group theory.

The theorem below is fairly obvious, but nonetheless worth putting on record.

Theorem 2.3 *The identity element of a group G is unique, and each member of G has exactly one inverse.*

Proof. Since any group is automatically a semigroup, Theorems 2.1 and 2.2 are applicable.

Our next two results deal with some of the properties of inverses in groups.

Theorem 2.4 *If G is a group, then $(a^{-1})^{-1} = a$ for all $a \in G$.*

Proof. By definition, $(a^{-1})^{-1}$ is the inverse of a^{-1}:

$$(a^{-1})^{-1}(a^{-1}) = (a^{-1})(a^{-1})^{-1} = e.$$

Since $aa^{-1} = a^{-1}a = e$, the element a is also an inverse of a^{-1}. But we know that inverses of group elements are unique, and so $(a^{-1})^{-1} = a$.

Theorem 2.5 *If G is a group, then $(ab)^{-1} = b^{-1}a^{-1}$ for all $a, b \in G$; in words, the inverse of a product of two group elements is the product of their inverses in reverse order.*

Proof. It suffices to show that

$$(ab)(b^{-1}a^{-1}) = (b^{-1}a^{-1})(ab) = e.$$

This would tell us that $b^{-1}a^{-1}$ is an inverse of ab, and hence by the uniqueness of inverses the conclusion would follow. By the associative property of the group operation we get

$$(ab)(b^{-1}a^{-1}) = ((ab)b^{-1})a^{-1} = (a(bb^{-1}))a^{-1}$$
$$= (ae)a^{-1} = aa^{-1} = e.$$

A similar computation leads to $(b^{-1}a^{-1})(ab) = e$.

To be sure, if the group operation is commutative, then we do have $(ab)^{-1} = a^{-1}b^{-1}$ for all $a, b \in G$. But in the absence of commutativity the inverse of a product need not be equal to the product of the inverses in direct order. Since this is a frequent source of error, it may be instructive to look at a concrete illustration. As was seen earlier, the set G of ordered pairs of real numbers having a nonzero first component forms a group under the operation $*$ defined by

$$(a, b) * (c, d) = (ac, bc + d).$$

The pair $(1, 0)$ serves as the identity element for this group; and the inverse of an element $(a, b) \in G$ is $(1/a, -b/a)$. A quick calculation indicates that

$$((1, 2) * (-1, 3))^{-1} = (-1, 1)^{-1} = (-1, 1)$$

while, at the same time,

$$(-1, 3)^{-1} * (1, 2)^{-1} = (-1, 3) * (1, -2) = (-1, 1).$$

Thus the equation $((1, 2) * (-1, 3))^{-1} = (-1, 3)^{-1} * (1, 2)^{-1}$ holds, as guaranteed by Theorem 2.5. Now let us compute the product of the inverses in the opposite order; one finds that

$$(1, 2)^{-1} * (-1, 3)^{-1} \neq (1, -2) * (-1, 3) = (-1, 5)$$

and so $((1, 2) * (-1, 3))^{-1} \neq (1, 2)^{-1} * (-1, 3)^{-1}$. This inequality is not a complete surprise, since the group under consideration happens to be noncommutative.

Among the distinctive types of elements that a given system may or may not possess are those that are cancellable: An element a of the system S is called *left-cancellable* or is said to satisfy the *left-cancellation law* if $ax = ay$ implies that $x = y$ for all x, y in S; *right-cancellable* elements are defined in a similar manner. We speak of a as being *cancellable* if it is both left- and right-cancellable.

THEOREM 2.6 *Every element of a group G is cancellable.*

Proof. Consider the equation $ax = ay$, where $a, x, y \in G$. Since a^{-1} exists in G, we may multiply both sides of this equation on the left by a^{-1} to arrive at $a^{-1}(ax) = a^{-1}(ay)$. Use of the associative law gives $(a^{-1}a)x = (a^{-1}a)y$ or $ex = ey$; hence $x = y$, which makes a a left-cancellable element. A similar argument shows that cancellation on the right is also permissible.

COROLLARY *In any group G the only element satisfying $xx = x$ is $x = e$.*

Proof. The conclusion is immediate from the cancellation laws and the fact that $xx = xe$.

As for terminology, an element x of the system S is said to be *idempotent* with respect to the given binary operation provided that $xx = x$. The preceding corollary asserts that the sole idempotent in a group is the identity element. For an illustration of a system in which every element is idempotent, the reader is referred to Problem 5 of Section 2.2.

Another feature of groups is that they permit us to solve what might be called "linear equations." Formally stated,

THEOREM 2.7 *If a, b are given elements of the group G, then the equations $ax = b$ and $ya = b$ have unique solutions in G.*

Proof. First, observe that $x = a^{-1}b$ satisfies the equation $ax = b$; for $a(a^{-1}b) = (aa^{-1})b = eb = b$. This shows that at least one solution exists in G. To see that the solution is unique, suppose that there is some other element $x' \in G$ for which $ax' = b$. Then $ax' = a(a^{-1}b)$, and so we may left-cancel to get $x' = a^{-1}b = x$. The second part of the theorem is proved in an analogous manner.

From this we may derive the following corollary:

COROLLARY *In the multiplication table for a group G, every element appears exactly once in each row and column.*

Proof. For a proof by contradiction, assume that the element b occurred twice in the row headed by a. Then there would exist elements $x_1, x_2 \in G$, with $x_1 \neq x_2$, such that $ax_1 = b$ and $ax_2 = b$. But this situation is incompatible with the above theorem, which assures us that there can be one and only one solution of the equation $ax = b$. A proof for columns is obtained by imitating the argument for rows.

It can be shown that the solution of linear equations provides a characterization of groups. Here is the exact statement: A semigroup S is a group if and only if the equations $ax = b$ and $ya = b$ have solutions in S for all elements $a, b \in S$. The reader might enjoy proving this.

We have stressed the fact that the associative law gives the product abc unambiguous meaning, even though it may be formed in two distinct ways—by combining ab first or bc first. When more than three elements are involved, the situation becomes increasingly complicated, owing to the variety of ways of inserting parentheses so as to make the result computable. The next theorem asserts that the value of a product of finitely many group elements is independent of the position of the parentheses (provided, of course, that the elements are parenthesized in a meaningful manner). More precisely, any two products each involving the same finite set of elements in the same order are equal.

We intend to prove this result for semigroups, rather than groups; the extra generality presents no problem and will be useful later. First, let us introduce some auxiliary notation: Suppose that S is a (multiplicative) semigroup and that the elements $a_1, a_2, \ldots, a_n$ belong to S, where n is an integer greater than 1. Their *standard product*, symbolized by $a_1 a_2 \cdots a_n$, is defined recursively as

$$a_1 a_2 \cdots a_n = (a_1 a_2 \cdots a_{n-1}) a_n.$$

When $n = 4$, for instance, $a_1 a_2 a_3 a_4$ will turn out to be

$$a_1 a_2 a_3 a_4 = (a_1 a_2 a_3) a_4 = ((a_1 a_2) a_3) a_4.$$

In essence, we are using induction to define a particular grouping of parentheses and therefore a particular element of S.

These remarks launch us into a proof of

THEOREM 2.8 (General Associative Law) *Let S be a semigroup and a_1, $a_2, \ldots, a_n$ be a set of $n \geqslant 3$ elements of S (not necessarily distinct). Then all possible products of $a_1, a_2, \ldots, a_n$, taken in this order and arrived at by placing parentheses in any meaningful positions, yield one and the same element.*

Proof. The strategy is to prove that all products of the elements $a_1, a_2, \ldots, a_n$ are equal to the standard product $a_1 a_2 \cdots a_n$ and hence are equal to each other; our argument proceeds by induction on n, where $n \geqslant 3$. In the case $n = 3$ the associativity of the operation implies the stated conclusion

$$a_1(a_2 a_3) = (a_1 a_2) a_3 = a_1 a_2 a_3.$$

Next assume that the assertion holds for all products of m factors, where $3 \leqslant m < n$, and let x be any product involving n factors.

Since x is obtained by successive applications of the binary operation, there must be a final multiplication between two expressions having less than n factors. By the induction hypothesis, each of these two expressions equals the standard product, so x may be written as

$$x = (a_1 a_2 \cdots a_k)(a_{k+1} a_{k+2} \cdots a_n), \qquad 1 \leqslant k < n.$$

By definition of the standard product, $a_{k+1}a_{k+2} \cdots a_n = (a_{k+1}a_{k+2} \cdots a_{n-1})a_n$, whence

$$x = (a_1 a_2 \cdots a_k)[(a_{k+1}a_{k+2} \cdots a_{n-1})a_n].$$

The usual associative law now yields

$$x = [(a_1 a_2 \cdots a_k)(a_{k+1}a_{k+2} \cdots a_{n-1})]a_n.$$

Applying first our induction assumption to the expression in brackets—which certainly contains less than n factors—and then the definition of the standard product again, we infer that

$$x = (a_1 a_2 \cdots a_{n-1})a_n = a_1 a_2 \cdots a_n.$$

This completes the induction step and therefore the proof of the theorem.

Given four elements a_1, a_2, a_3, $a_4 \in S$, parentheses can be legally inserted and the elements multiplied (in the given order) five different ways. Theorem 2.8 permits us to conclude that all these products are equal:

$$\begin{aligned}((a_1a_2)a_3)a_4 &= (a_1(a_2a_3))a_4 = a_1((a_2a_3)a_4) \\ &= a_1(a_2(a_3a_4)) = (a_1a_2)(a_3a_4).\end{aligned}$$

In practice, it will be our tendency to omit parentheses in writing products; the exception to this will be found on these occasions when we wish to emphasize the associative law or a certain grouping of elements.

There is also a generalized commutative law which is worth noting.

THEOREM 2.9 (General Commutative Law) *If $a_1, a_2, \ldots, a_n$ ($n \geqslant 2$) are elements of the commutative semigroup S, then the product $a_1 a_2 \cdots a_n$ is independent of the order of the factors.*

Proof. From Theorem 2.8 we have

$$\begin{aligned}a_1a_2 \cdots a_n &= a_1 \cdots a_{k-1}(a_k a_{k+1})a_{k+2} \cdots a_n \\ &= a_1 \cdots a_{k-1}(a_{k+1}a_k)a_{k+2} \cdots a_n.\end{aligned}$$

Thus the value of the product is unaltered if two consecutive factors are interchanged. By a succession of interchanges of adjacent elements all possible orders of the factors are obtainable.

The multiplicative notation for the operation in a group provides a convenient shorthand for representing the repeated multiplication of an element by itself. If G is any group and a is an element of G, we shall define a^n inductively as follows: Let $a^1 = a$ and if k is a positive integer such that a^k is defined put $a^{k+1} = a^k a$. Thus one has $a^2 = a^1 a = aa$, $a^3 = a^2 a = aaa$,

$a^4 = a^3a = aaaa$, and so on (in view of the General Associative Law, the right-hand sides are uniquely defined without parentheses). We shall speak of a^n as being an *integral power of* a and the integer n as the *exponent* of this power of a. Our definition of the expression a^n may be extended to include zero and negative integers by requiring that

$$a^0 = e \text{ and } a^{-n} = (a^{-1})^n, \qquad \text{where } n \in Z_+.$$

Since the elements a and a^{-1} commute, Theorem 2.9 yields

$$e = e^n = (aa^{-1})^n = a^n(a^{-1})^n$$

for any positive integer n. The implication is that $(a^{-1})^n$ is the inverse of a^n or, in symbols, $(a^n)^{-1} = (a^{-1})^n = a^{-n}$. Let us next assume that n is a negative integer, so that $-n > 0$. By what we have just proved (as well as the definition of negative exponents),

$$a^{-n} = [(a^{-n})^{-1}]^{-1} = [(a^{-1})^{-n}]^{-1} = [a^{-(-n)}]^{-1} = (a^n)^{-1}$$

while, at the same time,

$$a^{-n} = [(a^{-1})^{-1}]^{-n} = (a^{-1})^{-(-n)} = (a^{-1})^n.$$

The point we wish to make is that the formula

$$(a^n)^{-1} = (a^{-1})^n = a^{-n}$$

holds for every integer n, whether it be positive, negative, or zero.

Our aim is to prove that the familiar "laws of exponents" have their counterpart in group theory. We begin with the following theorem:

THEOREM 2.10 *If a is an element of the group G, then $a^m a^n = a^{m+n}$ for all integers m and n.*

Proof. The proof is divided into various cases. If $m = 0$, $n = 0$, or $m + n = 0$, the theorem is obvious, so let us begin by assuming that both $m > 0$ and $n > 0$. For each such m we prove the formula $a^m a^n = a^{m+n}$ by induction on n. If $n = 1$, then clearly $a^m a^1 = a^m a = a^{m+1}$ by definition of integral powers. Assuming that $a^m a^k = a^{m+k}$, we have

$$a^m a^{k+1} = a^m(a^k a) = (a^m a^k)a = a^{m+k}a = a^{m+(k+1)}.$$

This being so, the induction principle assures us that the formula in question is true for every positive integer n.

Now consider the situation where $m < 0$ and $n < 0$. The case already discussed may be applied to give

$$\begin{aligned} a^m a^n &= a^{-(-m)}a^{-(-n)} = (a^{-n}a^{-m})^{-1} \\ &= (a^{-n-m})^{-1} = (a^{-(n+m)})^{-1} = a^{n+m} \end{aligned}$$

as required.

Next let m and n have opposite signs; say $m < 0$ and $n > 0$, with $m + n > 0$. Then

$$a^{-m}a^{m+n} = a^{(-m)+(m+n)} = a^n,$$

which, when we multiply both sides by a^m, leads to $a^{m+n} = a^m a^n$. Since the case where $m < 0$, $n > 0$, and $m + n < 0$ may be treated similarly, we are finished.

Parallel to Theorem 2.10, we have the following result.

THEOREM 2.11 *If a is an element of the group G, then* $(a^m)^n = a^{mn}$ *for all integers m and n.*

Proof. Given a fixed $m \in Z$, let us show that $(a^m)^n = a^{mn}$ for all positive integers n by inducting on n. The case $n = 1$ is immediate, since $(a^m)^1 = a^m = a^{m1}$. As our induction assumption, we suppose that $(a^m)^k = a^{mk}$; then

$$(a^m)^{k+1} = (a^m)^k a^m = a^{mk}a^m = a^{mk+m} = a^{m(k+1)}.$$

Thus the theorem holds for $n = k + 1$ and, in consequence, for every positive integer n.

This leaves only the case where n is a negative integer for consideration (if $n = 0$, both sides of the equation clearly equal e). But then $-n > 0$, and so

$$(a^m)^n = ((a^m)^{-n})^{-1} = (a^{m(-n)})^{-1} = (a^{-(mn)})^{-1} = a^{mn},$$

completing the argument.

As a final comment on exponents, we should caution the reader that the property $(ab)^n = a^n b^n$ is not to be expected for arbitrary elements of a group. In fact, this condition holds for all $a, b \in G$ and $n \in Z$ if and only if G is a commutative group.

It can be demonstrated that all groups with fewer than six elements are commutative; thus a noncommutative group must necessarily contain at least six elements. The proof of this fact is somewhat lengthy, although the actual details are not by themselves particularly difficult. The subsequent lemma will serve to isolate the most tedious aspect of the theorem.

LEMMA *If a and b are noncommuting elements of a group G, then the elements of the set* $\{e, a, b, ab, ba\}$ *are all distinct.*

Proof. The underlying idea of the proof is to examine the members of the set $\{e, a, b, ab, ba\}$ two at a time and show that each of the ten possible equalities leads to a contradiction of the hypothesis that $ab = ba$. On several occasions the cancellation law is used without explicit reference. The reasoning runs as follows:

1. $a = e$ implies that $ab = eb = b = be = ba$; a similar argument holds if $b = e$.
2. $ab = e$ implies that $ae = ea = (ab)a = a(ba)$; hence $e = ba$, and so $ab = ba$. A similar argument holds if $ba = e$.
3. $a = b$ implies that $ab = a^2 = ba$.
4. $a = ab$ or $a = ba$ implies that $b = e$, reducing to case 1.
5. $b = ab$ or $b = ba$ implies that $a = e$, reducing to case 1.
6. $ab = ba$ contradicts our hypothesis.

With this lemma established, the main theorem is easily accessible.

THEOREM 2.12 *Any noncommutative group has at least six elements.*

Proof. If G is a noncommutative group, then it must contain a pair of noncommuting elements a and b. As seen in the lemma, the set $\{e, a, b, ab, ba\}$ consists of distinct members. We next establish that one of the group elements a^2 or aba is distinct from these five; however, it is not possible to specify whether this sixth distinct element will be a^2 or aba.

With the aid of the lemma we begin by showing that a^2 is different from each member of $\{a, b, ab, ba\}$. Observe that

1. $a^2 = a$ implies that $a = e$, reducing to case 1 of the lemma;
2. $a^2 = b$ implies that $ab = a^3 = ba$, a contradiction;
3. $a^2 = ab$ or $a^2 = ba$ implies that $a = b$, reducing to case 3 of the lemma.

Thus either $a^2 \neq e$, in which event a^2 is the sixth distinct element of G, or else $a^2 = e$.

In this latter event it can be established that aba is different from each member of $\{e, a, b, ab, ba\}$ and consequently is the required sixth element, since

4. $aba = e$ implies that $ab = (ab)e = (ab)(aba) = (aba)(ba) = e(ba) = ba$, a contradiction;
5. $aba = a$ implies that $ba = e$, reducing to case 2 of the lemma;
6. $aba = b$ implies that $ab = a(aba) = a^2(ba) = e(ba) = ba$, a contradiction;
7. $aba = ab$ or $aba = ba$ implies that $a = e$, reducing to case 1 of the lemma.

This settles our result.

Theorem 2.12 invites a short comment. The reader should not infer more from the theorem than what is actually said. It does not assert that every group with six or more elements is noncommutative (simply consider the additive group Z of integers); all the theorem maintains is that the noncommutative groups will be found among those groups having at least six elements.

We close this section by mentioning some of the applications of group theory, which is widely employed beyond the boundaries of mathematics—for instance, in such areas as crystallography, quantum mechanics, and particle physics. The early 1890's saw the first direct application of the newly emerging theory to the natural and physical sciences. As a mathematical tool, the group concept is peculiarly suited to investigations of symmetry, one of the elemental phenomena of nature. An object or figure is said to have a symmetry if some operation—a rotation, a translation, an inversion, a mirror reflection, or some combination of these—brings it to a position that is essentially indistinguishable from its original one. An operation of this kind is called a *symmetry operation.* To obtain a useful characterization of the physical properties of some object, the focus of study sometimes shifts from the object itself to the groups of symmetry operations upon it.

The notion of symmetry permeates all of crystallography; without it, one can neither study nor understand the structure of crystals such as quartz, diamond, or salt. These substances are composed of atoms that occupy positions with definite geometric relationships to each other that repeat themselves throughout the crystal. For example, in the simplest case the atoms may be thought of as being located at the vertices of a cube. Each arrangement in which atoms are positioned within a three-dimensional crystal corresponds to a group of symmetry operations known as a *space group.* A complete enumeration of the space groups was carried out by the Russian crystallographer E. S. Federov (1853–1919) during the years 1885–1890. The work was done a little later, quite independently, by the German mathematician Artur Schoenflies (1853–1928). Both showed that the number of space groups, though large, is limited and that in all there are 230 such groups. Federov's articles had a slight priority of publication but were almost concealed behind the barrier of the Russian language; as a result, they did not become known generally to crystallographers of other nations until after the appearance in 1891 of Schoenflies's work "Crystal Systems and Crystal Structures." The theoretical classification of the different kinds of symmetry arrangements possible within crystalline matter predated the experimental results; it was 20 years before experimental means were available for analyzing the crystals themselves. This latter milestone in science came in 1912, when Max von Laue (1879–1960), in the first demonstration of the diffraction of X-rays by a crystal, passed a narrow beam of X-rays through a crystal of copper sulfate and recorded the resulting pattern of fuzzy spots on a photographic plate.

The application of the theory of groups to chemistry and theoretical physics has expanded enormously with the advent of quantum mechanics. Today, group theory is a central tool in the study of the atom's internal structure. Starting in the 1950's, a new generation of particle accelerators produced a bewildering variety of subatomic objects, a multitude that physicists began calling a "particle zoo." To inject order into the zoo, theorists turned to the group of symmetries of the associated quantum fields. A striking consequence of the group-theoretic approach to modern elementary particle physics occurred in 1962, when Murray Gell-Mann and Yuval Ne'eman, working independently, became sufficiently convinced of the existence of a new particle to predict the values of its mass, charge, and spin. Since nine particles in a tenfold symmetry scheme of nucleons had recently been detected, there were compelling grounds for postulating a tenth nucleon, which they christened "Omega-Minus." Theoretical argument requires proof by experiment; and in a race to isolate an Omega-Minus a variety of experiments were begun at Brookhaven National Laboratory and CERN (the European Laboratory for Particle Physics on the outskirts of Geneva). In spite of strenuous efforts, convincing evidence was not found for another two years. The hunt officially ended in February 1964, when the Brookhaven physicists announced that after a detailed examination of some 150,000 bubble chamber photographs they had unambiguously identified the tracks of the elusive Omega-Minus.

PROBLEMS 2.3

1. Show that a group G is commutative if and only if $a^{-1}b^{-1}ab = e$ for all a, $b \in G$.

2. Suppose that $a \neq e$ and b are elements of a certain group G and that $a^4b = ba^5$. Prove that $ab \neq ba$.

3. Given elements a, b, c of the group G, establish the following:

a) cac^{-1} commutes with cbc^{-1} if and only if a and b commute.

b) $aca^{-1} = bcb^{-1}$ if and only if $a^{-1}b$ commutes with c.

4. Find a specific example of a group G and elements a, $b \in G$ for which $(ab)^2 \neq a^2b^2$.

5. If a is an element of a group G such that $a^3 = e$, show that a is a square; that is, $a = b^2$ for some $b \in G$.

6. Let G be a group that satisfies the "mixed-cancellation" law; in other words, whenever a, b, $c \in G$ with $ab = ca$, then $b = c$. Prove that G is a commutative group.

7. a) Assume that a and b are elements of a certain group G such that $a^2 = e$, $b^6 = e$, and $ab = b^4a$. Verify that $(ab)^6 = e$. [*Hint*: Begin by showing that $(ab)^2 = b^5$; then $(ab)^3 = a$.]

b) If a group G has elements a, b, c satisfying the relations $a^2 = e$, $b^2 = e$, and $bc = cab$, prove that $ab = ba$. [*Hint*: Note that $bc = cba$.]

8. Given any $a \in R^{\#}$, show that the matrices

$$\begin{bmatrix} 1 & 0 \\ a & 0 \end{bmatrix} \quad \text{and} \quad \begin{bmatrix} 1 & 0 \\ -a & 0 \end{bmatrix}$$

are both idempotent elements of $M_2(R^{\#})$ with respect to matrix multiplication. Does this violate the corollary to Theorem 2.6?

9. Prove that if a and b are elements of a group G satisfying $(ab)^3 = a^3b^3$, then $(ba)^2 = a^2b^2$.

10. For each $a \in R^{\#}$, define the translation mapping $\tau_a: R^{\#} \to R^{\#}$ by $\tau_a(x) = x + a$. Prove that

a) the set $G = \{\tau_a | a \in R^{\#}\}$ forms a group, using the operation of functional composition;

b) if $n \in Z$, then $\tau_n = \tau_1^n$.

11. Given a group G, use induction to establish each of the following assertions:

a) $e^n = e$ for any $n \in Z$.

b) If $a_1, a_2, \ldots, a_r \in G$, then $(a_1 a_2 \cdots a_r)^{-1} = a_r^{-1} \cdots a_2^{-1} a_1^{-1}$.

c) If $ab = ba$, then $(ab)^n = a^n b^n$ for any $n \in Z_+$. [*Hint*: First show that $ab^n = b^n a$ for $n \in Z_+$.]

d) $(aba^{-1})^n = ab^n a^{-1}$ for any $n \in Z_+$.

e) If $aba = e$, then $(ab)^{2n} = b^n$ for any $n \in Z_+$.

12. Let $G = \{a_1, a_2, \ldots, a_n\}$ be a group with n elements. Prove that

a) If n is even, then G contains an element $a \neq e$ for which $a^2 = e$. [*Hint*: If $a \neq a^{-1}$ for all $a \neq e$, then G possesses an odd number of elements.]

b) If n is odd, then the product $a_1 a_2 \cdots a_n = e$.

13. For any group G, prove that the following conditions are equivalent.

a) G is commutative;

b) $(ab)^{-1} = a^{-1}b^{-1}$ for all $a, b \in G$;

c) $(ab)^2 = a^2b^2$ for all $a, b \in G$.

14. Show that every element of the group G_s of symmetries of the square can be written in the form $r_1^n h^m$ for suitable integers $n = 0, 1, 2, 3$ and $m = 0, 1$.

15. If $a^2 = e$ for each element of the group G, verify that G is a commutative group.

2.4 TWO IMPORTANT GROUPS

This section is devoted to an examination of two important and frequently used groups: the group of integers modulo *n* and the group of permutations of the elements of a set, the so-called symmetric group. While these groups are of some interest per se, our main purpose in introducing them is to provide two more concrete examples to illustrate concepts that will be developed later.

We begin with an investigation of the notion of congruence, in terms of which the group of integers modulo *n* will be formulated. The concept, and the notation that makes it such a powerful tool, was first introduced by the German mathematician Karl Friedrich Gauss (1777–1855) in his *Disquisitiones Arithmeticae*; this monumental work, which appeared in 1801 when Gauss was 24 years old, laid the foundations of modern number theory. "It is really astonishing," said Leopold Kronecker, "to think that a single man of such young years was able to bring to light such a wealth of results, and above all to present such a profound and well-organized treatment of an entirely new discipline."

Gauss was initially undecided on becoming a mathematician because he excelled in languages, being fluent from an early age in classical Greek and Latin as well as French and his native German. But this vacillation between philology and mathematics came to an abrupt end when, still a month short of his nineteenth birthday, Gauss made a brilliant discovery. The problem of constructing regular polygons, using only a straightedge and a compass, had long been laid aside in the belief that the ancients had exhausted all possible constructions. In 1796, Gauss showed that the 17-sided regular polygon is so constructible, the first advance in this direction since Euclid's time. He valued this feat so highly that he expressed the wish that his tombstone should bear the figure of a regular polygon of 17 sides; in due time the wish was carried out on a monument erected at Brunswick to honor Gauss. Gauss's doctoral thesis of 1799 is noteworthy for providing the first rigorous proof of the Fundamental Theorem of Algebra (the result that an algebraic equation of degree *n* has *n* complex roots). His great work *Disquisitiones Arithmeticae*, which at once placed him in the front rank of mathematicians, followed two years later.

The most extraordinary achievement of Gauss was more in the realm of theoretical astronomy than of mathematics. On the opening night of the nineteenth century, January 1, 1801, the Italian astronomer Piazzi discovered the first of the so-called minor planets (planetoids or asteroids), later called Ceres. But after the course of this newly found body, visible only by telescope, passed the sun, neither Piazzi nor any other astronomer could locate it again. Piazzi's observations extended over a period of 41 days, during which the orbit swept out an angle of only nine degrees. From the scanty data available, Gauss was able to calculate Ceres's orbit with amazing accuracy: the elusive planet was rediscovered at the end of the year in almost exactly the position he had forecast. This success brought Gauss worldwide fame and led to his appointment as director of Göttingen Observatory.

By the middle of the nineteenth century, mathematics had grown into an enormous and unwieldy structure, divided into a large number of fields in which only the specialist knew a path. Gauss was the last complete mathematician, and it is no exaggeration to say that he was in some degree connected with nearly every aspect of the subject. His contemporaries regarded him as *Princeps Mathematicorum* (Prince of Mathematicians), on a par with Archimedes and Isaac Newton. The title has not been used since.

Gauss placed the definition of congruence in the first section of the *Disquisitiones Arithmeticae*, at the head of his whole interpretation of number theory. (He explains that he was induced to adopt the symbol $\equiv$ because of its close analogy with ordinary equality.)

DEFINITION 2.12 Let n be a fixed positive integer. Two integers a and b are said to be *congruent modulo n*, written

$$a \equiv b \pmod{n},$$

if and only if the difference $a - b$ is divisible by n. That is, $a \equiv b \pmod{n}$ if and only if $a - b = kn$ for some integer k.

For instance, if $n = 7$, we have

$$3 \equiv 24 \pmod{7},$$
$$-5 \equiv 2 \pmod{7},$$
$$-8 = -50 \pmod{7}, \qquad \text{etc.}$$

If $a - b$ is not divisible by n, we say that a is incongruent to b modulo n and, in this case, write $a \not\equiv b \pmod{n}$.

It is noteworthy that every pair of integers is congruent modulo 1, while a pair of integers is congruent modulo 2 provided that they are both even or both odd.

Our first theorem provides a useful characterization of congruence modulo n in terms of remainders on division by n.

THEOREM 2.13 *Let n be a fixed positive integer and a, b be arbitrary integers. Then $a \equiv b \pmod{n}$ if and only if a and b have the same nonnegative remainder when divided by n.*

Proof. Suppose first that $a \equiv b \pmod{n}$, so that $a = b + kn$ for some integer k. On division by n, b leaves a certain remainder r:

$$b = qn + r, \qquad \text{where} \qquad 0 \leqslant r < n.$$

Thus $a = b + kn = (q + k)n + r$, which shows that a has the same remainder as b.

On the other hand, let $a = q_1 n + r$ and $b = q_2 n + r$, with the same remainder $r (0 \leqslant r < n)$. Then

$$a - b = (q_1 n + r) - (q_2 n + r) = (q_1 - q_2)n,$$

with $q_1 - q_2$ an integer. Hence n is a factor of $a - b$ and so $a \equiv b \pmod{n}$.

Congruence may be viewed as a type of "equality" in the sense that its behavior with respect to addition and multiplication is reminiscent of ordinary equality. Some of the elementary properties of equality that also carry over to congruences are listed in the next theorem.

THEOREM 2.14 *Let n be a fixed positive integer and a, b, c be arbitrary integers. Then*

a) $a \equiv a \pmod{n}$;
b) if $a \equiv b \pmod{n}$, *then* $b \equiv a \pmod{n}$;
c) if $a \equiv b \pmod{n}$ *and* $b \equiv c \pmod{n}$, *then* $a \equiv c \pmod{n}$;
d) if $a \equiv b \pmod{n}$ *and* $c \equiv d \pmod{n}$,
then $a + c \equiv b + d \pmod{n}$, $ac \equiv bd \pmod{n}$;
e) if $a \equiv b \pmod{n}$, *then* $ac \equiv bc \pmod{n}$;
f) if $a \equiv b \pmod{n}$, *then* $a^k \equiv b^k \pmod{n}$ *for every positive integer* k.

Proof. For any integer a, $a - a = 0n$, so $a \equiv a \pmod{n}$ by Definition 2.12. If $a \equiv b \pmod{n}$, then $a - b = kn$ for some integer k. Hence

$$b - a = (-k)n,$$

where $-k$ is an integer. This yields property (b).

To obtain property (c), suppose that $a \equiv b \pmod{n}$ and $b \equiv c \pmod{n}$. Then

$$a - b = kn \qquad \text{and} \qquad b - c = hn$$

for some integers k, h. Therefore

$$a - c = (a - b) + (b - c) = kn + hn = (k + h)n,$$

which implies that $a \equiv c \pmod{n}$.

Similarly, if $a \equiv b \pmod{n}$ and $c \equiv d \pmod{n}$, then there exist integers k_1, k_2 such that

$$a - b = k_1 n \qquad \text{and} \qquad c - d = k_2 n.$$

Consequently,

$$\begin{aligned}(a + c) - (b + d) &= (a - b) + (c - d) = k_1 n + k_2 n \\ &= (k_1 + k_2)n\end{aligned}$$

or

$$a + c \equiv b + d \pmod{n}.$$

Also,

$$ac = (b + k_1 n)(d + k_2 n) = bd + (bk_2 + dk_1 + k_1 k_2 n)n.$$

Since $bk_2 + dk_1 + k_1 k_2 n$ is an integer, $ac - bd$ is divisible by n, so

$$ac \equiv bd \pmod{n}.$$

This establishes property (d).

Property (e) follows directly from the second part of (4), since $c \equiv c \pmod{n}$.

Finally, we prove property (f) by an inductive argument. The statement is certainly true for $k = 1$. Assuming that it holds for an arbitrary k, we must show that it also holds for $k + 1$. But this is immediate from property (d), since $a^k \equiv b^k \pmod{n}$ and $a \equiv b \pmod{n}$ imply $a^k a \equiv b^k b \pmod{n}$, or $a^{k+1} \equiv b^{k+1} \pmod{n}$.

In the foregoing theorem we saw that if $a \equiv b \pmod{n}$, then $ca \equiv cb \pmod{n}$ for any integer c. It is interesting to note that the converse of this statement fails to be true. For instance, $2 \cdot 4 \equiv 2 \cdot 1 \pmod{6}$, but $4 \not\equiv 1 \pmod{6}$. To put it another way, one cannot unrestrictedly apply the cancellation law in the algebra of congruences. The only positive assertion that can be made in this regard is embodied in the following theorem.

THEOREM 2.15 *If $ca \equiv cb \pmod{n}$ and the integer c is relatively prime to n, then $a \equiv b \pmod{n}$.*

Proof. By hypothesis, $c(a - b) = kn$ for some integer k. Since c is prime to n, it follows that n must divide $a - b$ [Theorem 1.15]; hence $a \equiv b \pmod{n}$.

Using the facts embodied in Theorem 2.14, we can prove a result having to do with congruences involving polynomials with integral coefficients.

THEOREM 2.16 *Let $P(x) = \sum_{k=0}^{m} c_k x^k$ be a polynomial function of x with integral coefficients c_k. If $a \equiv b \pmod{n}$, then $P(a) \equiv P(b) \pmod{n}$.*

Proof. Since $a \equiv b \pmod{n}$, part (f) of Theorem 2.14 can be applied to give $a^k \equiv b^k \pmod{n}$ for $k = 0, 1, \ldots, m$. Therefore

$$c_k a^k \equiv c_k b^k \pmod{n}$$

for all such k. Adding these $m + 1$ congruences, we conclude that

$$\sum_{k=0}^{m} c_k a^k \equiv \sum_{k=0}^{m} c_k b^k \pmod{n}$$

or equivalently, $P(a) \equiv P(b) \pmod{n}$.

If $P(x)$ is a polynomial with integral coefficients, one says that a is a *solution* of the congruence $P(x) \equiv 0 \pmod{n}$ if $P(a) \equiv 0 \pmod{n}$.

COROLLARY *If a is a solution of $P(x) \equiv 0 \pmod{n}$ and $a \equiv b \pmod{n}$, then b is also a solution.*

Proof. From the last theorem it is known that $P(a) \equiv P(b) \pmod{n}$. Hence if a is a solution of $P(x) \equiv 0 \pmod{n}$, then $P(b) \equiv P(a) \equiv 0 \pmod{n}$, making b a solution.

As an application of Theorem 2.16, we derive a criterion for divisibility by 9. The test is this: A positive integer is divisible by 9 if and only if the sum of its digits is divisible by 9.

THEOREM 2.17 *Let $a = a_m 10^m + a_{m-1} 10^{m-1} + \cdots + a_1 10 + a_0$ be the decimal representation of the positive integer a, $0 \leqslant a_k < 10$, and consider the sum $s = a_0 + a_1 + \cdots + a_m$. Then $9|a$ if and only if $9|s$.*

Proof. It is evident that $a = P(10)$, where the polynomial $P(x) = \sum_{k=0}^{m} a_k x^k$. Since $10 \equiv 1 \pmod{9}$, the preceding corollary implies that $P(10) \equiv 0 \pmod{9}$ if and only if $P(1) \equiv 0 \pmod{9}$. But $P(1) = a_0 + a_1 + \cdots + a_m = s$. Thus $a \equiv 0 \pmod{9}$ if and only if $s \equiv 0 \pmod{9}$, which is what we wanted to show.

The relation $\equiv \pmod{n}$ will now be used to construct a new group, with exactly n elements, from the additive group $(Z, +)$ of integers. Before attempting to set out this material in an orderly fashion, let us introduce some additional notation and terminology.

DEFINITION 2.13 For an arbitrary integer a, let $[a]$ denote the set of all integers congruent to a modulo n:

$$\begin{aligned}[a] &= \{x \in Z \mid x \equiv a \pmod{n}\} \\ &= \{x \in Z \mid x = a + kn \text{ for some integer } k\}.\end{aligned}$$

We call $[a]$ the *congruence class*, modulo n, determined by a and refer to a as a *representative* of this class.

By way of illustration, suppose that we are dealing with congruence modulo 3. Then

$$[0] = \{x \in Z | x = 3k \text{ for some } k \in Z\}$$
$$= \{\ldots, -9, -6, -3, 0, 3, 6, 9, \ldots\}.$$

Also

$$[1] = \{x \in Z | x = 1 + 3k \text{ for some } k \in Z\}$$
$$= \{\ldots, -8, -5, -2, 1, 4, 7, 10, \ldots\}.$$

Similarly,

$$[2] = \{\ldots, -7, -4, -1, 2, 5, 8, 11, \ldots\}.$$

Observe that every integer lies in one of these three classes. Integers in the same congruence class are congruent modulo 3, while integers in different classes are incongruent modulo 3.

A particular congruence class may be designated in a variety of ways by merely changing its representative. In the above illustration, for instance,

$$[-7] = [2] = [11] = [35] = \cdots$$

It suffices to remark that the characteristic feature of these various representatives is that in this case they all differ from each other by multiples of 3, and in general they differ by multiples of n. For convenience, one often selects the smallest nonnegative integer from each congruence class to represent it: in practice, we shall adhere to this notational convention.

To return to the general case of congruence modulo n, let

$$Z_n = \{[0], [1], [2], \ldots, [n-1]\}.$$

Several properties of the collection Z_n that we shall require later appear in the next theorem.

THEOREM 2.18 *Let n be a positive integer and Z_n be as defined above. Then*

a) for each $[a] \in Z_n$, $[a] \neq \varnothing$;

b) if $[a] \in Z_n$ and $b \in [a]$, then $[b] = [a]$; that is, any element of the congruence class $[a]$ determines the class;

c) for any $[a], [b] \in Z_n$ where $[a] \neq [b]$, $[a] \cap [b] = \varnothing$;

d) $\cup\{[a] | [a] \in Z_n\} = Z$.

Proof. The first three parts of Theorem 2.14 state that the relation $a \equiv b \pmod{n}$ forms an equivalence relation in the set Z of integers. Indeed the congruence classes as defined in Definition 2.13 are simply the equivalence classes for this equivalence relation. Viewed in this light, the statement of Theorem 2.18 is a translation of Theorem 1.8 into the language of "congruence modulo n."

The set Z_n, whose elements are the congruences classes modulo n, is traditionally known as the set of *integers modulo n*. It may strike the reader that this terminology is somewhat inappropriate because, precisely speaking, the elements of Z_n are not single integers, but rather infinite sets of integers. Moreover, the set Z_n is not infinite, like the integers, but is a finite set with n elements. Despite this violation of our intuition, we bow to long-standing custom and shall continue to refer to Z_n as the integers modulo n.

By a partition of the set S, we mean a family of nonempty subsets of S which are pairwise disjoint and whose union is all of S. It follows from Theorem 2.18 that, for each $n \in Z_+$, the integers modulo n constitute a partition of the set Z.

DEFINITION 2.14 A binary operation $+_n$ may be defined on Z_n as follows: for each $[a], [b] \in Z_n$, let $[a] +_n [b] = [a + b]$.

Definition 2.14 asserts that the modular sum of two congruence classes $[a]$ and $[b]$ is the unique member of Z_n that contains the ordinary sum $a + b$. However, there is a subtle problem connected with this definition. Inasmuch as addition of congruence classes in Z_n is defined in terms of representatives from these classes, we must show that the operation $+_n$ does not depend on the two representatives chosen. It must be proved formally that if $[a'] = [a]$ and $[b'] = [b]$, then $[a'] +_n [b'] = [a] +_n [b]$, or rather $[a' + b'] = [a + b]$. Now $a' \in [a'] = [a]$ and $b' \in [b'] = [b]$, which implies that

$$a' \equiv a \pmod{n} \qquad \text{and} \qquad b' \equiv b \pmod{n}.$$

By virtue of Theorem 2.14(d) it follows that

$$a' + b' \equiv a + b \pmod{n} \qquad \text{or} \qquad a' + b' \in [a + b].$$

Theorem 2.18(b) then indicates that $[a' + b'] = [a + b]$, as desired. Thus the operation $+_n$ is unambiguously defined, independent of the arbitrary choice of representatives.

Example 2.24

Suppose we consider congruence modulo 7 and the typical addition

$$[3] +_7 [6] = [3 + 6] = [9].$$

Since $[3] = [10]$ and $[6] = [-15]$, the same answer should be obtained if one used

$$[10] +_7 [-15] = [10 - 15] = [-5].$$

Although these results appear superficially different, both congruence classes [9] and [−5] may be expressed more simply as [2]. Thus although written in terms of different representatives, either modular addition gives the same sum, [2]. Other possible choices $[-4] +_7 [-8] = [-12]$, $[17] +_7 [6] = [23]$, $[3] +_7 [13] = [16]$ also yield the sum [2].

We are now in a position to prove one of the principal theorems of this section.

THEOREM 2.19 *For each positive integer n the system $(Z_n, +_n)$ forms a commutative group, known as the group of integers modulo n.*

Proof. The associativity and commutativity of the operation $+_n$ are a direct consequence of the same properties of the integers under ordinary addition. Indeed, if $[a], [b], [c] \in Z_n$, then

$$\begin{aligned} [a] +_n ([b] +_n [c]) &= [a] +_n [b + c] \\ &= [a + (b + c)] \\ &= [(a + b) + c] \\ &= [a + b] +_n [c] \\ &= ([a] +_n [b]) +_n [c]. \end{aligned}$$

Similarly,

$$[a] +_n [b] = [a + b] = [b + a] = [b] +_n [a].$$

By definition of $+_n$ it is clear that [0] is the identity element. Finally, if $[a] \in Z_n$, then $[n - a] \in Z_n$, and

$$[a] +_n [n - a] = [a + (n - a)] = [0],$$

so that $[a]^{-1} = [n - a]$. This completes the proof that $(Z_n, +_n)$ is a commutative group.

Incidentally, Theorem 2.19 also shows that for every positive integer n there exists at least one commutative group with n elements.

If we adopt the convention of designating each congruence class by its smallest nonnegative representative, then the operation table for, say, $(Z_4, +_4)$ looks like

$+_4$	[0]	[1]	[2]	[3]
[0]	[0]	[1]	[2]	[3]
[1]	[1]	[2]	[3]	[0]
[2]	[2]	[3]	[0]	[1]
[3]	[3]	[0]	[1]	[2]

For simplicity it is convenient to remove the brackets in the designation of the congruence classes of Z_n. Thus we often write $Z_n = \{0, 1, 2, \ldots, n-1\}$. With this notation the above operation table assumes the form

$+_4$	0	1	2	3
0	0	1	2	3
1	1	2	3	0
2	2	3	0	1
3	3	0	1	2

For the second of our two examples we turn to the study of permutation groups. Permutation groups are of particular importance because they provide us with a concrete representation of all finite groups (Theorem 2.59). To start, suppose that N is a finite set having n distinct elements which, for simplicity, we take to be the first n natural numbers; that is to say,

$$N = \{1, 2, \ldots, n\}.$$

DEFINITION 2.15 By a *permutation* of the set N is meant any one-to-one mapping of N onto itself. The totality of all permutations of the set N will be denoted by the symbol S_n.

Since N is a finite set, each function carrying N onto itself is automatically one-to-one. Likewise, a one-to-one mapping from N into N must be onto. Thus Definition 2.15 could be stated in an equivalent way by requiring a permutation to be any one-to-one (hence onto) function from N into itself.

If $\alpha \in S_n$, then there are n possibilities for $\alpha(1)$, since it may be any one of the integers $1, 2, \ldots, n$. Once $\alpha(1)$ has been specified, there are only $n-1$ choices for the image of 2—$\alpha(2) = \alpha(1)$ would violate the one-to-one property of α—making $n(n-1)$ choices for $\alpha(1)$ and $\alpha(2)$. Continuing in this manner, we find that α may be defined in $n!$ different ways, so that S_n must contain $n!$ distinct elements.

Any permutation $\alpha \in S_n$ may be described as a set of ordered pairs

$$\alpha = \{(1, \alpha(1)), (2, \alpha(2)), \ldots, (n, \alpha(n))\}.$$

Although this is acceptable functional notation, it will prove to be more convenient to represent α in a two-line form

$$\alpha = \begin{pmatrix} 1 & 2 & \cdots & n \\ \alpha(1) & \alpha(2) & \cdots & \alpha(n) \end{pmatrix}$$

where the corresponding images appear below each integer. The order of the elements in the top row of this symbol is immaterial, since the columns may be rearranged without affecting the nature of the function. Precisely speaking, if β is an arbitrary permutation of the integers $1, 2, \ldots, n$, then α could equally well be given by

$$\alpha = \begin{pmatrix} \beta(1) & \beta(2) & \cdots & \beta(n) \\ \alpha(\beta(1)) & \alpha(\beta(2)) & \cdots & \alpha(\beta(n)) \end{pmatrix}$$

From this we infer that each of the $n!$ permutations of S_n may itself be written in $n!$ different ways. For instance, the following two symbols both represent the same element of S_4:

$$\begin{pmatrix} 1 & 2 & 3 & 4 \\ 2 & 4 & 3 & 1 \end{pmatrix}, \quad \begin{pmatrix} 2 & 1 & 4 & 3 \\ 4 & 2 & 1 & 3 \end{pmatrix}.$$

Permutations, being functions, may be multiplied under the operation of functional composition. What is important is that the resulting product is not just a function; it is also a permutation of N (if this is not clear, the reader should supply a proof). Thus for permutations $\alpha, \beta \in S_n$,

$$\begin{aligned} \alpha \circ \beta &= \begin{pmatrix} 1 & 2 & \cdots & n \\ \alpha(1) & \alpha(2) & \cdots & \alpha(n) \end{pmatrix} \circ \begin{pmatrix} 1 & 2 & \cdots & n \\ \beta(1) & \beta(2) & \cdots & \beta(n) \end{pmatrix} \\ &= \begin{pmatrix} \beta(1) & \beta(2) & \cdots & \beta(n) \\ \alpha(\beta(1)) & \alpha(\beta(2)) & \cdots & \alpha(\beta(n)) \end{pmatrix} \circ \begin{pmatrix} 1 & 2 & \cdots & n \\ \beta(1) & \beta(2) & \cdots & \beta(n) \end{pmatrix} \\ &= \begin{pmatrix} 1 & 2 & \cdots & n \\ \alpha(\beta(1)) & \alpha(\beta(2)) & \cdots & \alpha(\beta(n)) \end{pmatrix}. \end{aligned}$$

What we have done is to rearrange the columns of the first (left) permutation until its top row is the same as the bottom row of the second (right) permutation; the product $\alpha \circ \beta$ is then the permutation whose top row is the top row of the second factor and whose bottom row is the bottom row of the first factor. With a little practice, one can evaluate products without having to write out this intermediate preparation. Many authors prefer to carry out the multiplication of permutations in the opposite order—that is, they apply the factors in a product from left to right—and the student should be particularly watchful for this.

Before stating a theorem that indicates the algebraic nature of S_n, let us clarify some of the foregoing points with an example.

Example 2.25

Consider the three-element set $N = \{1, 2, 3\}$. There are $3! = 6$ permutations of N, namely,

$$\alpha_1 = \begin{pmatrix} 1 & 2 & 3 \\ 1 & 2 & 3 \end{pmatrix}, \quad \alpha_2 = \begin{pmatrix} 1 & 2 & 3 \\ 2 & 3 & 1 \end{pmatrix}, \quad \alpha_3 = \begin{pmatrix} 1 & 2 & 3 \\ 3 & 1 & 2 \end{pmatrix},$$

$$\alpha_4 = \begin{pmatrix} 1 & 2 & 3 \\ 1 & 3 & 2 \end{pmatrix}, \quad \alpha_5 = \begin{pmatrix} 1 & 2 & 3 \\ 2 & 1 & 3 \end{pmatrix}, \quad \alpha_6 = \begin{pmatrix} 1 & 2 & 3 \\ 3 & 2 & 1 \end{pmatrix}.$$

A typical product, say $\alpha_4 \circ \alpha_6$, may be described by saying what it does to the elements of N. The effect of $\alpha_4 \circ \alpha_6$ on N is

$$(\alpha_4 \circ \alpha_6)(1) = \alpha_4(\alpha_6(1)) = \alpha_4(3) = 2$$

$$(\alpha_4 \circ \alpha_6)(2) = \alpha_4(\alpha_6(2)) = \alpha_4(2) = 3$$

$$(\alpha_4 \circ \alpha_6)(3) = \alpha_4(\alpha_6(3)) = \alpha_4(1) = 1$$

whence

$$\alpha_4 \circ \alpha_6 = \begin{pmatrix} 1 & 2 & 3 \\ 2 & 3 & 1 \end{pmatrix} = \alpha_2.$$

This result could be calculated just as rapidly by "rearranging columns":

$$\begin{aligned} \alpha_4 \circ \alpha_6 &= \begin{pmatrix} 1 & 2 & 3 \\ 1 & 3 & 2 \end{pmatrix} \circ \begin{pmatrix} 1 & 2 & 3 \\ 3 & 2 & 1 \end{pmatrix} \\ &= \begin{pmatrix} 3 & 2 & 1 \\ 2 & 3 & 1 \end{pmatrix} \circ \begin{pmatrix} 1 & 2 & 3 \\ 3 & 2 & 1 \end{pmatrix} = \begin{pmatrix} 1 & 2 & 3 \\ 2 & 3 & 1 \end{pmatrix} = \alpha_2. \end{aligned}$$

A similar analysis shows that

$$\begin{aligned} \alpha_6 \circ \alpha_4 &= \begin{pmatrix} 1 & 2 & 3 \\ 3 & 2 & 1 \end{pmatrix} \circ \begin{pmatrix} 1 & 2 & 3 \\ 1 & 3 & 2 \end{pmatrix} \\ &= \begin{pmatrix} 1 & 3 & 2 \\ 3 & 1 & 2 \end{pmatrix} \circ \begin{pmatrix} 1 & 2 & 3 \\ 1 & 3 & 2 \end{pmatrix} = \begin{pmatrix} 1 & 2 & 3 \\ 3 & 1 & 2 \end{pmatrix} = \alpha_3, \end{aligned}$$

so that multiplication of permutations in S_3 is not commutative.

THEOREM 2.20 *The system $(S_n, \circ)$ forms a group, known as the symmetric group on n symbols, which is noncommutative for $n \geqslant 3$.*

The proof of this fact is not too difficult, but we prefer to put off its presentation until a more general version of the theorem can be given. At this point it is necessary to note only that the identity element for $(S_n, \circ)$ is the permutation

$$\iota = \begin{pmatrix} 1 & 2 & \cdots & n \\ 1 & 2 & \cdots & n \end{pmatrix},$$

while the multiplicative inverse of any permutation $\alpha \in S_n$ is described by

$$\alpha^{-1} = \begin{pmatrix} \alpha(1) & \alpha(2) & \cdots & \alpha(n) \\ 1 & 2 & \cdots & n \end{pmatrix}.$$

Parenthetically, we might observe that the group of symmetries of the square (Example 2.22) can be subsumed under the theory of permutation groups; these symmetries really do nothing more than map the four vertices of the square in a one-to-one fashion onto themselves. This particular group is obviously not equal to the symmetric group $(S_4, \circ)$, since the set S_4 must contain $4! = 24$ elements. Indeed, one can easily find permutations in S_4 that do not correspond to any symmetry of the square. Rather, we have an example of what is called a "subgroup" of the group $(S_4, \circ)$.

In the following somewhat technical definition we introduce a special type of permutation called a cycle.

DEFINITION 2.16 Let $i_1, i_2, \ldots, i_k$ be k distinct integers between 1 and n. If a permutation $\alpha \in S_n$ is such that

$$\alpha(i_1) = i_2, \qquad \alpha(i_2) = i_3, \qquad \ldots, \qquad \alpha(i_{k-1}) = i_k, \qquad \alpha(i_k) = i_1$$

while $\alpha(i) = i$ for $i \notin \{i_1, i_2, \ldots, i_k\}$, then α is said to be a *k-cycle*, or a *cycle of length k*.

Simply put, a cycle replaces i_1 by i_2, i_2 by $i_3, \ldots,$ and finally i_k by i_1, while leaving all other elements fixed. For cycles a more condensed notation than the usual two-line form is to write $(i_1 \quad i_2 \cdots i_k)$, indicating that each integer is to be replaced by its successor on the right and the last integer by the first. An obvious drawback to this notation is that, unless S_n is specified, there is no way of telling where the cycle lies. In the symmetric group $(S_5, \circ)$, for example, we have

$$\begin{pmatrix} 1 & 2 & 3 & 4 & 5 \\ 1 & 5 & 2 & 4 & 3 \end{pmatrix} = (2 \quad 5 \quad 3).$$

Of course, each integer that is omitted in the cycle notation is presumed to map onto itself under the permutation. A given cycle may clearly be described in more than one way, since any of its elements can be put in the first position of the one-line form, as with

$$(2 \quad 5 \quad 3) = (5 \quad 3 \quad 2) = (3 \quad 2 \quad 5).$$

Any cycle that consists of a single element merely indicates that this element—as well as all others—is left unaltered; hence all 1-cycles necessarily equal the identity permutation.

We shall find it convenient to speak of two or more cycles as being *disjoint* if they have no integer in common; that is, if they move disjoint sets of elements (a permutation α *moves* i or *fixes* i according as $\alpha(i) \neq i$ or $\alpha(i) = i$). For instance, the cycles $(1 \quad 3 \quad 6)$ and $(2 \quad 4 \quad 5)$ are disjoint. Disjoint cycles α and β commute, since each cycle affects only elements left fixed by the other. Indeed, depending on whether α, β or neither moves i, both $(\alpha \circ \beta)(i)$ and $(\beta \circ \alpha)(i)$ take on the value $\alpha(i)$, $\beta(i)$, or i.

The next theorem shows why cycles are so important in the theory of permutations.

THEOREM 2.21 *Every permutation $\alpha \neq \iota$ in S_n can be written uniquely (apart from the order of the factors) as a product of disjoint cycles.*

Proof. Assuming that $\alpha(1) \neq 1$, consider the sequence of images of 1 under successive powers of α, in other words, the sequence

$$1 = \alpha^0(1),\ \alpha(1),\ \alpha^2(1),\ \alpha^3(1),\ \ldots$$

(If $\alpha(1) = 1$, we start with the first integer i for which $\alpha(i) \neq i$.) Since the range of α is finite, these elements cannot all be distinct. Let $\alpha^k(1)$ be the first term that has appeared previously in this sequence. More exactly, k is the least positive integer such that $1, \alpha(1), \alpha^2(1), \ldots, \alpha^{k-1}(1)$ are all distinct, but $\alpha^k(1) = \alpha^j(1)$ for some j, with $0 \leqslant j < k$. Then $\alpha^{k-j}(1) = 1$, which is a contradiction to our choice of k, unless $j = 0$. This gives rise to a cycle

$$\alpha_1 = (1 \quad \alpha(1) \quad \alpha^2(1) \quad \cdots \quad \alpha^{k-1}(1)).$$

Note that α_1 has the same effect on the k elements in it as does α.

If the integer 2 does not appear in the cycle α_1, repeat the foregoing argument to obtain a second cycle

$$\alpha_2 = (2 \quad \alpha(2) \quad \alpha^2(2) \quad \cdots \quad \alpha^{j-1}(2))$$

where j is the smallest positive integer such that $\alpha^j(2) = 2$. It will be found that the cycles α_1 and α_2 are disjoint. Otherwise, $\alpha^t(1) = \alpha^s(2)$ with, say, $1 \leqslant s < t \leqslant k - 1$; then $\alpha^{t-s}(1) = 2$, which puts 2 in the cycle α_1, contrary to assumption.

Needless to say, if the integer 2 had appeared in the cycle α_1, we would have formed α_2 by using the smallest integer not in α_1.

This process terminates in at most n such steps with a cycle α_r; of course, some of the α_i may be 1-cycles and hence equal to the identity permutation. Because the cycles α_i are disjoint, the product $\alpha_1 \circ \alpha_2 \circ \cdots \circ \alpha_r$ has the same effect on the elements of $\{1, 2, \ldots, n\}$ as α does, so that $\alpha = \alpha_1 \circ \alpha_2 \circ \cdots \circ \alpha_r$.

To illustrate Theorem 2.21, let us consider the permutation $\alpha \in S_6$ given by

$$\alpha = \begin{pmatrix} 1 & 2 & 3 & 4 & 5 & 6 \\ 5 & 6 & 1 & 4 & 3 & 2 \end{pmatrix}.$$

Starting with any of the six integers, say 1, we apply α repeatedly to get

$$\alpha(1) = 5, \qquad \alpha^2(1) = \alpha(5) = 3, \qquad \alpha^3(1) = \alpha(3) = 1.$$

Thus one of the disjoint cycles will be $(1 \quad 5 \quad 3)$. Next select any integer not appearing in $(1 \quad 5 \quad 3)$, for instance, 2. Evidently,

$$\alpha(2) = 6, \qquad \alpha^2(2) = \alpha(6) = 2,$$

yielding the cycle (2 6). This takes care of all the integers except 4, which remains fixed by α. We thus conclude that

$$\alpha = (1 \quad 5 \quad 3) \circ (2 \quad 6).$$

A particularly simple type of permutation is the 2-cycle; for these cycles, just interchange two integers while leaving the remaining $n - 2$ fixed. Such cycles are customarily referred to as *transpositions*. Each transposition is its own inverse, since $(i \quad j) \circ (i \quad j) = \iota$.

Not only is each element $\alpha \in S_n$ a product of cycles, but α can also be represented as a succession of transpositions. We prove this next.

THEOREM 2.22 *Every permutation in $S_n (n \geq 2)$ can be expressed as a product of transpositions.*

Proof. The identity permutation can be written as $\iota = (1 \quad 2) \circ (1 \quad 2)$. For $\alpha \neq \iota$, it suffices to prove that each cycle of α is a suitable product of transpositions. Given a k-cycle $(a_1 \quad a_2 \ldots a_k)$, this can be done in the following way:

$$(a_1 \quad a_2 \ldots a_k) = (a_1 \quad a_k) \circ (a_1 \quad a_{k-1}) \circ \cdots \circ (a_1 \quad a_2).$$

If $a \notin \{a_1, a_2, \ldots, a_k\}$, both sides of the displayed equation leave a fixed. If $i \neq k$, then $(a_1 \quad a_2 \ldots a_k)a_i = a_{i+1}$, while

$$\begin{aligned}(a_1 \quad a_k) \circ (a_1 \quad a_{k-1}) \circ \cdots \circ (a_1 \quad a_2)a_i &\\ &= (a_1 \quad a_k) \circ \cdots \circ (a_1 \quad a_i)a_i \\ &= (a_1 \quad a_k) \circ \cdots \circ (a_1 \quad a_{i+1})a_1 = a_{i+1}.\end{aligned}$$

Finally, $(a_1 \quad a_2 \ldots a_k)a_k = a_1$, while

$$(a_1 \quad a_k) \circ (a_1 \quad a_{k-1}) \circ \cdots \circ (a_1 \quad a_2)a_k = (a_1 \quad a_k)a_k = a_1.$$

The right- and left-hand sides of the equation have the same effect on each integer, and so the permutations are equal.

This new factorization lacks some of the advantages of the factorization of a permutation into cycles. First of all, the transpositions may not be disjoint and hence need not commute. One also loses uniqueness in this case. Another factorization of the k-cycle $(a_1 \quad a_2 \ldots a_k)$ is

$$(a_1 \quad a_2 \ldots a_k) = (a_1 \quad a_2) \circ (a_2 \quad a_3) \circ \cdots \circ (a_{k-1} \quad a_k).$$

Inasmuch as $(a_i \quad a_j) \circ (a_i \quad a_j) = \iota$, we could equally well insert the transposition $(a_i \quad a_j)$ twice in this expression without altering its value.

Although the factors in the representation of a given permutation as a product of transpositions are not uniquely determined, all such representations have one aspect in common, namely, the parity of the number of transpositions. (When we refer to the *parity* of an integer, we mean its classification as odd or even.) To put the matter more precisely, if $\alpha \in S_n$ can be expressed as both a product of j and k transpositions, then $j \equiv k \pmod 2$.

The notation of the *signature* or *sign function* is especially suitable for treating this question of parity. The signature of a permutation α in S_n is defined to be

$$\operatorname{sgn} \alpha = \prod_{i<j}^{n} \frac{\alpha(j) - \alpha(i)}{j - i}$$

where it is understood that the symbol $\prod$ denotes usual multiplication of real numbers. The condition $i < j$ implies that every pair of distinct integers between 1 and n (inclusive) occurs in exactly one factor on the right-hand side, so that $\operatorname{sgn} \alpha$ is the product of $n(n-1)/2$ factors.

To look at an actual example, let us choose $\alpha \in S_3$ to be the permutation

$$\alpha = \begin{pmatrix} 1 & 2 & 3 \\ 2 & 3 & 1 \end{pmatrix}.$$

Then

$$\operatorname{sgn} \alpha = \frac{\alpha(2) - \alpha(1)}{2 - 1} \frac{\alpha(3) - \alpha(1)}{3 - 1} \frac{\alpha(3) - \alpha(2)}{3 - 2}$$

$$= \frac{3 - 2}{2 - 1} \frac{1 - 2}{3 - 1} \frac{1 - 3}{3 - 2} = 1.$$

Likewise, if

$$\beta = \begin{pmatrix} 1 & 2 & 3 \\ 3 & 2 & 1 \end{pmatrix},$$

then

$$\operatorname{sgn} \beta = \frac{\beta(2) - \beta(1)}{2 - 1} \frac{\beta(3) - \beta(1)}{3 - 1} \frac{\beta(3) - \beta(2)}{3 - 2}$$

$$= \frac{2 - 3}{2 - 1} \frac{1 - 3}{3 - 1} \frac{1 - 2}{3 - 2} = -1.$$

This is typical of the general situation, for it is not hard to show that the value of $\operatorname{sgn} \alpha$ is always 1 or -1; in fact, except for signs, the numerators in the formula for $\operatorname{sgn} \alpha$ are exactly the same as the denominators.

From the point of view of the product of permutations the basic property of sgn is that the signature of a product is the product of the signatures; this is taken up next.

LEMMA *For any two permutations $\alpha, \beta \in S_n$,*

$$\operatorname{sgn}(\alpha \circ \beta) = (\operatorname{sgn} \alpha)(\operatorname{sgn} \beta).$$

Proof. The definition of sgn quickly reveals that

$$\begin{aligned}\operatorname{sgn}(\alpha\circ\beta) &= \prod_{i<j}^{n} \frac{\alpha(\beta(j)) - \alpha(\beta(i))}{j-i} \\ &= \prod_{i<j}^{n} \frac{\alpha(\beta(j)) - \alpha(\beta(i))}{\beta(j)-\beta(i)} \prod_{i<j}^{n} \frac{\beta(j)-\beta(i)}{j-i} \\ &= (\operatorname{sgn}\alpha)(\operatorname{sgn}\beta)\end{aligned}$$

What makes the argument work is that as i and j range over all possible pairs of distinct integers from 1 to n, so do $\beta(i)$ and $\beta(j)$.

This lemma launches us into a proof of the following theorem.

THEOREM 2.23 *Let the permutation $\alpha \in S_n$ and let*

$$\alpha = \tau_1 \circ \tau_2 \circ \cdots \circ \tau_j = \tau_1' \circ \tau_2' \circ \cdots \circ \tau_k'$$

be two factorizations of α into a product of transpositions. Then the integers j and k have the same parity.

Proof. We make use of the fact that if $\tau \in S_n$ is a transposition, then $\operatorname{sgn}\tau = -1$. To verify this, suppose that $\tau = (r\quad s)$, with $r < s$. Exactly one factor of $\operatorname{sgn}\tau$ involves both r and s, the factor

$$\frac{\tau(s)-\tau(r)}{s-r} = \frac{r-s}{s-r} = -1.$$

Every factor of $\operatorname{sgn}\tau$ that contains neither r nor s is equal to 1, since τ leaves the integers involved unchanged. All other factors—that is, those containing either r or s, but not both—can be paired to form one of the following products:

$$\frac{\tau(t)-\tau(r)}{t-r}\,\frac{\tau(t)-\tau(s)}{t-s} = \frac{t-s}{t-r}\,\frac{t-r}{t-s}, \qquad t > s$$

$$\frac{\tau(t)-\tau(r)}{t-r}\,\frac{\tau(s)-\tau(t)}{s-t} = \frac{t-s}{t-r}\,\frac{r-t}{s-t}, \qquad r < t < s$$

$$\frac{\tau(r)-\tau(t)}{r-t}\,\frac{\tau(s)-\tau(t)}{s-t} = \frac{s-t}{r-t}\,\frac{r-t}{s-t}, \quad t < r$$

for some integer t. As each of these pairs has value 1, we find that $\operatorname{sgn}\tau = -1$.

Now let α be an arbitrary permutation and express α as a product of transpositions,

$$\alpha = \tau_1 \circ \tau_2 \circ \cdots \circ \tau_j.$$

Using the lemma above (together with an obvious induction), we obtain

$$\operatorname{sgn} \alpha = (\operatorname{sgn} \tau_1)(\operatorname{sgn} \tau_2) \cdots (\operatorname{sgn} \tau_j) = (-1)^j.$$

Consequently, if there is another factorization of α into a product of transpositions, say

$$\alpha = \tau'_1 \circ \tau'_2 \circ \cdots \circ \tau'_k,$$

then $(-1)^j = (-1)^k$, or $(-1)^{j+k} = 1$. This can only happen if $j + k$ is an even integer, that is, if both j and k are even integers or both odd.

COROLLARY *If $\alpha \in S_n$ is written as a product of transpositions, $\alpha = \tau_1 \circ \tau_2 \circ \cdots \circ \tau_k$, then k is an even or odd integer according as sgn $\alpha = 1$ or sgn $\alpha = -1$.*

We have demonstrated that for $\alpha \in S_n$ the number of factors occurring in any factorization of α into transpositions is always even or always odd. This invariance of parity justifies the definition below.

DEFINITION 2.17 A permutation $\alpha \in S_n$ is called *even* or *odd* according as α can be written as a product of an even or odd number of transpositions.

Clearly, if α and β are both even or both odd permutations, then $\alpha \circ \beta$ is even, while if one of α or β is even and the other is odd, then $\alpha \circ \beta$ is odd. It should not escape notice that any transposition is odd and, since $\iota = (1\ \ 2) \circ (1\ \ 2)$, the identity permutation is always even. As regards notation, the subset of S_n consisting of all even permutations is traditionally designated by the symbol A_n:

$$A_n = \{\alpha \in S_n | \alpha \text{ is even}\}.$$

An interesting aspect of S_n is that the number of even permutations is the same as the number of odd permutations, both values being equal to $n!/2$. To see that this is so, let us temporarily denote the set of odd permutations in S_n by B_n. Then $S_n = A_n \cup B_n$ and $A_n \cap B_n = \varnothing$ where the latter equality is the content of Theorem 2.23. For a fixed transposition τ, define the function $f_\tau\colon A_n \to B_n$ by $f_\tau(\alpha) = \tau \circ \alpha$. Since α is even, the permutation $\tau \circ \alpha$ can be expressed as a product of $2k + 1$ transpositions and so $\tau \circ \alpha$ indeed belongs to B_n. There is no particular trouble in showing that f_τ is a one-to-one mapping of A_n onto B_n; we leave this to the reader. What is significant is that the elements of A_n and B_n are in one-to-one correspondence (via f_τ); hence each set contains half of the $n!$ permutations in S_n.

Here is another useful observation about cycles: A k-cycle is even or odd according as k is an odd or even integer, respectively. This stems from the fact that any k-cycle can be written as

$$(i_1 \quad i_2 \ldots i_k) = (i_1 \quad i_k) \circ (i_1 \quad i_{k-1}) \circ \cdots \circ (i_1 \quad i_2),$$

a product of $k - 1$ transpositions. Thus the parity of a permutation may be quickly calculated once its disjoint cycle representation is known. A good illustration is furnished by the permutation

$$\alpha = \begin{pmatrix} 1 & 2 & 3 & 4 & 5 & 6 & 7 \\ 7 & 1 & 2 & 6 & 4 & 5 & 3 \end{pmatrix}.$$

Since α admits the decomposition

$$\alpha = (1 \quad 7 \quad 3 \quad 2) \circ (4 \quad 6 \quad 5),$$

it must be odd, being the product of a 4-cycle (odd permutation) and a 3-cycle (even permutation).

The notion of permutation as presented in Definition 2.15 is not sufficiently general for all our purposes, since it restricts us to finite sets. It would seem more natural to define this concept for arbitrary sets, finite or infinite, as follows:

DEFINITION 2.18 If G is any nonempty set, then a *permutation* of G is a one-to-one function from the set G onto itself.

We represent the set of all permutations of G by the symbol, Sym G—the reason for this designation will be clear in a moment. To be sure, if G is a finite set having n elements, then Sym $G = S_n$.

In this general framework it is possible to prove a structure theorem from which Theorem 2.20 will follow as a special case.

THEOREM 2.24 *For a nonempty set G the system* (Sym G, $\circ$) *constitutes a group, called the symmetric group of G.*

Proof. The first thing to be done is to show that $\circ$ is actually a binary operation on Sym G. Given arbitrary permutations $f, g \in$ Sym G, the composition $f \circ g$ is obviously a function from G into G. If the element $c \in G$, then there exists some $b \in G$ such that $f(b) = c$; similarly, there is an element a in G for which $g(a) = b$. Hence

$$(f \circ g)(a) = f(g(a)) = f(b) = c,$$

which means that $f \circ g$ is an onto function.

By definition, both f and g are one-to-one functions; we would like to conclude that the composition $f \circ g$ also has this property. For this purpose, consider arbitrary elements $a, b \in G$ with $a \neq b$. Since g is one-to-one, the images $g(a)$ and $g(b)$ are necessarily distinct in G. But then, the one-to-one character of f implies $f(g(a)) \neq f(g(b))$. In other words, $(f \circ g)(a) \neq (f \circ g)(b)$, which establishes the desired one-to-oneness of $f \circ g$. From the preceding remarks we infer that the product $f \circ g \in \text{Sym}\, G$ whenever f and g are in Sym G, making $\circ$ a binary operation on the set Sym G.

The associativity of functional composition follows from Theorem 1.5. Plainly, the identity function i_G is a permutation of G and is such that

$$f \circ i_G = f = i_G \circ f,$$

for any $f \in \text{Sym}\, G$, showing i_G to be an identity element for the system (Sym G, $\circ$). If $f \in \text{Sym}\, G$, the inverse function f^{-1} exists, is a one-to-one function and maps the set G onto itself. Moreover, f^{-1} is the inverse of f with respect to composition. All of this justifies our assertion that (Sym G, $\circ$) is a group.

In this section, two more groups—one commutative and one not—have been added to our arsenal of standard examples. Both these groups will be seen again in the sequel as we illustrate the various concepts that arise.

PROBLEMS 2.4

1. If $a \equiv b \pmod{n}$ and $m|n$, prove that $a \equiv b \pmod{m}$.
2. Assuming that $a^2 \equiv b^2 \pmod{p}$, where p is a prime number, show that either $a \equiv b \pmod{p}$ or $a \equiv -b \pmod{p}$.
3. Verify that if $a \equiv b \pmod{n_1}$ and $a \equiv b \pmod{n_2}$, then $a \equiv b \pmod{n}$, where the integer $n = \text{lcm}(n_1, n_2)$. Hence if n_1 and n_2 are relatively prime, then $a \equiv b \pmod{n_1 n_2}$.
4. For any integer a, establish that $a^3 \equiv a \pmod{3}$, $a^5 \equiv a \pmod{5}$, and $a^5 \equiv a \pmod{15}$. [*Hint*: The last congruence follows from Problem 3.]
5. Describe the partition of Z determined by the relation congruence modulo 5.
6. Show that $(a + b)^p \equiv a^p + b^p \pmod{p}$, where p is a prime. [*Hint*: First prove that, for $1 \leqslant k \leqslant p - 1$, $p|\binom{p}{k}$.]
7. Let the decimal representation of the positive integer a be $a = \sum_{k=0}^{n} a_k 10^k$. Prove that
   a) $a \equiv 0 \pmod{3}$ if and only if $\sum_{k=0}^{n} a_k \equiv 0 \pmod{3}$;
   b) $a \equiv 0 \pmod{11}$ if and only if $\sum_{k=0}^{n} (-1)^k a_k \equiv 0 \pmod{11}$. [*Hint*: Use the fact that $10 \equiv -1 \pmod{11}$.]
8. If $a \equiv b \pmod{n}$, prove that $\gcd(a, n) = \gcd(b, n)$.
9. Prove the following generalizations of Theorem 2.15:
   a) If $ac \equiv bd \pmod{n}$, $c \equiv d \pmod{n}$, and $\gcd(c, n) = 1$, then $a \equiv b \pmod{n}$.

b) If $ac \equiv bc \pmod{n}$, $d = \gcd(c, n)$, and $n = n_1 d$, then $a \equiv b \pmod{n_1}$.

10. For any integer a, prove that either $a^2 \equiv 0 \pmod{4}$ or $a^2 \equiv 1 \pmod{4}$.

11. Show that if a is an odd integer, then $a^{2^n} \equiv 1 \pmod{2^{n+2}}$.

12. Use congruence theory to show that for every integer $n \geqslant 1$,

a) $3 | 4^{n+1} + 5^{2n-1}$;
[*Hint*: $4^{n+1} + 5^{2n-1} \equiv 1^{n+1} + (-1)^{2n-1} \pmod{3}$.]

b) $5 | 7 \cdot 16^n + 3$;

c) $8 | 5^{2n} + 7$;

d) $27 | 2^{5n+1} + 5^{n+2}$.

13. Find the last digit of 7^{99}. [*Hint*: Work modulo 10, noting that $7^4 \equiv 1 \pmod{10}$.]

14. Show that $4! \equiv -1 \pmod{5}$ and $6! \equiv -1 \pmod{7}$.

15. Given that $495 | 273x49y5$, obtain the digits x and y. [*Hint*: See Problem 7.]

16. a) Prove that if $p_1, p_2, \ldots, p_r$ are distinct primes and if $a \equiv b \pmod{p_k}$ for each k, then $a \equiv b \pmod{p_1 p_2 \cdots p_r}$.

b) Use part (a) to verify that $2^{30} \equiv 4 \pmod{15}$ and $24^{12} \equiv 1 \pmod{35}$.

17. Establish that $2^{32} \equiv -1 \pmod{641}$ and hence $641 | 2^{2^5} + 1$. This disproves a once-famous conjecture that integers of the form $2^{2^m} + 1$ are all prime. [*Hint*: Raise $5 \cdot 2^7 \equiv -1 \pmod{641}$ to the fourth power.]

18. If a and b are both odd integers, prove that $a^2 \equiv b^2 \pmod{8}$.

19. a) Given that $\gcd(a, n) = 1$, verify that the linear congruence $ax \equiv b \pmod{n}$ possesses a unique solution modulo n. [*Hint*: Apply Theorem 1.14.]

b) Find the solutions of $5x \equiv 2 \pmod{26}$ and $7x \equiv 4 \pmod{12}$.

20. Construct the operation tables for the groups $(Z_8, +_8)$ and $(Z_9, +_9)$.

21. Show that the system $(\{0, 4, 8, 12\}, +_{16})$ forms a commutative group.

22. Represent each of the symmetries of the square (Example 2.22) as a permutation of the vertices of the square.

23. If X is a finite set, prove that a function $f: X \to X$ is one-to-one if and only if it maps onto X.

24. Demonstrate that the group $(S_3, \circ)$, and hence any larger symmetric group S_n, is noncommutative by considering the permutations

$$\begin{pmatrix} 1 & 2 & 3 \\ 2 & 3 & 1 \end{pmatrix} \quad \text{and} \quad \begin{pmatrix} 1 & 2 & 3 \\ 3 & 2 & 1 \end{pmatrix}.$$

25. In S_n for $n \geqslant 4$, show that

a) $(1\ 2\ 3) \circ (2\ 3\ 4) = (1\ 2) \circ (3\ 4)$;

b) $(1\ 3) = (1\ 2) \circ (2\ 3) \circ (1\ 2)$;

c) $(1\ 2\ 3) = (1\ 2\ 4) \circ (1\ 3) \circ (1\ 2\ 4)^{-1} \circ (1\ 3)^{-1}$.

26. Suppose that α and β are disjoint cycles in S_n, say,

$$\alpha = (a_1 \quad a_2 \quad \cdots \quad a_r),$$
$$\beta = (b_1 \quad b_2 \quad \cdots \quad b_s).$$

Prove each of the statements below:

a) $\alpha \circ \beta \circ (b_s \quad a_r) = (a_1 \cdots a_r \quad b_1 \cdots b_s)$;

b) $(\alpha \circ \beta)^k = \alpha^k \circ \beta^k$ for all $k \in Z_+$;

c) if $\alpha \circ \beta = \iota$, then $\alpha = \iota$ and $\beta = \iota$;

d) if $\gamma \in S_n$, then $\gamma \circ \alpha \circ \gamma^{-1} = (\gamma(a_1) \quad \gamma(a_2) \quad \cdots \quad \gamma(a_r))$;

e) $\gamma \circ \alpha \circ \gamma^{-1} = \beta$ for some $\gamma \in S_n$.

27. Express the following permutations as (a) products of disjoint cycles and (b) products of transpositions:

$$\begin{pmatrix} 1 & 2 & 3 & 4 & 5 & 6 \\ 3 & 6 & 4 & 1 & 2 & 5 \end{pmatrix}, \quad \begin{pmatrix} 1 & 2 & 3 & 4 & 5 & 6 & 7 \\ 5 & 7 & 6 & 2 & 1 & 3 & 4 \end{pmatrix},$$
$$\begin{pmatrix} 1 & 2 & 3 & 4 & 5 & 6 & 7 & 8 & 9 & 10 \\ 10 & 5 & 6 & 4 & 9 & 7 & 8 & 3 & 2 & 1 \end{pmatrix}.$$

28. Show that the cycle $(1 \quad 2 \quad 3 \quad 4 \quad 5)$ may be written as a product of 3-cycles.

29. Given the k-cycle $(1 \quad 2 \ldots k)$, establish the following relations:

a) $(1 \quad 2 \ldots k) = (1 \quad 2 \ldots j) \circ (j \quad j+1 \ldots k), \qquad 1 < j < k$

b) $(1 \quad 2 \ldots k) = (1 \quad k) \circ (1 \quad 2 \ldots k-1), \qquad k \geqslant 2$

c) $(1 \quad 2 \ldots k) = (1 \quad 2) \circ (2 \quad 3) \circ \cdots \circ (k-1 \quad k), \qquad k \geqslant 2$

d) $(1 \quad 2 \ldots k) \circ (k-1 \ldots 2 \quad 1) = (1 \quad k)$

30. If $n < 4$, show that every permutation in S_n is a cycle.

31. Assuming that $\alpha \in S_n$ is a k-cycle, verify that

a) the inverse of α is also a k-cycle; [*Hint*: $(i_1 \quad i_2 \ldots i_k)^{-1} = (i_k \quad i_{k-1} \ldots i_1)$.]

b) whenever k is an odd integer, α^2 is a k-cycle;

c) $\alpha^k = \iota$.

32. Consider the set G consisting of the four permutations

$$\begin{pmatrix} 1 & 2 & 3 & 4 \\ 1 & 2 & 3 & 4 \end{pmatrix}, \begin{pmatrix} 1 & 2 & 3 & 4 \\ 2 & 1 & 4 & 3 \end{pmatrix}, \begin{pmatrix} 1 & 2 & 3 & 4 \\ 3 & 4 & 1 & 2 \end{pmatrix}, \begin{pmatrix} 1 & 2 & 3 & 4 \\ 4 & 3 & 2 & 1 \end{pmatrix}.$$

Show that $(G, \circ)$ constitutes a commutative group.

33. Form the set $G = \{f, f^2, f^3, f^4, f^5, f^6\}$, where f is the permutation

$$f = \begin{pmatrix} 1 & 2 & 3 & 4 & 5 & 6 \\ 2 & 3 & 4 & 5 & 6 & 1 \end{pmatrix}.$$

Prove that the pair $(G, \circ)$ is a commutative group.

34. By an *inversion* of a permutation $\alpha \in S_n$ is meant a pair (i, j) of integers, where $1 \leqslant i < j \leqslant n$, such that $\alpha(i) > \alpha(j)$. If $\pi(\alpha)$ denotes the total number of inversions performed by α, show that $\operatorname{sgn} \alpha = (-1)^{\pi(\alpha)}$.

35. Given two permutations $\alpha, \beta \in S_n$, prove the following:

a) α is even if and only if in its disjoint cycle representation there occur an even number of cycles of even length.

b) α and α^{-1} have the same parity.

c) α and $\beta \circ \alpha \circ \beta^{-1}$ have the same parity.

36. Find all even permutations in S_3 and S_4.

37. Let G be an infinite set and H be the set of all permutations $\alpha \in \operatorname{Sym} G$ that move only a finite number of elements of G. Prove that the pair $(H, \circ)$ forms a group.

2.5 SUBGROUPS

Two standard techniques are used for attacking the problem of the structure of a particular group. One method calls for finding all the subgroups of the group, with the hope of decomposing the parent group into simpler groups whose structure is known. The other approach is to determine all operation-preserving functions from the given group into a more familiar group; the idea here is that the images will reflect some of the algebraic properties of the original group. On closer scrutiny we shall see that although these lines of investigation aim in different directions, they are not entirely unrelated, but rather are different aspects of the same problem. For the moment, however, our attention is focused on analyzing a group by means of its subgroups; the question of structure-preserving mappings is a more subtle matter and will be deferred to a later section.

From various examples and exercises the reader may have noticed that certain subsets of the elements of a group lead to new groups when one restricts the group operation to these subsets. It is this situation that now commands our attention.

DEFINITION 2.19 Let H be a nonempty subset of the group G. Then H is said to be a *subgroup* of G if H is itself a group under the same operation as that of G.

Any group G has two obvious subgroups. For the set $\{e\}$ consisting only of the identity element of G satisfies Definition 2.19, as does G itself. These two subgroups are often referred to as the *trivial subgroups* of G; subgroups that are of genuine interest lie somewhere between these opposite extremes. A subgroup H is termed *proper* if it does not exhaust G, that is, if $H \subset G$.

Example 2.26
If Z_e and Z_o denote the sets of even and odd integers, respectively, then $(Z_e, +)$ forms a subgroup of the additive group $(Z, +)$, but $(Z_o, +)$ does not. (Why?)

Example 2.27
Consider $(Z_9, +_9)$, the group of integers modulo 9. If $H = \{0, 3, 6\}$, it will be found that $(H, +_9)$ is a subgroup of $(Z_9, +_9)$. The identity element of this subgroup is 0, while 3 and 6 are inverses of each other.

Example 2.28
As a further illustration, let us list the nontrivial subgroups of the group $(G_s, \circ)$ of symmetries of the square (see Example 2.22). We leave it to the reader to check that the following eight sets comprise the elements of these subgroups:

$$\{e, r_1, r_2, r_3\}, \quad \{e, r_2, h, v\}, \quad \{e, r_2, d_1, d_2\},$$
$$\{e, r_2\}, \quad \{e, h\}, \quad \{e, v\}, \quad \{e, d_1\}, \quad \{e, d_2\}.$$

It is sometimes helpful to visualize the relationships between the subgroups of a group by means of a lattice diagram. In such a diagram a line segment joining two subgroups indicates that the lower subgroup is contained in the upper one. The lattice of subgroups of G_s is

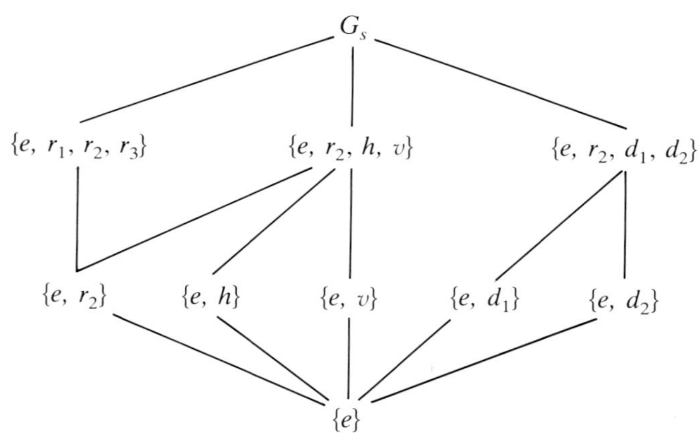

Suppose H is a subgroup of the group G. Since the identity element of H satisfies the equation $x^2 = x$, it must be the same as the identity of the larger group G; otherwise, we would have two idempotent elements in G, contrary to Theorem 2.7. The identity element of a group therefore serves as the identity element for any of its subgroups. Moreover, the uniqueness of the inverse elements in a group implies the inverse of an element $h \in H$ in the subgroup H is the same as its inverse in the whole group G.

To establish that a given subset H of G, along with the induced operation of G, constitutes a subgroup, we must verify that all the requirements of Definition 2.11 are met. However, the associativity of the operation in H is an immediate consequence of its associativity in G, since $H \subseteq G$. It is necessary then to show only the following:

1. $a, b \in H$ implies $ab \in H$ (closure condition),
2. $e \in H$, where e is the identity element of G,
3. $a \in H$ implies $a^{-1} \in H$, where a^{-1} is the inverse of a in G.

Needless to say, the saving in not having to check the associative law is considerable.

A theorem that establishes a single convenient criterion for determining subgroups is given below.

THEOREM 2.25 *Let G be a group and $\varnothing \neq H \subseteq G$. Then H is a subgroup of G if and only if $a, b \in H$ implies $ab^{-1} \in H$.*

Proof. If H is a subgroup and $a, b \in H$, then $b^{-1} \in H$, and so $ab^{-1} \in H$ by the closure condition. Conversely, suppose that H is a nonempty subset of G that contains the element ab^{-1} whenever $a, b \in H$. Since H contains at least one element a, we may take a equal to b to see that $bb^{-1} = e \in H$. Also, $b^{-1} = eb^{-1} \in H$ for every b in H, applying the hypothesis to the pair $e, b \in H$. Finally, if a and b are any two members of the set H, then by what was just proved, b^{-1} also belongs to H, so $ab = a(b^{-1})^{-1} \in H$. In other words, the set H is closed with respect to the group operation. Because H inherits the associative law as a subset of G, all the group axioms are satisfied, and H is therefore a subgroup of G.

In much the same way, one can show that a nonempty subset H of a group G is a subgroup of G if and only if $a^{-1}b \in H$ whenever $a, b \in H$.

Example 2.29

In Section 2.4 we discussed the subset A_n of S_n consisting of all even permutations on the set $\{1, 2, \ldots, n\}$. If both $\alpha, \beta \in A_n$, then each can be expressed as a product of an even number of transpositions; say,

$$\alpha = \tau_1 \circ \tau_2 \circ \cdots \circ \tau_{2n}, \qquad \beta = \bar{\tau}_1 \circ \bar{\tau}_2 \circ \cdots \circ \bar{\tau}_{2m}$$

where τ_i, $\bar{\tau}_j$ are transpositions. But then

$$\begin{aligned}\alpha \circ \beta^{-1} &= (\tau_1 \circ \cdots \circ \tau_{2n}) \circ (\bar{\tau}_1 \circ \cdots \circ \bar{\tau}_{2m})^{-1} \\ &= \tau_1 \circ \cdots \circ \tau_{2n} \circ \bar{\tau}_{2m}^{-1} \circ \cdots \circ \bar{\tau}_1^{-1}\end{aligned}$$

is also expressible as a product of an even number of transpositions, the inverse of a transposition being a transposition. That is, $\alpha \circ \beta^{-1} \in A_n$, which makes A_n a subgroup of the symmetric group S_n. This group is traditionally called the *alternating group on n symbols.*

Let us remark that a group G is termed *finite* if the set G of its elements is a finite set. The *order* of a finite group G, denoted by $o(G)$, is the number of elements in the group. The symmetric group S_n, for instance, is a finite group of order $n!$. Analogously, a group that contains infinitely many elements is said to be an *infinite group* and to have *infinite order*. For finite groups there is an even simpler test for subgroups; this is brought out by the following theorem.

THEOREM 2.26 *Let H be a finite (nonempty) subset of the group G. If H is closed under multiplication, then H is a subgroup of G.*

Proof. It suffices to show that $a^{-1} \in H$ whenever $a \in H$; after doing this the hypothesized closure condition makes an appeal to Theorem 2.25 legitimate. Choose any $a \in H$. Then the elements a, $a^2, \ldots, a^k, \ldots$ lie in H and, since H is finite, cannot all be distinct. Suppose that $a^m = a^n$ for integers $m > n > 0$. Canceling a^n, we get $a^{m-n} = e$ or

$$a(a^{m-n-1}) = (a^{m-n-1})a = e,$$

which signifies that $a^{-1} = a^{m-n-1} \in H$.

COROLLARY *A nonempty subset H of a finite group G is a subgroup of G if and only if $a, b \in H$ implies that $ab \in H$.*

We should mention that if G is a commutative group, then any subgroup H of G is commutative (in other words, the property of being commutative is *hereditary*); elements of H are also elements of G, in which the commutative law certainly holds.

As an illustration of the use of Theorem 2.25, let us examine the behavior of subgroups under the set-theoretic operations of intersection and union.

THEOREM 2.27 *If $\{H_i\}$ is an indexed collection of subgroups of the group G, then $\cap H_i$ is also a subgroup of G.*

Proof. Since the sets H_i all contain the identity element of G, the intersection $\cap H_i \neq \varnothing$. Next, suppose that a and b are any two elements of $\cap H_i$; then $a, b \in H_i$ where i ranges over the index set. The set H_i being a subgroup, it follows that the product ab^{-1} also belongs to H_i. As this is true for every value of the index i, $ab^{-1} \in \cap H_i$, which implies that $\cap H_i$ is a subgroup of G.

In general, without further restriction on the subgroups H_i, it is not true that $\cup H_i$ will again be a subgroup of G. One simply cannot guarantee that $\cup H_i$ will contain products whose factors come from different H_i. To give a concrete illustration, both $\{0, 6\}$ and $\{0, 4, 8\}$ are subgroups of Z_{12}, yet on taking their union, $\{0, 4, 6, 8\}$ fails to be so. The difficulty in this case is that the modular sums $4 +_{12} 6$ and $6 +_{12} 8$ do not belong to the set $\{0, 4, 6, 8\}$.

By way of an analog to Theorem 2.27 we have the following.

THEOREM 2.28 *Let $\{H_i\}$ be an indexed collection of subgroups of the group G. Suppose that $\{H_i\}$ has the property that for any two of its members H_i and H_j there exists a set H_k (depending on i and j) in $\{H_i\}$ such that $H_i \subseteq H_k$ and $H_j \subseteq H_k$. Then $\cup H_i$ is also a subgroup of G.*

Proof. By now the pattern of proof should be clear. We assume that a and b are arbitrary elements of $\cup H_i$ and show that $ab^{-1} \in \cup H_i$. If $a, b \in \cup H_i$, then there exist subsets H_i, H_j containing a and b, respectively. According to our hypothesis, $H_i \subseteq H_k$ and $H_j \subseteq H_k$ for some choice of H_k in $\{H_i\}$. Since H_k is a subgroup and both $a, b \in H_k$, it follows that the product ab^{-1} belongs to H_k. Accordingly, $ab^{-1} \in \cup H_i$, as was claimed at the beginning.

As a particular case of the foregoing result, consider just two subgroups H_1 and H_2. Theorem 2.28 may be interpreted as asserting that $H_1 \cup H_2$ will again be a subgroup of G provided that either $H_1 \subseteq H_2$ or $H_2 \subseteq H_1$. What is rather interesting is that this condition is necessary, as well as sufficient. The next theorem gives the details.

THEOREM 2.29 *Let H_1 and H_2 be subgroups of the group G. Then $H_1 \cup H_2$ is also a subgroup of G if and only if $H_1 \subseteq H_2$ or $H_2 \subseteq H_1$.*

Proof. In view of the preceding remarks, it is enough to show that if $H_1 \cup H_2$ is a subgroup of G, then one of the sets H_1 or H_2 must be contained in the other. Suppose to the contrary that this assertion were false, that is, $H_1 \not\subseteq H_2$ and $H_2 \not\subseteq H_1$. Then there would exist elements $a \in H_1 - H_2$ and $b \in H_2 - H_1$. Now, if the product ab were a member of the set H_1, we could infer that $b = a^{-1}(ab) \in H_1$, which is clearly not true. On the other hand, the possibility $ab \in H_2$ yields the equally false conclusion that $a = (ab)b^{-1} \in H_2$. That is, the elements a, $b \in H_1 \cup H_2$, but $ab \notin H_1 \cup H_2$. This conclusion is obviously untenable, since it contradicts the fact that $H_1 \cup H_2$ is a group. Now that we have arrived at a contradiction, the proof is complete.

COROLLARY *A group G cannot be the union of two of its proper subgroups.*

Proof. Let $G = H_1 \cup H_2$, where H_i $(i = 1, 2)$ are proper subgroups. By the foregoing theorem, either $H_1 \subseteq H_2$ or $H_2 \subseteq H_1$; let us say that $H_1 \subseteq H_2$. Then $H_2 = H_1 \cup H_2 = G$, a contradiction.

An interesting subset of any group is its center, defined below.

DEFINITION 2.20 The *center* of a group G, denoted by cent G, is the set

$$\text{cent } G = \{c \in G \mid cx = xc \text{ for all } x \in G\}.$$

Simply put, cent G consists of those elements that commute with every element of G. In the case of the group of symmetries of the square, for example, we have cent $G_s = \{e, r_2\}$. Given the multiplication table for a finite group G, there is no problem in finding its center. An element $a \in \text{cent } G$ if and only if the entries in the row headed by a appear in the same order as the entries in the column headed by a. In particular, G is a commutative group if and only if the entire table is symmetric about the main diagonal (upper left to lower right).

THEOREM 2.30 *The center of any group G is a subgroup of G.*

Proof. We first observe that cent G is nonempty, for at the very least $e \in \text{cent } G$. Now consider any two elements $a, b \in \text{cent } G$. By the definition of center we know that $ax = xa$ and $bx = xb$ for every element x of G. Thus if $x \in G$,

$$\begin{aligned}
(ab^{-1})x &= a(b^{-1}x) \\
&= a(x^{-1}b)^{-1} \\
&= a(bx^{-1})^{-1} \\
&= a(xb^{-1}) \\
&= (ax)b^{-1} \\
&= (xa)b^{-1} = x(ab^{-1}),
\end{aligned}$$

which implies that $ab^{-1} \in \text{cent } G$. According to Theorem 2.25, this is a sufficient condition for cent G to be a subgroup of G.

The center furnishes a measure of the departure from commutativity; the larger the center, the closer the group is to being commutative. The reader may already have deduced that a group G is commutative if and only if cent $G = G$. The antithesis of the commutative group would be a group G for which cent $G = \{e\}$, that is, a group with *trivial center* (the phrase "centerless group" is also current). Such groups exist, as is evidenced by the following example.

Example 2.30
With matrix multiplication as the operation the set

$$G = \left\{ \begin{bmatrix} a & 0 \\ b & 1 \end{bmatrix} \middle| a, b \in R^{\#}; \quad a \neq 0 \right\}$$

is a noncommutative group. To determine its center, let

$$\begin{bmatrix} a & 0 \\ b & 1 \end{bmatrix}$$

be in cent G. Since this matrix must commute with all elements of G, it follows in particular that

$$\begin{bmatrix} a & 0 \\ b & 1 \end{bmatrix} \begin{bmatrix} 1 & 0 \\ -1 & 1 \end{bmatrix} = \begin{bmatrix} 1 & 0 \\ -1 & 1 \end{bmatrix} \begin{bmatrix} a & 0 \\ b & 1 \end{bmatrix}.$$

Carrying out the multiplication leads to the equation

$$\begin{bmatrix} a & 0 \\ b-1 & 1 \end{bmatrix} = \begin{bmatrix} a & 0 \\ -a+b & 1 \end{bmatrix},$$

which holds only if $b - 1 = -a + b$, or $a = 1$. Similarly, in order that

$$\begin{bmatrix} a & 0 \\ b & 1 \end{bmatrix} \begin{bmatrix} -1 & 0 \\ 0 & 1 \end{bmatrix} = \begin{bmatrix} -1 & 0 \\ 0 & 1 \end{bmatrix} \begin{bmatrix} a & 0 \\ b & 1 \end{bmatrix},$$

we must have

$$\begin{bmatrix} -a & 0 \\ -b & 1 \end{bmatrix} = \begin{bmatrix} -a & 0 \\ b & 1 \end{bmatrix},$$

forcing $b = 0$. These calculations imply that cent G contains the single matrix

$$\begin{bmatrix} 1 & 0 \\ 0 & 1 \end{bmatrix},$$

the identity element of the group.

To give another instance of a centerless group, let us prove the following basic theorem.

THEOREM 2.31 *For $n \geqslant 3$ the center of the symmetric group S_n consists only of the identity permutation.*

Proof. Consider a typical permutation $\alpha \in S_n$, with $\alpha \neq \iota$. An observation of some importance is that if $\beta = (i_1 \quad i_2 \ldots i_r)$ is any cycle in S_n, then $\alpha \circ \beta \circ \alpha^{-1}$ is the cycle obtained by replacing each symbol in β

by its image under α:

$$\alpha \circ \beta \circ \alpha^{-1} = (\alpha(i_1) \quad \alpha(i_2) \ldots \alpha(i_r)).$$

This follows from the fact that, for $1 \leqslant j < k$,

$$(\alpha \circ \beta \circ \alpha^{-1})(\alpha(i_j)) = \alpha(\beta(i_j)) = \alpha(i_{j+1})$$

while

$$(\alpha \circ \beta \circ \alpha^{-1})(\alpha(i_k)) = \alpha(\beta(i_k)) = \alpha(i_1).$$

Now express $\alpha = \alpha_1 \circ \alpha_2 \circ \cdots \circ \alpha_r$ as a product of disjoint cycles where, let us say, the first cyclic $\alpha_1 = (i_1 \quad i_2 \ldots i_k)$. If $k \geqslant 3$, we consider the transposition $\tau = (i_1 \quad i_2)$. By the remark of the preceding paragraph,

$$\alpha \circ \tau \circ \alpha^{-1} = (\alpha(i_1) \quad \alpha(i_2)) = (i_2 \quad i_3) \neq \tau.$$

On the other hand, should it happen that $k = 2$, let $\beta = (i_1 \quad i_2 \quad i_3)$. Then

$$\alpha \circ \beta \circ \alpha^{-1} = (\alpha(i_1) \quad \alpha(i_2) \quad \alpha(i_3)) = (i_2 \quad i_1 \quad \alpha(i_3))$$

where $\alpha(i_3) \neq i_1, i_2$. Regardless of the choice of $\alpha(i_3)$, we have $\alpha \circ \beta \circ \alpha^{-1} \neq \beta$. Our argument tells us that $\alpha \notin \operatorname{cent} S_n$; in other words, no $\alpha \neq \iota$ lies in the center of S_n, which is the assertion of the theorem.

The next topic of interest concerns cyclic subgroups. This requires a few preliminary remarks on generating sets.

DEFINITION 2.21 If G is an arbitrary group and $\varnothing \neq S \subseteq G$, then the symbol (S) will represent the set

$$(S) = \cap\{H \mid S \subseteq H;\ H \text{ is a subgroup of } G\}.$$

The set (S) clearly exists, for G itself is a member of the family appearing on right; that is, G is a trivial subgroup of itself and $S \subseteq G$. In addition, since S is contained in each of the sets being intersected, we always have the inclusion $S \subseteq (S)$. Theorem 2.27 leads immediately to the following theorem.

THEOREM 2.32 *Let S be a nonempty subset of the group G. Then (S) forms a subgroup of T, known as the subgroup generated by the set S.*

Definition 2.21 implies that whenever H is a subgroup of the group G with $S \subseteq H$, then $(S) \subseteq H$. For this reason, one speaks informally of (S) as being the smallest subgroup that contains the set S. Of course, it may well happen that $(S) = G$, and in such a situation the group G is said to be *generated* by the subset S. For example, it is easy to see that the additive group Z is generated by Z_o, the set of odd integers.

We shall give an alternative description of the subset (S) that is frequently easier to work with than Definition 2.21. In what follows, the symbol S^{-1} is used to indicate the collection of inverses of elements in S; that is, $S^{-1} = \{a^{-1} | a \in S\}$. The result that we propose to obtain is that

$$(S) = \{a_1 a_2 \cdots a_n | a_1, \ldots, a_n \in S \cup S^{-1};\ n \in Z_+\}.$$

Although the notation is self-explanatory, it is helpful to point out explicitly that the set on the right consists of all finite products whose factors are either elements of S or inverses of elements in S. Let us temporarily designate this set of products by $[S]$.

An abbreviated proof of the assertion above might run as follows: The set $[S]$ is a subgroup of the group G with the property that $S \subseteq [S]$. Since (S) is the smallest such subgroup, it follows that $(S) \subseteq [S]$. The reverse inclusion is justified by the fact that any subgroup that contains the set S must necessarily contain all the elements of $[S]$.

To see that $[S]$ is actually a subgroup of G, let x, y be typical elements of $[S]$. Then

$$x = a_1 a_2 \cdots a_n, \qquad y = b_1 b_2 \cdots b_m$$

where $a_i, b_j \in S \cup S^{-1}$ for $i = 1, 2, \ldots, n$ and $j = 1, 2, \ldots, m$. It follows that

$$xy^{-1} = a_1 \cdots a_n b_m^{-1} \cdots b_1^{-1}$$

with all $a_i, b_j^{-1} \in S \cup S^{-1}$. By definition, $xy^{-1} \in [S]$, which makes $[S]$ a subgroup of G.

There is another way of characterizing the subgroup of G generated by the set S. By the preceding remarks, any element $x \in (S)$ can be written as $x = x_1 x_2 \cdots x_r$, where each x_i is either an element of S or the inverse of an element of S. If adjacent factors are equal, these may be combined into a single integral power, so that we end up with an expression for x of the form $x = a_1^{k_1} a_2^{k_2} \cdots a_n^{k_n}$, with $k_i \in Z$. Accordingly, (S) consists of finite products of powers of elements in S (powers of a given element may occur several times in the same product):

$$(S) = \{a_1^{k_1} a_2^{k_2} \cdots a_n^{k_n} | a_i \in S;\ k_i \in Z;\ n \geqslant 1\}.$$

This description of (S) allows us to conclude, for instance, that the symmetric group S_n is generated by the transpositions it contains.

When the subgroup of G generated by a (nonempty) subset S is the whole of G, we say that S is a *set of generators* of G. A group G is *finitely generated* if there exists a finite set of generators of G.

A single example will suffice for purposes of illustration.

Example 2.31

For $n \geqslant 2$ the symmetric group S_n is finitely generated, with the set of transpositions

$$\{(1\quad 2), (1\quad 3), \ldots, (1\quad n)\}$$

forming a set of generators. We have previously learned that every permutation in S_n is a product of cycles. Thus it is enough to factor any k-cycle into a product of elements of the given set. This is done in the following way: If the cycle in question contains the integer 1, then

$$(1 \quad a_1 \quad a_2 \ldots a_k) = (1 \quad a_k) \circ (1 \quad a_{k-1}) \circ \cdots \circ (1 \quad a_2) \circ (1 \quad a_1);$$

while for any cycle not containing 1,

$$(a_1 \quad a_2 \ldots a_k) = (1 \quad a_1) \circ (1 \quad a_k) \circ \cdots \circ (1 \quad a_2) \circ (1 \quad a_1).$$

A case of special importance arises when S consists of a single element a of G. In this situation it is usual to write (a) instead of $(\{a\})$ and refer to the subgroup (a) as the *cyclic subgroup generated by* a. The subgroup (a) is rather easy to describe; since all its products involve either the element a or its inverse, (a) simply reduces to the integral powers of a:

$$(a) = \{a^n | n \in Z\}.$$

It is entirely possible for a group G to be equal to one of its cyclic subgroups. This prompts the following definition.

DEFINITION 2.22 A group G is *cyclic* with generator a if $G = (a)$ for some $a \in G$.

Thus to say that a group is cyclic means that each of its members can be expressed as an integral power of some fixed element of the group. Any cyclic group $G = (a)$ is commutative, since

$$a^n a^m = a^{n+m} = a^{m+n} = a^m a^n$$

for arbitrary integers n and m. Note too that a given cyclic group may possess several different generators; indeed, we always have $(a) = (a^{-1})$.

To make these notions clearer, let us pause to examine several examples.

Example 2.32

First consider the additive group Z of integers. In this case, since the group operation is that of addition, the abstract product $aa \cdots a$ is replaced by the ordinary sum $a + a + \cdots + a$. Accordingly, the subset generated by the element a takes the form

$$(a) = \{na | n \in Z\}.$$

Using this notation, $\{0\} = (0)$, $Z = (1)$, while $Z_e = (2)$. We may thus conclude that both the additive groups Z and Z_e are cyclic, with the integers 1 and 2 as their respective generators.

Example 2.33
Another illustration is furnished by Z_{12}, the group of integers modulo 12. Note that we now write $a +_{12} a +_{12} \cdots +_{12} a$ for $aa \cdots a$. The cyclic subgroup generated by, say, 3 is $\{0, 3, 6, 9\}$; here

$$(3) = \{3n \pmod{12} | n \in Z\} = \{0, 3, 6, 9\}.$$

As $(1) = Z_{12}$, the group Z_{12} is itself cyclic; other possible generators of Z_{12} are 5, 7, and 11. It is not difficult to see that, in general, the group Z_n of integers modulo n is cyclic with 1 as a generator and thus there exists a cyclic group of order n for each $n \in Z_+$.

Example 2.34
The group G_s of symmetries of the square is not cyclic, since cyclic groups are necessarily commutative. Save for the identity, the distinct powers of any of its elements form a proper cyclic subgroup. The rotation r_1, in particular, generates the subgroup whose elements are $\{e, r_1, r_2, r_3\}$.

In the following theorems we shall see that finite cyclic groups have a particularly simple structure. For one thing, the elements of a finite cyclic group with generator a are just $e, a, a^2, \ldots, a^{n-1}$, where n is the order of the group. All other powers of a are superfluous, merely repeating these.

THEOREM 2.33 *If (a) is a finite cyclic group of order n, then the elements $e, a, \ldots, a^{n-1}$ are all distinct and*

$$(a) = \{e, a, a^2, \ldots, a^{n-1}\}.$$

Proof. As the set (a) is finite, not all the powers of the generator a are distinct. There must be some repetition $a^i = a^j$ with $i < j$. On multiplying this equation by $a^{-i} = (a^i)^{-1}$ it follows that $a^{j-i} = e$. Thus the set of positive integers k for which $a^k = e$ is nonempty. By the Well-Ordering Principle there is a smallest positive integer m with this property; that is, $a^m = e$, while $a^k \neq e$ for $0 < k < m$.

The set $S = \{e, a, a^2, \ldots, a^{m-1}\}$ consists of distinct elements of (a). For $a^r = a^s$, with $0 \leqslant r < s \leqslant m - 1$, implies that $a^{s-r} = e$, contrary to the minimality of m. To complete the proof, it remains to show each member a^k of the group (a) is equal to an element of S. Now by the Division Algorithm we may write $k = qm + r$ for some integers q and r with $0 \leqslant r < m$. Hence

$$a^k = (a^m)^q a^r = ea^r = a^r \in S.$$

This means the set $(a) \subseteq S$, yielding $(a) = \{e, a, a^2, \ldots, a^{m-1}\}$; since there are n elements in (a), the equality $m = n$ follows.

COROLLARY *$G = (a)$ is a finite cyclic group of order n if and only if n is the least positive integer such that $a^n = e$.*

We add a further definition that fits into the picture: If a is an element of the group G, the *order of a* is the order of the cyclic subgroup generated by a; the order of a will be denoted by $o(a)$. Clearly, an element of G has order 1 if and only if it is the identity element. Theorem 2.33 tells us that an element $a \in G$ has finite order n if and only if n is the smallest positive integer for which $a^n = e$.

As an illustration, consider the group Z_4; here, the element 0 has order 1, 1 has order 4, 2 has order 2, while 3 is of order 4. In the additive group Z, every nonzero element is of infinite order.

We shall have occasion to refer to the following fact a number of times:

THEOREM 2.34 Let a be an element of the group G. If $o(a) = n$, and $a^m = e$, then $n|m$.

Proof. By the Division Algorithm there exist unique integers q and r such that $m = qn + r$, where $0 \leqslant r < n$. Thus

$$e = a^m = a^{qn+r} = (a^n)^q a^r = a^r.$$

Since n is assumed to be the least positive integer for which $a^n = e$, this forces $r = 0$; hence $m = qn$ or equivalently $n|m$.

COROLLARY *Let $G = (a)$ be a finite cyclic group of order n. Then $a^k = a^j$ if and only if $k \equiv j \pmod{n}$.*

The next theorem provides a criterion for an element to serve as a generator of a cyclic group. First, a technical lemma is required.

LEMMA *Let $G = (a)$ be a finite cyclic group of order n. If $d = \gcd(n, m)$, where $m \in Z_+$, then $o(a^m) = n/d$.*

Proof. By Problem 8(c) of Section 1.3 we may write $n = rd$ and $m = sd$, with $\gcd(r, s) = 1$. Then

$$(a^m)^{n/d} = (a^{sd})^{n/d} = (a^n)^s = e^s = e.$$

Setting $q = o(a^m)$, Theorem 2.34 tells us that $q|n/d$. On the other hand, $a^{mq} = (a^m)^q = e$ and so $n|mq$. Therefore $rd|sdq$ or $r|sq$. Since r and s are relatively prime, the implication is that $r|q$, or rather $n/d|q$. Taken together, the conditions $q|n/d$ and $n/d|q$ yield $n/d = q = o(a^m)$.

COROLLARY *Under the hypothesis of the theorem, $o(a^m) = n$ if and only if $\gcd(n, m) = 1$.*

This puts us in a position to describe the generators of a cyclic group.

THEOREM 2.35 *Let G be a cyclic group with generator a,* $G = (a)$.

a) *If G has finite order n, then* a^m *is a generator of G if and only if* $\gcd(n, m) = 1$.
b) *If G has infinite order, then a and* a^{-1} *are the only generators of G.*

Proof. Suppose $o(G) = n$. Then the element a^m will act as a generator of G if and only if $o(a^m) = n$. In view of the corollary above, this latter condition holds if and only if m and n are relatively prime.

To establish the second assertion, let G be of infinite order and assume that $G = (a^m)$, where $m \neq \pm 1$. Then $a = a^{mk}$ for some integer k, hence $a^{mk-1} = e$. Since the equation $a^x = e$ has a positive solution, the argument used in Theorem 2.33 shows that G must be finite, a contradiction. Therefore a and a^{-1} are the only generators of G.

In certain cases—unfortunately, far too few—it is possible completely to characterize the subgroups of a given group. To cite one instance, in the additive group of integers Z the subgroups are all of the form (n) for some nonnegative integer n. Actually, the situation is somewhat more general, for it can be shown that any subgroup of a cyclic group is again cyclic; we shall discuss this next.

THEOREM 2.36 *Every subgroup of a cyclic group is cyclic.*

Proof. Let (a) be a cyclic group generated by the element a, and let H be one of its subgroups. If $H = \{e\}$, the theorem is trivially true, for $\{e\}$ is the cyclic subgroup generated by the identity element. We may thus suppose that the set $H \neq \{e\}$. If $a^m \in H$, where $m \neq 0$, then a^{-m} is also an element of H; hence H must contain positive powers of a. Let n be the smallest positive integer such that $a^n \in H$. We proposed to show that $H = (a^n)$.

To establish the inclusion $H \subseteq (a^n)$, let a^k be an arbitrary element in the set H. The Division Algorithm implies that there exist integers q and r for which $k = qn + r$, $0 \leqslant r < n$. Since both a^n and a^k are elements of H,

$$a^r = a^{k-qn} = a^k(a^n)^{-q} \in H.$$

Were $r > 0$, we would have a contradiction to the assumption that a^n is the minimal positive power of a in H. Accordingly, $r = 0$ and $k = qn$. Thus only powers of a^n lie in H, indicating that $H \subseteq (a^n)$.

On the other hand, since the set H is closed under the group operation, any power of a^n must again be a member of H. Consequently, $(a^n) \subseteq H$. The two inclusions demonstrate that $H = (a^n)$.

COROLLARY *If H is a subgroup of (a) and $H \neq \{e\}$, then $H = (a^n)$ where n is the least positive integer such that $a^n \in H$.*

We shall return to a further discussion of cyclic groups at the appropriate place in the sequel. For the moment, though, let us indicate another useful method for manufacturing new subgroups from given ones. Some preliminaries are needed:

DEFINITION 2.23 Let G be a group and H, K be nonempty subsets of G. *The product of H and K is the set*

$$HK = \{hk \mid h \in H,\ k \in K\}.$$

As a comment on notation, it should be remarked that the usual practice is to write the product HH as H^2 and, if one of the sets consists of the single element a, to simplify $\{a\}H$ to aH.

At first sight, the reader might reasonably conjecture that whenever H and K are both subgroups of G, then HK will also be a subgroup. The group of symmetries of the square, however, shows that such a simple outcome is not to be expected. Here, it is enough to consider the subgroups having elements $H = \{e, d_1\}$ and $K = \{e, v\}$. A quick check establishes that there is no subgroup whose members comprise the product set

$$HK = \{e^2, ev, d_1e, d_1v\} = \{e, v, d_1, r_1\}.$$

In fact, the set HK isn't even closed under the group operation. One need not be overwhelmed by this state of affairs, for an additional assumption on the subsets H and K readily overcomes the difficulty.

THEOREM 2.37 *Let H and K be subgroups of the group G. Then HK is a subgroup of G if and only if $HK = KH$.*

Proof. Innocuous as the equality $HK = KH$ appears, it is nontheless the source of some difficulty. The notation does not mean each element of H commutes with each element of K; all it signifies is that whenever h and k are arbitrary members of H and K, then there exist elements $h' \in H$, $k' \in K$ for which $hk = k'h'$. Bearing this in mind, let us proceed with the proof proper.

To start, assume that $HK = KH$ and let $a, b \in HK$ be arbitrary elements. (We observe that the set HK is nonempty, for $e = e^2 \in HK$.) Then $a = hk$ and $b = h_1k_1$ for some choice of $h, h_1 \in H$ and $k, k_1 \in K$. As usual, our aim in what follows is to show that the product ab^{-1} lies in HK. This is achieved through first noting that

$$ab^{-1} = (hk)(h_1k_1)^{-1} = h(kk_1^{-1})h_1^{-1}.$$

Since K is closed, the element kk_1^{-1} belongs to K; consequently,

$$(kk_1^{-1})h_1^{-1} \in KH.$$

By virtue of the condition $KH = HK$ there exist elements $h_2 \in H$ and $k_2 \in K$ satisfying $(kk_1^{-1})h_1^{-1} = h_2k_2$. We may thus conclude that

$$ab^{-1} = h(h_2k_2) = (hh_2)k_2 \in HK,$$

since the closure of the set H ensures that hh_2 also is a member of it. To complete the argument that HK is a subgroup, it suffices to invoke Theorem 2.25.

Going in the other direction, let us suppose that HK forms a subgroup of G and that $h \in H$, $k \in K$. Then the product $kh = (h^{-1}k^{-1})^{-1}$, being the inverse of an element in HK, must also lie in HK. This results in the inclusion $KH \subseteq HK$. Next, if the element $a \in HK$, then $a^{-1} \in HK$ and so can be written as $a^{-1} = h_1k_1$ for suitable $h_1 \in H$, $k_1 \in K$. In consequence, $a = k_1^{-1}h_1^{-1} \in KH$, yielding $HK \subseteq KH$ and equality follows.

COROLLARY *If H and K are subgroups of the commutative group G, then HK is also a subgroup of G.*

The utility of Theorem 2.37 lies in the fact that it permits another characterization of the subgroup generated by a union of sets. What happens is this: If HK forms a subgroup of G, it must in fact be that subgroup generated by $H \cup K$; in symbols,

$$HK = (H \cup K).$$

Let us briefly outline the argument involved. To start with, the two inclusions

$$H = He \subseteq HK \qquad \text{and} \qquad K = eK \subseteq HK$$

indicate $H \cup K \subseteq HK$. We have already observed that the subgroup generated by $H \cup K$ is the smallest subgroup to contain this union. Thus whenever HK is a subgroup, it follows that $(H \cup K) \subseteq HK$. On the other hand, the set $(H \cup K)$ by definition contains all products of the form hk with $h \in H$, $k \in K$. This results in the reverse inclusion $HK \subseteq (H \cup K)$ and the subsequent equality $HK = (H \cup K)$.

The significant point in this discussion is that we have gained a great deal of insight into the structure of the group generated by the union $H \cup K$, where H and K are both subgroups of the group G. To be specific, in the event the condition $HK = KH$ holds, each member of $(H \cup K)$ is expressible as the product of an element of H with an element of K. This statement obviously applies when the group G is commutative.

Example 2.35
For purposes of illustrating the above remarks, let us return to the commutative group Z_{12} and the two subgroups $\{0, 6\}$ and $\{0, 4, 8\}$. To obtain the smallest subgroup that contains $\{0, 6\}$ and $\{0, 4, 8\}$, it suffices merely to compute the product of these subsets:

$$\{0, 6\} +_{12} \{0, 4, 8\} = \{0 +_{12} 0,\ 0 +_{12} 4,\ 0 +_{12} 8,\ 6 +_{12} 0,\ 6 +_{12} 4,\ 6 +_{12} 8\}$$

$$= \{0, 4, 8, 6, 10, 2\}.$$

Hence the subgroup of Z_{12} generated by the union $\{0, 6\} \cup \{0, 4, 8\}$ is just the set $\{0, 2, 4, 6, 8, 10\}$.

PROBLEMS 2.5

1. For each of the following sets H, establish that $(H, \cdot)$ is a subgroup of the group $(G, \cdot)$:
   a) $G = \{1, -1, i, -i\}$; $H = \{1, -1\}$ (where $i^2 = -1$)
   b) $G = Q - \{0\}$; $H = \{2^n | n \in Z\}$
   c) $G = R^{\#} - \{0\}$; $H = \{a + b\sqrt{2} | a, b \in Q;\ a^2 + b^2 \neq 0\}$
   d) $G = Q - \{0\}$; $H = \{(1 + 2n)/(1 + 2m) | n, m \in Z\}$
2. For a fixed prime p, consider the set $H = \{m/p^n | m, n \in Z\}$. Show that H is a subgroup of the additive group Q of rational numbers.
3. If H is a subgroup of the group G and K is a subgroup of H, verify that K is also a subgroup of G.
4. In the symmetric group S_n, prove that:
   a) If H is the set of permutations that leave the integer n fixed, that is,
   $$H = \{\alpha \in S_n | \alpha(n) = n\},$$
   then H is a subgroup of S_n.
   b) If the set K is defined by
   $$K = \{\alpha \in S_n | \alpha(n) = n \text{ or } \alpha(n) = n - 1\},$$
   then K does not form a subgroup of S_n.
5. For a finite group G, show that there exists a positive integer n such that $a^n = e$ for all $a \in G$. [*Hint:* First show that there is an integer $n(a)$, depending on a, for which $a^{n(a)} = e$.]
6. Suppose that G is a group and $a \in G$. Let $C(a)$ denote the set of all elements of G that commute with a; that is,
   $$C(a) = \{x \in G | ax = xa\}.$$

Establish each of the assertions below.

a) $C(a)$ forms a subgroup of G, known as the *centralizer of a (in G)*;

b) cent $G \subseteq C(a)$ for all $a \in G$;

c) cent $G = \{a \in G | C(a) = G\} = \bigcap_{a \in G} C(a)$.

7. Given that H is a subgroup of the commutative group G, prove that the set

$$K = \{x \in G | x^n \in H \text{ for some } n \in Z_+\}$$

also forms a subgroup of G.

8. Using the sets G and H below and the operation of matrix multiplication, verify that H is a subgroup of G:

a) $G = \left\{ \begin{bmatrix} a & b \\ 0 & c \end{bmatrix} \middle| a, b, c \in R^{\#};\ ac \neq 0 \right\};\ H = \left\{ \begin{bmatrix} 1 & b \\ 0 & 1 \end{bmatrix} \middle| b \in R^{\#} \right\}$

b) $G = \left\{ \begin{bmatrix} a & b \\ 0 & 1 \end{bmatrix} \middle| a, b \in R^{\#};\ a \neq 0 \right\};\ H = \left\{ \begin{bmatrix} 1 & n \\ 0 & 1 \end{bmatrix} \middle| n \in Z \right\}$

c) $G = \left\{ \begin{bmatrix} a & 0 \\ 0 & 1 \end{bmatrix} \middle| a \in R^{\#} - \{0\} \right\};\ H = \left\{ \begin{bmatrix} 2^n & 0 \\ 0 & 1 \end{bmatrix} \middle| n \in Z \right\}$

9. Prove that in a commutative group G the set

$$H = \{x \in G | x = x^{-1}\}$$

forms a subgroup of G.

10. Given a group G, let H be the set of squares of elements of G; that is,

$$H = \{x \in G | x = y^2 \text{ for some } y \in G\}.$$

Establish that if G is commutative, then H is a subgroup of G, but that this need not be true when G is noncommutative. [*Hint:* For a counterexample, consider the alternating group A_4.]

11. Find the center of the group $M_2^*(R^{\#})$ of 2×2 invertible matrices with real entries. [*Hint:* Any element in the center commutes with $\begin{bmatrix} 1 & 0 \\ 1 & 1 \end{bmatrix}$ and $\begin{bmatrix} 1 & 1 \\ 0 & 1 \end{bmatrix}$.]

12. Consider the group $G = \{(a, b) | a, b \in R^{\#};\ a \neq 0\}$, where multiplication is defined by

$$(a, b) * (c, d) = (ac, bc + d).$$

Show that the center of G consists of the identity element alone.

13. If the set X has at least two elements, prove that the group $(P(X), \Delta)$ is not cyclic where Δ is the symmetric difference operation.

14. Prove the following:

a) If G is a finite commutative group of order 6 that contains an element of order 3, then G is cyclic. [*Hint*: Let $o(a) = 3$ and b be an element of G that is not a power of a. Then $G = \{e, a, a^2, b, ab, a^2b\}$. If $b^2 = a$ or $b^2 = a^2$, then $G = (b)$.]

b) Any commutative group of order 6 is cyclic.

15. Let G be the multiplicative group of 2×2 invertible matrices with entries from $R^{\#}$. If

$$a = \begin{bmatrix} 0 & -1 \\ 1 & 0 \end{bmatrix} \quad \text{and} \quad b = \begin{bmatrix} 1 & 1 \\ 0 & 1 \end{bmatrix},$$

verify that a has finite order, while b has infinite order.

16. Verify the following statements.

a) The additive group Z of integers is generated by the set $S = \{2, 3\}$.

b) The additive group Q of rational numbers is generated by the set $S = \{1/p \mid p \text{ a prime}\}$.

c) The group G_s of symmetries of the square is generated by the set $S = \{r_1, h\}$.

17. Show that the additive group Q of rational numbers is not finitely generated.

18. Prove each of the following.

a) If S and T are nonempty subsets of the group G such that $G = S \cup T$, then either $(S) = G$ or $(T) = G$. [*Hint:* Use the Corollary to Theorem 2.29.]

b) If H is a proper subgroup of the group G, then the subgroup generated by the complement $G - H$ is G itself.

c) The set of odd integers generates the additive group of integers.

19. In G_s, the group of symmetries of the square, determine the subgroup generated by the set of elements of order 2.

20. Assuming that G is a group and $a, b \in G$, establish the following facts concerning the order of an element.

a) $o(a) = o(a^{-1}) = o(bab^{-1})$

b) $a = a^{-1}$ if and only if $o(a) = 1$ or 2

c) $o(ab) = o(ba)$ [*Hint:* Write $ab = a(ba)a^{-1}$.]

d) If $ab = ba$, then $o(ab)$ divides $o(a)o(b)$.

e) If G has exactly one element of order 2, say a, then $a \in \text{cent } G$. [*Hint:* Use part (a).]

21. Show that the elements of finite order in a commutative group G form a subgroup of G; this subgroup is called the *torsion subgroup* of G.

22. Suppose that H is a nonempty subset of a group G and H is closed under multiplication. If every element of H has finite order, prove that H is a subgroup of G.

23. Let G be a commutative group. For a fixed positive integer n, define H_n to be the set of all elements of G whose orders divide n:

$$H_n = \{a \in G \mid o(a) \mid n\}.$$

Show that H_n forms a subgroup of G.

24. Prove that any noncommutative group contains nontrivial subgroups.

25. Let H be a subgroup of the symmetric group S_n. Prove that either $H \subseteq A_n$ or else H contains as many odd as even permutations.

26. For any permutation $\alpha \in S_n$, verify each of the following:

a) if α is a k-cycle, then the order of α is k;

b) if $\alpha = \beta_1 \circ \beta_2 \circ \cdots \circ \beta_s$, where $\beta_1, \beta_2, \ldots, \beta_s$ are disjoint cycles of lengths $k_1, k_2, \ldots, k_s$, then the order of α is $\operatorname{lcm}(k_1, k_2, \ldots, k_s)$;

c) if α is a p-cycle, with p a prime number, then any power of α is either a p-cycle or the identity permutation.

27. Show that the symmetric group S_n contains elements of each order $1, 2, \ldots, n$.

28. a) Show that the number of distinct generators of a cyclic group of order n is equal to the number of k such that $1 \leqslant k < n$ and $\gcd(k, n) = 1$.

b) Use part (a) to find the number of generators of Z_{p^k} and Z_{pq}, where p and q are distinct primes.

29. If H is a nonempty subset of the group G, verify that the following conditions are equivalent.

a) H is a subgroup of G

b) $H^2 \subseteq H$ and $H^{-1} \subseteq H$

c) $HH^{-1} \subseteq H$

d) $hH = H$ for all $h \in H$

30. Let H and K be subgroups of the commutative group G with orders m and n, respectively. If $H \cap K = \{e\}$, prove that $o(HK) = mn$.

31. Consider the group G_s of symmetries of the square. Use Theorem 2.37 to obtain the subgroup generated by $H \cup K$, where $H = \{e, r_2\}$ and $K = \{e, d_1\}$.

32. Let G be a commutative group.

a) Prove that the set $H^{(n)} = \{a^n \mid a \in G\}$ forms a subgroup of G for any $n \geqslant 1$.

b) Determine whether $\bigcup_{n \geqslant 1} H^{(n)}$ is also a subgroup of G.

33. Let a, b be elements of the group G of orders m and n, respectively, such that $ab = ba$. If $\gcd(m, n) = 1$, show that $o(ab) = mn$. [*Hint:* Set $k = o(ab)$. To see that $mn \mid k$, raise both sides of $a^k = b^{-k}$ to the mth power and, similarly, to the nth power.]

34. In $M_2^*(R^\#)$, the multiplicative group of real 2×2 matrices, show that the following sets form subgroups:

a) $H = \left\{ \begin{bmatrix} a & b \\ 0 & a \end{bmatrix} \middle| a, b \in R^{\#};\ a \neq 0 \right\}$;

b) $H = \left\{ \begin{bmatrix} a & 0 \\ 0 & a^{-1} \end{bmatrix} \middle| a \in R^{\#} - \{0\} \right\}$;

c) $H = \left\{ \begin{bmatrix} a & a^{-1} - a \\ 0 & a^{-1} \end{bmatrix} \middle| a \in R^{\#} - \{0\} \right\}$.

35. For a subgroup H of the group G and any element $a \in G$, define

$$aHa^{-1} = \{aha^{-1} | h \in H\}.$$

Prove each of the statements below:

a) the set aHa^{-1} is a subgroup of G;

b) if H is a commutative group, then aHa^{-1} is also commutative;

c) if H is a cyclic group, then aHa^{-1} is also cyclic.

36. In the multiplicative group of matrices

$$G = \left\{ \begin{bmatrix} a & 0 \\ b & 1 \end{bmatrix} \middle| a, b \in R^{\#};\ a \neq 0 \right\},$$

show that

a) the cyclic subgroup generated by the matrix $\begin{bmatrix} 1 & 0 \\ 1 & 1 \end{bmatrix}$ is

$$H_1 = \left\{ \begin{bmatrix} 1 & 0 \\ n & 1 \end{bmatrix} \middle| n \in Z \right\};$$

b) the centralizer of the matrix $\begin{bmatrix} 1 & 0 \\ 1 & 1 \end{bmatrix}$ is

$$H_2 = \left\{ \begin{bmatrix} 1 & 0 \\ b & 1 \end{bmatrix} \middle| b \in R^{\#} \right\}.$$

2.6 NORMAL SUBGROUPS AND QUOTIENT GROUPS

Although we have derived some interesting results concerning subgroups, this concept, if unrestricted, is too general for many purposes. To obtain certain highly desirable conclusions (most notably, the Fundamental Homomorphism Theorem), additional assumptions that go beyond Definition 2.19 must be imposed.

Thus in the present section we narrow the field and focus attention on a restricted class of subgroups, which we shall refer to as normal subgroups. While such groups are "normal" in the sense that not every subgroup need be of this type, they occur nonetheless with considerable frequency. It will soon become apparent that for the major part of our work the significant aspect of this class of subgroups is that they permit the construction of algebraic structures known as quotient groups.

Having already divulged some of the content of this section, let us now proceed to develop these ideas in detail. As a starting point, we prove several theorems leading to the conclusion that each subgroup induces a decomposition of the elements of the parent group into disjoint subsets known as cosets.

DEFINITION 2.24 Let H be a subgroup of the group G and let $a \in G$. The set

$$aH = \{ah | h \in H\}$$

is called a *left coset of H in G*. The element a is a *representative* of aH.

In a similar fashion we can define the right cosets Ha of H. The right cosets of a subgroup are in general different from its left cosets. If the group operation of G is commutative, then clearly $aH = Ha$ for all $a \in G$. In the subsequent discussions we will generally consider only left cosets of a subgroup. It is obvious that a parallel theory for right cosets could be developed.

Remark. If the group operation is written additively, then we shall denote the left cosets of H in $(G, +)$ by the symbol $a + H$:

$$a + H = \{a + h | h \in H\}.$$

Before proceeding to an example, we shall make several simple observations. First, if e is the identity element of G, then

$$eH = \{eh | h \in H\} = \{h | h \in H\} = H,$$

so that H itself is a left coset of H. Moreover, since $e \in H$, we have

$$a = ae \in aH;$$

that is, every element a of G belongs to some left coset of H and, more specifically, to the coset aH. We shall presently make use of this fact.

We note further that there is a one-to-one correspondence between the elements of H and those of any coset of H. Indeed, if aH is a left coset of H, we may define a mapping $f : H \to aH$ by $f(h) = ah$. This function maps H onto aH, since every element of aH is of the form ah for some choice of $h \in H$. In addition f is a one-to-one function, for if $ah_1 = ah_2$, where $h_1, h_2 \in H$, then the cancellation law yields $h_1 = h_2$. That is, $f(h_1) = f(h_2)$ implies $h_1 = h_2$. If the group G has a finite number of elements, we may conclude that any two left cosets of H have the same number of elements, to wit, the number of elements in H.

Example 2.36

Returning to the group G_s of symmetries of the square, let us select the subgroup $H = \{e, v\}$. The task of computing the left cosets of H in G_s is straightforward; using the multiplication table on page 64, we see that

$$\begin{aligned} eH &= \{e^2, ev\} = \{e, v\} \\ r_1H &= \{r_1e, r_1v\} = \{r_1, d_2\} \\ r_2H &= \{r_2e, r_2v\} = \{r_2, h\} \\ r_3H &= \{r_3e, r_3v\} = \{r_3, d_1\} \\ hH &= \{he, hv\} = \{h, r_2\} \\ vH &= \{ve, v^2\} = \{v, e\} \\ d_1H &= \{d_1e, d_1v\} = \{d_1, r_3\} \\ d_2H &= \{d_2e, d_2v\} = \{d_2, r_1\} \end{aligned}$$

A quick inspection shows that there are only four distinct left cosets, namely

$$\{e, v\}, \qquad \{r_1, d_2\}, \qquad \{r_2, h\}, \qquad \{r_3, d_1\}.$$

These cosets are disjoint, and their union is the underlying group; as we shall see, this is always the case. For this subgroup a given element does not necessarily determine the same left and right coset, since

$$d_1H = \{d_1, r_3\} \neq \{d_1, r_1\} = Hd_1.$$

In fact, d_1H is not equal to any right coset of H, the right cosets being

$$\{e, v\}, \qquad \{r_1, d_1\}, \qquad \{r_2, h\}, \qquad \{r_3, d_2\}.$$

Associated with each subgroup there is a natural equivalence relation and hence a partition of the underlying group. Let H be a subgroup of the group G and let a, b be elements of G. If it happens that the product $a^{-1}b$ lies in H, then we say that a and b are *left-congruent modulo H* and write $a \equiv_l b \pmod{H}$; that is,

$$a \equiv_l b \pmod{H} \qquad \text{if and only if} \qquad a^{-1}b \in H.$$

The properties of this relation of congruence are revealed in the following theorem.

THEOREM 2.38 *If H is a subgroup of the group G, then the relation* $\equiv_l (\text{mod } H)$ *is an equivalence relation on G; the equivalence classes are the left cosets of H in G.*

Proof. First we have $a \equiv_l a \, (\text{mod } H)$, since $a^{-1}a = e \in H$ for each $a \in G$. If $a \equiv_l b \, (\text{mod } H)$, so that $a^{-1}b \in H$, then $b^{-1}a = (a^{-1}b)^{-1} \in H$; this just says that $b \equiv_l a \, (\text{mod } H)$. Finally, suppose that $a \equiv_l b \, (\text{mod } H)$ and $b \equiv_l c \, (\text{mod } H)$; or equivalently, $a^{-1}b \in H$, $b^{-1}c \in H$. The closure of H under multiplication gives $a^{-1}c = (a^{-1}b)(b^{-1}c) \in H$, in consequence of which $a \equiv_l c \, (\text{mod } H)$. This establishes that left congruence modulo H is an equivalence relation on G.

For $a \in G$ the equivalence class of a with respect to the relation $\equiv_l (\text{mod } H)$ is the set $[a] = \{x \in G \mid x \equiv_l a \, (\text{mod } H)\}$. If $b = ah$ is any member of the left coset aH, then $a^{-1}b = h \in H$. By definition of $\equiv_l (\text{mod } H)$ this implies that $b \in [a]$, and so $aH \subseteq [a]$. On the other hand, if $x \in [a]$, we must have $a^{-1}x = h$ for some h in H, whence $x = ah \in aH$. Thus the inclusion $[a] \subseteq aH$ also holds, and the equality $[a] = aH$ follows.

One benefit of Theorem 2.38 is that all the results on equivalence relations are now available to us. These tell us, for instance, that the set of left cosets of H in G partition G; that is to say, G is the union of all left cosets of H and distinct left cosets are disjoint. Also Theorem 1.8 provides a condition for the equality of two equivalence classes when a representative of each is known. In the notation of the present section this becomes

THEOREM 2.39 *Let H be a subgroup of the group G. Then aH = bH if and only if* $a^{-1}b \in H$*; in particular, aH = H if and only if* $a \in H$.

For those who would prefer to see Theorem 2.39 proved directly, without recourse to Theorem 1.8, the argument proceeds as follows: If $aH = bH$, then $b \in bH = aH$, and hence $b = ah$ for some $h \in H$; this implies that $a^{-1}b = h \in H$. For the converse, let us suppose that $a^{-1}b \in H$, say, $a^{-1}b = h_1$, where $h_1 \in H$. Then, for any element $bh_2 \in bH$, we have

$$bh_2 = (ah_1)h_2 = a(h_1h_2) \in aH.$$

As a result, $bH \subseteq aH$. The opposite inclusion is obtained similarly, leading to the equality $aH = bH$.

Needless to say, there is a parallel theory with the results phrased on the right. If one starts with the relation $a \equiv_r b \, (\text{mod } H)$ defined by $ab^{-1} \in H$, then this turns out to be an equivalence relation on G whose equivalence classes are simply the right cosets of H. The implication is that any subgroup H induces a partition of the group G into right cosets. In working with right cosets the criterion for equality assumes the form $Ha = Hb$ if and only if $ab^{-1} \in H$. For future reference we record these remarks as follows.

THEOREM 2.40 *If H is a subgroup of the group G, then the left (right) cosets of H in G form a partition of G.*

Before proceeding to examples, there is another consequence of Theorem 2.39 that is worth our attention, namely, that any member of a coset can serve as a representative of that coset. More precisely, if H is a subgroup of the group G and the element b belongs to the coset aH, then $bH = aH$. Obviously, $b \in aH$ implies that $b = ah$ for suitable $h \in H$. This being true, we have $a^{-1}b = h \in H$; and Theorem 2.39 leads immediately to $aH = bH$. The point that we wish to make is that a given coset aH may be described by as many different names as there are elements in it.

Now let us look at two illustrations of these ideas.

Example 2.37
Consider Z_{12}, the group of integers modulo 12. If we take $H = \{0, 4, 8\}$ for our subgroup, then the (left) cosets of H in Z_{12} are

$$\begin{aligned} 0 +_{12} H &= \{0, 4, 8\} = 4 +_{12} H = 8 +_{12} H, \\ 1 +_{12} H &= \{1, 5, 9\} = 5 +_{12} H = 9 +_{12} H, \\ 2 +_{12} H &= \{2, 6, 10\} = 6 +_{12} H = 10 +_{12} H, \\ 3 +_{12} H &= \{3, 7, 11\} = 7 +_{12} H = 11 +_{12} H. \end{aligned}$$

In this case the coset decomposition of Z_{12} relative to the given subgroup is just

$$Z_{12} = \{0, 4, 8\} \cup \{1, 5, 9\} \cup \{2, 6, 10\} \cup \{3, 7, 11\}.$$

A convenient way of depicting the situation is shown in Figure 2.5.

Example 2.38
Another illustration is provided by the multiplicative group of matrices

$$G = \left\{ \begin{bmatrix} a & 0 \\ b & 1 \end{bmatrix} \middle| a, b \in R^{\#};\ a \neq 0 \right\}.$$

In this group the subset

$$H = \left\{ \begin{bmatrix} 1 & 0 \\ c & 1 \end{bmatrix} \middle| c \in R^{\#} \right\}$$

forms a subgroup. The details are that

$$\begin{bmatrix} 1 & 0 \\ c & 1 \end{bmatrix} \begin{bmatrix} 1 & 0 \\ d & 1 \end{bmatrix}^{-1} = \begin{bmatrix} 1 & 0 \\ c & 1 \end{bmatrix} \begin{bmatrix} 1 & 0 \\ -d & 1 \end{bmatrix} = \begin{bmatrix} 1 & 0 \\ c - d & 1 \end{bmatrix} \in H,$$

and so an appeal to Theorem 2.25 is legitimate.

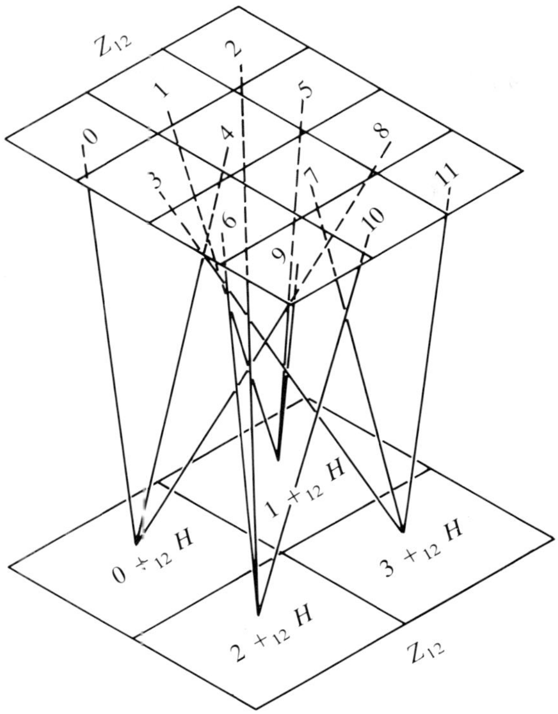

Figure 2.5

Take an arbitrary element $\begin{bmatrix} a & 0 \\ b & 1 \end{bmatrix}$ in G. Then the coset

$$\begin{aligned}\begin{bmatrix} a & 0 \\ b & 1 \end{bmatrix} H &= \left\{ \begin{bmatrix} a & 0 \\ b & 1 \end{bmatrix}\begin{bmatrix} 1 & 0 \\ c & 1 \end{bmatrix} \middle| c \in R^{\#} \right\} \\ &= \left\{ \begin{bmatrix} a & 0 \\ b+c & 1 \end{bmatrix} \middle| c \in R^{\#} \right\} \\ &= \left\{ \begin{bmatrix} a & 0 \\ d & 1 \end{bmatrix} \middle| d \in R^{\#} \right\}.\end{aligned}$$

The elements in this coset are just the matrices of G with the real number a in the upper left corner. This makes it clear that there will be as many distinct cosets of H as there are choices for such a matrix entry. The subgroup H induces, in a natural way, a partition of G into infinitely many cosets.

If H is a subgroup of the group G, then the *index of H in G* is defined to be the number of distinct left cosets of H; this integer will be denoted by $[G:H]$. In light of Problem 3(e) at the end of this section there is no necessity of distinguishing between a right and left index. Note that if we take $H = \{e\}$, then $o(G) = [G:\{e\}]$.

In 1770, Joseph Louis Lagrange (1736–1813), the greatest mathematician of the generation between Euler and Gauss, published an extensive paper called "*Réflexions sur la Résolution Algébrique des Equations*." This memoir first applies what Lagrange calls the "*calcul des combinaisons*" to the solvability of algebraic equations, thus beginning the research into the study of permutations. In it there appears, in an imperfect form, a result that some have described as "the most important theorem in finite group theory," namely, the statement that the order of a finite group is divisible by the order of each of its subgroups. Today this elementary theorem is known as Lagrange's theorem, though it is scarcely necessary to add that Lagrange did not develop it in its abstract form. Confining himself entirely to permutations, he observed that (in modern terminology) the order of a subgroup of the symmetric group S_n divides $n!$. The general result itself was probably first proved by Evariste Galois in 1830.

THEOREM 2.41 (Lagrange) *The order and index of any subgroup of a finite group divide the order of the group.*

Proof. Suppose now that G is a finite group, say of order n, and H is a subgroup of G of order k. We can then decompose the set G into a union of a finite number of disjoint left cosets of H:

$$G = a_1H \cup a_2H \cup \cdots \cup a_rH.$$

Since each of the r cosets in the above decomposition has k elements, the set G itself must have rk elements; hence $n = rk$ or, in symbols,

$$o(G) = [G:H]o(H).$$

There is a corollary to Theorem 2.41 that is of some intrinsic interest.

COROLLARY *If G is a group of order n, then the order of any element $a \in G$ divides n; in addition, $a^n = e$.*

Proof. Let the element a have order k. By definition, the cyclic subgroup (a) generated by a must also be of order k. According to the conclusion of Lagrange's Theorem, k is a divisor of n; that is, $n = rk$ for some $r \in Z_+$. Hence

$$a^n = a^{rk} = (a^k)^r = e^r = e,$$

completing the proof of both assertions.

From Lagrange's Theorem we are able to conclude that any finite group of prime order has no nontrivial subgroups. Actually, a stronger statement can be made:

THEOREM 2.42 *If G is a finite group of composite order, then G has nontrivial subgroups.*

Proof. If the group G is not cyclic, any element $e \neq a \in G$ generates a nontrivial cyclic subgroup (a). Thus it suffices to consider cyclic groups of composite order. To this end, suppose that $G = (a)$, where the generator a has order $nm(n, m \neq 1)$. Then $(a^n)^m = e$, while $(a^n)^{m'} \neq e$ for $0 < m' < m$. From this it is obvious that (a^n) is a nontrivial cyclic subgroup of G with order m.

COROLLARY *Every group G of prime order is cyclic.*

Proof. Consider the cyclic subgroup (a) generated by any $a \in G$, with $a \neq e$. Now, the order of (a) must divide the order of G, a prime; since (a) contains more than one element, $o((a)) = o(G)$, whence $(a) = G$.

As a further application of Lagrange's Theorem, we give the following.

THEOREM 2.43 *Any noncommutative group has at least six elements.*

Proof. A group of prime order, being a cyclic group, is necessarily commutative. Accordingly, any group having order 2, 3, or 5 will be commutative. Suppose next that G is a group of order 4. By Lagrange's Theorem, each element of G distinct from the identity has order 2 or 4. If one of them has order 4, then G is a cyclic group of order 4 and therefore commutative. On the other hand, a group each of whose elements other than the identity has order 2 must be commutative by Problem 15 of Section 2.3. This argument establishes that all groups of order less than 6 are commutative groups.

Incidentally, the implication of Lagrange's Theorem cannot be reversed; that is to say, a group of order n need not have a subgroup of order k, where k is a divisor of n. To be more specific, it will be shown that the alternating group A_4—a group of order 12—contains no subgroup of order 6.

We take this opportunity, however, to develop a point brought out by the last proof. Let G be a noncyclic group of order 4. Then every element of G, other than the identity, will have order 2. If a and b are two such elements of G, $a^2 = b^2 = e$, hence $a = a^{-1}$ and $b = b^{-1}$. The implication is that the product $ab \neq e$. But ab cannot be equal to either a or b; were $ab = a$, then the cancellation law would produce the contradiction that $b = e$. Thus ab is the fourth member of G and, as a result, the multiplication table for G must be as shown:

$\cdot$	e	a	b	ab
e	e	a	b	ab
a	a	e	ab	b
b	b	ab	e	a
ab	ab	b	a	e

This particular group is usually called the *Klein 4-group*, in honor of Felix Klein, and is denoted by K_4. To sum up, the only noncyclic group of order 4 is the Klein 4-group.

Felix Klein (1849–1925) received his doctorate from Bonn and later taught at the universities of Erlangen, Munich, Leipzig, and Göttingen. The inaugural lecture that he gave upon his appointment as full professor at Erlangen in 1872 outlined how the group concept could serve as a means of classifying the various geometries that had come into existence. This celebrated address, which has become known as the *Erlanger Programm*, outlined a scheme for classifying any geometry by specifying which properties of geometric figures remain unchanged when the figures are subjected to a particular group of transformations. For example, plane Euclidean geometry is characterized by the study of properties that remain invariant under the group of rigid motions (translations, rotations, reflections, and combinations of these), such as length, area, congruence, and parallelism. By viewing geometry as embedded within this more general framework, Klein did much to spread and popularize group theory in the late nineteenth century.

We return to an arbitrary group G and subgroup H of G. In general, the collection of left cosets of H in G is not the same as the collection of right cosets of H in G (a glance at Example 2.36 will confirm this). Those subgroups for which every left coset turns out to be a right coset are of sufficient importance to warrant a definition:

DEFINITION 2.25 A subgroup H of a group G is said to be *normal* in G if every left coset of H in G is also a right coset of H in G.

Thus if H is a normal subgroup of G and aH is a typical left coset of H, there exists some element $b \in G$ for which $aH = Hb$. Since a lies in the coset aH, this means that $a \in Hb$; let us say, $a = hb$. Then $ab^{-1} = h \in H$ and so, by the analog of Theorem 2.39 for right cosets, $Ha = Hb$. The situation is now this: If aH happens to be a right coset of H, then it must be the coset Ha.

The observation just made suggests reformulating Definition 2.25 as follows.

DEFINITION 2.26 A subgroup H is *normal* in the group G if and only if $aH = Ha$ for every $a \in G$.

For a normal subgroup H we may thus speak simply of the cosets of H in G without specifying right or left. The trivial subgroups G and $\{e\}$ are obviously normal. Every subgroup of a commutative group is a normal subgroup; in particular, every subgroup of a cyclic group is normal.

We shall sometimes speak of a *simple group*, meaning thereby that it has no normal subgroups other than the two trivial ones. For instance, the finite cyclic groups of prime order are simple groups.

Care must be taken to interpret the normality condition correctly. The statement $aH = Ha$ means that for any $h \in H$ there exists an element $h' \in H$ (depending on h) such that $ah = h'a$; it does not assert that $ah = ha$.

It would be gratifying to have a less cumbersome procedure than the computation of cosets for determining whether a given subgroup is normal. Just such a criterion is given in the next theorem, and we shall have frequent occasion to make use of it.

THEOREM 2.44 *The subgroup H is a normal subgroup of the group G if and only if $aHa^{-1} \subseteq H$ for each element $a \in G$.*

Proof. First, assume that $aHa^{-1} \subseteq H$ for any $a \in G$. We must prove that in this event $aH = Ha$. Pick an arbitrary element ah of aH. Since $aHa^{-1} \subseteq H$, it follows that $aha^{-1} = h_1$ for some $h_1 \in H$. Thus

$$ah = (aha^{-1})a = h_1 a.$$

The product $h_1 a$ lies in the right coset Ha, and so we conclude that $aH \subseteq Ha$. We obtain the opposite inclusion, $Ha \subseteq aH$, by a similar argument upon observing that our hypothesis also implies that

$$a^{-1}Ha = a^{-1}H(a^{-1})^{-1} \subseteq H.$$

Conversely, suppose that H is normal in G so that $aH = Ha$ for each $a \in G$. Let ah_1a^{-1} be a typical element in aHa^{-1}. Since $aH = Ha$, there exists an element $h_2 \in H$ such that $ah_1 = h_2 a$. Consequently,

$$ah_1a^{-1} = (h_2 a)a^{-1} = h_2,$$

which yields $aHa^{-1} \subseteq H$.

The assertion $aHa^{-1} \subseteq H$ for all $a \in G$ is easily seen to be equivalent to the statement $aHa^{-1} = H$ for all $a \in G$. If we assume that $aHa^{-1} \subseteq H$ for any $a \in G$, then when we choose a^{-1} as the particular element of G, $a^{-1}Ha = a^{-1}H(a^{-1})^{-1} \subseteq H$, whence

$$H = a(a^{-1}Ha)a^{-1} \subseteq aHa^{-1} \subseteq H,$$

which signifies that $H = aHa^{-1}$. On the other hand, if $aHa^{-1} = H$ for every $a \in G$, then plainly $aHa^{-1} \subseteq H$.

It follows from our next theorem that the center of any group is a normal subgroup.

THEOREM 2.45 *If H is a subgroup of the group G such that $H \subseteq$ cent G, then H is a normal (commutative) subgroup of G.*

Proof. The proof is short: It must be shown that if $h \in H$ and a is arbitrary in G, then $aha^{-1} \in H$; that is, $aHa^{-1} \subseteq H$. Since $h \in$ cent G, we quickly get $aha^{-1} = haa^{-1} = h \in H$.

COROLLARY *Every subgroup of a commutative group is normal.*

Example 2.39

Let us take G to be the noncommutative group of order 6 presented in Example 2.21. The reader may recall that the elements of G are functions $f_1, f_2, \ldots, f_6$, while the group operation is functional composition. For convenience the operation table is reproduced below.

$\circ$	f_1	f_2	f_3	f_4	f_5	f_6
f_1	f_1	f_2	f_3	f_4	f_5	f_6
f_2	f_2	f_1	f_4	f_3	f_6	f_5
f_3	f_3	f_5	f_1	f_6	f_2	f_4
f_4	f_4	f_6	f_2	f_5	f_1	f_3
f_5	f_5	f_3	f_6	f_1	f_4	f_2
f_6	f_6	f_4	f_5	f_2	f_3	f_1

If we let H be the subset $\{f_1, f_4, f_5\}$, then it is easily calculated that H is a normal subgroup of G. The coset decomposition for the subgroup in question is

$$f_k \circ H = \{f_1, f_4, f_5\} = H \circ f_k \qquad \text{for } k = 1, 4, 5;$$

$$f_k \circ H = \{f_2, f_3, f_6\} = H \circ f_k \qquad \text{for } k = 2, 3, 6.$$

On the other hand, the subgroup $\{f_1, f_2\}$ is not normal; a short computation indicates that the criterion of Theorem 2.44 fails to be satisfied:

$$f_4 \circ f_2 \circ f_4^{-1} = f_4 \circ f_2 \circ f_5 = f_3 \notin \{f_1, f_2\}.$$

Example 2.40

Consider once again the symmetric group S_n $(n \geqslant 2)$. We assert that the alternating group A_n is normal in S_n. For suppose that $\beta \in A_n$ and α is arbitrary in S_n. Then each of these can be written as a product of transpositions, with β requiring an even number of factors:

$$\alpha = \tau_1 \circ \tau_2 \circ \cdots \circ \tau_r, \qquad \beta = \bar{\tau}_1 \circ \bar{\tau}_2 \circ \cdots \circ \bar{\tau}_{2n}.$$

Now α and α^{-1} together involve an even number of transpositions, whether α be even or odd; thus the product

$$\alpha \circ \beta \circ \alpha^{-1} = (\tau_1 \circ \cdots \circ \tau_r) \circ (\bar{\tau}_1 \circ \cdots \circ \bar{\tau}_{2n}) \circ (\tau_r^{-1} \circ \cdots \circ \tau_1^{-1})$$

is expressible as a product of an even number of transpositions, namely, $2r + 2n$ transpositions. This places $\alpha \circ \beta \circ \alpha^{-1}$ in A_n, whence the announced normality of A_n.

Example 2.41

For our last illustration of these ideas, let us fall back on the multiplicative group of matrices

$$G = \left\{ \begin{bmatrix} a & 0 \\ b & 1 \end{bmatrix} \middle| a, b \in R^{\#};\ a \neq 0 \right\}.$$

A straightforward calculation will reveal that

$$H = \left\{ \begin{bmatrix} a & 0 \\ 0 & 1 \end{bmatrix} \middle| a \in R^{\#} - \{0\} \right\}$$

is a subgroup of G. Consider the particular matrix

$$x = \begin{bmatrix} 1 & 0 \\ 1 & 1 \end{bmatrix}$$

and take $\begin{bmatrix} a & 0 \\ 0 & 1 \end{bmatrix}$ to be an arbitrary member of H. Then the product

$$\begin{bmatrix} 1 & 0 \\ 1 & 1 \end{bmatrix}\begin{bmatrix} a & 0 \\ 0 & 1 \end{bmatrix}\begin{bmatrix} 1 & 0 \\ 1 & 1 \end{bmatrix}^{-1} = \begin{bmatrix} 1 & 0 \\ 1 & 1 \end{bmatrix}\begin{bmatrix} a & 0 \\ 0 & 1 \end{bmatrix}\begin{bmatrix} 1 & 0 \\ -1 & 1 \end{bmatrix}$$

$$= \begin{bmatrix} a & 0 \\ a & 1 \end{bmatrix}\begin{bmatrix} 1 & 0 \\ -1 & 1 \end{bmatrix} = \begin{bmatrix} a & 0 \\ a-1 & 1 \end{bmatrix} \notin H.$$

The argument leaves us with $xHx^{-1} \not\subseteq H$, and in view of Theorem 2.44 we can assert that H is not a normal subgroup of G.

As a further fact of interest, notice that the set

$$xHx^{-1} = \left\{ \begin{bmatrix} a & 0 \\ a-1 & 1 \end{bmatrix} \middle| a \in R^{\#} - \{0\} \right\}$$

forms a subgroup of G. This serves to illustrate and partly to suggest the following general result: If H is a subgroup of the group G and $x \in G$, then xHx^{-1} is also a subgroup of G. This will be proven later as Theorem 3.12.

Assume that H is a normal subgroup of G and that aH and bH are two cosets of H. Cosets of H are subsets of G, and so, as a special case of Definition 2.23, we may form their product

$$(aH)(bH) = \{(ah)(bh') | h, h' \in H\}.$$

We assert that the set on the right is actually the coset abH. To demonstrate this, let $h, h' \in H$ be arbitrarily given. Inasmuch as $hb \in Hb = bH$, there exists some element $h_1 \in H$ such that $hb = bh_1$. In consequence,

$$(ah)(bh') = a(hb)h' = a(bh_1)h' = (ab)(h_1 h') \in abH,$$

leading to the inclusion $(aH)(bH) \subseteq abH$. On the other hand,

$$(ab)h = (ae)(bh) \in (aH)(bH)$$

for any h in H. This proves that $abH \subseteq (aH)(bH)$, which, combined with the inclusion obtained before, yields $(aH)(bH) = abH$.

Curiously enough, the converse also holds. That is, if the equation $(aH)(bH) = abH$ is satisfied for all $a, b \in G$, then H must be a normal subgroup of G. Suppose that $h \in H$ and $a \in G$; then

$$aha^{-1} = (ah)(a^{-1}e) \in (aH)(a^{-1}H) = aa^{-1}H = eH = H,$$

which signifies that H is normal.

We ought to note formally the result of our labors.

LEMMA If H is a subgroup of the group G, then H is a normal subgroup of G if and only if $(aH)(bH) = abH$ for all $a, b \in G$.

If H is a normal subgroup of the group G, the symbol G/H will be used to designate the collection of left cosets of H in G:

$$G/H = \{aH | a \in G\}.$$

Since these are also right cosets—H being normal guarantees that $aH = Ha$ for every $a \in G$—we may hereafter speak of the cosets of H without ambiguity.

The notion of a product of sets has permitted us to introduce a coset multiplication in G/H. Because our definition is phrased in terms of coset representatives, the question arises as to whether the formula $(aH)(bH) = abH$ is independent of the choice of a and b. Put differently, does it follow that if $aH = a_1H$ and $bH = b_1H$, then $abH = a_1b_1H$? According to Theorem 2.39, all one really needs to establish is that the product $(ab)^{-1}(a_1b_1) \in H$. Now the equalities $aH = a_1H$ and $bH = b_1H$ tell us that $a^{-1}a_1, b^{-1}b_1 \in H$. The fact that H is taken to be a normal subgroup and not merely a subgroup implies that $b^{-1}(a^{-1}a_1)b$ lies in H. Knowing this, we see that the closure of H gives

$$(ab)^{-1}(a_1b_1) = b^{-1}a^{-1}a_1b_1 = (b^{-1}(a^{-1}a_1)b)(b^{-1}b_1) \in H,$$

as desired. This argument indicates that coset multiplication is a well-defined binary operation on G/H, in the sense that the product of two cosets depends only on the cosets involved and in no way on the representative elements chosen from them; any other choice would have yielded the same product.

Having thus prepared the way, we are now ready to tackle the principal result of this section.

THEOREM 2.46 *If H is a normal subgroup of the group G, then G/H forms a group with respect to the operation $(aH)(bH) = abH$; its order is $[G:H]$.*

Proof. First let us observe that the associativity of the operation in G/H is inherited from that of the operation in G:

$$\begin{aligned}[(aH)(bH)](cH) &= (abH)(cH)\\ &= (ab)cH\\ &= a(bc)H\\ &= (aH)(bcH) = (aH)[(bH)(cH)].\end{aligned}$$

The coset $H = eH$ containing the identity of G serves as the identity element for G/H, since

$$(aH)(eH) = aeH = aH = eaH = (eH)(aH).$$

It is just as easy to show that the inverse of the coset aH is $a^{-1}H$, where a^{-1} indicates the inverse of a in G; this follows from

$$(aH)(a^{-1}H) = aa^{-1}H = eH = a^{-1}aH = (a^{-1}H)(aH),$$

and so all is proved.

The group G/H just obtained is called the *quotient group of G by H* or, sometimes, the *factor group of G by H*. The usual practice is to denote the group operation in G/H by the same symbol as that used for the original operation in G. In the case of an additive group $(G, +)$, for instance, one would write the quotient group operation as $(a + H) + (b + H) = a + b + H$.

Before turning to some examples, let us stress that only normal subgroups can be used to define a quotient group of a group G. In order that the product of aH and bH be the coset abH, we must take H normal in G; in fact, for $(aH)(bH)$ to be equal to any left coset of H, H has to be normal.

Example 2.42

Once again we fall back on the familiar group G_s of symmetries of the square. Here the subgroup $H = \{e, r_2\}$ is normal, being the center of the group. The distinct cosets of H comprise the elements of the set

$$G_s/H = \{\{e, r_2\}, \{r_1, r_3\}, \{h, v\}, \{d_1, d_2\}\}.$$

A typical coset multiplication would procced as follows:

$$\begin{aligned}\{d_1, d_2\} \cdot \{r_1, r_3\} &= (d_1 H)(r_1 H) \\ &= d_1 r_1 H = hH = \{h, v\}.\end{aligned}$$

To multiply two cosets in G_s/H, all we need to do is select an arbitrary representative from each coset, multiply these elements under the original group operation, and determine to which coset the resulting product belongs.

With this in mind, the operation table for G_s/H can be filled in as shown below.

$\cdot$	$\{e, r_2\}$	$\{r_1, r_3\}$	$\{h, v\}$	$\{d_1, d_2\}$
$\{e, r_2\}$	$\{e, r_2\}$	$\{r_1, r_3\}$	$\{h, v\}$	$\{d_1, d_2\}$
$\{r_1, r_3\}$	$\{r_1, r_3\}$	$\{e, r_2\}$	$\{d_1, d_2\}$	$\{h, v\}$
$\{h, v\}$	$\{h, v\}$	$\{d_1, d_2\}$	$\{e, r_2\}$	$\{r_1, r_3\}$
$\{d_1, d_2\}$	$\{d_1, d_2\}$	$\{h, v\}$	$\{r_1, r_3\}$	$\{e, r_2\}$

Incidentally, this example indicates that it is possible for a noncommutative group to have a commutative quotient group.

Example 2.43 A useful illustration to keep in mind when working with quotient groups comes from the additive group Z of integers. We already know that the (normal) subgroups of Z are the cyclic subgroups (n), where n is a nonnegative integer. Assuming that $n > 1$, the cosets of (n) in Z take the form

$$a + (n) = \{a + kn | k \in Z\} = [a].$$

In other words, the cosets of the subgroup (n) are precisely the congruence classes modulo n, which were defined in Definition 2.13. Coset addition in $Z/(n)$ is given by the rule

$$(a + (n)) + (b + (n)) = a + b + (n),$$

or, with an appropriate change of notation, by $[a] + [b] = [a + b]$. The point that we wish to make is that the group of integers modulo n is a perfect disguise for the quotient group of Z by (n):

$$(Z_n, +_n) = (Z/(n), +).$$

It is a simple matter to see that any quotient group of a commutative group is necessarily commutative. A natural question is: Under what conditions will a noncommutative group possess commutative quotient groups? As the next topic in this section, we investigate this particular situation. Our analysis begins with a basic definition.

DEFINITION 2.27 *Given a group G and a pair of elements $a, b \in G$, the commutator of a and b is defined to be the product $aba^{-1}b^{-1}$.*

To simplify matters, the symbol $[a, b]$ will be used to represent the commutator of two elements a and b; any other symbol would do as well, but this notation is fairly standard. Inasmuch as $[a, b]$ satisfies the identity $ab = [a, b]ba$, one may view the commutator of a and b as a measure of the extent to which ab differs from ba. Indeed, the elements a and b commute if and only if $[a, b] = e$.

In general, the commutators do not by themselves form the elements of a subgroup, since the product of two commutators fails to be a commutator. The usual procedure for bypassing this difficulty is to work instead with the subgroup that is generated by the set of all the commutators. This subgroup is known as either the *derived subgroup* or the *commutator subgroup of* G and may be denoted by $[G, G]$.

Now the inverse of a commutator is again a commutator; specifically, $[a, b]^{-1} = [b, a]$. There is no need then to explicitly consider inverses in the definition of the set $[G, G]$; its elements consist merely of products of finitely many commutators of G. That is,

$$[G, G] = \{\Pi[a_i, b_i] \mid a_i, b_i \in G\},$$

where the symbol Π should be interpreted as representing a finite product with one or more terms.

With these preparatory remarks out of the way, we proceed to establish some of the special properties of the commutator subgroup.

THEOREM 2.47 *If H is a subgroup of the group G such that $[G, G] \subseteq H$, then H is normal in G.*

Proof. The argument proceeds along the usual line. Namely, it must be shown that, for $a \in G$ and $h \in H$, aha^{-1} lies in H. But

$$aha^{-1} = (aha^{-1}h^{-1})h = [a, h]h.$$

The commutator $[a, h]$ belongs to $[G, G] \subseteq H$, by hypothesis, whence closure implies $[a, h]h \in H$.

COROLLARY *The commutator subgroup $[G, G]$ is a normal subgroup of G.*

The quotient group $G/[G, G]$, which exists by virtue of the corollary above, is called the *commutator quotient group* or *abelianized group*. The motivation for this latter choice of terminology will become apparent after the next result.

THEOREM 2.48 *Let H be a normal subgroup of the group G. The quotient group G/H is commutative if and only if $[G, G] \subseteq H$.*

Proof. Suppose aH and bH are two arbitrary elements in G/H. Since the coset $H = eH$ is the identity element of G/H, the group operation in G/H will be commutative if and only if

$$H = [aH, bH] = (aH)(bH)(aH)^{-1}(bH)^{-1},$$

or, what amounts to the same thing, $H = (aba^{-1}b^{-1})H$. But Theorem 2.39 tells us that a necessary and sufficient condition for the last equality to hold is that $[a, b] = aba^{-1}b^{-1} \in H$. In other words, commutativity of the group G/H is equivalent to requiring that the subgroup H contain all the commutators of G. As $[G, G]$ is, by definition, the smallest subgroup with this property, the latter condition may be replaced by $[G, G] \subseteq H$.

A special case, but of interest in itself, occurs on taking $H = [G, G]$:

COROLLARY *For any group G, the commutator quotient group $G/[G, G]$ is commutative.*

The foregoing theorem says, in effect, that the commutator subgroup is the smallest—again, in the sense of inclusion—normal subgroup having a commutative quotient group. The transition from a group to its commutator quotient group is referred to as the *abelianization* of the group and provides a convenient means of manufacturing commutative groups from noncommutative ones.

Casual reference has already been made to the notion of a simple group (recall that this is a group $G \neq \{e\}$ that has no nontrivial normal subgroup). In the commutative setting, Lagrange's Theorem tells us that the only simple groups are the cyclic groups of prime order; these are "simple" groups both in the technical sense and in the ordinary sense of having an uncomplicated structure. The concept of a simple group has a venerable history, going back to the studies of Evariste Galois (1811–1832) on the unsolvability of the general fifth-degree equation. On the eve of his death, Galois wrote a letter to a friend in which he set forth the idea of a simple group, as well as the fact (but not the proof) that the alternating group A_5 is the noncommutative simple group of smallest order. The proof that A_5 is simple appears in Camille Jordan's monumental treatise of 665 pages, *Traité des Substitutions* (1870).

We might pause here to mention a famous conjecture of the English algebraist William Burnside (1852–1927). One of the earliest textbooks on group theory was Burnside's popular *The Theory of Groups of Finite Order*, the first edition of which appeared in 1897. It was the first treatment of the subject in English and the first to develop the theory from the modern standpoint of abstract groups rather than by way of permutation groups. In his book, Burnside remarks, "No noncommutative simple group of odd order is at the present known to exist." He devoted much effort in the late 1800's and early 1900's to proving that there is no noncommutative simple group of odd order less than 40,000. By the time the second edition of his textbook was published in 1911, Burnside's efforts had convinced him to write:

> The contrast that these results shew [sic] between groups of odd and of even order suggests inevitably that noncommutative simple groups of odd order do not exist.

This conjecture stood as an open question for more than 50 years. Burnside's prophecy was at last fulfilled in 1963 by two Americans, John Thompson and Walter Feit, in a paper entitled "Solvability of Groups of Odd Order." Their proof of the "Odd Order Theorem" required a full 255-page issue of the *Pacific Journal of Mathematics*. For this achievement and subsequent work, Thompson was awarded the Fields Medal at the 1970 International Congress of Mathematicians. The Fields Medal is considered to be the equivalent in mathematics of a Nobel Prize. (It is generally believed that Alfred Nobel's bad relations with the Swedish mathematician Gösta Mittag-Leffler was the reason that Nobel did not establish a prize in mathematics.)

The goal during the remaining pages of this section is to provide the reader with an extensive class of simple groups, the alternating groups A_n $(n \neq 4)$. Let us begin with an observation on the generators of A_n. (In what follows, we shall omit the small circle and write $\alpha\beta$ for the composition of two permutations α and β.)

LEMMA *For $n \geq 3$ the alternating group A_n is generated by its 3-cycles.*

Proof. Every element of A_n can be written as a product of an even number of transpositions. Thus it is enough to show that a product of two transpositions is a product of 3-cycles. If $(a \quad b)$ and $(c \quad d)$ are disjoint transpositions, then $(a \quad b)(c \quad d) = (a \quad c \quad b)(a \quad c \quad d)$. If $(a \quad b)$ and $(c \quad d)$ have a letter in common, say $a = c$, then $(a \quad b)(a \quad d) = (a \quad b \quad d)$, which proves the lemma.

THEOREM 2.49 *Let H be a normal subgroup of the alternating group A_n, where $n \geq 4$. If H contains a 3-cycle, then $H = A_n$.*

Proof. There is no loss in denoting the 3-cycle by $(1\ \ 2\ \ 3)$. We first treat the case of A_4. The normality of H in A_4 implies that H contains

$$(2\ \ 4\ \ 3)(1\ \ 2\ \ 3)(2\ \ 4\ \ 3)^{-1} = (1\ \ 4\ \ 2)$$
$$(2\ \ 3\ \ 4)(1\ \ 2\ \ 3)(2\ \ 3\ \ 4)^{-1} = (1\ \ 3\ \ 4)$$
$$(1\ \ 2\ \ 4)(1\ \ 2\ \ 3)(1\ \ 2\ \ 4)^{-1} = (2\ \ 4\ \ 3)$$

as well as the squares

$$(1\ \ 2\ \ 3)^2 = (1\ \ 3\ \ 2), \qquad (1\ \ 4\ \ 2)^2 = (1\ \ 2\ \ 4)$$
$$(1\ \ 3\ \ 4)^2 = (1\ \ 4\ \ 3), \qquad (2\ \ 4\ \ 3)^2 = (2\ \ 3\ \ 4).$$

Thus all the 3-cycles of S_4 lie in H; but these 3-cycles generate A_4, so $H = A_4$.

We now consider A_n, where $n \geqslant 5$. Given an arbitrary 3-cycle $(a\ \ b\ \ c)$, there is a permutation $\alpha \in S_n$ such that $\alpha(1) = a$, $\alpha(2) = b$ and $\alpha(3) = c$. If α is even, set $\beta = \alpha$; while if α is odd, set $\beta = (c\ \ d)\alpha$, where c and d are distinct integers in $\{4, 5, \ldots, n\}$. In either event, $\beta \in A_n$ and

$$\beta(1\ \ 2\ \ 3)\beta^{-1} = \alpha(1\ \ 2\ \ 3)\alpha^{-1}$$
$$= (\alpha(1)\ \ \alpha(2)\ \ \alpha(3)) = (a\ \ b\ \ c).$$

Since H is normal in A_n, $(a\ \ b\ \ c) \in H$. Again, H contains all 3-cycles, whence $H = A_n$.

With this in hand, it is an easy matter to show that the converse of Lagrange's Theorem does not hold in general; that is, if k is a divisor of the order of a finite group G, there need be no subgroup of G of order k. We consider A_4, a group of order 12.

COROLLARY *The alternating group A_4 contains no subgroup of order 6.*

Proof. Assume that H is a subgroup of A_4 of order 6. Then $[A_4 : H] = 2$, so H is a normal subgroup of A_4 (Problem 9 at the end of this section). Since 8 of the 12 elements of A_4 are 3-cycles, H must contain a 3-cycle. It follows from the theorem that $H = A_4$, which is absurd. This contradiction forces us to accept that there is no subgroup H with $o(H) = 6$.

We now give the promised proof of the simplicity of A_n, $n \neq 4$.

THEOREM 2.50 *The alternating group A_n is a simple group for $n \geqslant 5$.*

Proof. Let $H \neq \{\imath\}$ be a normal subgroup of A_n. We argue that H must contain a 3-cycle; then Theorem 2.49 can be applied to conclude that $H = A_n$.

Choose any permutation $\alpha \neq \iota$ in H and express α as a product $\alpha_1\alpha_2 \cdots \alpha_r$ of disjoint pairwise commutative cycles α_i. The proof will be broken up into four cases.

Case 1 Assume that α has some cycle of length $k \geqslant 4$ as a factor, say $\alpha_1 = (1 \quad 2 \quad 3 \ldots k)$. The normality of H in A_n implies that the product $(1 \quad 2 \quad 3)\alpha(1 \quad 3 \quad 2) = (1 \quad 2 \quad 3)\alpha(1 \quad 2 \quad 3)^{-1} \in H$. Since $\alpha_2 \ldots \alpha_r$ is a product of mutually disjoint cycles, each of which is a disjoint from $(1 \quad 2 \quad 3)$ (and hence from $(1 \quad 3 \quad 2)$), it follows that $(1 \quad 3 \quad 2)$ commutes with $\alpha_2 \ldots \alpha_r$:

$$\begin{aligned}(1 \quad 2 \quad 3)\alpha(1 \quad 3 \quad 2) &= (1 \quad 2 \quad 3)(1 \quad 2 \quad 3 \ldots k)(1 \quad 3 \quad 2)\alpha_2 \ldots \alpha_r \\ &= (1 \quad 4 \ldots k \quad 2 \quad 3)\alpha_2 \ldots \alpha_r.\end{aligned}$$

As a result, H must contain the permutation

$$\begin{aligned}(1 \quad 4 \ldots k \quad 2 \quad 3)\alpha_2 \ldots \alpha_r\alpha^{-1} &= (1 \quad 4 \ldots k \quad 2 \quad 3)(k \ldots 3 \quad 2 \quad 1) \\ &= (1 \quad 2 \quad 4).\end{aligned}$$

We next investigate those cases in which the lengths of the cycles α_i are not greater than 3.

Case 2 Assume that α has at least two 3-cycles as factors, for instance, $\alpha_1 = (1 \quad 2 \quad 3)$ and $\alpha_2 = (4 \quad 5 \quad 6)$. Then

$$\begin{aligned}(3 \quad 4 \quad 5)\alpha(3 \quad 5 \quad 4) &= (3 \quad 4 \quad 5)(1 \quad 2 \quad 3)(4 \quad 5 \quad 6)(3 \quad 5 \quad 4)\alpha_3 \ldots \alpha_r \\ &= (1 \quad 2 \quad 4)(3 \quad 6 \quad 5)\alpha_3 \ldots \alpha_r\end{aligned}$$

belongs to H, as does

$$\begin{aligned}(1 \quad 2 \quad 4)(3 \quad 6 \quad 5)\alpha_3 \ldots \alpha_r\alpha^{-1} &= (1 \quad 2 \quad 4)(3 \quad 6 \quad 5)(1 \quad 3 \quad 2)(4 \quad 6 \quad 5) \\ &= (1 \quad 6 \quad 3 \quad 4 \quad 5).\end{aligned}$$

This means that H contains a cycle of length 5 and so, by Case 1, contains a 3-cycle.

Case 3 In this case, assume that α has exactly one 3-cycle as a factor, all other factors being transpositions; say, $\alpha_1 = (1 \quad 2 \quad 3)$. Then

$$\alpha^2 = (1 \quad 2 \quad 3)^2(\alpha_2 \ldots \alpha_r)^2 = (1 \quad 3 \quad 2)$$

is the desired 3-cycle in H.

Case 4 The last case occurs when each factor of α is a transposition, with $\alpha_1 = (1 \quad 2)$ and $\alpha_2 = (3 \quad 4)$. Arguing as in the previous cases, it follows that

$$(1 \quad 2 \quad 4)\alpha(1 \quad 4 \quad 2)\alpha^{-1} = (1 \quad 2 \quad 4)(1 \quad 2)(3 \quad 4)(1 \quad 4 \quad 2) = (1 \quad 4)(2 \quad 3)$$

lies in H. Now consider the permutation $(2 \quad 3 \quad 5)$; this is possible, since

$n \geq 5$. Then

$$(2\ 5\ 3)(1\ 4)(2\ 3)(2\ 3\ 5) = (1\ 4)(2\ 5)$$

is an element of H, which puts the 3-cycle

$$(1\ 4)(2\ 5)(2\ 3)(1\ 4) = (2\ 3\ 5)$$

in H. This finally completes the proof.

COROLLARY *The alternating group A_n is simple for $n \neq 4$.*

Proof. Since $o(A_2) = 1$ and $o(A_3) = 3$, the assertion is obvious for $n = 2, 3$. If $n \geq 5$, then the foregoing theorem shows A_n to be simple. This leaves only the case of the alternating group A_4. The reader may check that A_4 contains a normal subgroup having as its elements the permutations

$$\begin{pmatrix} 1 & 2 & 3 & 4 \\ 1 & 2 & 3 & 4 \end{pmatrix}, \quad \begin{pmatrix} 1 & 2 & 3 & 4 \\ 2 & 1 & 4 & 3 \end{pmatrix}, \quad \begin{pmatrix} 1 & 2 & 3 & 4 \\ 3 & 4 & 1 & 2 \end{pmatrix}, \quad \begin{pmatrix} 1 & 2 & 3 & 4 \\ 4 & 3 & 2 & 1 \end{pmatrix}.$$

This seems a good point at which to inflict yet another proof on the reader; this one has at any rate the merit of brevity.

THEOREM 2.51 *For $n \geq 5$, the center of A_n consists of the identity permutation alone.*

Proof. Since cent A_n is a normal subgroup of A_n, the hypothesized simplicity of A_n implies that either cent $A_n = A_n$ or cent $A_n = \{\iota\}$. But the first possibility is ruled out by the fact that cent A_n is a commutative group, while A_n is not.

In the last 20 years the outstanding problem of algebra has been that of finding some reasonable description or classification of all finite simple groups. A classification theorem concerning such groups would assert that an arbitrary finite simple group is abstractly equivalent to one of the groups on a specified list of simple groups. The Feit-Thompson "Odd-Order Theorem," by showing that a finite (noncommutative) simple group must have odd order, brought about a tremendous surge of interest in this problem. Research activity in the field of finite groups greatly intensified and broadened in the 1960's and 1970's, and a large number of talented young mathematicians were attracted to the subject. In February 1981 the classification of the finite simple groups was completed, representing what has been called "one of the most remarkable achievements in the history of mathematics." The full classification proof has enormous length; involving the joint efforts of several hundred mathematicians from around the world, it runs close to 10,000 journal pages spread over as many as 500 individual papers.

Without going into detail we mention that until 1955 no finite simple groups were disclosed that were not known to algebraists by 1905. Relatively little progress was made in classifying finite simple groups during this entire half-century, an extraordinarily long period in contemporary mathematical history. In 1955, Claude Chevalley introduced a construction that yielded several new infinite families of finite simple groups. Variations on Chevalley's approach produced several further families, and by the early 1960's it was widely presumed that the list of finite simple groups would prove to be complete.

It was something of a mixed blessing for researchers when a new finite simple group, a so-called *sporadic group*, was detected in 1965. William Burnside gave them this name because they did not fit in a natural way into any known infinite family of simple groups, such as the alternating groups A_n for $n \geqslant 5$. The first five of these apparently isolated simple groups had been described a long time ago, in 1861 and 1873 to be exact, so the discovery in 1965 of the next sporadic group took almost 100 years. After 1965 the search for sporadic simple groups picked up, with a discovery rate of at least one per year. For a while it was thought that there might well exist an additional family of finite simple groups of which the sporadic groups provided a few examples; but we know now that there are no more to be found. There are exactly 26 of these exceptional groups. The final step in the classification theorem was the establishment of the uniqueness (that is, its nonequivalence to any existing simple group) of the 26th sporadic simple group, originally dubbed "the monster"; the order of this group is about 10^{54}, which does indeed make it monstrous

PROBLEMS 2.6

1. Describe the left coset decomposition of the following groups with respect to the indicated subgroup H:
   a) The group Z_{24} of integers modulo 24, $H = \{0, 6, 12, 18\}$.
   b) The group G_s of symmetries of the square, $H = \{e, d_1\}$.
   c) The symmetric group S_3, $H = A_3$.
   d) The additive group Z of integers, $H = (3)$.
   e) The multiplicative group of matrices

$$\begin{bmatrix}1 & 0\\0 & 1\end{bmatrix}, \begin{bmatrix}0 & 1\\1 & 0\end{bmatrix}, \begin{bmatrix}1 & 0\\-1 & -1\end{bmatrix}, \begin{bmatrix}0 & 1\\-1 & -1\end{bmatrix},$$

$$\begin{bmatrix}-1 & -1\\1 & 0\end{bmatrix}, \begin{bmatrix}-1 & -1\\0 & 1\end{bmatrix},$$

H consisting of the two matrices

$$\begin{bmatrix} 1 & 0 \\ 0 & 1 \end{bmatrix}, \quad \begin{bmatrix} 0 & 1 \\ 1 & 0 \end{bmatrix}.$$

f) The symmetric group S_4, H consisting of the four permutations

$$\begin{pmatrix} 1 & 2 & 3 & 4 \\ 1 & 2 & 3 & 4 \end{pmatrix}, \quad \begin{pmatrix} 1 & 2 & 3 & 4 \\ 2 & 1 & 4 & 3 \end{pmatrix}, \quad \begin{pmatrix} 1 & 2 & 3 & 4 \\ 3 & 4 & 1 & 2 \end{pmatrix}, \quad \begin{pmatrix} 1 & 2 & 3 & 4 \\ 4 & 3 & 2 & 1 \end{pmatrix}.$$

2. If G is a finite group, prove that $a^{o(G)} = e$ for all $a \in G$.

3. Assuming that H is a subgroup of the group G, establish each of the assertions below:

a) The equation $caH = cbH$ implies that $aH = bH$.

b) If $aH = Ha$, then $a^{-1}H = Ha^{-1}$.

c) If $a^2 = e$, then aH consists of the inverses of the elements in Ha.

d) All the cosets of H in G (both right and left) have the same number of elements.

e) The number of left cosets of H in G is equal to the number of right cosets of H in G. [*Hint:* Consider the mapping that sends aH into Ha^{-1}.]

4. Let H be a subgroup of the group G. If $G - H$ is finite and nonempty, prove that G is also finite.

5. Given that H and K are subgroups of the group G, show that

a) if K is a subgroup of H, then $[G:K] = [G:H][H:K]$;

b) if $H \cap K \neq \{e\}$ and $o(K)$ is a prime number, then $K \subseteq H$.

6. If $\{H_i\}$ is an indexed collection of normal subgroups of the group G, prove that $\cap H_i$ is also normal in G.

7. Suppose that H and K are normal subgroups of the group G such that $H \cap K = \{e\}$. By considering products of the form $hkh^{-1}k^{-1}$, show that the elements of H commute with those of K.

8. Let H be a subgroup of the group G. Prove that

a) H has index 2 if and only if $a^{-1}b \in H$ for all $a, b \in G - H$.

b) If H has index 2, then H is a normal subgroup of G. [*Hint:* For any $a \in G - H$, one has $H \cup aH = G = H \cup Ha$.]

c) If H has index 2, then $a^2 \in H$ for all $a \in G$.

9. Show that the alternating group A_n is a normal subgroup of S_n by proving that $[S_n : A_n] = 2$.

10. Let G be a finite group of odd order n. For each $a \in G$, show that the equation $x^2 = a$ has a solution in G. [*Hint*: Since $\gcd(n, 2) = 1$, there exist u, v satisfying $nu + 2v = 1$; consider $x = a^v$.]

11. In the group G_s of symmetries of the square, verify that

a) $H_1 = \{e, r_2\}$, $\quad H_2 = \{e, r_1, r_2, r_3\}$, $\quad H_3 = \{e, r_2, h, v\}$,

$H_4 = \{e, r_2, d_1, d_2\}$ are the only nontrivial normal subgroups. [*Hint:* Problem 8(b) is helpful.]

b) $\{e, d_1\}$ is normal in $\{e, r_2, d_1, d_2\}$, which in its turn is normal in G_s, but $\{e, d_1\}$ is not normal in G_s.

12. Let H be a subgroup of the group G. Prove that H is normal in G if and only if $ab \in H$ implies that $ba \in H$ for all $a, b \in G$. [*Hint:* Since $a^{-1}(ah) \in H$, the condition gives $(ah)a^{-1} \in H$.]

13. a) Show that $H = \{1, -1, i, -i\}$ is a normal subgroup of the multiplicative group

$$G = \left\{ \pm 1,\ \pm i,\ \frac{\pm 1 \pm \sqrt{3}i}{2},\ \frac{\pm i \pm \sqrt{3}}{2} \right\}.$$

b) For the multiplicative group G whose elements are the matrices

$$\begin{bmatrix} 1 & 0 \\ 0 & 1 \end{bmatrix}, \begin{bmatrix} 0 & 1 \\ 1 & 0 \end{bmatrix}, \begin{bmatrix} 1 & 0 \\ -1 & -1 \end{bmatrix}, \begin{bmatrix} 0 & 1 \\ -1 & -1 \end{bmatrix},$$
$$\begin{bmatrix} -1 & -1 \\ 1 & 0 \end{bmatrix}, \begin{bmatrix} -1 & -1 \\ 0 & 1 \end{bmatrix}$$

show that

$$H = \left\{ \begin{bmatrix} 1 & 0 \\ 0 & 1 \end{bmatrix}, \begin{bmatrix} 0 & 1 \\ -1 & -1 \end{bmatrix}, \begin{bmatrix} -1 & -1 \\ 1 & 0 \end{bmatrix} \right\}$$

forms a normal subgroup of G, but

$$K = \left\{ \begin{bmatrix} 1 & 0 \\ 0 & 1 \end{bmatrix}, \begin{bmatrix} 0 & 1 \\ 1 & 0 \end{bmatrix} \right\}$$

does not.

14. For a subgroup H of the group G, define the set

$$N(H) = \{a \in G \mid aHa^{-1} = H\}.$$

Prove each of the statements below:

a) $N(H)$ forms a subgroup of G, called the *normalizer of H in G.*

b) $N(H)$ is the largest (in the sense of inclusion) subgroup of G in which H is normal.

c) H is normal in G if and only if $N(H) = G$.

15. Assume that H and K are subgroups of the group G and that one of these is normal in G. Establish that HK is a subgroup of G. When both H and K are normal in G, show that the group HK is also normal.

16. Consider a group G and an element $a \in G$. Show that

a) The cyclic subgroup (a) is normal in G if and only if for each $x \in G$ there exists some $n \in Z$ satisfying $ax = xa^n$.

b) If a has order 2, then the subgroup $H = \{e, a\}$ is normal in G if and only if $H \subseteq \text{cent } G$.

17. In each of the following, H is a normal subgroup of the group G. List the elements of the quotient group G/H and make the multiplication table for G/H.

a) $H = \{1, -1\}$, $G = \{1, -1, i, -i\}$.

b) $H = \{0, 4, 8, 12\}$, $G = Z_{16}$.

c) $H = \left\{\begin{pmatrix}1 & 2 & 3\\ 1 & 2 & 3\end{pmatrix}, \begin{pmatrix}1 & 2 & 3\\ 3 & 1 & 2\end{pmatrix}, \begin{pmatrix}1 & 2 & 3\\ 2 & 3 & 1\end{pmatrix}\right\}$, $G = S_3$.

d) $H = \left\{\begin{bmatrix}1 & 0\\ 0 & 1\end{bmatrix}, \begin{bmatrix}-1 & 0\\ 0 & -1\end{bmatrix}\right\}$, $G = \left\{\begin{bmatrix}\pm 1 & 0\\ 0 & \pm 1\end{bmatrix}, \begin{bmatrix}0 & \pm 1\\ \pm 1 & 0\end{bmatrix}\right\}$.

e) $H = Z_e$, $G = Z$.

18. Prove that every element of G/H is its own inverse if and only if $a^2 \in H$ for all $a \in G$.

19. Given a cyclic group $G = (a)$ of order 15 and a subgroup $H = (a^3)$, make the multiplication table for the quotient group G/H.

20. Let Q be the additive group of rational numbers and Z be the subgroup of integers. Show that every element of the quotient group Q/Z has finite order.

21. For any group G, determine the quotient groups of the trivial normal subgroups $\{e\}$ and G.

22. Let H be a normal subgroup of the group G. Prove that if G is commutative (cyclic), then the quotient group G/H is also commutative (cyclic).

23. If H is a normal subgroup of the finite group G, prove that

a) $o(G/H) = o(G)/o(H)$

b) For all $a \in G$, $a^n \in H$, where $n = [G:H]$

c) The order of aH in G/H is a divisor of the order of a in G.

24. Let G be a group and T be the set of elements in G of finite order. Assuming that T is a subgroup of G, verify the following facts.

a) T forms a normal subgroup of G.

b) The quotient group G/T is *torsion-free*; that is, the torsion subgroup of G/T consists of the identity element alone.

c) If H is a normal subgroup of G for which G/H is torsion-free, then $T \subseteq H$.

25. Prove that if the quotient group $G/\text{cent}\ G$ is cyclic, then G must be commutative.

26. Let H be a subgroup of the commutative group G. If H and G/H are both finitely generated, show that G is also finitely generated.

27. Show that a group G is commutative if and only if $[G, G] = \{e\}$.

28. For a group G, let S be the subgroup generated by the set of squares of elements of G. Establish the following.

a) S is a normal subgroup of G [*Hint:* $as^2a^{-1} = (asa^{-1})^2$.]

b) The quotient group G/S is commutative. [*Hint:* The inclusion $[G, G] \subseteq S$ holds, since $[a, b] = (ab)^2(b^{-1}a^{-1}b)^2b^{-2}$.]

29. Prove that the commutator subgroup of S_3 is A_3. [*Hint:* For any $\alpha \in S_3$, $\alpha\beta\alpha^{-1}\beta^{-1}$ is even.]

30. Show that the symmetric group S_n, $n \geqslant 2$, can be generated by the following sets.

a) The n transpositions $(1\ \ 2), (1\ \ 3), \ldots, (1\ \ n)$

b) The $n - 1$ transpositions $(1\ \ 2), (2\ \ 3), \ldots, (n-1\ \ n)$. [*Hint:* Use (a) and the fact that $(2\ \ 3)(1\ \ 2)(2\ \ 3) = (1\ \ 3)$, $(3\ \ 4)(1\ \ 3)(3\ \ 4) = (1\ \ 4), \ldots$.]

c) The two cycles $\alpha = (1\ \ 2\ \ 3 \ldots n)$ and $\tau = (1\ \ 2)$. [*Hint:* Use (b) and the fact that $\alpha^k\tau\alpha^{-k} = (k\ \ k+1)$ for $1 \leqslant k \leqslant n-1$.]

31. Given that $n \geqslant 3$, prove that the alternating group A_n is generated by the 3-cycles $(1\ \ 2\ \ 3), (1\ \ 2\ \ 4), \ldots, (1\ \ 2\ \ n)$. [*Hint:* Each $\alpha \in A_n$ is a product of an even number of pairs $(1\ \ j)(1\ \ k)$. If $j, k \geqslant 3$ and $j \neq k$, then $(1\ \ j)(1\ \ k) = (1\ \ 2\ \ j)^2(1\ \ 2\ \ k)(1\ \ 2\ \ j)$ while $(1\ \ 2)(1\ \ k) = (1\ \ 2\ \ k)^{-1}$.]

32. Prove that, for $n \geqslant 5$, the only normal subgroup of the symmetric group S_n is A_n. [*Hint:* If H is normal in S_n, then $H \cap A_n$ is normal in A_n; hence $H \cap A_n = \{\imath\}$ or A_n.]

2.7 HOMOMORPHISMS

Up to this point in the text we have not considered mappings from one group to another; indeed, any knowledge of functions was irrelevant to most of the topics considered. They now enter in an essential way because we wish to introduce a concept, the idea of *algebraically indistinguishable systems*, that will be of fundamental importance throughout the remainder of the book and that is, in fact, one of the most important notions in mathematics.

For a simple illustration of just what we mean by algebraically indistinguishable systems, let us look at the group $(Z_2, +_2)$ of integers modulo 2 and the multiplicative group $(\{1, -1\}, \cdot)$. Their respective multiplication tables are

$+_2$	0	1
0	0	1
1	1	0

$\cdot$	1	-1
1	1	-1
-1	-1	1

If in the multiplication table for Z_2, 0 is replaced by 1 and 1 by -1, then we obtain the table for $\{-1, 1\}$. That is, the tables are "carbon copies" of each other, differing only in the symbols chosen for their elements. But the multiplication table tells us everything there is to know about the group. Since $(Z_2, +_2)$ and $(\{-1, 1\}, \cdot)$ have matching tables, it is natural to regard them as being the same group. In his *Traité des Substitutions* (1870), Camille Jordan used the word "isomorphic" to describe two (permutation) groups that look superficially different but behave in essentially the same way. The term is derived from the two Greek words *isos* meaning "equal" and *morphe* meaning "form."

When comparing groups, mappings that preserve the operations, and hence preserve the algebraic structure, play an essential role. These are called *group homomorphisms* or simply *homomorphisms* (from the Greek *homos*, "like"), terminology introduced by Felix Klein in 1893.

DEFINITION 2.28 Let $(G, \cdot)$ and $(G', \cdot')$ be two groups, not necessarily distinct. A *homomorphism* from $(G, \cdot)$ into $(G', \cdot')$ is a mapping $f : G \to G'$ such that

$$f(a \cdot b) = f(a) \cdot' f(b)$$

for arbitrary $a, b \in G$. If, in addition, f is one-to-one, then f is said to be an *isomorphism*.

Before considering any examples, some comments on this definition are in order. Observe first that on the left-hand side of the above equation the product $a \cdot b$ is computed in G, while on the right-hand side the product $f(a) \cdot' f(b)$ is that of elements of G'. Homomorphisms have the property of carrying products (in G) into products (in G'), a situation that is conveniently described by the phrase "the image of a product of two elements is equal to the product of their images."

Another interpretation is perhaps helpful. The condition $f(a \cdot b) = f(a) \cdot' f(b)$ asserts that if we start with a pair of elements $a, b \in G$ and move then to G' by either of two routes—by first forming the product $a \cdot b$ and then applying f to it, or by first obtaining the images $f(a)$ and $f(b)$ and then taking their product—the result will be the same. This idea is illustrated by the "commutative" diagram below:

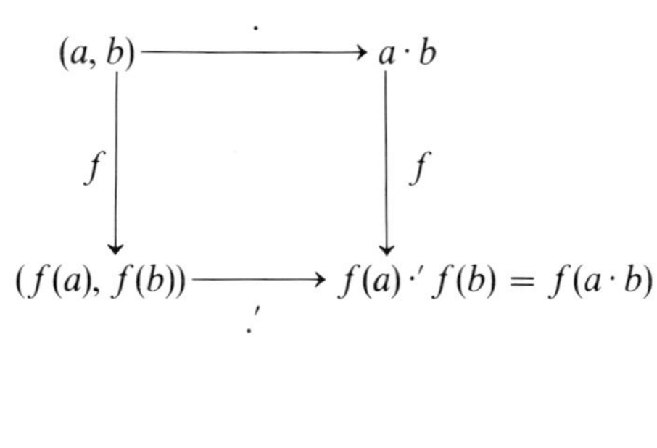

(In rough terms, a diagram of mappings is *commutative* if, whenever there are two sequences of arrows in the diagram leading from one group to another, the composition of mappings along either of these paths produces the same function.)

If the operations of the two groups G and G' are both indicated by juxtaposition, then Definition 2.28 takes the following form: a homomorphism from G into G' is a function $f: G \to G'$ such that $f(ab) = f(a)f(b)$ for all $a, b \in G$. When $G = G'$, so that both groups are the same, we shall speak of f as being a homomorphism from G into itself. In this connection a homomorphism from G into itself is frequently referred to as an *endomorphism* of the group G or, if an isomorphism onto G, an *automorphism* of G.

Without further ado, we pass on to an inspection of several examples.

Example 2.44
Given an arbitrary group $(G, \cdot)$, take f to be the identity mapping on G; that is, $f = i_G$. The verification that f meets the requirement of a homomorphism is immediate:

$$f(a \cdot b) = a \cdot b = f(a) \cdot f(b).$$

Example 2.45
Suppose that $(G, \cdot)$ and $(G', \cdot')$ are two groups with identity elements e and e', respectively. The function $f: G \to G'$ given by $f(a) = e'$ for all $a \in G$ turns out to be a homomorphism, since

$$f(a \cdot b) = e' = e' \cdot' e' = f(a) \cdot' f(b).$$

This particular mapping (the so-called *trivial homomorphism*) is the only constant function that satisfies Definition 2.28.

Example 2.46
Let $(R^{\#}, +)$ be the additive group of real numbers and $(R^{\#} - \{0\}, \cdot)$ be the multiplicative group of nonzero reals. For $a \in R^{\#}$, define the function f by $f(a) = 2^a$. To show that f is a homomorphism, we must verify that $f(a + b) = f(a) \cdot f(b)$. But this presents no problem, since

$$f(a + b) = 2^{a+b} = 2^a \cdot 2^b = f(a) \cdot f(b).$$

Example 2.47
Consider the two additive groups $(Z, +)$ and $(Z_n, +_n)$. Define $f: Z \to Z_n$ by setting $f(a) = [a]$; in other words, send each integer into the congruence class containing it. That the function just cited preserves the group operation follows from the definition of addition modulo n:

$$f(a + b) = [a + b] = [a] +_n [b] = f(a) +_n f(b).$$

Let us label the set of all homomorphisms from the group G into the group G' by the symbol Hom(G, G'). In the event that $G = G'$, the simpler notation Hom G will be used in place of Hom(G, G). Some authors prefer to write End G, for endomorphism; both notations have a certain suggestive power, and it reduces to a matter of personal preference.

Interestingly enough, the set Hom G can be endowed with an algebraic structure:

LEMMA *With the operation of functional composition,* Hom G *forms a semigroup with identity.*

Proof. For the proof, which is quite elementary, it must first be shown that the composition $f \circ g$ of two functions $f, g \in \text{Hom } G$ also preserves the group operation. This is easily accomplished by noting that whenever $a, b \in G$,

$$\begin{aligned}(f \circ g)(ab) &= f(g(ab)) \\ &= f(g(a)g(b)) \\ &= f(g(a))f(g(b)) = (f \circ g)(a)(f \circ g)(b).\end{aligned}$$

As we saw in an earlier section, composition of functions is associative. Finally, Example 2.41 indicates that the identity mapping i_G, the identity element for composition, is itself operation-preserving.

It seems reasonable to inquire whether there exists any subset of Hom G that forms a group under composition. For inverses to exist, one must plainly single out the one-to-one functions. Moreover, in order that the domain of f^{-1} be the set G, consideration should be further restricted to those functions that map onto G. Thus a natural undertaking is to investigate the collection of all isomorphisms from the group G onto itself; as a matter of notation, we shall designate this latter set of mappings by the symbol Aut G, for automorphism.

The elements of Aut G are now restricted to the extent that (Aut G, $\circ$) does indeed have the agreeable property of being a group. Let us give some details.

THEOREM 2.52 *For any group G,* Aut G *is a subgroup of the symmetric group* Sym G.

Proof. Given functions $f, g \in \text{Aut } G$, we already know that the composition $f \circ g$ is in Sym G. In conjunction with the last lemma, this shows that $f \circ g$ belongs to Aut G, as does the identity map i_G. It only remains to verify here that whenever a function $f \in \text{Aut } G$, its inverse f^{-1}—which clearly is a member of Sym G—is likewise a homomorphism. If $x, y \in G$, the onto character of f implies that $x = f(a)$, $y = f(b)$ for some choice of $a, b \in G$. Therefore

$$\begin{aligned} f^{-1}(xy) &= f^{-1}(f(a)f(b)) \\ &= f^{-1}(f(ab)) \\ &= ab = f^{-1}(x)f^{-1}(y), \end{aligned}$$

which establishes the result claimed.

We now proceed to analyze the behavior of homomorphisms in detail. As a starting point, it will be shown that such functions "preserve" identities and inverses.

THEOREM 2.53 *If f is a homomorphism from the group G into the group G', then*

a) *f maps the identity element e of G onto the identity element e' of G'; in symbols, $f(e) = e'$; and*
b) *f maps the inverse of any element a of G onto the inverse of $f(a)$; in symbols, $f(a^{-1}) = f(a)^{-1}$.*

Proof. To prove the first assertion, observe that

$$f(a)e' = f(a) = f(ae) = f(a)f(e)$$

for each $a \in G$. On left-canceling $f(a)$, we obtain $f(e) = e'$. (One could also use the argument that $f(e)$ is idempotent in G', hence equal to $e' : f(e)^2 = f(e^2) = f(e)$.)

Turning to the second part of the theorem, let a be a typical element of G. By what we have just proved,

$$f(a)f(a^{-1}) = f(aa^{-1}) = f(e) = e'.$$

Likewise, $f(a^{-1})f(a) = e'$, so that $f(a^{-1})$ is an inverse of $f(a)$. Since inverses are unique in G', it may be concluded that $f(a^{-1}) = f(a)^{-1}$.

Example 2.48

As an immediate application of these ideas, we propose to establish that for each real number $r \neq 0$ there is exactly one homomorphism f from the additive group Z into the multiplicative group $R^{\#} - \{0\}$ for which $f(1) = r$. The existence of such a function is trivial, for we need consider only the mapping $f(n) = r^n$, $n \in Z$.

To prove that there can be at most one function satisfying the indicated conditions provides a more challenging problem. The basic idea is simple enough: Assume that there are two homomorphisms $f, g : Z \to R^{\#} - \{0\}$ which carry 1 onto r, and show that they are actually the same. Now each positive integer n may be written as

$$n = 1 + 1 + \cdots + 1 \qquad (n \text{ summands}).$$

The operation-preserving character of f and g implies that

$$f(n) = f(1)^n = r^n = g(1)^n = g(n), \qquad n \in Z_+.$$

On the other hand, if n is a nonzero negative integer, then $-n \in Z_+$. Hence

$$f(n) = f(-(-n)) = f(-n)^{-1} = g(-n)^{-1} = g(-(-n)) = g(n).$$

The crucial equality, $f(-n) = g(-n)$, is justified by the fact that f and g are already known to agree on the positive integers. By the first part of Theorem 2.53, $f(0) = 1 = g(0)$, so $f(n) = g(n)$ for every integer n; therefore $f = g$.

Our immediate goal is to ascertain the algebraic nature of direct and inverse images of subgroups under homomorphisms. We shall see, among other things, that if f is a homomorphism from the group G into the group G', then $f(G)$ forms a subgroup of G'. (Concerning terminology, $f(G)$ is called the *homomorphic image* of G under f.) The complete story is told below.

THEOREM 2.54 *Let f be a homomorphism from the group G into the group G'. Then*

a) for each subgroup H of G, $f(H)$ is a subgroup of G', and

b) for each subgroup H' of G', $f^{-1}(H')$ is a subgroup of G.

Proof. To obtain the first part of the theorem, recall that the image $f(H) = \{f(h) | h \in H\}$. Now suppose that $f(a)$ and $f(b)$ are arbitrary elements of $f(H)$. Then both a and b belong to the set H, as does the product ab^{-1}—H being a subgroup of G. Hence

$$f(a)f(b)^{-1} = f(a)f(b^{-1}) = f(ab^{-1}) \in f(H).$$

According to Theorem 2.25, this is a sufficient condition for $f(H)$ to be a subgroup of G'.

The proof of the second assertion proceeds along similar lines. First, remember that $f^{-1}(H') = \{a \in G | f(a) \in H'\}$. Thus if a, $b \in f^{-1}(H')$, their images $f(a)$ and $f(b)$ must be members of H'. Since H' is a subgroup of G', it follows at once that

$$f(ab^{-1}) = f(a)f(b^{-1}) = f(a)f(b)^{-1} \in H'.$$

This means that ab^{-1} lies in $f^{-1}(H')$, from which we deduce that $f^{-1}(H')$ forms a subgroup of G.

Left unresolved is the matter of replacing the term "subgroup" in Theorem 2.54 by "normal subgroup." It is not difficult to show that part (b) of the theorem remains true under such a substitution. More precisely, if H' is a normal subgroup of G', then the subgroup $f^{-1}(H')$ is normal in G. By way of proof, suppose that $h \in f^{-1}(H')$, so that $f(h) \in H'$ and let a be an arbitrary element of G. Then $f(aha^{-1}) = f(a)f(h)f(a)^{-1} \in H'$ by the normality of H'; in other words, the product $aha^{-1} \in f^{-1}(H')$, which is enough to make $f^{-1}(H')$ a normal subgroup of G.

Without further restriction it cannot be inferred that the image $f(H)$ will form a normal subgroup of G', whenever H is normal in G. One would need to know that $a'f(h)(a')^{-1} \in f(H)$ for all $h \in H$ and $a' \in G'$. In general, there is no justification for replacing a' by some $f(a)$ in order to exploit the fact that H is a normal subgroup of G. The solution is obvious: Just take f to be an onto mapping.

Summarizing these remarks, we now state the following.

COROLLARY

a) *For each normal subgroup H' of G', the subgroup $f^{-1}(H')$ is normal in G.*
b) *If $f(G) = G'$, then for each normal subgroup H of G, the subgroup $f(H)$ is normal in G'.*

Before we go further, a new idea needs to be introduced.

DEFINITION 2.29 Let f be a homomorphism from the group G into the group G'. The *kernel* of f, denoted by ker f, consists of those elements in G that are mapped by f onto the identity element of the group G':

$$\ker f = \{a \in G \mid f(a) = e'\}.$$

Theorem 2.53 indicates that ker f is a nonempty subset of G, since, if nothing else, the identity $e \in \ker f$. As the reader is aware, the trivial subgroup $\{e'\}$ forms a normal subgroup of the group G'. Since $\ker f = f^{-1}(e')$, it follows from the preceding corollary that the kernel of f will be a normal subgroup of G. This observation is recorded as follows.

THEOREM 2.55 *If f is a homomorphism from the group G into the group G', then* ker f *is a normal subgroup of G.*

The proof of Theorem 2.55, as we have given it, is deceptively simple; it hides a good deal of machinery. An alternative approach would be to fall back on first principles and let the elements $a, b \in \ker f$. Then

$$f(ab^{-1}) = f(a)f(b)^{-1} = e'(e')^{-1} = e',$$

whence $ab^{-1} \in \ker f$. As a result, $\ker f$ is a subgroup of the group G. For normality, select $k \in \ker f$ and $a \in G$. It is readily seen that

$$f(aka^{-1}) = f(a)f(k)f(a)^{-1} = f(a)e'f(a)^{-1} = e'$$

and so aka^{-1} lies in $\ker f$; by Theorem 2.44, $\ker f$ is a normal subgroup of G.

Among the homomorphisms of any group G are the identity map on G and the trivial homomorphism that sends every element of G to e. The corresponding kernels are $\{e\}$ and G, the trivial normal subgroups of G.

Example 2.49
As a simple example of the foregoing theorems, consider the additive group Z of integers and multiplicative group $R^{\#} - \{0\}$. The mapping $f: Z \to R^{\#} - \{0\}$ defined by

$$f(n) = \begin{cases} 1 & \text{if } n \in Z_e \\ -1 & \text{if } n \in Z_o \end{cases}$$

is a homomorphism, as the reader may verify by checking the various possibilities that could arise. Here $\ker f = \{n \in Z \mid f(n) = 1\} = Z_e$, while the direct image $f(Z)$ is given by $f(Z) = \{1, -1\}$. It is not particularly difficult to show that Z_e is a normal subgroup of Z and that $\{1, -1\}$ is a subgroup of $R^{\#} - \{0\}$.

The kernel of a homomorphism may be viewed as a measure of the extent to which the homomorphism fails to be one-to-one and hence fails to be an isomorphism. Otherwise stated:

THEOREM 2.56 *A homomorphism f from a group G into a group G' is one-to-one if and only if* $\ker f = \{e\}$.

Proof. To see this, let f be a one-to-one function and $a \in \ker f$. Then $f(a) = e' = f(e)$, whence $a = e$; that is, $\ker f = \{e\}$. On the other hand, suppose that the kernel consists only of e. If $f(a) = f(b)$, then

$$f(ab^{-1}) = f(a)f(b)^{-1} = f(a)f(a)^{-1} = e',$$

which entails that $ab^{-1} \in \ker f$. Since $\ker f = \{e\}$, we must have $ab^{-1} = e$, or equivalently $a = b$, making f a one-to-one function.

In Theorem 2.55 we saw that certain normal subgroups occur as kernels of homomorphisms. Let us now demonstrate that every normal subgroup arises in this manner.

THEOREM 2.57 *Let H be a normal subgroup of the group G. Then the mapping* $\mathrm{nat}_H : G \to G/H$ *defined by* $\mathrm{nat}_H(a) = aH$ *is a homomorphism of G onto the quotient group G/H; the kernel of* nat_H *is just H itself.*

Proof. The fact that nat_H is a homomorphism follows directly from the manner in which multiplication is defined in the quotient group:

$$\text{nat}_H(ab) = abH = (aH)(bH) = \text{nat}_H(a)\,\text{nat}_H(b).$$

That nat_H sends G onto G/H is all but obvious; indeed, every element of G/H is a coset aH, with $a \in G$, and so by definition $\text{nat}_H(a) = aH$.

Inasmuch as the coset $H = eH$ serves as the identity element for the group G/H, we necessarily have

$$\begin{aligned}\ker(\text{nat}_H) &= \{a \in G \mid \text{nat}_H(a) = H\} \\ &= \{a \in G \mid aH = H\} = H.\end{aligned}$$

The last equality was achieved with the help of Theorem 2.39.

It is possible—and sometimes convenient—to recast this theorem so that no reference is made to the notion of a quotient group:

COROLLARY *Let H be a normal subgroup of the group G. Then there exists a group G' and a homomorphism from G onto G' whose kernel is H.*

In the future we shall speak of the function nat_H, which assigns to each element of G the coset in G/H of which it is the representative, as being the *natural* or *canonical mapping* of G onto the quotient group G/H. When there is no danger of confusion, the subscript H will be omitted in writing this mapping.

Let us pause for a moment to interpret Theorem 2.57 in the case of the additive group Z of integers. We already know that its normal subgroups are the cyclic groups (n), where n is a nonnegative integer. Moreover, the quotient group corresponding to any fixed $n \in Z_+$ is simply Z_n, the group of integers modulo n; that is, $Z/(n) = Z_n$. It is fairly evident from this that the natural mapping $\text{nat}: Z \to Z_n$ does nothing more than send each integer into its congruence class modulo n; in short, $\text{nat}(a) = [a]$.

We mentioned earlier that there are two standard techniques for investigating the structure of a particular group. One method calls for finding all the subgroups and, more pointedly, all the normal subgroups of the group, in the hope of gaining information about the group through its local structure. The other approach is to determine all homomorphisms from the given group into a simpler group; the idea here is that the homomorphic images will tend to reflect some of the algebraic properties of the original group. (The reader is warned to proceed with some care, since, for example, it is quite possible for multiplication to be commutative in the image group without the given group being commutative.) Although these lines of attack aim in different directions, Theorems 2.55 and 2.57 show that for all practical purposes these are the same: Every normal subgroup determines a homomorphism, and every homomorphism determines a normal subgroup.

Isomorphisms play a particularly prominent role in group theory, and we now focus our attention on these.

DEFINITION 2.30 A group G is said to be *isomorphic* to a group G', denoted symbolically by $G \simeq G'$, if there exists an isomorphism from G onto G'.

Although this definition is unsymmetric, in that it makes mention of a function from one particular group to another, let us remark that if f is an isomorphism of G onto G', then the inverse mapping, f^{-1}, will map G' isomorphically onto G. We may therefore ignore the apparent lack of symmetry and merely speak of the two groups G and G' as being isomorphic, without specifying one group as isomorphic to the other; notationally, this situation will be recognized by writing either $G \simeq G'$ or $G' \simeq G$.

Isomorphic groups share the same structural properties, even though they may differ in the notation for and nature of their elements and operations. Although not formally identical, two such groups are the same from the point of view of abstract group theory; indeed, any statement that can be made about one using only the group operation will be translated by an isomorphism into a corresponding statement about the other.

A group G' is said to be an *isomorphic* (*homomorphic*) *image* of a group G if there exists an isomorphism (homomorphism) from G onto G'. The crucial distinction is this: The properties of a group are reflected with complete precision in any isomorphic image, whereas homomorphic images do not necessarily inherit all the properties of the original group.

So much for the technical framework. Let us look at some specific examples.

Example 2.50
Consider the additive group Z_4 and the multiplicative group $G = \{1, -1, i, -i\}$, where $i^2 = -1$. For convenience the operation tables for these two groups are reproduced below.

$+_4$	0	1	2	3
0	0	1	2	3
1	1	2	3	0
2	2	3	0	1
3	3	0	1	2

$\cdot$	1	-1	i	$-i$
1	1	-1	i	$-i$
-1	-1	1	$-i$	i
i	i	$-i$	-1	1
$-i$	$-i$	i	1	-1

We wish to prove that the groups Z_4 and G are abstractly equal. To do so, we must produce a one-to-one homomorphism f from Z_4 onto G.

Since the preservation of identity elements is a general feature of any homomorphism, f must be such that $f(0) = 1$. Let us suppose for the moment that we were to define $f(1) = -1$. The image of the inverse of an element must equal the inverse of its image. We would then have

$$f(3) = f(1^{-1}) = f(1)^{-1} = (-1)^{-1} = -1,$$

or $f(3) = f(1)$. This, however, would prevent f from being one-to-one.

A more appropriate choice, in the sense that it avoids the above difficulty, is to take $f(1) = i$. The condition on inverses then implies that $f(3) = -i$. Since f is further required to preserve modular addition,

$$f(2) = f(1 +_4 1) = f(1)f(1) = i^2 = -1.$$

We are thus led in a natural way to consider the function defined by

$$f(0) = 1, \qquad f(1) = i, \qquad f(2) = -1, \qquad f(3) = -i.$$

Clearly, this function is a one-to-one mapping of the set Z_4 onto the set G. To establish that f is actually a homomorphism, we must consider all possible sums of two elements of Z_4 and show that each such sum is mapped onto the product of the images by f. Merely to verify one instance, observe that

$$f(1 +_4 2) = f(3) = -i = i(-1) = f(1)f(2).$$

Granting the other cases, we have $Z_4 \simeq G$.

Loosely speaking, two finite groups are isomorphic if it is possible to obtain each multiplication table from the other by merely renaming the elements. The nature of the function f suggests the appropriate rearrangement of the table for G is as follows.

$\cdot$	1	i	-1	$-i$
1	1	i	-1	$-i$
i	i	-1	$-i$	1
-1	-1	$-i$	1	i
$-i$	$-i$	1	i	-1

Apart from the particular symbols used, this group table is identical with that of Z_4, for corresponding elements appear at the same place in each table. Both groups are simply disguises for the same abstract system.

In passing, we might note that Z_4 is also isomorphic to G under the function $g: Z_4 \to G$, whereby

$$g(0) = 1, \qquad g(1) = -i, \qquad g(2) = -1, \qquad g(3) = i.$$

A standard procedure for showing that two groups are not isomorphic is to find some property of one that is not possessed by the other, which by its nature would necessarily be shared if these groups were actually isomorphic. For instance, the group Z_4 and the Klein 4-group K_4 are differentiated by the fact that the former is a cyclic group, whereas the latter is not.

Example 2.51

Let us show that the additive group Z is not isomorphic to the multiplicative group $Q - \{0\}$ of nonzero rationals. To see this, suppose that there exists a one-to-one onto function $f: Z \to Q - \{0\}$ with the property $f(a + b) = f(a)f(b)$ for all $a, b \in Z$. If x denotes the element of Z such that $f(x) = -1$, then

$$f(2x) = f(x + x) = f(x)f(x) = (-1)(-1) = 1.$$

In accordance with Theorem 2.53 the identity element of Z is the unique element of Z corresponding to the identity of $Q - \{0\}$, so that $2x = 0$ or $x = 0$. Consequently, both $f(0) = 1$, and $f(0) = -1$, contradicting the fact that f is a function. This argument shows that Z cannot be isomorphic to $Q - \{0\}$, for no function satisfying Definition 2.30 can exist.

Example 2.52

For an instructive example in connection with the additive group Z of integers, consider the following assertion: The only functions under which Z is isomorphic to itself are the identity mapping and its negative. A fairly succinct description of all this is that $\text{Aut } Z = \{i_Z, -i_Z\}$.

Perhaps the quickest proof of the above assertion consists of showing that if $f \in \text{Aut } Z$, then the cyclic subgroup $(f(1))$ generated by $f(1)$ is the group Z itself. Since the inclusion $(f(1)) \subseteq Z$ trivially holds, our aim would be achieved on establishing $Z \subseteq (f(1))$. But this is a straightforward matter. If n is an arbitrary integer, $n = f(m)$ for some $m \in Z$—recall f is a mapping onto Z—so that

$$\begin{aligned} n &= f(1 + 1 + \cdots + 1) \qquad (m \text{ times}) \\ &= f(1) + f(1) + \cdots + f(1) \\ &= mf(1), \end{aligned}$$

whence $n \in (f(1))$.

Since 1 and -1 are the only two elements that individually generate Z, either $f(1) = 1$ or $f(1) = -1$. The argument of the previous paragraph can be used to show that $f(n) = nf(1)$ for each $n \in Z$; from this it is evident that $f = i_Z$ or $f = -i_Z$ according as $f(1) = 1$ or $f(1) = -1$.

We now resume our discussion of the general theory by proving that the groups $(Z_n, +_n)$ and $(Z, +)$ are the prototypes of all finite and infinite cyclic groups, respectively.

THEOREM 2.58 *Every finite cyclic group of order $n > 0$ is isomorphic to the additive group Z_n of integers modulo n; every infinite cyclic group is isomorphic to the additive group Z of integers.*

Proof. To start, suppose that the cyclic group $G = (a)$ is of finite order n. In this case we know from Theorem 2.33 that

$$G = \{e, a, a^2, \ldots, a^{n-1}\}.$$

It seems natural then to investigate the mapping $f: G \to Z_n$ given by the rule $f(a^k) = [k]$, $0 \leqslant k < n$. This function plainly sends G onto the set Z_n. Next observe that f is one-to-one; if $f(a^k) = f(a^j)$, then $k \equiv j (\mathrm{mod}\, n)$, so that $a^k = a^j$. Finally, for any elements a^k, a^j in G we have

$$f(a^k a^j) = f(a^{k+j}) = [k + j] = [k] +_n [j] = f(a^k) +_n f(a^j).$$

This shows that the function f preserves the group operation and completes the proof of the isomorphism $G \simeq Z_n$.

For the second part of the theorem the cyclic group $G = (a)$ is assumed to be of infinite order. Here the choice of a mapping f between G and Z is the obvious one: Simply take $f(a^k) = k$. It is immediate that f, so defined, is an onto mapping. Moreover, this function acts in a one-to-one fashion on G; if $f(a^k) = f(a^j)$, then $k = j$, giving $a^k = a^j$. To complete the picture, we need to show that f is a homomorphism; but this is easy:

$$f(a^k a^j) = f(a^{k+j}) = k + j = f(a^k) + f(a^j).$$

The outcome is that G is isomorphic to the additive group Z via f.

As a by-product of the above theorem, we have

COROLLARY *Any two cyclic groups of the same order are isomorphic.*

Proof. This follows from the fact that the relation "is isomorphic to" is an equivalence relation on any collection of groups and hence is transitive.

As another application of the isomorphism concept, let us show that any group can be represented in concrete terms as a group of permutations on some set. To make things more specific, suppose that G is an arbitrary group and a is a fixed element of G. We define the function $f_a: G \to G$ by setting

$$f_a(x) = ax \qquad (x \in G);$$

this is called the *left-multiplication function induced by a*. Given $x \in G$,

$$x = a(a^{-1}x) = f_a(a^{-1}x),$$

so f_a maps G onto itself. It is also one-to-one; if $x, y \in G$, with $f_a(x) = f_a(y)$, then $ax = ay$ and, by the cancellation law, $x = y$. In brief, f_a is a permutation of the set G or, to put it another way, an element of the symmetric group Sym G.

The next stage in the proceedings is to prove that the set $H = \{f_a | a \in G\}$, with the usual composition of functions, forms a subgroup of Sym G. For any $x \in G$ we have

$$\begin{aligned}(f_a \circ f_b)(x) &= f_a(f_b(x)) = f_a(bx)\\ &= a(bx) = (ab)x = f_{ab}(x),\end{aligned}$$

implying that $f_a \circ f_b = f_{ab}$. The last relation tells us that H is closed under multiplication. If e is the identity element of G, then

$$f_a \circ f_e = f_{ae} = f_a = f_{ea} = f_e \circ f_a;$$

hence the permutation $f_e = \iota$ acts as the identity for H. Finally, observe that

$$f_a \circ f_{a^{-1}} = f_{aa^{-1}} = f_e = f_{a^{-1}a} = f_{a^{-1}} \circ f_a,$$

in consequence of which $(f_a)^{-1} = f_{a^{-1}} \in H$. All of this adds up to the fact that H is a subgroup of the symmetric group Sym G.

There still remains the matter of showing that G is isomorphic to this group H. What could be more natural than to map an element of G onto the left-multiplication function induced by it? In other words, we consider the mapping $\Phi: G \to H$ given by $\Phi(a) = f_a$. That Φ carries G onto H is perfectly clear. If $f_a = f_b$, then $f_a(e) = f_b(e)$ or $a = b$, which makes Φ one-to-one. Since

$$\Phi(ab) = f_{ab} = f_a \circ f_b = \Phi(a) \circ \Phi(b),$$

Φ preserves the group operation and so is an isomorphism of G onto H. We have proved all the ingredients of the following theorem:

THEOREM 2.59 (Cayley) *Any group G is isomorphic to a subgroup of the symmetric group* Sym G.

When G is finite, this result takes the following form.

COROLLARY *Any finite group of order n is isomorphic to a subgroup of the symmetric group* S_n.

Example 2.53
By way of illustrating Cayley's Theorem, let us determine the left-multiplication functions induced by the elements of the Klein 4-group K_4. As we have observed before, the multiplication table for this group is

$\cdot$	e	a	b	c
e	e	a	b	c
a	a	e	c	b
b	b	c	e	a
c	c	b	a	e

where $c = ab$. A typical left-multiplication function is described by telling what it does to the elements of K_4. The effect of, say, f_c is given by

$$f_c(e) = ce = c, \qquad f_c(a) = ca = b,$$
$$f_c(b) = cb = a, \qquad f_c(c) = c^2 = e,$$

so that

$$f_c = \begin{pmatrix} e & a & b & c \\ c & b & a & e \end{pmatrix}.$$

The set of all such functions is

$$f_e = \begin{pmatrix} e & a & b & c \\ e & a & b & c \end{pmatrix} = \iota;$$

$$f_a = \begin{pmatrix} e & a & b & c \\ a & e & c & b \end{pmatrix} = (ea)(bc);$$

$$f_b = \begin{pmatrix} e & a & b & c \\ b & c & e & a \end{pmatrix} = (eb)(ac);$$

$$f_c = \begin{pmatrix} e & a & b & c \\ c & b & a & e \end{pmatrix} = (ec)(ab).$$

On sending each element of K_4 to the appropriate function we get an isomorphism of K_4 onto a subgroup of S_4 (in fact, a subgroup of A_4), namely,

$$\{\iota, (12)(34), (13)(24), (14)(23)\}.$$

In bringing our discussion of these matters to a close, we might add that the concept of isomorphism is applicable to all types of mathematical systems, since it seems reasonable to treat two systems as abstractly equal when they have exactly the same properties. The essence of the notion is that we can always find a one-to-one mapping between the elements of the two systems that in some way preserves whatever structure we are interested in studying.

PROBLEMS 2.7

1. Determine whether the indicated function f is a homomorphism from the first group into the second group.
   a) $f(a) = -a$, $(R^{\#}, +)$, $(R^{\#}, +)$
   b) $f(a) = 1/a$, $(R^{\#} - \{0\}, \cdot)$, $(R^{\#} - \{0\}, \cdot)$
   c) $f(a) = a + 1$, $(Z, +)$, $(Z, +)$
   d) $f(a) = a^2$, $(R^{\#} - \{0\}, \cdot)$, $(R^{\#}_{+}, \cdot)$
   e) $f(a) = a/q$ (q a fixed nonzero integer), $(Z, +)$, $(Q, +)$
   f) $f(a) = na$ (n a fixed integer), $(Z, +)$, $(Z, +)$
   g) $f(a) = |a|$, $(Q - \{0\}, \cdot)$, $(Q_{+}, \cdot)$
2. Given the permutations

$$\alpha_1 = \begin{pmatrix} 1 & 2 & 3 \\ 1 & 2 & 3 \end{pmatrix}, \quad \alpha_2 = \begin{pmatrix} 1 & 2 & 3 \\ 2 & 3 & 1 \end{pmatrix}, \quad \alpha_3 = \begin{pmatrix} 1 & 2 & 3 \\ 3 & 1 & 2 \end{pmatrix},$$

$$\alpha_4 = \begin{pmatrix} 1 & 2 & 3 \\ 1 & 3 & 2 \end{pmatrix}, \quad \alpha_5 = \begin{pmatrix} 1 & 2 & 3 \\ 2 & 1 & 3 \end{pmatrix}, \quad \alpha_6 = \begin{pmatrix} 1 & 2 & 3 \\ 3 & 2 & 1 \end{pmatrix},$$

   define the function $f: S_3 \to Z_2$ by setting

$$f(\alpha_i) = 0 \quad \text{for } i = 1, 2, 3; \qquad f(\alpha_i) = 1 \quad \text{for } i = 4, 5, 6.$$

   Show that f is a homomorphism from S_3 onto Z_2.
3. For any group G, prove that if G is cyclic, then Aut G is commutative.
4. Prove that the following conditions on a group G are equivalent:
   a) G is commutative.
   b) The mapping $f(a) = a^{-1}$ is a homomorphism of G onto itself.
   c) The mapping $f(a) = a^2$ is a homomorphism of G into itself.
5. Let f be a homomorphism from the group G into itself and let H denote the set of elements that are left fixed by f:

$$H = \{a \in G \mid f(a) = a\}.$$

   Establish that H forms a subgroup of G.
6. Assuming that f is a homomorphism from the group G into the group G', verify the assertions below:
   a) $f(\text{cent } G) \subseteq \text{cent } f(G)$.

b) If S is a set of generators for G, then $f(S)$ generates $f(G)$.
c) If $o(a)$ is finite, then the order of $f(a)$ divides the order of a.
d) Whenever G' is commutative, $[G, G] \subseteq \ker f$.
e) Two elements $a, b \in G$ have the same image in G' if and only if they belong to the same (left) coset of $\ker f$.
f) If G' is commutative, then any subgroup of G containing $\ker f$ is normal.

7. Let Z_8 be the group of integers modulo 8 and $G = (a)$ be any cyclic group of order 12. Consider the mapping $f: Z_8 \to G$ defined by $f(n) = a^{3n}$; that is,

$$f(0) = f(4) = e, \qquad f(1) = f(5) = a^3,$$
$$f(2) = f(6) = a^6, \qquad f(3) = f(7) = a^9.$$

Prove that f is a homomorphism from Z_8 into G and determine its kernel.

8. Suppose that G is an arbitrary multiplicative group and the element $a \in G$. Show that the function $f: Z \to G$ given by $f(n) = a^n$ is a homomorphism from the additive group Z into G. At the same time, describe the subgroups $f(Z)$ and $\ker f$.

9. Verify that the following properties of a group G are passed on to any homomorphic image of G: the property of being
a) commutative
b) cyclic
c) simple
d) a torsion group (that is, every element of the group has finite order)

10. Let G be a finite cyclic group of order n and $k \in Z_+$. If $\gcd(k, n) = 1$, establish that the mapping $f: G \to G$ defined by $f(x) = x^k$ is an isomorphism.

11. a) Show that the additive groups Z and Z_e are isomorphic (hence a group may be isomorphic to one of its proper subgroups).
b) If $n < m$, show that the symmetric group S_m contains a subgroup isomorphic to S_n.

12. Prove that the permutations

$$\begin{pmatrix} 1 & 2 & 3 & 4 \\ 1 & 2 & 3 & 4 \end{pmatrix}, \quad \begin{pmatrix} 1 & 2 & 3 & 4 \\ 2 & 3 & 4 & 1 \end{pmatrix}, \quad \begin{pmatrix} 1 & 2 & 3 & 4 \\ 3 & 4 & 1 & 2 \end{pmatrix}, \quad \begin{pmatrix} 1 & 2 & 3 & 4 \\ 4 & 1 & 2 & 3 \end{pmatrix}$$

form a subgroup of S_4 which is isomorphic to the multiplicative group $G = \{1, -1, i, -i\}$, where $i^2 = -1$.

13. Let G be the set of all matrices in $M_2(R^\#)$ of the form $\begin{bmatrix} 1 & a \\ 0 & 1 \end{bmatrix}$. Prove that

a) G is a commutative group under matrix multiplication;

b) G is isomorphic to the additive group $R^{\#}$.

14. Given that f is a homomorphism from a simple group G onto a group G', establish that either $G \simeq G'$ or else f is the trivial homomorphism.

15. Verify that the following groups are all isomorphic:

a) the group Z_4 of integers modulo 4;

b) the subgroup $\{e, r_1, r_2, r_3\}$ of the group of symmetries of the square;

c) the group of four permutations

$$\begin{pmatrix}1 & 2 & 3 & 4\\1 & 2 & 3 & 4\end{pmatrix}, \quad \begin{pmatrix}1 & 2 & 3 & 4\\2 & 3 & 4 & 1\end{pmatrix}, \quad \begin{pmatrix}1 & 2 & 3 & 4\\3 & 4 & 1 & 2\end{pmatrix}, \quad \begin{pmatrix}1 & 2 & 3 & 4\\4 & 1 & 2 & 3\end{pmatrix};$$

d) the multiplicative group of matrices

$$\begin{bmatrix}1 & 0\\0 & 1\end{bmatrix}, \quad \begin{bmatrix}-1 & 0\\0 & -1\end{bmatrix}, \quad \begin{bmatrix}i & 0\\0 & -i\end{bmatrix}, \quad \begin{bmatrix}-i & 0\\0 & i\end{bmatrix}.$$

16. Show that the multiplicative group

$$G = \left\{1, \frac{-1 + i\sqrt{3}}{2}, \frac{-1 - i\sqrt{3}}{2}\right\},$$

where $i^2 = -1$, is isomorphic to the additive group Z_3.

17. Confirm that the multiplicative group

$$G = \{a + b\sqrt{2} | a, b \in Q;\ a \text{ and } b \text{ not both zero}\}$$

is isomorphic to itself under the mapping $f(a + b\sqrt{2}) = a - b\sqrt{2}$.

18. Prove that $(R^{\#}, +) \not\simeq (R^{\#} - \{0\}, \cdot)$.

19. For the multiplicative group G whose elements are the matrices

$$\begin{bmatrix}1 & 0\\0 & 1\end{bmatrix}, \quad \begin{bmatrix}-1 & 0\\0 & -1\end{bmatrix}, \quad \begin{bmatrix}1 & 0\\0 & -1\end{bmatrix}, \quad \begin{bmatrix}-1 & 0\\0 & 1\end{bmatrix},$$

$$\begin{bmatrix}0 & 1\\1 & 0\end{bmatrix}, \quad \begin{bmatrix}0 & -1\\-1 & 0\end{bmatrix}, \quad \begin{bmatrix}0 & 1\\-1 & 0\end{bmatrix}, \quad \begin{bmatrix}0 & -1\\1 & 0\end{bmatrix},$$

establish that the mapping $f(A) = \det A$ is a homomorphism of G onto $\{1, -1\}$, and determine its kernel.

20. Verify that the following groups are all isomorphic:

a) the Klein 4-group K_4;

b) the subgroup $\{e, r_2, h, v\}$ of the group of symmetries of the square;

c) the group of four permutations

$$\begin{pmatrix}1 & 2 & 3 & 4\\1 & 2 & 3 & 4\end{pmatrix}, \quad \begin{pmatrix}1 & 2 & 3 & 4\\2 & 1 & 3 & 4\end{pmatrix}, \quad \begin{pmatrix}1 & 2 & 3 & 4\\1 & 2 & 4 & 3\end{pmatrix}, \quad \begin{pmatrix}1 & 2 & 3 & 4\\2 & 1 & 4 & 3\end{pmatrix};$$

d) the multiplicative group of matrices

$$\begin{bmatrix}1 & 0\\ 0 & 1\end{bmatrix}, \quad \begin{bmatrix}-1 & 0\\ 0 & -1\end{bmatrix}, \quad \begin{bmatrix}1 & 0\\ 0 & -1\end{bmatrix}, \quad \begin{bmatrix}-1 & 0\\ 0 & 1\end{bmatrix};$$

e) the group of functions $g_i (i = 1, 2, 3, 4)$ from $R^{\#} - \{0\}$ onto itself, where $g_1(x) = x$, $g_2(x) = -x$, $g_3(x) = 1/x$, $g_4(x) = -1/x$.

21. Prove that $(R^{\#}, +) \simeq (R^{\#}_{+}, \cdot)$ but that $(Q, +) \not\simeq (Q_{+}, \cdot)$. [*Hint:* If $(Q, +) \simeq (Q_{+}, \cdot)$ using $f: Q_{+} \to Q$ and if $f(q) = 3$, then $f^2(q/2) = 3$.]

22. Verify that the following groups are all isomorphic:

a) the symmetric group S_3;

b) the multiplicative group of matrices

$$\begin{bmatrix}1 & 0\\ 0 & 1\end{bmatrix}, \quad \begin{bmatrix}0 & 1\\ 1 & 0\end{bmatrix}, \quad \begin{bmatrix}0 & 1\\ -1 & -1\end{bmatrix},$$

$$\begin{bmatrix}-1 & -1\\ 0 & 1\end{bmatrix}, \quad \begin{bmatrix}-1 & -1\\ 1 & 0\end{bmatrix}, \quad \begin{bmatrix}1 & 0\\ -1 & -1\end{bmatrix},$$

c) The group of functions $f_i (i = 1, 2, \ldots, 6)$ from $R^{\#} - \{0\}$ onto itself, where

$$f_1(x) = x, \qquad f_2(x) = 1/x, \qquad f_3(x) = 1 - x,$$

$$f_4(x) = \frac{1}{1-x}, \qquad f_5(x) = \frac{x-1}{x}, \qquad f_6(x) = \frac{x}{1-x}.$$

23. If $R^{\#}_{+}$ denotes the positive real numbers, then the product $R^{\#}_{+} \times \{1, -1\}$ becomes a group when multiplication is defined componentwise; that is,

$$(a_1, b_1)(a_2, b_2) = (a_1 a_2, b_1 b_2).$$

Establish the isomorphism

$$R^{\#}_{+} \times \{1, -1\} \simeq R^{\#} - \{0\}.$$

[*Hint:* Consider the function $f(r, 1) = r$, $f(r, -1) = -r$.]

24. Determine whether the following two multiplicative groups of matrices are isomorphic:

$$G_1 = \left\{\begin{bmatrix}1 & 0\\ 0 & 1\end{bmatrix}, \begin{bmatrix}-1 & 0\\ 0 & -1\end{bmatrix}, \begin{bmatrix}1 & 0\\ 0 & -1\end{bmatrix}, \begin{bmatrix}-1 & 0\\ 0 & 1\end{bmatrix}\right\},$$

$$G_2 = \left\{\begin{bmatrix}1 & 0\\ 0 & 1\end{bmatrix}, \begin{bmatrix}-1 & 0\\ 0 & -1\end{bmatrix}, \begin{bmatrix}1 & -2\\ 0 & -1\end{bmatrix}, \begin{bmatrix}-1 & 2\\ 0 & 1\end{bmatrix}\right\}.$$

25. Consider an arbitrary group G, and let j be the mapping of G onto itself defined by the rule $j(a) = a^{-1}$. Prove the following:

a) The function j is an automorphism of G if and only if G is commutative.

b) If G is an infinite cyclic group, then $\text{Aut}\, G = \{i_G, j\}$.

26. Show by example that a subgroup of a simple group need not be simple. [*Hint:* A_5 is simple, but $H = \{\alpha \in A_5 | \alpha(5) = 5\}$ is a subgroup isomorphic to A_4.]

27. With matrix multiplication as the operation, prove that the set

$$G = \left\{ \begin{bmatrix} a & 0 \\ b & 1 \end{bmatrix} \middle| a, b \in R^{\#};\, a \neq 0 \right\}$$

forms a group that is isomorphic to that of Example 2.19.

28. Let T be the set of all functions $f_n : Z \to Z$ defined by

$$f_n(x) = x + n, \qquad x \in Z.$$

a) Taking functional composition as the operation, show that T forms a cyclic group with generator f_1.

b) Use the technique in Cayley's theorem to obtain an isomorphism from the additive group Z onto T.

2.8 THE CLASSICAL ISOMORPHISM THEOREMS

In this section we shall discuss a number of significant results concerning the relationship between homomorphisms and quotient groups. Of these, Theorem 2.61, commonly known as the *Fundamental Homomorphism Theorem for Groups*, is perhaps the most crucial. The importance of this result would be hard to overemphasize, for it stands as the cornerstone upon which much of group theory rests.

We begin with a proof of the Factorization Theorem, stated here in a form best suited to our immediate needs.

THEOREM 2.60 (Factorization of Homomorphisms) *Let f be a homomorphism from the group G onto the group G', and let H be a normal subgroup of G such that $H \subseteq \ker f$. Then there exists a unique homomorphism $\bar{f}: G/H \to G'$ with the property that $f = \bar{f} \circ \text{nat}_H$.*

Proof. Before going into the details of the proof, let us remember that the symbol nat_H designates the natural mapping of G onto G/H; that is, $\text{nat}_H : G \to G/H$ with $\text{nat}_H(a) = aH$.

To start with, we define a function $\bar{f} : G/H \to G'$ called the *induced mapping*, by taking

$$\bar{f}(aH) = f(a), \qquad a \in G.$$

The first question to be raised concerns whether or not $\bar{f}$ is actually well defined. In other words, it must be established that this function depends only on the cosets of H and in no way on the particular representatives used. In order to do so, let $aH = bH$. Because the elements a and b lie in the same coset, $ab^{-1} \in H \subseteq \ker f$. The implication is that

$$f(a) = f(ab^{-1}b) = f(ab^{-1})f(b) = e'f(b) = f(b),$$

and so, by the manner in which $\bar{f}$ was specified, $\bar{f}(aH) = \bar{f}(bH)$. Thanks to this, $\bar{f}$ is a well-defined mapping.

The fact that $\bar{f}$ is a homomorphism is scarcely more than a restatement of the definition of multiplication in G/H:

$$\bar{f}((aH)(bH)) = \bar{f}(abH) = f(ab) = f(a)f(b) = \bar{f}(aH)\bar{f}(bH).$$

In this connection, note that for each element $a \in G$,

$$f(a) = \bar{f}(aH) = \bar{f}(\text{nat}_H(a)) = (\bar{f} \circ \text{nat}_H)(a),$$

whence the equality $f = \bar{f} \circ \text{nat}_H$. It remains only to establish the uniqueness of $\bar{f}$. Suppose also that $f = g \circ \text{nat}_H$ for some other function $g: G/H \to G'$. Then

$$\bar{f}(aH) = f(a) = (g \circ \text{nat}_H)(a) = g(aH)$$

for all $a \in G$, and so $\bar{f} = g$. The induced mapping $\bar{f}$ is thus the only function from the quotient group G/H into G' satisfying the equation $f = \bar{f} \circ \text{nat}_H$.

COROLLARY *The induced mapping $\bar{f}$ is an isomorphism if and only if* $\ker f \subseteq H$.

Proof. What is needed here is an explicit description of the kernel of $\bar{f}$, that is,

$$\begin{aligned} \ker \bar{f} &= \{aH \mid \bar{f}(aH) = e'\} \\ &= \{aH \mid f(a) = e'\} \\ &= \{aH \mid a \in \ker f\} = \text{nat}_H(\ker f) \end{aligned}$$

where, of course, e' indicates the identity of the group G'. With reference to Theorem 2.56, a necessary and sufficient condition for $\bar{f}$ to be an isomorphism is that $\ker \bar{f} = eH = H$. This amounts to the demand that $\text{nat}_H(\ker f) = H$, which in its turn is equivalent to the inclusion $\ker f \subseteq H$.

In view of the equality $f = \bar{f} \circ \text{nat}_H$, the conclusion of Theorem 2.60 is sometimes expressed by saying that the homomorphism f can be factored through the quotient group G/H or, alternatively, that f can be factored by nat_H. What we have proved, in a technical sense, is that there exists precisely one function $\bar{f}$ that makes the following triangle of mappings commutative:

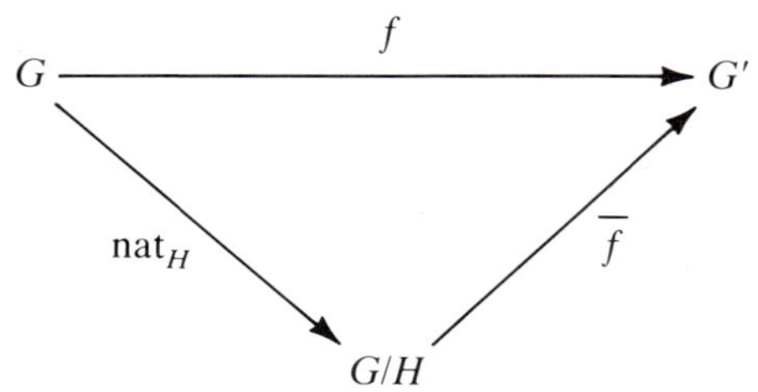

Example 2.54
Let us cite a specific instance of the Factorization Theorem. Our contention is that whenever the group G' is commutative, the homomorphism $f : G \to G'$ can be factored through the commutator quotient group $G/[G, G]$; otherwise stated,

$$f = \bar{f} \circ \text{nat}_{[G,G]}.$$

For this, consider any commutator $[a, b] = aba^{-1}b^{-1}$, where $a, b \in G$. As a consequence of the commutativity of G', we have

$$f([a, b]) = [f(a), f(b)] = e',$$

so that $[a, b] \in \ker f$. Now the commutator subgroup $[G, G]$ is, so to speak, the smallest subgroup of G to contain all the commutators. Thus we must conclude that $[G, G] \subseteq \ker f$. Having this inclusion, it is only necessary to apply Theorem 2.60 to obtain the desired factorization.

Although we did not expressly require the information in the proof of Theorem 2.60, it might be pointed out that the induced mapping $\bar{f}$ carries G/H onto the set G'. This is readily obtainable from the definition of $\bar{f}$ and the fact that f itself is an onto function. To be more specific, if $a' \in G'$, then $a' = f(a)$ for some element a in G; hence

$$a' = f(a) = \bar{f}(aH).$$

A rather simple observation, with far-reaching implications, is that whenever $H = \ker f$, so that both the Factorization Theorem and its corollary are applicable, f induces a mapping $\bar{f}$ under which G/H and G' are isomorphic groups. These considerations are formalized in the theorem below, a result that will be invoked on a number of occasions in the sequel.

THEOREM 2.61 (Fundamental Homomorphism Theorem) *If f is a homomorphism from the group G onto the group G', then $G/\ker f \simeq G'$.*

Remark. If in the statement of the theorem the word "onto" is replaced by "into," the conclusion takes the form $G/\ker f \simeq f(G)$.

Theorem 2.61 states that the images of G under homomorphisms can be duplicated, to within isomorphism, by quotient groups of G; to put it another way, every homomorphism of G is "essentially" a natural mapping. Thus the problem of finding all homomorphic images of a given group is reduced to the determination of its nonisomorphic quotient groups.

Another observation concerning Theorem 2.61 helps to shed some light on the nature of the quotient group $G/\ker f$. Specifically, the sets of elements in G that map onto particular elements of G' form the cosets of $\ker f$. To see this, suppose that $a, b \in G$ have the same image under f. Taking $f(a) = f(b)$ leads to

$$f(a^{-1}b) = f(a)^{-1}f(b) = e'$$

and, as a result, $a^{-1}b \in \ker f$; therefore $b = ak$ for some element $k \in \ker f$. This simply means that b lies in the coset $a \ker f$. On the other hand, choose $b \in a \ker f$; then $b = ak$ where $k \in \ker f$. We find without much trouble that

$$f(b) = f(ak) = f(a)f(k) = f(a)e' = f(a).$$

The point that we wish to make is that two elements of G have the same image in G' if and only if they belong to the same coset of $\ker f$. This can be described pictorially by Figure 2.6.

A few examples should furnish the best explanation of this last theorem.

Example 2.55

We saw earlier that the mapping $f: Z \to \{1, -1\}$, in which

$$f(n) = \begin{cases} 1 & \text{if } n \in Z_e \\ -1 & \text{if } n \in Z_o \end{cases}$$

is a homomorphism from the additive group Z of integers onto the multiplicative group $\{1, -1\}$. The kernel of this homomorphism is just the set Z_e, so that

$$Z/\ker f = Z/Z_e = \{Z_e, Z_o\}.$$

The Fundamental Theorem then guarantees that $\{Z_e, Z_o\}$ and $\{1, -1\}$ are isomorphic groups—indeed, this is fairly evident from an inspection

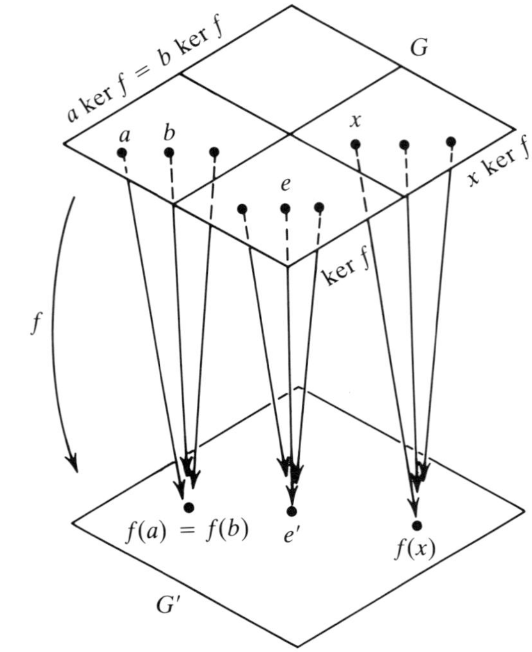

Figure 2.6

of their multiplication tables.

$\cdot$	Z_e	Z_o
Z_e	Z_e	Z_o
Z_o	Z_o	Z_e

$\cdot$	1	-1
1	1	-1
-1	-1	1

Further, the proof of the theorem indicates that the function $\bar{f}$ that actually accomplishes this isomorphism (the induced mapping) is given by

$$\bar{f}(Z_e) = f(0 + Z_e) = f(0) = 1,$$
$$\bar{f}(Z_o) = f(1 + Z_e) = f(1) = -1.$$

Example 2.56

For a more penetrating example, consider an arbitrary multiplicatively written group G and a fixed element $a \in G$. Define the mapping $f: Z \to G$ by the rule $f(n) = a^n$, $n \in Z$. It is not difficult to check that f, so defined, is a homomorphic mapping from the additive group Z of integers onto the cyclic subgroup (a). Hence by virtue of Theorem 2.61,

$$Z/\ker f \simeq (a),$$

where, in the situation at hand,

$$\ker f = \{n \in Z \mid a^n = e\}.$$

Now two possibilities arise according to the magnitude of the kernel, the first being that $\ker f = \{0\}$; in other words, $a^n = e$ implies $n = 0$. Under this circumstance, $Z/\ker f$ is just the group Z itself. On the other hand, if $\ker f \neq \{0\}$, there exists some least positive integer n for which $a^n = e$. One can easily deduce from this that $\ker f = (n)$, so we must have $Z/\ker f = Z_n$.

In summary, the preceding discussion reveals that (1) if the generator a is of infinite order, then $(a) \simeq Z$, and (2) if a is of finite order n, then $(a) \simeq Z_n$. These facts are already familiar; we have merely furnished an alternative approach to Theorem 2.58.

Example 2.57

We next pick an illustration from matrix theory. Under matrix multiplication the set G consisting of all 2×2 diagonal matrices with nonzero real entries forms a group; in symbols,

$$G = \left\{ \begin{bmatrix} a & 0 \\ 0 & b \end{bmatrix} \middle| a, b \in R^{\#} - \{0\} \right\}.$$

There is a homomorphism $f: G \to R^{\#} - \{0\}$ from G onto the multiplicative group $R^{\#} - \{0\}$ defined by

$$f\left(\begin{bmatrix} a & 0 \\ 0 & b \end{bmatrix}\right) = a.$$

For this mapping the kernel is

$$\begin{aligned} \ker f &= \left\{ \begin{bmatrix} a & 0 \\ 0 & b \end{bmatrix} \in G \middle| f\left(\begin{bmatrix} a & 0 \\ 0 & b \end{bmatrix}\right) = 1 \right\} \\ &= \left\{ \begin{bmatrix} a & 0 \\ 0 & b \end{bmatrix} \in G \middle| a = 1 \right\} \\ &= \left\{ \begin{bmatrix} 1 & 0 \\ 0 & b \end{bmatrix} \middle| b \in R^{\#} - \{0\} \right\}. \end{aligned}$$

Now a typical coset in the quotient group $G/\ker f$ looks like

$$\begin{aligned} \begin{bmatrix} r & 0 \\ 0 & s \end{bmatrix} \ker f &= \begin{bmatrix} r & 0 \\ 0 & s \end{bmatrix} \left\{ \begin{bmatrix} 1 & 0 \\ 0 & b \end{bmatrix} \middle| b \in R^{\#} - \{0\} \right\} \\ &= \left\{ \begin{bmatrix} r & 0 \\ 0 & x \end{bmatrix} \middle| x \in R^{\#} - \{0\} \right\}, \end{aligned}$$

which could alternatively be described as

$$\begin{bmatrix} r & 0 \\ 0 & 1 \end{bmatrix} \ker f.$$

The latter representation suggests an obvious mapping (in fact, an isomorphism) $\Phi : G/\ker f \to R^{\#} - \{0\}$, namely,

$$\Phi\left(\begin{bmatrix} r & 0 \\ 0 & 1 \end{bmatrix} \ker f\right) = r.$$

Of course, there is nothing particularly new here; Φ is just the induced mapping of Theorem 2.60.

Example 2.58

We cannot resist using the Fundamental Theorem to prove that A_n is a normal subgroup of S_n of order $n!/2$. To this purpose, let G denote the multiplicative group whose elements are 1 and -1. Define a function $f : S_n \to G$ by taking $f(\alpha) = 1$ if α is an even permutation and $f(\alpha) = -1$ if α is odd. Then f is a homomorphism from S_n onto G. Since

$$\ker f = \{\alpha \in S_n | f(\alpha) = 1\} = A_n,$$

the subgroup A_n is normal in S_n (Theorem 2.55). Falling back on the Fundamental Theorem, we get

$$S_n/A_n \simeq G = \{1, -1\}.$$

The implication is that the index of A_n in S_n is 2. An appeal to Lagrange's Theorem now completes the argument:

$$o(A_n) = o(S_n)/[S_n : A_n] = n!/2.$$

The next theorem not only provides further evidence of the power of the Fundamental Theorem, but is of independent interest, since it gives additional insight into the structure of the quotient group $G/\text{cent}\, G$. To prepare the way, it is necessary to digress for a moment.

Given a fixed element a of the group G, define the mapping $\sigma_a : G \to G$ by taking

$$\sigma_a(x) = axa^{-1}$$

for every x in G. Let us obtain a few of the special properties of this function. First, σ_a turns out to be a homomorphism: If $x_1, x_2 \in G$, then

$$\begin{aligned} \sigma_a(x_1x_2) &= a(x_1x_2)a^{-1} \\ &= (ax_1a^{-1})(ax_2a^{-1}) = \sigma_a(x_1)\sigma_a(x_2). \end{aligned}$$

The next thing to notice is that σ_a maps the set G onto itself; specifically, for any element $x \in G$, $\sigma_a(a^{-1}xa) = x$. Finally, it can be proved that σ_a is actually a one-to-one function. For this purpose, assume that $\sigma_a(x_1) = \sigma_a(x_2)$, so that $ax_1a^{-1} = ax_2a^{-1}$. The superfluous elements may be removed through the cancellation law, allowing us to conclude that $x_1 = x_2$. All these observations may be conveniently summarized by saying that $\sigma_a \in \text{Aut } G$.

Functions of the form σ_a, with $a \in G$, are usually called *inner automorphisms* of the group G; to be more precise, σ_a is the *inner automorphism induced by the element a*. For brevity we label the set of functions arising in this way by Inn G:

$$\text{Inn } G = \{\sigma_a | a \in G\}.$$

In the case of a commutative group, Inn G reduces to the identity mapping $i_G = \sigma_e$. Thus it is only when G is noncommutative that the notion becomes meaningful.

LEMMA *The set* Inn G *forms a group under functional composition, known as the group of inner automorphisms of G; in fact,* Inn G *is a normal subgroup of* Aut G.

Proof. The proof that Inn G is a group presents no difficulties; it consists of nothing more than noticing that, whenever $\sigma_a, \sigma_b \in \text{Inn } G$,

$$\sigma_a \circ \sigma_b = \sigma_{ab}.$$

In addition to disposing of the closure condition, this relation also indicates that σ_e is the identity element for the system and that $\sigma_a^{-1} = \sigma_{a^{-1}}$.

Regarding the second assertion, it suffices to show that if $f \in \text{Aut } G$, the product $f \circ \sigma_a \circ f^{-1}$ is an inner automorphism. The argument proceeds as follows: For each element x in G,

$$\begin{aligned}(f \circ \sigma_a \circ f^{-1})(x) &= f(\sigma_a(f^{-1}(x))) \\ &= f(af^{-1}(x)a^{-1}) \\ &= f(a)f(f^{-1}(x))f(a^{-1}) \\ &= f(a)xf(a)^{-1} = \sigma_{f(a)}(x).\end{aligned}$$

The reasons for each of these steps are reasonably self-evident, and the reader should make sure of understanding them. What is significant is that

$$f \circ \sigma_a \circ f^{-1} = \sigma_{f(a)} \in \text{Inn } G,$$

as was to be proved.

We are now in a position to obtain the result that has been our goal, namely, that the groups $G/\text{cent } G$ and Inn G are the same in all algebraic respects.

THEOREM 2.62 *For any group G, $G/\text{cent } G \simeq \text{Inn } G$.*

Proof. To begin, consider the mapping $g: G \to \text{Inn } G$, whereby $g(a) = \sigma_a$. That this function maps G onto the set Inn G is quite plain. It is equally easy to see that g is a homomorphism, since

$$g(ab) = \sigma_{ab} = \sigma_a \circ \sigma_b = g(a) \circ g(b), \qquad a, b \in G.$$

The crucial aspect of the proof is to show that $\ker g = \text{cent } G$; once this is finished, the desired conclusion will follow immediately from the Fundamental Theorem. In the case at hand, the inner automorphism σ_e serves as the identity element for the group Inn G. Accordingly, the kernel of the function g is defined by

$$\ker g = \{a \in G | g(a) = \sigma_e\} = \{a \in G | \sigma_a = i_G\}.$$

From the definition of equality of functions, we conclude that $a \in \ker g$ if and only if $axa^{-1} = x$ for every x in G. But this is the same as insisting that $ax = xa$ for all $x \in G$ or, in equivalent terms, that $a \in \text{cent } G$. Therefore $\ker g = \text{cent } G$, leading to the stated conclusion.

It is worth calling the reader's attention to the fact that a subgroup H of the group G is normal if and only if H is mapped into itself by every inner automorphism of G. Indeed, the condition that $aHa^{-1} \subseteq H$ for all $a \in G$ is no different from requiring that $\sigma_a(H) \subseteq H$ for all $\sigma_a \in \text{Inn } G$.

This may be a good place to pause and list a number of characterizations of normality:

THEOREM 2.63 *For any subgroup H of the group G, the following statements are equivalent:*

a) H is normal in G.
b) $aH = Ha$ for every $a \in G$.
c) $aHa^{-1} \subseteq H$ for every $a \in G$.
d) H is the kernel of some homomorphism of G.
e) $\sigma_a(H) \subseteq H$ for any inner automorphism σ_a of G.

Coming to the next item on the program, we again let f be an arbitrary homomorphism from the group G onto the group G'. Theorem 2.54 ensures that every subgroup H of the group G determines a subgroup $f(H)$ of the group G'. It goes without saying that group theory would be considerably simplified if the subgroups of G were in a one-to-one correspondence with those of G' in this manner. Unfortunately, this need not be the case.

The difficulty is reflected in the fact that if H and K are two subgroups of G with $H \subseteq K \subseteq H(\ker f)$, then $f(H) = f(K)$. The quickest way to see this is to note that

$$f(H) \subseteq f(K) \subseteq f(H \ker f) \subseteq f(H)f(\ker f) = f(H),$$

from which we infer that all the inclusions are actually equalities. In essence, we are observing that distinct subgroups of G may have the same homomorphic image in G'.

This disconcerting situation could be remedied either by demanding that $\ker f = \{e\}$ or else by narrowing our view to consider only subgroups H with $\ker f \subseteq H$. In either event the source of our trouble—the condition $H \subseteq K \subseteq H(\ker f)$—would reduce to the equality $H = K$. The first of the restrictions just cited has the effect of making the function f one-to-one, in which case G and G' are isomorphic groups; it then comes as no surprise to find their subgroups in one-to-one correspondence. The second possibility is the subject of our next theorem.

We first establish a preliminary lemma that will provide the key to later success.

LEMMA *Let f be a homomorphism from the group G onto the group G'. If H is any subgroup of G such that* $\ker f \subseteq H$, *then* $H = f^{-1}(f(H))$.

Proof. Pick a typical element $a \in f^{-1}(f(H))$, so that $f(a) \in f(H)$. Then $f(a) = f(h)$ for some choice of h in H. As a result, we will have $f(ah^{-1}) = e'$ or, equivalently, $ah^{-1} \in \ker f \subseteq H$. This implies that $a \in H$, yielding the inclusion $f^{-1}(f(H)) \subseteq H$. But the reverse inclusion always holds, and so the desired equality follows.

The relationship between the subgroups of G and those of any homomorphic image may now be recorded.

THEOREM 2.64 (Correspondence Theorem) *Let f be a homomorphism from the group G onto the group G'. Then there is a one-to-one correspondence between the subgroups H of G such that* $\ker f \subseteq H$ *and the set of all subgroups H' of G'; specifically, H' is given by $H' = f(H)$.*

Proof. Our first concern is to show that the indicated correspondence actually maps onto the set of all subgroups of G'. To put it another way, starting with an arbitrary subgroup H' of G', we must produce some subgroup H of the group G, with $\ker f \subseteq H$, such that $f(H) = H'$. The set $H = f^{-1}(H')$ certainly meets these requirements. By Theorem 2.54, $f^{-1}(H')$ forms a subgroup of G, and since $e' \in H'$,

$$\ker f = f^{-1}(e') \subseteq f^{-1}(H').$$

The corollary to Theorem 1.6 affirms that $f(f^{-1}(H')) = H'$, f being an onto mapping.

To finish off the proof, we argue that the correspondence in question is one-to-one. Suppose then that H and K are subgroups of G, with $\ker f \subseteq H$, $\ker f \subseteq K$ and $f(H) = f(K)$. Using the lemma above, we get

$$H = f^{-1}(f(H)) = f^{-1}(f(K)) = K,$$

as desired.

Before proceeding further, it should be remarked that the Correspondence Theorem remains valid if we replace the term "subgroup" throughout by "normal subgroup." That is to say, there is also a one-to-one correspondence between those normal subgroups of G that contain $\ker f$ and the set of all normal subgroups of G'. The additional argument needed to establish this fact is left for the reader to supply.

With regard to the incidence of subgroups in a quotient group, it should be noted that Theorem 2.64 applies in particular to the case in which we start with a normal subgroup H of the group G and take f to be the natural mapping $\text{nat}_H : G \to G/H$. Since $\ker(\text{nat}_H) = H$, the conclusion of the Correspondence Theorem is modified slightly:

COROLLARY *Let H be a normal subgroup of the group G. Then there is a one-to-one correspondence between the (normal) subgroups K of G such that $H \subseteq K$ and the set of all (normal) subgroups K' of the quotient group G/H; specifically, K' is given by $K' = \text{nat}_H K$.*

Example 2.59

By way of an application of these ideas, consider the following statement: If G is a finite cyclic group of order n, then G has exactly one subgroup of order m for each positive divisor m of n and no other subgroups. Although there is nothing very surprising about this assertion, our aim is to demonstrate its validity by using the corollary to the Correspondence Theorem.

First observe that, since the groups G and Z_n are isomorphic, there is no loss in generality in working with Z_n. Furthermore, as we have pointed out on several occasions, $Z_n = Z/(n)$. The corollary tells us that there is a one-to-one correspondence between those subgroups of the additive group Z that contain the set (n) and the subgroups of Z_n.

But the subgroups of Z are just the cyclic subgroups (m), where m is a non-negative integer. The outcome is that there is a one-to-one correspondence between the subgroups of Z_n and those subgroups (m) of Z with $(n) \subseteq (m)$; this last inclusion occurs if and only if m divides n.

The next two theorems are somewhat deeper results than usual and require the full force of our accumulated machinery. They are often called the

First and Second Isomorphism Theorems, and they have important applications in the sequel. The reader is cautioned that there seems to be no universally accepted agreement as to which of the two is the First Isomorphism Theorem.

THEOREM 2.65 *Let f be a homomorphism from the group G onto the group G' and let H be a normal subgroup of G. If* $\ker f \subseteq H$, *then* $G/H \simeq G'/f(H)$.

Proof. As a prefatory remark, we might point out that the corollary to Theorem 2.54 implies that $f(H)$ is a normal subgroup of the group G'; thus it makes sense to speak of the quotient group $G'/f(H)$.

Let us define the function $g: G \to G'/f(H)$ by $g = \text{nat}_{f(H)} \circ f$, where $\text{nat}_{f(H)}$ designates the usual natural mapping of G' onto $G'/f(H)$. In brief, g assigns to each element $a \in G$ the coset $f(a)f(H)$ in $G'/f(H)$. The following diagram may be of some help in visualizing this situation.

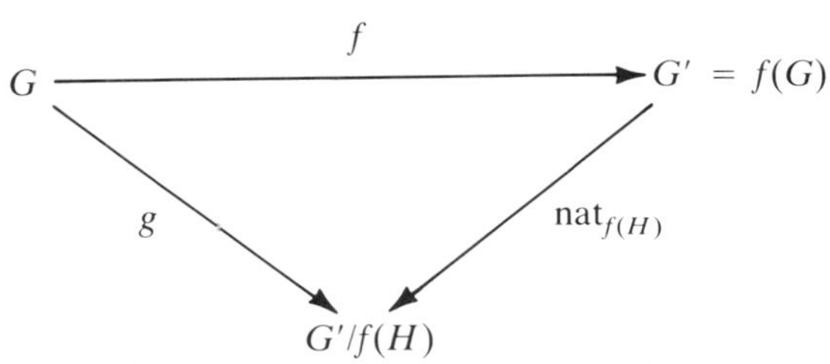

Since the functions f and $\text{nat}_{f(H)}$ are both onto homomorphisms, their combined effect is to produce a homomorphism from G onto the quotient group $G'/f(H)$.

The crux of the argument is to show that $\ker g = H$; then the desired conclusion would be an immediate consequence of the Fundamental Homomorphism Theorem. Since the identity element of $G'/f(H)$ is just the coset $e'f(H) = f(H)$, the kernel of g consists of those members of G that are mapped by g onto $f(H)$:

$$\begin{aligned}\ker g &= \{a \in G \mid g(a) = f(H)\}\\ &= \{a \in G \mid f(a)f(H) = f(H)\}\\ &= \{a \in G \mid f(a) \in f(H)\} = f^{-1}(f(H)).\end{aligned}$$

The hypothesis that $\ker f \subseteq H$ allows us to appeal to the lemma preceding Theorem 2.64, from which it may be inferred that $H = f^{-1}(f(H))$. But then $H = \ker g$, completing the proof.

When applying this result, one frequently starts with a normal subgroup of G' and utilizes inverse images rather than direct images. The theorem can then be rephrased as

COROLLARY 1 *Let f be a homomorphism from the group G onto the group G'. If H' is any normal subgroup of G', then $G/f^{-1}(H') \simeq G'/H'$.*

Proof. In compliance with the corollary to Theorem 2.54, $f^{-1}(H')$ forms a normal subgroup of G. Furthermore, $\ker f \subseteq f^{-1}(H')$, so that Theorem 2.65 leads directly to the isomorphism

$$G/f^{-1}(H') \simeq G'/f(f^{-1}(H')) = G'/H'.$$

Another special case, itself of interest, is the following.

COROLLARY 2 *Let H and K be two normal subgroups of the group G with $K \subseteq H$. Then H/K is a normal subgroup of G/K and $(G/K)/(H/K) \simeq G/H$.*

Proof. The corollary to Theorem 2.54 asserts that if H is a normal subgroup of the group G and f is any homomorphism of G, then $f(H)$ forms a normal subgroup of the homomorphic image $f(G)$. In the current setting, we take f to be the natural mapping $\text{nat}_K : G \to G/K$; then $\text{nat}_K H = H/K$ is a normal subgroup of the quotient group $G' = G/K$. (As the notation suggests, H/K is the collection of cosets of K having representatives in H.) Since $\ker(\text{nat}_K) = K \subseteq H$, Theorem 2.65 implies that G/H is isomorphic to $(G/K)/(H/K)$.

Our last isomorphism theorem concerns intersections and products of subgroups.

THEOREM 2.66 *If H and K are subgroups of the group G, with K normal in G, then*

$$H/(H \cap K) \simeq (HK)/K.$$

Proof. The first step should be to check that the quotient groups appearing in the statement of the theorem are actually defined. We leave to the reader the routine task of verifying that $H \cap K$ is a normal subgroup of H, that HK is a group, and that K is normal in HK.

The plan of action is to find a homomorphism f from H onto the quotient group $(HK)/K$ such that $\ker f = H \cap K$. Once this is accomplished, an appeal to the Fundamental Homomorphism Theorem would be in order. Our candidate for this function is defined by putting $f(a) = aK$, where $a \in H$. Since $H \subseteq HK$, f can be obtained by composing the injection map $i_H : H \to HK$ with the natural mapping $\text{nat}_K : HK \to (HK)/K$. To be quite explicit, $f = \text{nat}_K \circ i_H$ or, represented diagrammatically,

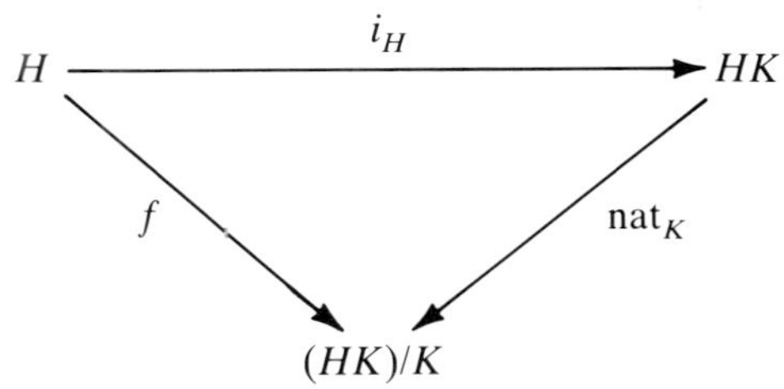

This factorization allows us to conclude that f is a homomorphism with image $f(H) = (HK)/K$. To ascertain that the kernel of f is precisely the set $H \cap K$, note that the coset $eK = K$ serves as the identity element of $(HK)/K$; in consequence,

$$\begin{aligned}\ker f &= \{a \in H \mid f(a) = K\} \\ &= \{a \in H \mid aK = K\} = \{a \in H \mid a \in K\} = H \cap K.\end{aligned}$$

All the pieces fit into place, and Theorem 2.61 can be applied to give the isomorphism

$$H/(H \cap K) = H/\ker f \simeq (HK)/K.$$

Example 2.60

For purposes of illustrating this last result, we fall back on the additive group Z of integers and the cyclic subgroups (3) and (4). Both these subgroups are normal in Z, the operation being commutative. It can be shown without too much difficulty that

$$(3) \cap (4) = (12) \qquad \text{and that} \qquad (3) + (4) = (1) = Z.$$

Knowing this, we can use Theorem 2.66 to secure the isomorphism $(3)/(12) \simeq Z/(4)$.

The notation tends to obscure the simplicity of our conclusion. The reader will doubtless recall that $Z/(4)$ is just the group of integers modulo 4. A closer examination of the cosets and operation in the quotient group $(3)/(12)$ reveals this to be none other than the group $(\{0, 3, 6, 9\}, +_{12})$. What we actually have is a disguised version of the isomorphism

$$(\{0, 3, 6, 9\}, +_{12}) \simeq (Z_4, +_4).$$

PROBLEMS 2.8

1. The following mapping f defines a homomorphism from the group G_s of symmetries of the square into the Klein 4-group K_4:

$$f(e) = f(r_2) = e, \qquad f(r_1) = f(r_3) = a$$

$$f(h) = f(v) = b, \qquad f(d_1) = f(d_2) = ab.$$

   a) Obtain the isomorphism $G_s/\text{cent } G_s \simeq K_4$.

   b) Write out the induced mapping $\bar{f}: G_s/\text{cent } G_s \to K_4$ that leads directly to this isomorphism.

2. Show that the Factorization Theorem implies that any group homomorphism can be expressed (nontrivially) as the composition of an onto function and a one-to-one function.

3. Let f be the mapping from the additive group Z onto the multiplicative group $\{1, -1, i, -i\}$ defined by

$$f(n) = i^n.$$

   Prove that $Z_4 \simeq Z/\ker f \simeq \{1, -1, i, -i\}$.

4. Suppose that G is the multiplicative group of 2×2 invertible matrices with real entries. If

$$H = \{A \in M_2(R^\#) | \det A = 1\},$$

   prove that G/H is isomorphic to the multiplicative group $R^\# - \{0\}$. [*Hint:* Consider the homomorphism $f(A) = \det A$.]

5. Consider the multiplicative group G whose elements are the six matrices

$$\begin{bmatrix} 1 & 0 \\ 0 & 1 \end{bmatrix}, \quad \begin{bmatrix} 0 & 1 \\ 1 & 0 \end{bmatrix}, \quad \begin{bmatrix} 1 & 0 \\ -1 & -1 \end{bmatrix},$$

$$\begin{bmatrix} 0 & 1 \\ -1 & -1 \end{bmatrix}, \quad \begin{bmatrix} -1 & -1 \\ 1 & 0 \end{bmatrix}, \quad \begin{bmatrix} -1 & -1 \\ 0 & 1 \end{bmatrix}$$

   and the set

$$H = \left\{\begin{bmatrix} 1 & 0 \\ 0 & 1 \end{bmatrix}, \begin{bmatrix} 0 & 1 \\ -1 & -1 \end{bmatrix}, \begin{bmatrix} -1 & -1 \\ 1 & 0 \end{bmatrix}\right\}.$$

   Establish that H is a normal subgroup of G and that $G/H \simeq \{1, -1\}$.

6. Find all the homomorphic images of the group G_s of symmetries of the square. [*Hint:* Obtain all groups G_s/H, where H is normal in G_s.]

7. The set

$$G = \{(a, b) \in R^\# \times R^\# | a^2 + b^2 \neq 0\}$$

   forms a group if multiplication is defined by

$$(a, b)(c, d) = (ac - bd, ad + bc).$$

Prove that if

$$H = \{(a, b) \in R^{\#} \times R^{\#} | a^2 + b^2 = 1\},$$

then the quotient group G/H is isomorphic to the multiplicative group $R^{\#}_{+}$ of positive real numbers.

8. Consider the additive group $R^{\#}$ and its cyclic subgroup (2π). Establish the isomorphism

$$R^{\#}/(2\pi) \simeq M,$$

where M is the multiplicative group of all 2×2 invertible matrices of the form

$$\begin{bmatrix} \cos x & \sin x \\ -\sin x & \cos x \end{bmatrix}, \qquad x \in R^{\#}.$$

9. If H is a subgroup of the group G, and no other subgroup of G has the same order as H, prove that H is normal in G. [*Hint:* H and $\sigma_a(H)$ have the same order.]

10. For each ordered pair (a, b) of real numbers, with $a \neq 0$, define a mapping $f_{(a,b)}$ of $R^{\#}$ into itself by $f_{(a,b)}(x) = ax + b$. If G denotes the set of all such functions, that is,

$$G = \{f_{(a,b)} | a, b \in R^{\#};\ a \neq 0\},$$

prove each of the statements below:

a) G forms a subgroup of Sym $R^{\#}$ that is isomorphic to the group of Example 2.19.

b) The set $H = \{f_{(1,b)} | b \in R^{\#}\}$ is a normal subgroup of G.

c) The quotient group G/H is isomorphic to the multiplicative group $R^{\#} - \{0\}$. [*Hint:* Consider the function Φ defined by $\Phi(f_{(a,b)}) = a$.]

11. Let G be the multiplicative group of nonzero real numbers and $H = \{1, -1\}$. Show that the quotient group G/H is isomorphic to the multiplicative group of positive real numbers. [*Hint:* Consider the homomorphism $f(a) = |a|$.]

12. Given a group G, establish the following facts concerning the inner automorphisms of G:

a) G is commutative if and only if Inn $G = \{i_G\}$.

b) If H is a subgroup of G, then $\sigma_a(H)$ is also a subgroup for all $a \in G$.

c) For any subgroup H of G, $K = \bigcap_{a \in G} \sigma_a(H)$ is the largest normal subgroup of G contained in H.

d) If the element $x \in G$ is such that $\sigma_a(x) = x$ for all $a \in G$, then the subgroup (x) is normal in G.

13. For $n \neq 4$, show that $S_n \simeq \text{Inn } S_n$.

14. Suppose that f is a homomorphism from the group G into itself which has the property of commuting with every inner automorphism of G; that is, $\sigma_a \circ f = f \circ \sigma_a$ for all $a \in G$. If the set H is defined by

$$H = \{x \in G \mid f(f(x)) = f(x)\},$$

prove the following statements:

a) H forms a normal subgroup of G.

b) The quotient group G/H is commutative. [*Hint:* $[G, G] \subseteq H$.]

15. For the group G_s of symmetries of the square:

a) Find all homomorphic images (to within isomorphism) of G_s.

b) Exhibit the correspondence between those subgroups H of G_s with cent $G_s \subseteq H$ and the set of all subgroups of G_s/cent G_s.

16. Given that f is a homomorphism from the group G onto the group G', verify the following assertions.

a) If H and K are subgroups of G with $\ker f \subseteq H \cap K$, then $f(H \cap K) = f(H) \cap f(K)$.

b) If $\ker f \subseteq [G, G]$, then $G/[G, G] \simeq G'/[G', G']$.

c) For any subgroup H of G, $f^{-1}(f(H)) = H(\ker f)$.

17. Let f be a homomorphism of the group G onto the group G', H a subgroup of G and H' a subgroup of G'. Prove that

a) If H' has finite index in G', then $[G : f^{-1}(H')] = [G' : H']$. [*Hint:* Suppose $x_1 H', \ldots, x_n H'$ are the distinct left cosets of H' in G'. Choose $a_i \in G$ such that $f(a_i) = x_i$ and show that $a_1 f^{-1}(H'), \ldots, a_n f^{-1}(H')$ are the distinct left cosets of $f^{-1}(H')$ in G.]

b) If H has finite index in G and $\ker f \subseteq H$, then $[G : H] = [G' : f(H)]$.

18. If $f: G \to Z_8$ is a homomorphism from a group G onto the group of integers modulo 8, establish that G has normal subgroups of index 2 and 4. [*Hint:* See Problem 17(a).]

19. Assume that H, K are subgroups of the group G, with K normal in G and $G = HK$. Verify that $G/K \simeq H$ if and only if $H \cap K = \{e\}$.

20. Let K be the normal subgroup of S_4 consisting of the four permutations

$$\begin{pmatrix} 1 & 2 & 3 & 4 \\ 1 & 2 & 3 & 4 \end{pmatrix}, \quad \begin{pmatrix} 1 & 2 & 3 & 4 \\ 2 & 1 & 4 & 3 \end{pmatrix}, \quad \begin{pmatrix} 1 & 2 & 3 & 4 \\ 3 & 4 & 1 & 2 \end{pmatrix}, \quad \begin{pmatrix} 1 & 2 & 3 & 4 \\ 4 & 3 & 2 & 1 \end{pmatrix}.$$

Derive the isomorphisms $S_4/K \simeq S_3$ and $A_4/K \simeq Z_3$. [*Hint:* For the first isomorphism, put $H = \{\alpha \in S_4 \mid \alpha(4) = 4\}$ and use Problem 19.]

21. a) Let H, K_1 and K_2 be subgroups of the group G, with $K_i (i = 1, 2)$ normal. If $H \cap K_1 = H \cap K_2$, prove that $(HK_2)/K_2 \simeq (HK_1)/K_1$.

b) Let H_1, H_2 and K be subgroups of the group G, with K normal. If $H_1 K = H_2 K$, prove that $H_1/(H_1 \cap K) \simeq H_2/(H_2 \cap K)$.

22. For a proof of Theorem 2.65 that does not depend on the Fundamental Homomorphism Theorem, define the function $g: G/H \to G'/f(H)$ by taking $g(aH) = f(a)f(H)$. Show that

a) g is a well-defined isomorphism from G/H onto $G'/f(H)$.

b) g is the unique mapping that makes the diagram below commutative:

$$\begin{array}{ccc} G & \xrightarrow{\;f\;} & G' = f(G) \\ {\scriptstyle \mathrm{nat}_H}\downarrow & & \downarrow{\scriptstyle \mathrm{nat}_{f(H)}} \\ G/H & \xrightarrow[\;g\;]{} & G'/f(H) \end{array}$$

3

Further Topics in Group Theory

As the prerogative of Natural Science is to cultivate a taste for observation, so that of Mathematics, almost from the starting point, is to stimulate the faculty of invention.

JAMES JOSEPH SYLVESTER

3.1 AUGUSTIN-LOUIS CAUCHY

Augustin-Louis Cauchy (1789–1857) was, along with Carl Friedrich Gauss, the major mathematician of the first decades of the nineteenth century. He played a central role in the shaping of permutation theory and, as a result, has some claim to be regarded as the founder of the theory of groups of finite order.

Cauchy acquired his early education from his father, a barrister and master of classical studies. In 1805 he entered the Ecole Polytechnique in Paris, proceeding two years later to the civil engineering school Ecole des Ponts et Chaussées (School of Bridges and Roads). Thereafter he worked briefly (1810–1813) as a military engineer at the harbor of Cherbourg, where Napoleon had started to build a naval base for his intended invasion of England. Cauchy first attracted the attention of the leading mathematicians of France, among them the aging Lagrange, by a series of brilliant memoirs that date from 1811, when he was but 22 years old. For reasons of health, Cauchy was persuaded by Lagrange to abandon the profession of engineering and to devote himself to pure science and mathematics. Cauchy returned to Paris in 1813 to accept an instructorship at the Ecole Polytechnique, rising to full professor in 1816. In the nineteenth century, scientific academies offered prizes for solving major outstanding problems. Cauchy won a grand prize from the French Academy of Sciences in 1816 for a paper on the propagation of waves at the surface of a liquid. Although he was worthy of admission to the Academy, Cauchy lost the respect of many scientists at this time by being appointed to the Academy by royal decree instead of being elected by its members. Moreover, the seat that became Cauchy's had been vacant because of the political expulsion of a noted geometer named Monge.

The changing political situation in France had unexpected effects on Cauchy's career. He was a religious conservative and an ardent royalist who supported the Bourbon monarchy almost to the point of perversity. When Louis-Philippe came to power following the Revolution of 1830, Cauchy refused, as a loyal supporter of the deposed King Charles X, to swear allegiance to the new government. Abandoning his academic post at the Ecole Polytechnique, Cauchy went into a self-imposed exile. He did not regain public employment for 18 years. Cauchy spent part of this time teaching at the University of Turin (1831–1833) and part in Prague (1833–1838), where the ex-monarch summoned Cauchy to tutor his grandson. In due time it pleased Charles X to make Cauchy a baron for his loyalty. When

Cauchy went back to France in 1838, he was permitted to resume his post at the Academy but not his professorship, since this involved taking an oath of allegiance. He had to be content to support himself by teaching at a Jesuit college. When Louis-Philippe was himself ousted in the Revolution of 1848 and oaths were no longer required, Cauchy was appointed professor of celestial mechanics at the Sorbonne, where he remained for the rest of his career.

Like Cayley's, Cauchy's mathematical output was prodigious and embraced almost all branches of the subject. In total he wrote eight full-length books as well as 789 papers, which in a modern edition fill 26 large volumes. Cauchy was so active that he had to establish a journal, the *Exercises de Mathematiques* (1826–1830), continued in a second series as *Exercises d'Analyse Mathematique et de Physique*, issues of which were often composed entirely of Cauchy's own work. In 1835 the Academy of Sciences began to publish a weekly bulletin, the *Comptes Rendus.* Cauchy inundated this journal with a succession of his papers to such an extent that the Academy became alarmed at the rapidly mounting printing bill; to cope with Cauchy's staggering productivity, a rule was passed (and is still in force today) limiting each contribution to a length of four pages. In all, the *Comptes Rendus* printed 589 articles by Cauchy.

Cauchy had universal mathematical interests. Perhaps his most brilliant achievement is the theory of functions of a complex variable, a field that from 1814 on he developed almost singlehandedly. Complex numbers had been used earlier by various mathematicians, but until Cauchy's work their legitimacy remained clouded. It was Cauchy's effort to give analysis an unimpeachable foundation that exerted the greatest influence and achieved for him the most recognition. After deriving an acceptable theory of limits, he rigorously defined—and proved theorems about—the central notions of continuity, convergence, differentiability, and integrability in terms of the limit concept. His definitions are essentially the ones found in modern textbooks on the calculus. Cauchy's foundations of calculus appear in the books based on the lectures he delivered at the Ecole Polytechnique, the most important one being his classic *Cours d'Analyse* of 1821. In the introduction, Cauchy himself said, "As to methods, I have tried to fill them with all the rigor one required in geometry." Trivial as the remark may appear today, it was revolutionary at the time; this new type of analytical reasoning, based on precise definitions and proofs for all the basic concepts, swept away the old foundations, which had relied largely on formal rules irrespective of any considerations of convergence. Beginning with the calculus, the critical spirit soon permeated all of analysis and, toward the end of the century, overhauled and recast the underpinnings of geometry itself. For initiating the movement to infuse rigor into mathematics, Cauchy has been called "the first of the moderns."

Cauchy can well be credited with making the theory of determinants into a formal branch of mathematics. In 1812 he published an 84-page memoir in which he placed this subject on a firm basis. The memoir brought the term "determinant" into popular use and contains the theorem relating to the product of two matrices: if A and B are both $n \times n$ matrices, then $\det(AB) = \det A \cdot \det B$.

In 1844, Cauchy published the earliest systematic account of permutation group theory as part of Volume 3 of his *Cours d'Analyse*. Filling some 100 pages, this presentation represented a synthesis of earlier results due to Cauchy, most notably a paper of 1815 in which he defined the terms "permutation" and "substitution." (Although Cauchy drew a distinction, most nineteenth century mathematicians tended to use these two words interchangeably.) Cauchy introduced the familiar two-row notation for a permutation—for example, $\begin{pmatrix} x & y & z \\ z & x & y \end{pmatrix}$—deleting the periods that he had placed between the letters in 1815. Already in 1815 he had defined the product of two permutations and recognized the identity mapping as the "identity permutation" (the term "inverse permutation" occurs only from 1844). Other lasting additions to the terminology, the words "cyclic permutation" and "transposition," date back to 1815. The cycle notation $(x \quad y \quad \cdots \quad z)$ appears first in the 1844 memoir, as does the representation of a permutation as a product of disjoint cycles. While the use of the word "group" in a nontechnical sense abounds in the earlier paper, it was only during his second period of activity along these lines that Cauchy introduced the phrase "a system of conjugate substitutions" when referring to a group; "conjugate" meant to him that the permutations were "joined together" in a group structure.

One of the fundamental results proved in the 1844 memoir is the assertion that every finite (permutation) group whose order is divisible by a prime p contains an element, and hence a subgroup, of order p. This result is usually called Cauchy's Theorem, and it constitutes the most difficult step in the proof of a series of results named for Ludwig Sylow and known as Sylow's Theorems. Cauchy established the first Sylow Theorem for the special case in which the group involved is the symmetric group S_n, to the extent of showing that if p^n is the highest power of p dividing $o(S_n)$, then there exists a subgroup of order p^n. Twenty-five years elapsed before Sylow published the general result without referring to the permutation group S_n.

Cauchy's work exerted considerable influence on the subsequent development of the theory of groups. By 1870, more than a dozen mathematicians were actively studying groups of permutations and had produced some 700 journal pages of articles on the subject. In 1870, Camille Jordan (1838–1922) brought out his classic *Traité des Substitutions*, the earliest separate textbook on (permutation) group theory. The *Traité* amassed the totality of knowledge in group theory at that time, including as well much new material of Jordan's own—for instance, a proof (in permutation-theoretic disguise) that the alternating group A_5 is simple. Since the *Traité* was written before the abstract theory was well established, there is no definition of an abstract group in it; nor does Cauchy's phrase "systems of conjugate substitutions" appear. For Jordan a "group" meant simply a finite set of permutations that is closed under multiplication.

Group theory remained a very specialized topic, known and understood by relatively few mathematicians, until the publication of the textbook *Substitutionentheorie* by Eugen Netto (1846–1919) in 1882. This work did much to place the theory of groups within reach of the mathematical public; aside from the journal articles of Cayley, the subject was little known in the United States until an English translation of the *Substitutionentheorie* appeared in 1892. The translation was the handiwork of a young algebraist from Harvard, Frank Nelson Cole (1861–1926), who had studied with Felix Klein at Leipzig. Cole later published the first American research paper on group theory (1894).

Augustin-Louis Cauchy (1789–1857)

3.2 DIRECT PRODUCTS OF GROUPS

In this section we pay special attention to finite groups, one of the richest and deepest branches of the whole of group theory. For no other general class of groups is the structure as completely known or as easily described. When one is setting up a structure theory, the overall strategy is to express—in some sense—the complicated algebraic systems in terms of those that are better behaved. We accomplish this in the present setting by introducing the concept of a direct product of groups. Our main theorem will then be to the effect that any finite commutative group can be split into the direct product of cyclic groups.

Let us begin by recalling that if H and K are subgroups of the group G, then the set HK consists of all products hk, with $h \in H$ and $k \in K$. While HK is not necessarily a subgroup of G, taking H and K to be normal in G is more than enough to assure that this will indeed be the case (Problem 15, Section 2.6). We are particularly interested in those situations in which HK is the entire group G. The significance of this assumption is that it enables us to write every member of G as a product of elements of H and K. If it also happens that H and K intersect *trivially* (in other words, $H \cap K = \{e\}$), then one speaks of G as being expressed as an internal direct product of these subgroups. To put the matter on a formal basis:

DEFINITION 3.1 Let H and K be normal subgroups of the group G. Then G is said to be the *internal direct product* of H and K, written $G = H \otimes K$, if $G = HK$ and $H \cap K = \{e\}$.

Before attempting to set out the theory of direct products in an orderly fashion, let us look at two examples.

Example 3.1
Consider the Klein 4-group $K_4 = \{e, a, b, ab\}$, where $a^2 = b^2 = e$ and $ab = ba$. Since K_4 is commutative, all its subgroups are normal. If H is the cyclic subgroup generated by the element a and K is the cyclic subgroup generated by b, then the conditions of Definition 3.1 are satisfied. Thus $K_4 = H \otimes K$ is the internal direct product of two cyclic subgroups of order 2.

Example 3.2
Some groups cannot be expressed as the internal direct product of two nontrivial normal subgroups (technically speaking, such groups are referred to as *indecomposable*). As an example of this situation, we mention any infinite cyclic group $G = (a)$. Suppose that $G = H \otimes K$, where H, K are nontrivial subgroups of G. Being subgroups of a cyclic group, H and K are themselves cyclic with, say, generators a^n and a^m,

respectively. Then the product $a^{nm} \in H \cap K$, showing that $H \cap K \neq \{e\}$, a contradiction to the definition of direct product. (The equality $a^{nm} = e$ is impossible; this would imply that G is finite.)

Several criteria exist for a group to be an internal direct product of its subgroups. We establish one such condition below.

THEOREM 3.1 *The group G is the internal direct product of its subgroups H and K if and only if*

1. *each element $x \in G$ can be uniquely expressed in the form $x = hk$, where $h \in H$ and $k \in K$; and*
2. *any element of H commutes with any element of K.*

Proof. As our starting point, suppose that $G = H \otimes K$. Since $G = HK$, any element x of G has the form $x = hk$, with $h \in H$ and $k \in K$. To see that this representation is uniquely determined by x, let $x = h_1 k_1$ for $h_1 \in H$, $k_1 \in K$. Then $hk = h_1 k_1$ or rather $h_1^{-1}h = k_1 k^{-1}$. Because $h_1^{-1}h \in H$ and $k_1 k^{-1} \in K$, each side of this last equation is simultaneously in H and K. From the definition of direct product, $H \cap K = \{e\}$, and so $h_1^{-1}h = e = k_1 k^{-1}$. Thus $h = h_1$ and $k = k_1$, proving condition 1. Now let $h \in H$ and $k \in K$ be arbitrarily given and consider the commutator $hkh^{-1}k^{-1}$. Knowing that K is a normal subgroup of G, we necessarily have $hkh^{-1} \in K$, and hence $(hkh^{-1})k^{-1} \in K$; similarly, the normality of H forces $h(kh^{-1}k^{-1}) \in H$. This puts the commutator $hkh^{-1}k^{-1}$ in $H \cap K = \{e\}$. As a result, $hkh^{-1}k^{-1} = e$, which leads to the required commutativity $hk = kh$.

For the other direction of the theorem we assume that the indicated conditions hold and prove that $G = H \otimes K$. It follows immediately from condition 1 that $G = HK$. Pick any $h \in H$ and $x \in G$. Then $x = h_1 k_1$ for some choice of $h_1 \in H$, $k_1 \in K$. Observe that

$$xhx^{-1} = (h_1 k_1)h(h_1 k_1)^{-1} = h_1 h h_1^{-1} \in H,$$

since the elements of H and K commute by condition 2. Thus the product xhx^{-1} belongs to H, in consequence of which H is normal in G. For reasons that are entirely similar to those just given, K is a normal subgroup of G. To complete the proof, we must show that $H \cap K = \{e\}$. Suppose that H and K have the element x in common; then x will have two representations as a member of HK:

$$x = xe \quad (x \in H,\ e \in K) \qquad \text{and} \qquad x = ex \quad (e \in H,\ x \in K).$$

The uniqueness of representation guaranteed by condition 1 would be violated unless $x = e$. Accordingly, $H \cap K = \{e\}$, and Theorem 3.1 is finished.

The next theorem exhibits the interplay between the notions of internal direct product and quotient group.

THEOREM 3.2 *If H and K are normal subgroups of the group G such that $G = H \otimes K$, then $G/H \simeq K$ and $G/K \simeq H$.*

Proof. Consider the mapping $f\colon G \to K$ defined as follows: If $x = hk$, with $h \in H$ and $k \in K$, then $f(x) = k$. It is to be noted that the uniqueness of this representation for x assures that f is well defined. We also see at once that f carries G onto K. Given two elements $x = hk$ and $y = h_1 k_1$ of G, the product

$$xy = (hk)(h_1 k_1) = (hh_1)(kk_1),$$

where the last equality makes use of (2) of Theorem 3.1. This entails that $f(xy) = kk_1 = f(x)f(y)$, which shows the mapping in question to be a homomorphism. Finally, we have

$$\begin{aligned} \ker f &= \{hk \in HK \mid f(hk) = e\} \\ &= \{hk \in HK \mid k = e\} = H. \end{aligned}$$

The Fundamental Homomorphism Theorem now enables us to assert that $G/H \simeq K$. In the same way the function g defined by $g(x) = g(hk) = h$ gives rise to the isomorphism $G/K \simeq H$.

An immediate consequence is

COROLLARY *If G is a finite group and if $G = H \otimes K$, then $o(G) = o(H)o(K)$.*

Proof. By Lagrange's Theorem the order of G is the product of the order of H and the index of H in G. The foregoing result shows that $[G:H] = o(K)$, so all is proved.

A quicker, but more sophisticated, route to Theorem 3.2 is via the Second Isomorphism Theorem; indeed, if $G = H \otimes K$, then

$$G/H = HK/H \simeq K/(H \cap K) = K/\{e\} = K.$$

Similarly, the quotient group $G/K \simeq H$.

If G is a finite group, the definition of internal direct product may be restated as the criterion of our next theorem. This result requires a "counting" lemma, which is of interest in its own right.

LEMMA *If H and K are finite subgroups of the group G, then*

$$o(HK) = \frac{o(H)o(K)}{o(H \cap K)}.$$

Proof. That the set HK contains at most $o(H)o(K)$ elements is clear from its definition. The problem is that there may be some duplication; to put the matter more precisely, it is entirely possible that a given element $hk \in HK$ might equal h_1k_1 for $h_1 \in H$, $k_1 \in K$, where $h_1 \neq h$ or $k_1 \neq k$. The idea is to show that this takes place as many times as there are elements in $H \cap K$.

Suppose that $x \in H \cap K$ and $hk \in HK$. We may then write $hk = (hx)(x^{-1}k)$, where hx belongs to H (since $x \in H$) and $x^{-1}k$ belongs to K (since $x \in K$). Thus the element hk is duplicated in HK at least $o(H \cap K)$ times. On the other hand, every duplication of hk arises in the manner just described. If $hk = h_1k_1$, then $x = h^{-1}h_1 = kk_1^{-1}$ lies in $H \cap K$; hence $h_1 = hx$, $k_1 = x^{-1}k$, and so $hk = (hx)(x^{-1}k)$. The implication is that hk appears in HK exactly $o(H \cap K)$ times. The number of distinct elements in HK is $o(H)o(K)$ divided by the number of times each element appears, namely, $o(H \cap K)$.

THEOREM 3.3 *Let H and K be normal subgroups of the finite group G with $o(G) = o(H)o(K)$. If either $G = HK$ or $H \cap K = \{e\}$, then $G = H \otimes K$.*

Proof. If $G = HK$, then the foregoing lemma implies that

$$o(G) = o(HK) = \frac{o(H)o(K)}{o(H \cap K)},$$

or rather $o(H \cap K)o(G) = o(H)o(K)$. But we are assuming that $o(G) = o(H)o(K)$. Hence $o(H \cap K) = 1$, and so $H \cap K = \{e\}$. This fact, together with $G = HK$, yields $G = H \otimes K$.

In the event that $H \cap K = \{e\}$, then the following equalities are obvious:

$$o(HK) = \frac{o(H)o(K)}{o(H \cap K)} = o(H)o(K) = o(G).$$

Since $HK \subseteq G$ with $o(HK) = o(G)$, one concludes that $HK = G$ and, again, $G = H \otimes K$.

To see how all this works in a specific instance, let us write the group Z_{10} as a direct product of subgroups. This particular group contains only two nontrivial (normal) subgroups, namely, $H = \{0, 5\}$ and $K = \{0, 2, 4, 6, 8\}$. Evidently, $o(Z_{10}) = 10 = 2 \cdot 5 = o(H)o(K)$. Since $H \cap K = \{0\}$, it follows from Theorem 3.3 that Z_{10} is decomposable into the internal direct product of H

and K. A quick check reveals that the elements of Z_{10} can be represented as

$$\begin{array}{ll} 0 = 0 +_{10} 0 & 5 = 5 +_{10} 0 \\ 1 = 5 +_{10} 6 & 6 = 0 +_{10} 6 \\ 2 = 0 +_{10} 2 & 7 = 5 +_{10} 2 \\ 3 = 5 +_{10} 8 & 8 = 0 +_{10} 8 \\ 4 = 0 +_{10} 4 & 9 = 5 +_{10} 4 \end{array}$$

where the right-hand sides are all in HK.

Thus far we have restricted our attention to the internal direct product of two subgroups. The definition lends itself to the following generalization:

DEFINITION 3.2 A group G is said to be the *internal direct product* of the normal subgroups H_1, H_2, ..., H_n, indicated by writing $G = H_1 \otimes H_2 \otimes \cdots \otimes H_n$ if

1. $G = H_1 H_2 \cdots H_n$; that is, every element $x \in G$ is a product $x = h_1 h_2 \cdots h_n$, with $h_i \in H_i$, and
2. $H_i \cap H_i' = \{e\}$, where $H_i' = H_1 \cdots H_{i-1} H_{i+1} \cdots H_n$, for each i.

In the interests of completeness we record the analog of Theorem 3.1; its proof is left as an exercise.

THEOREM 3.4 *The group G is the internal direct product of the subgroups $H_1, H_2, \ldots, H_n$ if and only if*

a) every element $x \in G$ is uniquely expressible in the form $x = h_1 h_2 \cdots h_n$, with $h_i \in H_i$;

b) $h_i \in H_i$ and $h_j \in H_j$ $(i \neq j)$ imply that $h_i h_j = h_j h_i$.

We digress a little to consider the following question: Given two groups G_1 and G_2, is there a group G that is the internal direct product of subgroups H and K isomorphic to G_1 and G_2? For an answer, take G to be the Cartesian product

$$G_1 \times G_2 = \{(a, b) | a \in G_1, b \in G_2\}.$$

We endow $G_1 \times G_2$ with the structure of a group by defining multiplication of elements componentwise; that is,

$$(a, b)(a', b') = (aa', bb'),$$

aa' being the product in G_1 and bb' that in G_2. The reader is left the straightforward verification that, in this way, $G_1 \times G_2$ really can be made into a group. This new group is called the *external direct product* of G_1 and G_2

and is denoted by $G_1 \times G_2$, the same notation as that used for the Cartesian product. (Since $G_1 \times G_2$ and $G_2 \times G_1$ are isomorphic as groups, we may speak of the direct product of two groups without specifying the order of the factors.) One finds that the identity element of $G_1 \times G_2$ is just (e_1, e_2), where e_i is the identity of G_i, and that the inverse of (a, b) is (a^{-1}, b^{-1}).

Next observe that $G_1 \times G_2$ contains normal subgroups that are isomorphic copies of G_1 and G_2, namely, the subgroups $H = G_1 \times \{e_2\}$ and $K = \{e_1\} \times G_2$. The only element of $G_1 \times G_2$ simultaneously of the forms (a, e_2) and (e_1, b) is the identity (e_1, e_2), whence H and K must intersect in the identity element. Furthermore, a typical member (a, b) of $G_1 \times G_2$ can be expressed as

$$(a, b) = (a, e_2)(e_1, b) \in HK,$$

which signifies that $G_1 \times G_2 = HK$. These remarks make it plain that $G_1 \times G_2$ is the internal direct product of its subgroups H and K:

$$G_1 \times G_2 = H \otimes K \qquad \text{with } G_1 \simeq H, \quad G_2 \simeq K.$$

For a somewhat different approach to the subject we now start with a group G that is decomposable into the internal direct product of two normal subgroups H and K: in symbols, $G = H \otimes K$. Construct the external direct product $H \times K$ of the groups H and K. Since $G = HK$, every element $x \in G$ is uniquely expressible in the form $x = hk$ with $h \in H$, $k \in K$. It is therefore possible to define a function $f\colon G \to H \times K$ by setting

$$f(x) = f(hk) = (h, k).$$

Our purpose is to show that f is an isomorphism of G onto $H \times K$, so that $G \simeq H \times K$.

Pursuing this aim, let $x = hk$ and $y = h'k'$ be any two members of G. In compliance with Theorem 3.1 the elements of H commute with those of K, and so $xy = (hk)(h'k') = (hh')(kk')$. This means that

$$f(xy) = (hh', kk') = (h, k)(h', k') = f(x)f(y),$$

confirming that f is a homomorphism. It is equally clear that $f(G) = H \times K$.

To verify that f is a one-to-one mapping, suppose that $x = hk$ belongs to the kernel of f. Then

$$f(x) = f(hk) = (h, k) = (e, e).$$

The criterion for equality of ordered pairs forces $h = k = e$, whence $x = e$. Accordingly, $\ker f = \{e\}$ and, as a result, f is one-to-one.

Altogether, we have proved the following theorem.

THEOREM 3.5 *Let G be a group with normal subgroups H and K such that $G = H \otimes K$. Then $G \simeq H \times K$.*

The notion of external direct product can easily be broadened to allow for three or more factors. Let $G_1, G_2, \ldots, G_n$ be a finite collection of groups (not necessarily distinct) and consider their Cartesian product

$$G = G_1 \times \cdots \times G_n = \{(a_1, \ldots, a_n) | a_i \in G_i\}.$$

Multiplication is defined in G by

$$(a_1, \ldots, a_n)(b_1, \ldots, b_n) = (a_1 b_1, \ldots, a_n b_n),$$

where $a_i b_i$ denotes the product in G_i. Under this binary operation G is a group, known as the external direct product of the set $\{G_i\}$. As before, G contains subgroups H_i that are isomorphic to G_i for $i = 1, 2, \ldots, n$ and such that the group $G = H_1 \otimes \cdots \otimes H_n$. We need only take H_i to be the set of elements of G having all components equal to the identity except for the ith; in other words, if e_i denotes the identity of G_i, let

$$H_i = \{(e_1, \ldots, e_{i-1}, a, e_{i+1}, \ldots, e_n) | a \in G_i\}.$$

To summarize, the external direct product G of the groups G_i $(i = 1, 2, \ldots, n)$ is also the internal direct product of the subgroups H_i, where for each i, $G_i \simeq H_i$. If one agrees to identify groups that are isomorphic, then the distinction between external and internal direct products disappears. In the future we shall use the term *direct product* (without the qualifying adjective) for both, leaving the context to clarify which one is meant.

Let us take advantage of these ideas to obtain a direct product decomposition of the group Z_n of integers modulo n. First, a collateral result:

LEMMA *If* $\gcd(m, n) = 1$, *then* $Z_{mn} \simeq Z_m \times Z_n$.

Proof. In the group Z_{mn} the integer n generates a cyclic subgroup

$$H = \{0, n, 2n, \ldots, (m-1)n\}$$

of order m. Likewise, the integer m in Z_{mn} generates a cyclic subgroup

$$K = \{0, m, 2m, \ldots, (n-1)m\},$$

having order n. Clearly, $o(Z_{mn}) = mn = o(H)o(K)$. If the integer $a \in H \cap K$, then $m|a$ and $n|a$. But m and n are relatively prime; hence $mn|a$, and consequently $a \equiv 0 \pmod{mn}$; viewed otherwise, $H \cap K = \{0\}$. An appeal to Theorem 3.3 being legitimate, we see in this way that

$$Z_{mn} = H \otimes K.$$

Theorem 2.58 now tells us that $H \simeq Z_m$ and $K \simeq Z_n$, from which the stated result is apparent. (Indeed, the reader will experience no difficulty in showing that if f and g are isomorphisms from H onto Z_m and K onto Z_n, respectively, then $H \otimes K \simeq Z_m \times Z_n$ via the mapping F, where $F(hk) = (f(h), g(k))$.)

This enables us to prove the following.

THEOREM 3.6 *If the integer* $n > 1$ *has the prime factorization* $n = p_1^{k_1} p_2^{k_2} \cdots p_r^{k_r}$ *(distinct primes), then*

$$Z_n \simeq Z_{n_1} \times Z_{n_2} \times \cdots \times Z_{n_r}$$

where $n_i = p_i^{k_i}$ for $i = 1, 2, \ldots, r$.

Proof. The assertion is trivial when $r = 1$. In general, we argue by induction on r, the number of distinct primes appearing in the prime factorization of n. Let $r > 1$ and assume that the theorem is true for all integers containing $r - 1$ distinct primes. From this we wish to prove that the result holds for $n = p_1^{k_1} \cdots p_{r-1}^{k_{r-1}} p_r^{k_r}$. Put $m_1 = p_1^{k_1} \cdots p_{r-1}^{k_{r-1}}$ and $m_2 = p_r^{k_r}$, so $n = m_1 m_2$ with $\gcd(m_1, m_2) = 1$. According to the previous lemma,

$$Z_n \simeq Z_{m_1} \times Z_{m_2}.$$

Our induction hypothesis states that

$$Z_{m_1} \simeq Z_{n_1} \times Z_{n_2} \times \cdots \times Z_{n_{r-1}}.$$

When we set $n_r = m_2$, the result is

$$Z_n \simeq Z_{n_1} \times Z_{n_2} \times \cdots \times Z_{n_r},$$

where the integers $n_i = p_i^{k_i}$ for every $i = 1, 2, \ldots, r$.

The task that lies ahead is to generalize this theorem from Z_n to arbitrary finite commutative groups. That is, we intend to prove that any commutative group of order $n = p_1^{k_1} p_2^{k_2} \cdots p_r^{k_r}$ splits into a direct product of r cyclic subgroups of orders $p_1^{k_1}, p_2^{k_2}, \ldots, p_r^{k_r}$. As a step toward this goal, it is expedient to introduce a new class of groups, the p-groups. These groups are not only essential to the whole theory of this section, but become all the more so in the next section when we develop the Sylow Theorems.

DEFINITION 3.3 Let p be a prime number. A group G is said to be a *p-group* if the order of each element of G is some power of p (not necessarily the same power).

Example 3.3
The group G_s of symmetries of the square is a p-group, with $p = 2$. Each of its elements has order 2, except r_1 and r_3, which have order 4.

Example 3.4
If p is a prime, then Z_{p^n} is a p-group for any $n > 0$. This follows directly from Lagrange's Theorem: Every element of Z_{p^n} must have order

dividing p^n, the order of the group; hence every element has order a power of p.

Example 3.5
Let G be a commutative group. For a fixed prime p, define

$$G_p = \{a \in G | o(a) = p^n \text{ for some } n\}.$$

Our assertion is that G_p forms a subgroup of G which, by its nature, is automatically a p-group. We establish that G_p is a subgroup by applying the usual subgroup criterion (quite possibly, $G_p = \{e\}$). Suppose then that a, b are elements of G_p with respective orders p^r and p^s. Since G is commutative,

$$(ab^{-1})^{p^{r+s}} = (a^{p^r})^{p^s}(b^{p^s})^{-p^r} = e.$$

Taking Theorem 2.34 into account, we infer that the order of ab^{-1} divides p^{r+s} and therefore will be a power of p. This puts ab^{-1} into G_p, which is the desired conclusion.

We should point out that this result is false for noncommutative groups; for instance, in the symmetric group S_3, the elements having order some power of 2 do not form a group.

The only additional remark called for here is that G_p is the largest (in the sense of inclusion) p-subgroup of G. Indeed, if H is an arbitrary p-subgroup, then every element of H has order a power of p; hence by the definition of G_p, each element of H must lie in G_p; that is to say, $H \subseteq G_p$.

In the course of the next section we shall characterize finite p-groups as those groups having p-power order. It will be convenient to pause and derive the commutative case now. For this a brief lemma is needed.

LEMMA *Let H be a normal subgroup of the group G. If G is a p-group, then both H and G/H are p-groups.*

Proof. The assertion is obvious for H, even without the normality requirement. That the quotient group G/H forms a p-group follows directly from the fact that $o(aH)$ divides $o(a)$ for all $a \in G$.

THEOREM 3.7 *A finite commutative group G is a p-group if and only if $o(G)$ is a power of p.*

Proof. From Lagrange's Theorem it is clear that any group of order p^n is a p-group. Concentrating on the less obvious direction of the theorem, let G be a p-group. We prove that G has p-power order by using induction on $o(G)$. That is, we assume this to be true for all finite commutative p-groups of orders less than that of G. The stated result is obvious for groups of order 1, which starts the inductive process. Let $o(G) > 1$. If G has no nontrivial subgroups, then G is cyclic, and any generator is an element of p-power order; hence G will have order a power of p. On the other hand, when G contains a nontrivial subgroup H, then both $o(H)$ and $o(G/H)$ are strictly less than $o(G)$. Apply the induction assumption (as one may by the previous lemma) to conclude that H and G/H each have p-power order. Since

$$o(G) = o(H)o(G/H),$$

it follows that the order of G is necessarily a power of p, and the proof is completed.

With the needed preliminaries out of the way, we are now ready to deal with the central topic of the present section, to wit, the description of all finite commutative groups. The first step in this program ties the structure of such groups to that of p-groups.

THEOREM 3.8 *Any finite commutative group $G \neq \{e\}$ is the direct product of p-groups.*

Proof. Let $o(G) = n$. We claim that if $n = p_1^{k_1} p_2^{k_2} \cdots p_r^{k_r}$ is the prime factorization of n, then G is the direct product of the subgroups G_{p_i} for $i = 1, 2, \ldots, r$ (notation as in Example 3.5). To support this contention, let us apply the criterion of Definition 3.2. First, it is necessary to show that an arbitrary element $a \in G$ can be represented in the form $a = a_1 a_2 \cdots a_r$, where $a_i \in G_{p_i}$. If a has order m, then the corollary to Theorem 2.41 asserts that $m|n$; hence $m = p_1^{n_1} p_2^{n_2} \cdots p_r^{n_r}$ for certain integers $n_i (0 \leqslant n_i \leqslant k_i)$. Put $m_i = m/p_i^{n_i}$. The m_i are relatively prime, and so integers u_i exist for which the sum $u_1 m_1 + u_2 m_2 + \cdots + u_r m_r = 1$. Then

$$a = a^{u_1 m_1 + \cdots + u_r m_r} = a^{u_1 m_1} a^{u_2 m_2} \cdots a^{u_r m_r}.$$

Setting $a_i = a^{u_i m_i}$ for $i = 1, 2, \ldots, r$, it follows that $a = a_1 a_2 \cdots a_r$, and if $v_i = p_i^{n_i}$, then

$$a_i^{v_i} = (a^{u_i m_i})^{v_i} = a^{u_i m} = (a^m)^{u_i} = e^{u_i} = e.$$

Being of p_i-power order, the element a_i necessarily lies in G_{p_i}. All in all, we have shown that

$$G = G_{p_1} G_{p_2} \cdots G_{p_r}.$$

The proof will be finished as soon as we can establish that each G_{p_i} has trivial intersection with the product of the remaining ones. To see that this is so, choose any element

$$a \in G_{p_i} \cap (G_{p_i} \cdots G_{p_{i-1}} G_{p_{i+1}} \cdots G_{p_r}).$$

On one hand, $a^{p_i^s} = e$ for some integer s. On the other hand, a can be expressed as a product $a = b_1 \cdots b_{i-1} b_{i+1} \cdots b_r$ where the order of b_j is a power of p_j; say, to be definite, that by raising b_j to the $p_j^{t_j}$th power one can produce the identity. Letting $t = p_1^{t_1} \cdots p_r^{t_r}$, we can get $a^t = e$. Furthermore, p_i^s and t are relatively prime, so $up_i^s + vt = 1$ for appropriate integers u and v. This forces

$$a = a^{up_i^s + vt} = (a^{p_i^s})^u (a^t)^v = e,$$

which proves what we wanted.

We can recast Theorem 3.8 in a form that gives more precise information:

COROLLARY *If the commutative group* $G \neq \{e\}$ *has order* $n = p_1^{k_1} p_2^{k_2} \cdots p_r^{k_r}$ *and if* $G_{p_i} = \{x \in G \mid o(x)$ *is a power of* $p_i\}$, *then*

$$G = G_{p_1} \otimes G_{p_2} \otimes \cdots \otimes G_{p_r}$$

where $o(G_{p_i}) = p_i^{k_i}$ *for every* i.

Proof. Only the last statement needs further justification. By Theorem 3.7, $o(G_{p_i}) = p_i^{j_i}$ for some nonnegative integer j_i. Since the order of the direct product of a finite number of subgroups is the product of their respective orders, we obtain

$$\begin{aligned} p_1^{k_1} p_2^{k_2} \cdots p_r^{k_r} &= o(G) = o(G_{p_1}) o(G_{p_2}) \cdots o(G_{p_r}) \\ &= p_1^{j_1} p_2^{j_2} \cdots p_r^{j_r}. \end{aligned}$$

The uniqueness of the prime factorization of $o(G)$ then implies that $j_i = k_i$ for every $i = 1, 2, \ldots, r$; hence $o(G_{p_i}) = p_i^{k_i}$, as asserted.

The preceding proof establishes that a finite commutative group G is the direct product of its subgroups G_p for all primes p such that $G_p \neq \{e\}$, which is simply to say for all primes p dividing $o(G)$. This reduces the study of arbitrary finite commutative groups to the study of finite commutative p-groups. The basic result on p-groups from which the whole structure theory can be pinned down is derived below.

THEOREM 3.9 *A finite commutative* p*-group* G *is the direct product of a finite number of cyclic* p*-subgroups.*

Proof. It is already known that $o(G) = p^n$ for some $n \geqslant 0$. Our method is to perform induction on n. The starting point is the case in which $n = 1$, the theorem being trivial when $n = 0$; here there is really nothing to prove, since G is cyclic of order p. Next, suppose that $n > 1$ and that the result in question holds for all groups of order p^m with $m < n$.

Since G is a p-group, the order of every element of G is necessarily a power of p. Let $a \in G$ be an element of maximal order, say $o(a) = p^k$ $(k \leqslant n)$, and consider the cyclic subgroup $H = (a)$ generated by a. If $k = n$, then $H = G$, so G itself is cyclic; in such an event the theorem is trivial. Thus we may assume that $k < n$. The implication is that the quotient group G/H will have order $p^{n-k} < o(G)$. By the hypothesis of our induction, G/H is the direct product

$$G/H = \bar{H}_1 \otimes \bar{H}_2 \otimes \cdots \otimes \bar{H}_r$$

of cyclic p-subgroups $\bar{H}_i$ $(i = 1, 2, \ldots, r)$.

Let these subgroups $\bar{H}_i$ be generated by the cosets b_iH and let their respective orders be p^{j_i}. The next step is to produce a representative c_i of b_iH such that the order of c_i in G is equal to the order of b_iH in G/H. As a temporary device, let us write b for b_i and j for j_i. Since $(bH)^{p^j} = H$, the element b^{p^j} lies in H and hence is some power of a, say $b^{p^j} = a^s$. By the maximal nature of k, $o(b) \leqslant p^k$, so $b^{p^k} = e$; as a result

$$e = b^{p^k} = (b^{p^j})^{p^{k-j}} = (a^s)^{p^{k-j}} = a^{sp^{k-j}}.$$

This reveals that $p^k | sp^{k-j}$. Thus there exists an element t with $tp^k = sp^{k-j}$, leading to $tp^j = s$. If we now set $c = ba^{-t}$, then $cb^{-1} \in H$ and

$$c^{p^j} = (ba^{-t})^{p^j} = b^{p^j}a^{-tp^j} = a^s a^{-s} = e.$$

If we restore the index i, the situation stands as follows: For each i $(i = 1, 2, \ldots, r)$ there is an element $c_i \in G$ and an integer $v_i = p^{j_i}$ with the properties

1. $c_iH = b_iH$ and
2. $c_i^{v_i} = e$.

We contend that if K is the subgroup generated by $c_1, c_2, \ldots, c_r$, then $G = H \otimes K$. Starting with an arbitrary $x \in G$, we can write the coset $xH \in G/H$ as

$$xH = (b_1H)^{n_1}(b_2H)^{n_2} \cdots (b_rH)^{n_r}$$

for suitable integers n_i. Since $b_iH = c_iH$ for $i = 1, 2, \ldots, r$, we obtain

$$xH = (c_1H)^{n_1}(c_2H)^{n_2} \cdots (c_rH)^{n_r}.$$

This is scarcely more than a statement that

$$x = c_1^{n_1} c_2^{n_2} \cdots c_r^{n_r} y$$

for some choice of y in H. In consequence the element $x \in HK$, thereby confirming that $G = HK$.

There remains the verification that $H \cap K = \{e\}$. Pick a typical element x in $H \cap K$. Because $x \in K$, integers m_i exist for which

$$x = c_1^{m_1} c_2^{m_2} \cdots c_r^{m_r},$$

and so, passing to the quotient group G/H,

$$xH = (c_1 H)^{m_1} (c_2 H)^{m_2} \cdots (c_r H)^{m_r}.$$

When we use the fact that $c_i H = b_i H$ (as well as the assumption that $x \in H$), this equation translates into

$$H = (b_1 H)^{m_1} (b_2 H)^{m_2} \cdots (b_r H)^{m_r}.$$

Since G/H is the direct product of the cyclic subgroups $\bar{H}_i$ with generators $b_i H$, the uniqueness of representation of an element implies that each factor on the right-hand side must equal the identity of G/H; that is, $(b_i H)^{m_i} = H$ for $i = 1, 2, \ldots, r$. The conclusion to be drawn is that p^{j_i}, the order of the coset $b_i H$, divides m_i. But this in its turn yields $c_i^{m_i} = e$ (recall that c_i raised to the p^{j_i}th power gives the identity) and hence our expression for x reduces to $x = e$.

Having exhibited that $G = H \otimes K$, our proof proceeds rapidly to its conclusion. The induction assumption tells us that K admits a direct product decomposition into cyclic p-subgroups. Inasmuch as H is itself cyclic, the net result is that G is expressible as a direct product of cyclic p-subgroups. This is the contention of the theorem.

Before we go further, it is worth remarking that a given finite commutative p-group G can usually be written as a direct product of cyclic subgroups in several ways. However, the number of subgroups and their orders are the same in all such representations; more precisely, if

$$G = (a_1) \otimes \cdots \otimes (a_n) = (b_1) \otimes \cdots \otimes (b_m),$$

then $n = m$ and, after renumbering, $o(a_i) = o(b_i)$. This implies that the decomposition of G is unique up to isomorphism of the factors.

Theorems 3.8 and 3.9 may be brought together to give a complete description of the groups under consideration. Without delay we present what is sometimes called the *Fundamental Theorem for Finite Commutative Groups*:

THEOREM 3.10 (Frobenius) *Every finite commutative group is the direct product of finitely many cyclic p-subgroups.*

Switching to external direct products and using the fact that any cyclic group of order k is isomorphic to Z_k, one could just as well express Theorem 3.10 as follows:

COROLLARY 1 *If G is a finite commutative group, then*

$$G \simeq Z_{p_1^{n_1}} \times Z_{p_2^{n_2}} \times \cdots \times Z_{p_r^{n_r}}$$

for suitable primes p_i, not necessarily distinct, and integers $n_i \in Z_+$.

So far, we have learned that any finite commutative group is a direct product of cyclic groups, each of which is a p-group. There can be no further factorization, since a finite cyclic group, say $G = (a)$, of p-power order is indecomposable. To clarify this point, let $o(G) = p^n$, $n \geqslant 1$, and suppose, in anticipation of contradiction, that G has the (nontrivial) representation $G = H \otimes K$. Then $o(H) = p^j$ and $o(K) = p^k$, where $j + k = n$ $(1 \leqslant j, k < n)$. If m is the larger of j and k, then certainly $m < n$. By writing the generator a as $a = hk$, with $h \in H$ and $k \in K$, it follows that

$$a^{p^m} = (hk)^{p^m} = h^{p^m}k^{p^m} = e.$$

This being so, $o(G) \leqslant p^m$, which is absurd. These considerations allow us to rephrase Theorem 3.10 in the following form, which we state as a corollary:

COROLLARY 2 *Any finite commutative group is the direct product of a finite number of indecomposable cyclic groups.*

We are now in a position to identify, up to isomorphism, all possible commutative groups of a given order. By Frobenius' Theorem, any finite commutative group G is the direct product of prime-power order cyclic groups. Thus it will be sufficient to consider the case wherein $o(G) = p^n$, p a prime. In terms of external direct products, Theorem 3.10 may be viewed as asserting that

$$G \simeq Z_{p^{n_1}} \times Z_{p^{n_2}} \times \cdots \times Z_{p^{n_r}}.$$

Since the order of G is the product of the orders of these $Z_{p^{n_i}}$, it follows that $p^n = p^{n_1 + n_2 + \cdots + n_r}$, or rather, $n = n_1 + n_2 + \cdots + n_r$.

For any positive integer n a summation $n = n_1 + n_2 + \cdots + n_r$, where $0 < n_i \leqslant n_j$ for $i < j$, is called a *partition* of n. The remarks of the last paragraph indicate that each commutative group of order p^n gives rise to a partition of n. From the opposite point of view, starting with a partition $n = n_1 + \cdots + n_r$, we can always produce a commutative group G of order p^n; simply take G to be the external direct product of $Z_{p^{n_1}}, \ldots, Z_{p^{n_r}}$. In brief, there are just as many nonisomorphic commutative groups of order p^n as there are ways to express n as a sum of positive integers. There are, to cite an example, only two commutative groups of order p^2, either Z_{p^2} or $Z_p \times Z_p$.

These groups are not isomorphic, since Z_{p^2} contains an element of order p^2, but every element of $Z_p \times Z_p$ other than the identity has order p.

In general, the number of nonisomorphic commutative groups of order $n = p_1^{k_1} p_2^{k_2} \cdots p_r^{k_r}$ (distinct primes) is $\pi(k_1)\pi(k_2)\cdots\pi(k_r)$, where $\pi(k_i)$ denotes the number of partitions of the integer k_i. As a typical illustration, there are three commutative groups of order $24 = 2^3 \cdot 3$; specifically,

$$Z_8 \times Z_3 \simeq Z_{24},$$

and

$$Z_4 \times Z_2 \times Z_3 \simeq Z_{12} \times Z_2,$$

$$Z_2 \times Z_2 \times Z_2 \times Z_3 \simeq K_4 \times Z_6.$$

There are other theorems that give insight into the structure of finite commutative groups. A second basic result states that any finite commutative group G is isomorphic to a direct product

$$Z_{n_1} \times Z_{n_2} \times \cdots \times Z_{n_r}$$

where $n_i | n_{i+1}$ for $1 \leqslant i \leqslant r - 1$. To arrive at this, all one needs to do is to write G as a direct product of cyclic groups of prime power order and assemble the latter judiciously; the formal proof is left as an exercise. Suppose, for example, that a group G of order 360 has the direct product decomposition

$$G \simeq Z_2 \times Z_2 \times Z_2 \times Z_3 \times Z_3 \times Z_5.$$

Rearranging factors, we can put this in the form

$$G \simeq Z_2 \times (Z_2 \times Z_3) \times (Z_2 \times Z_3 \times Z_5) \simeq Z_2 \times Z_6 \times Z_{30},$$

with $2|6$ and $6|30$.

Our final theorem provides an application of the results of this section.

THEOREM 3.11 *If $n > 1$ is a square-free integer (that is, n is not divisible by the square of any prime), then any commutative group of order n is isomorphic to the group Z_n.*

Proof. Since n is a square-free integer, we have $o(G) = n = p_1 p_2 \cdots p_r$, where the p_i are distinct primes. Now the corollary to Theorem 3.8 asserts that

$$G = G_{p_1} \otimes G_{p_2} \otimes \cdots \otimes G_{p_r}$$

with $o(G_{p_i}) = p_i$; hence $G_{p_i} \simeq Z_{p_i}$. This implies that

$$G \simeq Z_{p_1} \times Z_{p_2} \times \cdots \times Z_{p_r} \simeq Z_{p_1 p_2 \cdots p_r} = Z_n,$$

the second isomorphism being achieved by use of Theorem 3.6.

COROLLARY *If n is a square-free integer, then all commutative groups of order n are isomorphic.*

PROBLEMS 3.2

1. Establish the following assertions:
   a) The group Z_{15} is the internal direct product of $\{0, 5, 10\}$ and $\{0, 3, 6, 9, 12\}$.
   b) The group G_s of symmetries of the square is indecomposable.
   c) The additive groups Z and Q are both indecomposable.
2. Let $G = H \otimes K$ for subgroups H and K of the group G. Given that N is a normal subgroup of G, prove that either $N \subseteq \text{cent } G$ or N has nontrivial intersection with one of H or K.
3. Assume that H and K are normal subgroups of the finite group G. If $o(G) = o(H)o(K)$, where $o(H)$ and $o(K)$ are relatively prime, prove that $G = H \otimes K$.
4. Show that the symmetric group S_3 is indecomposable by proving that any direct product of subgroups of orders 2 and 3 must be cyclic.
5. Let the group $G = H_1 \otimes H_2 \otimes \cdots \otimes H_n$ for subgroups H_i. Consider the representation $a = a_1 a_2 \cdots a_n$, where $a_i \in H_i$, of an element $a \in G$. If the mappings $\pi_i: G \to H_i$ are defined by $\pi_i(a) = a_i$ for $i = 1, 2, \ldots, n$, verify that
   a) π_i is a homomorphism from G onto H_i, called the ith *projection.*
   b) A mapping $f: G' \to G$ from an arbitrary group G' into G is a homomorphism if and only if the composition $\pi_i \circ f: G' \to H_i$ is a homomorphism for each value of i.
6. Suppose that $f: G \to G'$ is a homomorphism from the group G onto the group G' and that H is a normal subgroup of G. If the restriction $f|H$ is an isomorphism from H onto G', prove that $G = H \otimes \ker f$.
7. Let H and K be normal subgroups of the group G. Verify that if the natural mapping $\text{nat}_K: G \to G/K$ induces an isomorphism of H onto G/K, then $G = H \otimes K$. [*Hint*: See Problem 6.]
8. Assuming that $G = H \otimes K$ for subgroups H and K of the group G, prove that:
   a) Any normal subgroup of H (or K) is also a normal subgroup of G.
   b) If N is a normal subgroup of H, then $G/N \simeq (H/N) \times K$.
9. Given that G_i $(i = 1, 2, 3)$ are groups, establish the following facts concerning (external) direct products:
   a) $G_1 \times G_2 \simeq G_2 \times G_1$
   b) $(G_1 \times G_2) \times G_3 \simeq G_1 \times (G_2 \times G_3)$

c) If $G_1 \simeq G_2$, then $G_1 \times G_3 \simeq G_2 \times G_3$.

d) $G_1 \times G_2$ is a commutative group if and only if G_1 and G_2 are both commutative.

10. Show that if G_1 and G_2 are finite groups and the element $(a, b) \in G_1 \times G_2$, then $o((a, b)) = \text{lcm}(o(a), o(b))$.

11. Prove the following:

a) The Klein 4-group K_4 is isomorphic to $Z_2 \times Z_2$.

b) Z_{n^2} is not isomorphic to $Z_n \times Z_n$ for any integer $n > 1$. [*Hint*: Z_{n^2} has only one subgroup of order n, but $Z_n \times Z_n$ has two such subgroups.]

12. Given that $G = G_1 \times G_2 \times \cdots \times G_n$, where the G_i are groups, verify that:

a) $\text{cent } G = \text{cent } G_1 \times \text{cent } G_2 \times \cdots \times \text{cent } G_n$.

b) $[G, G] = [G_1, G_1] \times [G_2, G_2] \times \cdots \times [G_n, G_n]$.

13. Let H_i be normal subgroups of the groups G_i $(i = 1, 2)$. Prove that $H_1 \times H_2$ is a normal subgroup of $G_1 \times G_2$ and that

$$(G_1 \times G_2)/(H_1 \times H_2) \simeq G_1/H_1 \times G_2/H_2.$$

14. Show that if H and K are normal subgroups of the group G, with $H \cap K = \{e\}$, then G is isomorphic to a subgroup of $G/H \times G/K$. [*Hint*: Use the mapping $f: G \to G/H \times G/K$ defined by $f(x) = (xH, xK)$.]

15. Show by example that the direct product of cyclic groups (p-groups) need not be a cyclic group (p-group).

16. Verify the following statements:

a) Any homomorphic image of a p-group is again a p-group.

b) A finite commutative p-group is generated by its elements of highest order.

c) If G is a cyclic p-group, then G is a finite p-group.

17. Let G and G' be finite commutative groups. Prove that:

a) If $f: G \to G'$ is a homomorphism, then $f(G_p) \subseteq G'_p$.

b) $G \simeq G'$ if and only if $G_p \simeq G'_p$ for all primes p.

18. Given that G is a finite cyclic group, prove that, for every prime p dividing $o(G)$, there are exactly $p - 1$ elements in G having order p. [*Hint*: If $p | o(G)$, then $G_p \simeq Z_{p^n}$ for some $n \in Z_+$.]

19. Show that a finite commutative group $G \neq \{e\}$ is cyclic if and only if

$$G \simeq Z_{n_1} \times Z_{n_2} \times \cdots \times Z_{n_r}$$

where $n_i = p_i^{k_i}$ for $i = 1, 2, \ldots, r$ (p_i distinct primes, $k_i \in Z_+$).

20. Show that there are three nonisomorphic commutative groups of order 8, namely, Z_8, $Z_4 \times Z_2$, and $K_4 \times Z_2$.

21. Determine, up to isomorphism, all commutative groups of the following orders: p^4 (p a prime), 144, and 360.

22. Prove that a cyclic group G is indecomposable if and only if G is either infinite or of prime power order.

23. Characterize those positive integers n for which the only commutative groups of order n are cyclic.

24. Prove that if G is a finite commutative group, then

$$G \simeq Z_{n_1} \times Z_{n_2} \times \cdots \times Z_{n_r},$$

where $n_i | n_{i+1}$ for $1 \leqslant i \leqslant r - 1$.

3.3 THE SYLOW THEOREMS

We recall that according to Lagrange's Theorem, the order of any subgroup of a finite group G is a divisor of the order of G. The converse, however, is false; we cannot be certain that a group G will have a subgroup of order n for each divisor n of its order. To be sure, under special circumstances (as, for example, in a finite cyclic group) it is true that such subgroups always exist. There are also partial converses of Lagrange's Theorem that assert the existence of subgroups of prescribed order in arbitrary finite groups. Cauchy proved (1845) that every finite group whose order is divisible by a prime number p must contain one or more subgroups of order p. In 1872 a Norwegian professor of mathematics, Ludwig Sylow (1832–1918), published a ten-page paper entitled "*Theoremes sur les Groupes de Substitutions*" in which a remarkable extension of Cauchy's result appeared. Our immediate goal is the first theorem of the article: If the order of the group G is divisible by p^k but not by p^{k+1}, with p a prime number, then G contains at least one subgroup of order p^k. This theorem, known as the "First Sylow Theorem" or simply as "Sylow's Theorem," has turned out to be the cornerstone of finite group theory.

As the title of his paper suggests, for Sylow the term "group" still meant a group of permutations (substitutions); and his proof of Sylow's Theorem is given in this context. Since any finite group can be represented as a group of permutations, the theorem holds for all finite groups. However, in view of the fundamental nature of Sylow's Theorem it seemed desirable to establish the result in the abstract, without explicit reference to permutation groups. This was done in 1887 by Georg Frobenius (1849–1917) of the University of Berlin; as he points out, "The elements of every finite group can be interpreted as substitutions. Nevertheless, considering the theorem to be proved, I specifically do not want to assume this interpretation in what follows." Frobenius thereupon proceeds from an abstract definition of a finite group, and it is essentially his argument that we shall present.

We begin with a survey of facts, some previously encountered in the exercises, arranged rather haphazardly around the notion of conjugacy. Despite the familiarity of the ground covered, the relevant definitions and results are collected here for the reader's convenience.

Starting with any subgroup H of the group G and an arbitrary element $a \in G$, consider the following subset of G:

$$aHa^{-1} = \{aha^{-1} | h \in H\}.$$

If x and y are members of aHa^{-1}, then there exist certain $h_1, h_2 \in H$ for which $x = ah_1a^{-1}$ and $y = ah_2a^{-1}$. Thus

$$xy^{-1} = (ah_1a^{-1})(ah_2a^{-1})^{-1} = a(h_1h_2^{-1})a^{-1},$$

where, since H is a subgroup of G, the product $h_1h_2^{-1}$ lies in H. But this means that $xy^{-1} \in aHa^{-1}$, so that aHa^{-1} is itself a subgroup of G. We are therefore led to the theorem below.

THEOREM 3.12 *If H is a subgroup of the group G and if $a \in G$, then aHa^{-1} is also a subgroup of G, called the conjugate subgroup of H induced by the element a.*

Not all the subgroups conjugate to H are necessarily distinct from H or, for that matter, distinct from each other. If $aHa^{-1} = H$ for some $a \in G$, we say that the subgroup H is *self-conjugate under a*. Any subgroup is, for example, self-conjugate under each one of its own elements. In terms of this definition the concept of a normal subgroup can be arrived at in an alternative way: The subgroup H is normal in G if and only if H is self-conjugate under every element of G.

Another point of interest is that the conjugate subgroups of H are all isomorphic to H and hence isomorphic to each other. Given a conjugate aHa^{-1} of H, define the mapping $f: H \to aHa^{-1}$ by taking $f(h) = aha^{-1}$; there is no particular trouble in showing that f is an isomorphism. One deduces from this that conjugate subgroups have the same order.

Now if K is any other subgroup of G, we may certainly consider those elements of K under which H is self-conjugate, in other words the set

$$N_K(H) = \{k \in K | kHk^{-1} = H\}.$$

The reader should prove the following theorem concerning the structure of $N_K(H)$.

THEOREM 3.13 *If H and K are two subgroups of the group G, then $N_K(H)$ is also a subgroup of G, known as the normalizer of H in K.*

When $K = G$, one customarily writes $N(H)$ in place of $N_G(H)$ and refers to the subgroup $N(H)$ simply as the *normalizer of H*. Although the subgroup H is not necessarily normal in G, a simple argument shows it to be normal in its normalizer; in fact, $N(H)$ is the largest subgroup of G in which H is normal. Thus H is a normal subgroup of G if and only if $N(H) = G$.

With this notation in mind we can now derive a formula for counting the conjugate subgroups aHa^{-1} of H.

THEOREM 3.14 *The number of distinct conjugates of a subgroup H of G induced by the elements of a subgroup K is equal to $[K:N_K(H)]$, the index of $N_K(H)$ in K.*

Proof. For arbitrary elements $a, b \in K$, we have $aHa^{-1} = bHb^{-1}$ if and only if $H = (a^{-1}b)H(a^{-1}b)^{-1}$, that is, if and only if $a^{-1}b \in N_K(H)$. This last relation holds, in its turn, if and only if $aN_K(H) = bN_K(H)$. Hence the number of distinct conjugate subgroups of H by elements of K is equal to the number of distinct cosets of $N_K(H)$ in K, namely, equal to the value $[K:N_K(H)]$.

COROLLARY *Let H be a subgroup of the finite group G. The number of distinct conjugate subgroups of H is precisely $[G:N(H)]$ and hence is a divisor of $o(G)$.*

We continue our discussion by considering a new class of subgroups.

DEFINITION 3.4 The *centralizer* of an element a of the group G, denoted by $C(a)$, is the set of all elements in G that commute with a:

$$C(a) = \{x \in G \mid xa = ax\}.$$

As the reader might suspect, the centralizer of an element is closely related to the center of a group. These concepts are not identical, since in the group G_s of symmetries of the square, we have $C(r_1) = \{e, r_1, r_2, r_3\}$, while cent $G_s = \{e, r_2\}$. The related definitions make it plain that the center of a group G could equally well be defined as the intersection of all the centralizers of the elements in G; expressed in symbols, cent $G = \bigcap_{a \in G} C(a)$. Note also that $C(a) = G$ if and only if $a \in \text{cent}\, G$.

The proof of the fact below is routine and is left as an exercise.

THEOREM 3.15 *For any element $a \in G$ the set $C(a)$ forms a subgroup of G.*

Our next theme, the development of the class equation of a group, needs some preparation. We begin by introducing a notion of conjugacy for group elements.

DEFINITION 3.5 Let G be a group and $a, b \in G$. The element b is said to be a *conjugate* of a (by x) if some $x \in G$ exists for which $b = xax^{-1}$.

In a manner similar to that in which Theorem 3.14 was obtained, one may prove that the number of distinct conjugate elements of a given element $a \in G$ is precisely the index of $C(a)$ in G. We shall simply state this as a theorem and leave its formal verification to the reader.

THEOREM 3.16 *If $a \in G$, the number of conjugate elements of a is equal to $[G:C(a)]$, the index of the centralizer of a in G.*

Given a, $b \in G$, we may define an equivalence relation $\sim$ in G by requiring that $a \sim b$ if and only if b is a conjugate of a. The relation $\sim$ induces a partition of the set G into equivalence classes, usually referred to as the *conjugacy classes of G*. Elements in the same conjugacy class are called *conjugates* of one another. By the preceding theorem there are $[G:C(a)]$ elements in the conjugacy class containing a.

Certain of these classes will consist of exactly one member; for example, if $a \in \text{cent } G$, then $xax^{-1} = a$ for every $x \in G$, so the conjugacy class containing a reduces to $\{a\}$. It should be clear, on the other hand, that $a \in \text{cent } G$ when the conjugacy class of a is $\{a\}$. From this we infer that a group is commutative if and only if each of its conjugacy classes is a singleton.

Example 3.6

There are three conjugacy classes of S_3, the symmetric group on three symbols. We leave to the reader the task of verifying that these classes are given by

$$\left\{\begin{pmatrix}1 & 2 & 3\\ 1 & 2 & 3\end{pmatrix}\right\},$$

$$\left\{\begin{pmatrix}1 & 2 & 3\\ 2 & 3 & 1\end{pmatrix}, \begin{pmatrix}1 & 2 & 3\\ 3 & 1 & 2\end{pmatrix}\right\},$$

$$\left\{\begin{pmatrix}1 & 2 & 3\\ 2 & 1 & 3\end{pmatrix}, \begin{pmatrix}1 & 2 & 3\\ 1 & 3 & 2\end{pmatrix}, \begin{pmatrix}1 & 2 & 3\\ 3 & 2 & 1\end{pmatrix}\right\}.$$

As each element of G lies in one and only one conjugacy class, we may determine the number of elements in G by counting them class by class. To do so, first observe that there are exactly $o(\text{cent } G)$ classes that contain one element. If we add to $o(\text{cent } G)$ the number of elements in the nonsingleton conjugacy classes, the elements of G will all have been counted. This leads to the important (conjugacy) *class equation*:

$$o(G) = o(\text{cent } G) + \sum [G:C(a)],$$

where the summation runs over a set of representatives for the distinct conjugacy classes having more than one member; that is to say, we choose a set of nonconjugate elements that are not in cent G.

In the group of symmetries of the square, for instance, the conjugacy classes are

$$\{e\}, \qquad \{r_2\}, \qquad \{r_1, r_3\}, \qquad \{d_1, d_2\}, \qquad \{h, v\},$$

so the class equation would read

$$o(G) = o(\text{cent } G) + [G:C(r_1)] + [G:C(d_1)] + [G:C(h)].$$

We now have the background to derive a partial converse of Lagrange's Theorem, to the effect that any finite group of order pn (p a prime) has a subgroup of order p. The following lemma serves as a starting point.

LEMMA *If G is a finite commutative group whose order is divisible by a prime p,* then *G contains an element of order p.*

Proof. We proceed by induction on the order of G. First, observe that if $o(G)$ is itself a prime, this prime must be p and every element other than the identity has order p; this takes care of the basis for the induction when $o(G) = 2$. Now, let $a \in G$, with $a \neq e$. Since G is a finite group, the element a has a finite order n. Either $p|n$ or $p\nmid n$. In case $p|n$, say $n = kp$, then $(a^k)^p = a^n = e$. The order of a^k therefore divides p, which implies, because p is prime, that $o(a^k) = p$.

Next, let us suppose that $p\nmid n$. Because G is commutative, the cyclic subgroup $H = (a)$ is normal in G, and we can form the quotient group G/H. Moreover, $o(H) > 1$, so $o(G/H) < o(G)$. From

$$o(G) = [G:H]o(H) = o(G/H)n$$

and the fact that $p\nmid n$ it follows that $o(G/H)$ is divisible by p. Using our induction assumption, G/H must contain some coset bH of order p. But the equation $b^pH = (bH)^p = eH$ implies that $b^p \in H$. As H is of order n, $(b^n)^p = (b^p)^n = e$. Thus either b^n has order p or $b^n = e$. Were $b^n = e$, then $(bH)^n = eH$ and we would be forced to conclude that $p|n$. Since this conflicts with our hypothesis, b^n is necessarily of order p. In either of the situations considered, we have produced an element in G of order p, as required.

It is instructive to present a second proof of this lemma using direct products. Since $p|o(G)$, the corollary to Theorem 3.8 asserts that the p-subgroup $G_p \neq \{e\}$. Thus by Theorem 3.9,

$$G_p \simeq Z_{p^{n_1}} \times Z_{p^{n_2}} \times \cdots \times Z_{p^{n_r}}$$

for suitable $n_i \in Z_+$. Inasmuch as the element

$$(p^{n_1-1}, p^{n_2-1}, \ldots, p^{n_r-1}) \in Z_{p^{n_1}} \times \cdots \times Z_{p^{n_r}}$$

has order p in the direct product, there exists an element in G_p (hence in G) of order p.

The next objective is to establish the preceding lemma without the hypothesis of commutativity. The class equation provides the key to the proof.

THEOREM 3.17 (Cauchy) *If G is a finite group whose order is divisible by a prime p, then G contains an element of order p.*

Proof. Use induction on the order of G. If $o(G) = 2$, then G is a commutative group and the theorem reduces to the lemma. Suppose, inductively, that the theorem holds true for all groups of order less than $o(G)$. Now consider the class equation for G,

$$o(G) = o(\text{cent } G) + \sum [G : C(a)],$$

where the summation, as noted previously, is made by choosing one representative from each conjugacy class with more than one element.

If $G = \text{cent } G$, then G is commutative, and we need only appeal to the lemma to complete the proof; in the contrary case there is some element $a \in G$ with $a \notin \text{cent } G$. For such an a, $[G : C(a)] > 1$, so by Lagrange's Theorem,

$$o(G) = [G : C(a)]o(C(a)) > o(C(a)).$$

Since the theorem is assumed to hold for the subgroup $C(a)$, if $p | o(C(a))$, then $C(a)$ has an element of order p, and we are through. On the other hand, if $p \nmid o(C(a))$ for every $a \notin \text{cent } G$, p must divide $[G : C(a)]$, since p is prime and divides $o(G) = [G : C(a)]o(C(a))$. But then, in the class equation, p divides each term of the summation and also divides $o(G)$, so that $p | o(\text{cent } G)$. As cent G is a commutative subgroup of G, it follows (again, from the lemma) that cent G contains an element of order p.

COROLLARY 1 *$G \neq \{e\}$ is a finite p-group if and only if $o(G) = p^k$, for some $k > 0$.*

Proof. Suppose that G is a p-group, but $q | o(G)$ for some prime $q \neq p$. Then by Cauchy's Theorem, G has an element of order q, contradicting the fact that G is a p-group. Thus p is the only prime divisor of $o(G)$, which implies that $o(G) = p^k$ $(k > 0)$. Conversely, if $o(G) = p^k$, then from Lagrange's Theorem, each element of G has a power of p as its order.

It is to be noted that our labors have also produced:

COROLLARY 2 *If the prime p divides the order of the finite group G, then G contains at least one subgroup of order p.*

Proof. The cyclic subgroup generated by an element of order p itself has order p.

While we have seen examples of groups that have a trivial center (for instance, the symmetric groups S_n with $n > 2$), finite p-groups always possess nontrivial centers. This is the content of an elegant little theorem due to Burnside.

THEOREM 3.18 (Burnside) *If G is a finite p-group with more than one element, then* $\text{cent}\, G \neq \{e\}$.

Proof. This follows directly from the class equation for G,

$$o(G) = o(\text{cent } G) + \sum [G:C(a)],$$

the summation being over certain elements a for which $[G:C(a)] \neq 1$. For $a \notin \text{cent}\, G$, $C(a)$ is a nontrivial subgroup of G; in fact, $C(a)$ is itself a p-group. (Any subgroup of a p-group is a p-group!) By the preceding corollary it follows that $o(C(a)) = p^k$ for some $k > 0$. Then

$$[G:C(a)] = o(G)/o(C(a))$$

must be divisible by p. Now p divides each term of the summation, as well as dividing $o(G)$, so $p | o(\text{cent}\, G)$. This implies that $o(\text{cent}\, G) > 1$.

COROLLARY 1 *A noncommutative group G of order p^k, where p is a prime and $k > 1$, is never simple.*

Proof. The center of G, which is normal in G, forms a nontrivial subgroup.

Perhaps the most interesting application of Burnside's Theorem is the following:

COROLLARY 2 *Any group G of order p^2 is commutative, where p is a prime.*

Proof. As we have just seen, $\text{cent}\, G \neq \{e\}$, so either $o(\text{cent}\, G) = p$ or $o(\text{cent}\, G) = p^2$. Suppose that $\text{cent}\, G$ has order p. Then $o(G/\text{cent}\, G) = p$, whence $G/\text{cent}\, G$ is a cyclic group. This is enough to ensure that G is commutative. To spell out the details, let $G/\text{cent}\, G$ be generated by the coset $a\, \text{cent}\, G$. Then a typical element of $G/\text{cent}\, G$ has the form $a^n \text{cent}\, G$ for some integer n. Given $x, y \in G$, we may write $x = a^n c$, $y = a^m c'$ for suitable choice of integers n and m, with c and c' in $\text{cent}\, G$. As a result,

$$xy = a^n c a^m c' = a^{n+m} cc' = a^{m+n} c'c = a^m c' a^n c = yx.$$

Every pair of elements of G commutes, making G a commutative group. The implication is that $G = \text{cent}\, G$ and so $\text{cent}\, G$ has order p^2, a contradiction. This leaves only the possibility that $o(\text{cent}\, G) = p^2$, which, of course, forces G to be commutative.

We should caution that the last result is peculiar to groups of order p^2 and need not hold for those having order p^n, with $n > 2$. Perhaps this can best be illustrated by the group of symmetries of the square, a non-commutative group of order 2^3.

We now shift our emphasis to the case in which G is not necessarily a p-group but does contain p-subgroups. Of particular interest are the "maximal" p-subgroups, the so-called Sylow p-subgroups.

DEFINITION 3.6 Let G be a finite group and p a prime. A subgroup P of G is said to be a *Sylow p-subgroup* if P is a p-group and is not properly contained in any other p-subgroup of G.

Example 3.7
The symmetric group S_3 has three Sylow 2-subgroups, specifically the subgroups

$$P_1 = \left\{\begin{pmatrix} 1 & 2 & 3 \\ 1 & 2 & 3 \end{pmatrix}, \begin{pmatrix} 1 & 2 & 3 \\ 2 & 1 & 3 \end{pmatrix}\right\},$$

$$P_2 = \left\{\begin{pmatrix} 1 & 2 & 3 \\ 1 & 2 & 3 \end{pmatrix}, \begin{pmatrix} 1 & 2 & 3 \\ 1 & 3 & 2 \end{pmatrix}\right\},$$

$$P_3 = \left\{\begin{pmatrix} 1 & 2 & 3 \\ 1 & 2 & 3 \end{pmatrix}, \begin{pmatrix} 1 & 2 & 3 \\ 3 & 2 & 1 \end{pmatrix}\right\}.$$

These subgroups are all conjugate; indeed, if

$$\alpha = \begin{pmatrix} 1 & 2 & 3 \\ 3 & 1 & 2 \end{pmatrix}, \qquad \beta = \begin{pmatrix} 1 & 2 & 3 \\ 2 & 3 & 1 \end{pmatrix},$$

then we have $P_1 = \alpha \circ P_2 \circ \alpha^{-1} = \beta \circ P_3 \circ \beta^{-1}$. Note too that the number of Sylow 2-subgroups is a divisor of the order of S_3. This is typical of the general situation.

Example 3.8
Consider the Sylow 2-subgroups of the alternating group A_4. They each have order 2 or 4, since these are the only powers of 2 dividing $o(A_4) = 12$. The candidates for elements of the Sylow 2-subgroups are to be found in the set

$$P = \left\{\begin{pmatrix} 1 & 2 & 3 & 4 \\ 1 & 2 & 3 & 4 \end{pmatrix}, \begin{pmatrix} 1 & 2 & 3 & 4 \\ 2 & 1 & 4 & 3 \end{pmatrix}, \begin{pmatrix} 1 & 2 & 3 & 4 \\ 3 & 4 & 1 & 2 \end{pmatrix}, \begin{pmatrix} 1 & 2 & 3 & 4 \\ 4 & 3 & 2 & 1 \end{pmatrix}\right\},$$

since all other members of A_4 have order 3. But the set P itself forms a group. Any 2-subgroup obviously will be contained in P, making it the unique Sylow 2-subgroup of A_4. As the reader may verify, there are four Sylow 3-subgroups in A_4.

Let us pick up a few loose threads. Glancing back at the last section, the reader will observe that when the group G is commutative the set G_p consisting of all elements of order a power of p forms a Sylow p-subgroup; in fact, G_p happens to be the only Sylow p-subgroup of G. We even proved that G is the direct product of its Sylow p-subgroups G_p for those primes p which divide $o(G)$. (This is just the corollary to Theorem 3.8 restated with a slight change of language.) Our goal is to develop enough of the Sylow theory to be able to derive the analog of this result for noncommutative groups; namely, if for every $p|o(G)$ there exists exactly one Sylow p-subgroup of G, then G is their direct product.

We pass on now to the question of existence of Sylow p-subgroups.

THEOREM 3.19 *For each prime p, the finite group G has a Sylow p-subgroup (possibly trivial).*

Proof. If $o(G) = 1$ or if $p \nmid o(G)$, then $\{e\}$ is, in a trivial sense, the required Sylow p-subgroup. On the other hand, whenever $p|o(G)$, Cauchy's Theorem guarantees that there exists at least one subgroup H of order p. If H is not already a Sylow p-subgroup, then it is properly contained in another p-subgroup of G. Since G is finite, we are led after a finite number of inspections to a Sylow p-subgroup containing H.

Because the order of an element is preserved under conjugation, it is clear that any conjugate of a p-group is again a p-group. Even more is true. The conjugates of a Sylow p-subgroup P are themselves Sylow p-subgroups (of course, some or all of these conjugate subgroups may be identical with P); this fact emerges from the following theorem:

THEOREM 3.20 *Let P be a Sylow p-subgroup of the group G. If f is any automorphism of G, then f(P) is also a Sylow p-subgroup of G.*

Proof. Being the homomorphic image of a p-group, $f(P)$ is necessarily a p-group (Problem 16, Section 3.2). For purposes of contradiction, let us assume that $f(P)$ is not a Sylow p-subgroup of G. Then $f(P)$ is properly contained in some p-subgroup, say P'. Since f^{-1} is itself an automorphism of G, the inverse image $f^{-1}(P')$ will be a p-subgroup containing P. This inclusion is proper, since there exists an element $a \in P'$ with $a \notin f(P)$; hence $f^{-1}(a) \in f^{-1}(P')$, but $f^{-1}(a) \notin P$. In short, we have shown that P is properly contained in a p-subgroup of G, contradicting the hypothesis that P is a Sylow p-subgroup.

COROLLARY *Any conjugate of a Sylow p-subgroup P of G is again a Sylow p-subgroup.*

Proof. Each conjugate of P is the image of P under a suitable inner automorphism of G.

In the following theorems our goal is to show that if P is a fixed Sylow p-subgroup of the finite group G, then any other Sylow p-subgroup of G must in fact be a conjugate of P. We will also want to determine, so far as is possible, the number of distinct Sylow p-subgroups of G. If, for example, there is a unique Sylow p-subgroup corresponding to a particular prime p, then this subgroup must be normal, since it is conjugate only to itself. Using the Sylow theorems, we can obtain some rather general results concerning the existence of subgroups of finite groups; for this, one needs to know only the prime factorization of the order of the group.

But we anticipate events. In the meantime, two preparatory lemmas are required.

LEMMA 1 *Let H be a normal subgroup of the group G. If H and G/H are both p-groups, then G itself is a p-group.*

Proof. For each a in G the coset $aH \in G/H$ has order some power of p, say order p^k. Then $a^{p^k}H = (aH)^{p^k} = eH$, so $a^{p^k} \in H$. By hypothesis the element a^{p^k} also has p-power order, say order p^j. This implies that $e = (a^{p^k})^{p^j} = a^{p^{k+j}}$, whence $o(a) | p^{k+j}$. Since the only divisors of p^{k+j} are powers p^i, where $0 \leqslant i \leqslant j + k$, it follows that $o(a)$ is a power of p.

LEMMA 2 *Let P be a Sylow p-subgroup of the group G and $a \in G$ be an element whose order is a power of p. If $aPa^{-1} = P$, then $a \in P$.*

Proof. Inasmuch as the condition $aPa^{-1} = P$ means that $a \in N(P)$, the problem is one of showing that there can be no element in $N(P) - P$ of p-power order. Suppose such an element a actually did exist. Now P is normal in its normalizer, so we may form the quotient group $N(P)/P$ and consider the coset aP. The order of a coset as an element in the quotient group divides the order of any of its representatives; hence $o(aP)$ is a power of p. This implies that the cyclic subgroup (aP) of $N(P)/P$ has order a power of p and, in consequence, (aP) is a p-group (see the corollary to Cauchy's Theorem). By the Correspondence Theorem, $N(P)$ possesses a subgroup K with $K \supseteq P$ and $K/P = (aP)$. Since we know that P and (aP) are both p-groups, Lemma 1 permits us to conclude that K must be a p-group. Moreover, since the element $a \notin P$, K properly contains P. The combined effect of these assertions is to violate the hypothesis that P is a Sylow p-subgroup of G. Accordingly, there can be no members of $N(P) - P$ whose order is a power of p.

This whole tangle of ideas can best be summed up by the following:

COROLLARY *If H is an arbitrary p-subgroup of the group G and if P is a Sylow p-subgroup of G, then $N_H(P) = H \cap P$.*

Proof. Since $P \subseteq N(P)$, it is clear that $H \cap P \subseteq H \cap N(P)$, while the opposite inclusion is the content of the preceding lemma. This results in the equality that $N_H(P) = H \cap N(P) = H \cap P$.

We are now ready to prove the first of the Sylow Theorems. This result provides information about the number of distinct Sylow p-subgroups of a given group.

THEOREM 3.21 (Sylow) *For a fixed prime p the number of Sylow p-subgroups of the finite group G is congruent to 1 modulo p and is a divisor of $o(G)$.*

Proof. Let P be any Sylow p-subgroup of G. We have already learned that each conjugate subgroup xPx^{-1} of P is likewise a Sylow p-subgroup and that there are $[G:N(P)]$ of them. Thus a natural starting point in the determination of the number of Sylow p-subgroups in G is to count the distinct conjugates $P_x = xPx^{-1}$ of P.

We begin by partitioning the set of conjugate subgroups of P into equivalence classes by means of the equivalence relation $\sim$ wherein $P_x \sim P_y$ if and only if $P_x = aP_ya^{-1}$ for some $a \in P$. Theorem 3.14 asserts that the number of subgroups in the equivalence class determined by P_x (in other words, the number of distinct conjugate subgroups of P_x induced by elements from P) is equal to $[P:N_P(P_x)]$. This leads to the formula

$$[G:N(P)] = \sum [P:N_P(P_x)],$$

where the sum extends over a set of representatives P_x, one from each equivalence class.

By virtue of the corollary above, we obtain

$$[P:N_P(P_x)] = [P:P \cap P_x] = o(P)/o(P \cap P_x).$$

Being p-groups, P and $P \cap P_x$ are both of p-power order, so $[P:N_P(P_x)]$ must be a suitable power of p, say p^k. Now the integer $k = 0$ if and only if $P_x = P$. This happens exactly once in the displayed summation. The only conjugate subgroup of P induced by elements of P is P itself; hence P is the only member (thus the only choice for representative) of its equivalence class. Except for this one case, $[P:N_P(P_x)]$ is a positive power of p, implying that the total number of distinct conjugates of P is

$$1 + p^{k_1} + \cdots + p^{k_r} = 1 + mp \equiv 1 \pmod{p}, \qquad k_i > 0.$$

At this stage an obvious question arises: Does G contain any Sylow p-subgroups that are not conjugate subgroups of P? To show that the answer is no, let us assume that such a subgroup R exists (that is, R is a Sylow p-subgroup not conjugate to P) and reach a contradiction. Following the procedure just outlined, we count the conjugates of R to get

$$[G:N(R)] = \sum [P:N_P(R_x)].$$

The essential point here is that no term of this sum reduces to 1; if this happened, R would turn out to be a conjugate of P, contrary to assumption. To fill in the details, suppose that

$$1 = [P:N_P(R_x)] = o(P)/o(P \cap R_x)$$

for some x. Then $o(P) = o(P \cap R_x)$; hence $P = P \cap R_x$ or, equivalently, $P \subseteq R_x$. The fact that P is a Sylow p-subgroup would imply that $P = R_x$ and so $R = P_{x^{-1}}$, which is impossible. The total number of conjugate subgroups of R is therefore found to be

$$p^{j_1} + p^{j_2} + \cdots + p^{j_m} \equiv 0 \pmod{p}, \qquad j_i > 0.$$

But as was established earlier in the proof, the number of conjugate subgroups of any Sylow p-subgroup (in this particular case, of R) is congruent to 1 modulo p. Since no integer can be congruent at once to $0 \pmod{p}$ and to $1 \pmod{p}$, we have arrived at the desired contradiction.

Our efforts have shown that the number of Sylow p-subgroups of G is equal to $[G:N(P)]$, the index of $N(P)$ in G. Since $N(P)$ is a subgroup of G, $[G:N(P)]$ evidently divides $o(G)$. This proves the theorem in its entirety.

The foregoing argument brought out several facts that are important for subsequent work. These are recorded as

Corollary 1 *Any two Sylow p-subgroups of G corresponding to the same prime p are conjugate and hence isomorphic.*

Corollary 2 *The number of distinct Sylow p-subgroups (for a fixed prime p) is* $[G:N(P)]$, *where P is any Sylow p-subgroup of G.*

We now tackle the second major result of this section; from this theorem we shall determine the order of the Sylow p-subgroups of a finite group.

Theorem 3.22 (Sylow) *Let G be a finite group of order* $p^k q$, *where the prime p does not divide q. A subgroup P of G is a Sylow p-subgroup if and only if* $o(P) = p^k$.

Proof. One direction of the theorem is immediate: If P is a subgroup of G of order p^k, it must be a Sylow p-subgroup; any p-subgroup of G properly containing P would necessarily be of a higher order of p than appears in $o(G)$.

Concentrating on the less obvious part of the theorem, suppose that P is any Sylow p-subgroup of G. Being a p-group, P will have order some power of p. Since

$$[G:P]o(P) = o(G) = p^k q,$$

if one knew that $[G:P]$ were divisible by no power of p, it would follow that $o(P) = p^k$. Thus to establish that $o(P) = p^k$, we need only show that $[G:P]$ has no p-power divisor.

The strategy that we employ is to prove that both factors on the right-hand side of the equation

$$[G:P] = [G:N(P)]o(N(P)/P)$$

are prime to p. First, observe that $[G:N(P)]$ is equal to the number of conjugate subgroups of P in G and, by our last theorem, this value is congruent to 1 modulo p. Therefore the possibility that p divides $[G:N(P)]$ cannot occur.

Assume for purposes of contradiction that $o(N(P)/P)$ has a p-power divisor. By Cauchy's Theorem there must exist an element $aP \in N(P)/P$ of order p. Then $a^pP = (aP)^p = eP$, which signifies that $a^p \in P$. Inasmuch as P is a p-group, this element has order a power of p; hence a itself has p-power order. Because $a \in N(P)$, the second lemma to Theorem 3.21 implies that $a \in P$ and so $aP = eP$. From this contradiction we infer that $p \nmid o(N(P)/P)$. Since the prime p divides neither of the factors of $[G:P]$, it cannot divide $[G:P]$, thereby completing the argument.

COROLLARY *Let G be a finite group, p a prime and p^k the highest power of p dividing the order of G. Then G has at least one subgroup of order p^k.*

Theorem 3.22 tells us that if G is a finite group of order n and if $n = p_1^{k_1}p_2^{k_2}\cdots p_r^{k_r}$ is the factorization of n into a product of distinct primes, then G has subgroups of orders $p_i^{k_i}$ $(i = 1, 2, \ldots, r)$. As a matter of fact, one can prove considerably more: For each prime p_i, G contains subgroups of orders $p_i, p_i^2, p_i^3, \ldots, p_i^{k_i}$. For instance, any group of order $12 = 2^2 \cdot 3$ will have subgroups of orders 2, 2^2, and 3; nothing can be concluded, however, regarding the existence of a subgroup of order 6.

Example 3.9

To see an application of the Sylow Theorems, let us show that there can be no simple group G of order 42. Since $42 = 2 \cdot 3 \cdot 7$, the preceding theorem implies that G contains a subgroup H of order 7. By Theorem 3.21 the number of conjugate subgroups of H is of the form $1 + 7k$, for some integer k, and divides 42. But $1 + 7k$ divides 42 if and only if $k = 0$, so there is only one Sylow 7-subgroup. Since H coincides with all its conjugates, it is a normal subgroup of G. In consequence, G cannot be simple.

THEOREM 3.23 *If H is a normal p-subgroup of the group G, then H is contained in every Sylow p-subgroup of G for the same prime p.*

Proof. Take P to be any Sylow p-subgroup of G. Since H is assumed to be normal in G, the product HP forms a group (Problem 15, Section 2.6). We actually contend more—that HP is a p-group. From the Second Isomorphism Theorem,

$$HP/H \simeq P/(H \cap P)$$

Now $P/(H \cap P)$ is itself a p-group, being the homomorphic image of the p-group P; this makes the quotient group HP/H a p-group. As H is by hypothesis a p-group, a glance at Lemma 1 on page 218 will confirm that HP must also be a p-group. The rest is now clear. Because HP is a p-subgroup of G containing P, the maximal nature of P forces $HP = P$, from which it follows that $H \subseteq P$.

COROLLARY *Given a particular prime p, a group G has exactly one Sylow p-subgroup if and only if it contains a normal Sylow p-subgroup.*

Proof. Suppose that P is the only Sylow p-subgroup of G. By virtue of Theorem 3.20 we know that the conjugates of P are all Sylow p-subgroups. Thus $xPx^{-1} = P$ for all $x \in G$, implying that P is normal in G. Conversely, let P be a normal Sylow p-subgroup of G. If P' is an arbitrary Sylow p-subgroup, then by the foregoing theorem, $P \subseteq P'$ and it may be concluded that $P = P'$. In consequence, P is the only Sylow p-subgroup of G.

An important deduction from this last corollary is that if G is a finite commutative group, then G has precisely one Sylow p-subgroup for each prime p dividing the order of G. This is not a new result, of course, but just a shorter proof of the results in Example 3.5.

Using the foregoing corollary, we are able to extend Theorem 3.8 to a more general setting.

THEOREM 3.24 *If the Sylow p-subgroups of a finite group $G \neq \{e\}$ are all normal, then G is the direct product of its Sylow p-subgroups for those primes p dividing $o(G)$.*

Proof. Suppose that $o(G) = p_1^{k_1} p_2^{k_2} \cdots p_r^{k_r}$, where the p_i are distinct primes. Since the Sylow p-subgroups are assumed to be normal, there is precisely one Sylow p_i-subgroup for $i = 1, 2, \ldots, r$; call this unique p_i-subgroup P_i. By Theorem 3.22, each P_i is of order $p_i^{k_i}$. Since the p_i are distinct primes, the intersection of P_i with the product of the remaining ones is always e; in particular, $P_i \cap P_j = \{e\}$ for $i \neq j$.

Consider the subgroup $P_1 P_2$. By the lemma preceding Theorem 3.3 its order is

$$\frac{o(P_1)o(P_2)}{o(P_1 \cap P_2)}.$$

But $P_1 \cap P_2 = \{e\}$, so

$$o(P_1 P_2) = o(P_1)o(P_2) = p_1^{k_1} p_2^{k_2}.$$

Continuing in this way, we obtain $o(P_1 P_2 \cdots P_r) = p_1^{k_1} p_2^{k_2} \cdots p_r^{k_r}$. Since the subgroup $P_1 P_2 \cdots P_r$ is contained in G and has order equal to that of G, $P_1 P_2 \cdots P_r$ must be identical with G; that is, $G = P_1 \otimes P_2 \otimes \cdots \otimes P_r$, as asserted.

Example 3.10

Our work up to now has given us the means of showing that any group G of order 15 is cyclic. Since $15 = 3 \cdot 5$, Theorem 3.22 implies that G possesses at least one Sylow 3-subgroup and at least one Sylow 5-subgroup. The number of Sylow 3-subgroups must, by Theorem 3.21, divide 15 and at the same time be congruent to 1 modulo 3. Since the only such integer is 1, there is a unique Sylow 3-subgroup; call it H. Theorem 3.23 allows us to assert that H is normal in G. Being of prime order, H is also cyclic. By the same reasoning, there is precisely one Sylow 5-subgroup of G, label it K, that is both cyclic and normal.

In the present circumstances an appeal to Theorem 3.24 is legitimate, and we thereby conclude that $G \simeq H \otimes K$; but on closer scrutiny,

$$H \otimes K \simeq Z_3 \times Z_5 \simeq Z_{15},$$

a cyclic group of order 15 (the equipment needed is Theorem 3.6).

By way of summary we reformulate some of the essential results of this section as

THEOREM 3.25 (Sylow) *Let G be a finite group of order $p^k q$, where p is a prime and $p \nmid q$. Then*

a) *G contains at least one subgroup of order p^k.*

b) *Any two subgroups of order p^k are conjugate.*

c) *The number of distinct subgroups of order p^k is congruent to* 1 *modulo* p *and divides* $o(G)$.

d) *Any* p*-subgroup of* G *is contained in a subgroup of order* p^k.

PROBLEMS 3.3

In the following set of problems, p always denotes a prime number.

1. Let H and K be two subgroups of the group G. Show that H is normal in K if and only if $H \subseteq K \subseteq N(H)$.

2. Suppose that H is a proper subgroup of the finite group G. Verify that G contains an element that belongs to no conjugate subgroup of H.

3. Let G be an arbitrary group and a be a fixed element of G. Define the function f by $f(xC(a)) = xax^{-1}$, $x \in G$. Show that f is a one-to-one mapping from the left cosets of $C(a)$ onto the set of distinct conjugates of a, thereby deducing Theorem 3.16.

4. Prove that a subgroup H of G is a normal subgroup if and only if the set H is a union of conjugacy classes of G.

5. If in a finite group G an element exists that has exactly two conjugates, prove that G contains a nontrivial normal subgroup.

6. Describe the class equation for the symmetric group S_3 and alternating group A_4.

7. Let G be a group such that $o(G) = p^n$, $n > 0$. Prove that

a) cent G contains a subgroup of order p;

b) G has at least one normal subgroup of order p^k for all $0 \leqslant k \leqslant n$. [*Hint:* Proceed by induction on n and utilize part (a).]

8. Establish that if a finite group G has exactly two conjugacy classes, then $o(G) = 2$.

9. Let H be a normal subgroup of the finite p-group G and suppose that $o(H) = p$. Establish the inclusion $H \subseteq \text{cent } G$. [*Hint:* If $H = (a)$, show that $[G : C(a)] = 1$.]

10. Show that if G is a group of order p^2, then either $G \simeq Z_{p^2}$ or else $G \simeq Z_p \times Z_p$. [*Hint:* Use the corollary to Theorem 3.18 and the remark on page 205.]

11. Prove that if G is a noncommutative group of order p^3, then $G/\text{cent } G$ has order p^2.

12. If $o(G) = p^n$, $n > 0$, verify that

a) $o(\text{cent } G) \neq p^{n-1}$;

b) any subgroup H of order p^{n-1} is normal in G. [*Hint:* Proceed by induction on n. Let $a \in \text{cent } G$ be an element of order p. If $a \in H$, consider the subgroup nat H of $G/(a)$.]

13. a) Determine all Sylow p-subgroups of the group Z_{24}.

b) Determine the Sylow 3-subgroups of the alternating group A_4.

14. Show that a p-subgroup P of a finite group G is a Sylow p-subgroup if and only if $p \nmid [G:P]$.

15. a) Assume that H is an arbitrary p-subgroup of the finite group G. If H is contained in exactly one Sylow p-subgroup P of G, verify the inclusion $N(H) \subseteq N(P)$.

b) If P is a Sylow p-subgroup of the finite group G, show that P is the only Sylow p-subgroup contained in $N(P)$.

16. Let G be a finite group and P be any Sylow p-subgroup of G. Given that H is any p-subgroup of G (for the same prime p), show that H is contained in a conjugate of P. [*Hint:* Every p-subgroup of G is contained in a Sylow p-subgroup.]

17. Let P be a Sylow p-subgroup of the finite group G. Prove that if H is any subgroup of G with $P \subseteq N(P) \subseteq H$, then $N(H) = H$; in particular, deduce that $N(N(P)) = N(P)$. [*Hint:* If $x \in N(H)$, then P and xPx^{-1} are both Sylow p-subgroups of H and hence are conjugate.]

18. Prove each of the following:

a) Any group of order $2p$ has a normal subgroup of order p (p an odd prime).

b) There are no simple groups of orders 20, 30, or 56.

c) There is exactly one Sylow 13-subgroup in any group of order 130.

d) Any group of order 12 must contain a normal subgroup of order 2 or 3.

19. Let G be a group of order pq, where $p < q$ are both primes.

a) Show that G has precisely one normal subgroup of order q.

b) If $q \not\equiv 1 \pmod{p}$, prove that G is a cyclic group.

20. Prove that the following conditions on a finite group G are equivalent.

a) Each Sylow p-subgroup of G is normal.

b) Any pair of elements in G with relatively prime orders commute.

c) G is the direct product of its Sylow p-subgroups.

21. If the Sylow p-subgroups of the finite group G are normal and cyclic, prove that G is itself cyclic.

22. Let G be a group of order p^2q, where $p < q$ are both primes. If $q \not\equiv 1 \pmod{p}$, prove that G is commutative. [*Hint:* G has a unique Sylow p-subgroup and a unique Sylow q-subgroup.]

23. State and prove a theorem ensuring that any group of order p^2q^2, where p and q are distinct primes, will be commutative.

24. Establish that if G is a group of order 6, then either $G \simeq Z_6$ or $G \simeq S_3$.

3.4 THE JORDAN-HOLDER THEOREM

As the section title indicates, the main purpose of this section is to establish the significant and historic Jordan-Holder Theorem. For the sake of simplicity we shall limit ourselves to the case of finite groups. Because the theorem is rather involved, it will be convenient to begin by introducing some special terminology.

DEFINITION 3.7 By a *chain* for a group G we mean any finite sequence of subgroups of G:

$$G = H_0 \supset H_1 \supset \cdots \supset H_{n-1} \supset H_n = \{e\}$$

descending with proper inclusions from G to $\{e\}$. The integer n is called the *length* of the chain.

What we are really interested in, and shall henceforth confine our attention to, are the so-called *normal chains*. These are chains in which each group H_i is a normal subgroup of its immediate predecessor H_{i-1}. For groups G with two or more elements there is always one normal chain, the trivial chain $G \supset \{e\}$; however, this may very well be the only such chain.

Example 3.11
In the group Z_{12} of integers modulo 12 the following chains are normal chains:

$$Z_{12} \supset (6) \supset \{0\}, \qquad Z_{12} \supset (2) \supset (4) \supset \{0\},$$

$$Z_{12} \supset (3) \supset (6) \supset \{0\}, \qquad Z_{12} \supset (2) \supset (6) \supset \{0\}.$$

All the subgroups involved are normal, since Z_{12} is a commutative group.

Among other things, this particular example indicates that a given chain may be lengthened or refined by the insertion of admissible subgroups. In technical terms a second chain

$$G = K_0 \supset K_1 \supset \cdots \supset K_{m-1} \supset K_m = \{e\}$$

is said to be a *refinement* of the chain

$$G = H_0 \supset H_1 \supset \cdots \supset H_{n-1} \supset H_n = \{e\}$$

provided that a one-to-one function f exists from $\{0, 1, \ldots, n\}$ into $\{0, 1, \ldots, m\}$ such that $H_i = K_{f(i)}$ for all i. What we are requiring, in effect, is that every H_i coincide with one of the K_j. The lengths of the foregoing chains clearly must satisfy the inequality $n \leqslant m$, equality being allowed because every chain is trivially a refinement of itself. A refinement of a chain is termed *proper* if the refinement contains a subgroup that is not in the original chain. A normal chain that has no proper refinement is customarily called a *composition chain.* We summarize all this in the following definition.

DEFINITION 3.8 In the group G the descending sequence of sets

$$G = H_0 \supset H_1 \supset \cdots \supset H_{n-1} \supset H_n = \{e\}$$

forms a *composition chain* for G provided that

1. H_i is a subgroup of G,
2. H_i is a normal subgroup of H_{i-1},
3. the inclusion $H_{i-1} \supseteq K \supseteq H_i$, where K is a normal subgroup of H_{i-1}, implies that either $K = H_{i-1}$ or $K = H_i$.

Before proceeding to develop some of the properties of composition chains, let us pause to consider several examples.

Example 3.12
In the group Z_{24} the normal chain

$$Z_{24} \supset (2) \supset (12) \supset \{0\}$$

is not a composition chain, since it may be refined one stage further by inserting either of the cyclic groups (4) or (6). On the other hand,

$$Z_{24} \supset (2) \supset (4) \supset (8) \supset \{0\}$$

and

$$Z_{24} \supset (3) \supset (6) \supset (12) \supset \{0\}$$

are both composition groups for Z_{24}. One way of verifying this is to check the orders of the subgroups concerned. For instance, to insert a subgroup between (2) and (4), there would have to exist a subgroup of Z_{24} of order n, where $6 < n < 12$, such that n divides 12 and is itself divisible by 6; clearly, no such subgroup exists.

Example 3.13
For a second illustration, consider the group G_s of symmetries of the square. In this case the following sequence of subgroups serves as a composition chain:

$$G_s \supset \{e, r_2, h, v\} \supset \{e, h\} \supset \{e\}.$$

It is worth noting that the subgroup $\{e, h\}$ is not normal in the original group G_s, but only in its immediate predecessor.

A normal chain for G_s that fails to be a composition chain is

$$G_s \supset \{e, r_2\} \supset \{e\}.$$

This chain admits any one of several refinements; among others there is

$$G_s \supset \{e, r_1, r_2, r_3\} \supset \{e, r_2\} \supset \{e\}.$$

Example 3.14

To see that not every group possesses a composition chain, merely consider the additive group of integers Z. We have previously observed that the normal subgroups of Z are the cyclic subgroups (n) where n is a nonnegative integer. Since the inclusion $(kn) \subseteq (n)$ holds for all $k \in Z_+$, a proper subgroup of any given group always exists. Accordingly, each chain for Z may be refined indefinitely.

Condition (3) of Definition 3.8 merits closer attention. Loosely speaking, it prohibits us from squeezing a normal subgroup between H_{i-1} and H_i. There is a special terminology for this situation.

DEFINITION 3.9 A normal subgroup H is called a *maximal subgroup* of the group G if $H \neq G$ and no normal subgroup K of G exists such that $H \subset K \subset G$.

In terms of this new notion a chain

$$G = H_0 \supset H_1 \supset \cdots \supset H_{n-1} \supset H_n = \{e\}$$

is a composition chain for G if each subgroup H_i is a maximal normal subgroup of H_{i-1}.

We should point out that Definition 3.9 does not imply that H is maximal in the sense of being the subgroup of G with the largest order. A given group may well have many distinct maximal normal subgroups of various orders. Indeed, in the group Z_{24}, the cyclic subgroups (2) and (3) are both maximal normal and have orders 12 and 8, respectively.

The next theorem develops the connection between a maximal normal subgroup and its associated quotient group. For this, recall that a group is said to be simple if the only normal subgroups are the two trivial ones.

THEOREM 3.26 *A normal subgroup H of the group G is maximal if and only if the quotient group G/H is simple.*

Proof. This result follows immediately from the Correspondence Theorem. We saw that to each normal subgroup K of G with $H \subseteq K$ there corresponds, in a one-to-one way, a normal subgroup K/H of the quotient group G/H. Hence H is maximal normal in G if and only if there are exactly two normal subgroups in G/H, the entire group and the identity subgroup.

Certainly all the quotient groups H_{i-1}/H_i of a composition chain are simple. Conversely, any normal chain whose quotient groups are all simple groups is not subject to further refinement and is therefore a composition chain.

Not every group contains a maximal normal subgroup, so we cannot always find a composition chain for a given group. However, in the case of finite groups there is the following result.

THEOREM 3.27 *Every finite group $G \neq \{e\}$ has a composition chain.*

Proof. If G is a simple group, then the trivial chain $G \supset \{e\}$ is a composition chain. Suppose G is not simple, so that there exists a nontrivial normal subgroup H. If H is maximal in G and $\{e\}$ is maximal in H, then

$$G \supset H \supset \{e\}$$

is a composition chain, and we are finished. When H is not already maximal normal, a larger normal subgroup K exists for which $H \subset K \subset G$. Now if H is maximal in K and K is maximal in G, $G \supset K \supset H \supset \{e\}$ is the desired composition chain. Continue to argue along these lines. Since G is finite, there are only a finite number of subgroups, and this process must ultimately terminate. The resulting chain of sets is obviously a composition chain for G.

The set of subgroups that make up a composition chain for a specific group G is not, in general, unique. For instance, the group Z_6 of integers modulo 6 has two composition chains,

$$Z_6 \supset (2) \supset \{0\} \qquad \text{and} \qquad Z_6 \supset (3) \supset \{0\}.$$

The subgroups involved in these chains are not only different, but they are not even isomorphic. The next theorem ensures that while the composition chains for a given group can differ, they cannot differ too widely. For this we need a method for comparing different chains.

DEFINITION 3.10 Two composition chains for the group G,

$$G = H_0 \supset H_1 \supset \cdots \supset H_{n-1} \supset H_n = \{e\}$$

and

$$G = K_0 \supset K_1 \supset \cdots \supset K_{m-1} \supset K_m = \{e\},$$

are termed *equivalent* if they have the same length (that is, $n = m$) and a permutation $f \in S_n$ exists such that

$$H_{i-1}/H_i \simeq K_{f(i)-1}/K_{f(i)} \qquad \text{for } i = 1, 2, \ldots, n.$$

In other words, two chains are equivalent if their associated quotient groups are isomorphic in some order.

With this terminology the Jordan-Holder Theorem asserts that any two composition chains for a given finite group are equivalent. Jordan proved in 1869 that if a finite group has two different composition chains, then the number of subgroups in them is the same, and the orders of the quotients formed from consecutive groups are the same, except possibly for the sequence in which these orders occur. To say it a different way, the indices of the subgroups in any composition chain are invariant. In 1889, Otto Holder (1859–1937), a professor at the University of Leipzig, extended Jordan's result on the uniqueness of orders to show that the quotient groups themselves are independent of the choices of the normal subgroups found in a composition chain; that is, to within isomorphism the same quotient groups would be present for any composition chain (although not necessarily in the same sequence). These two results are linked in the Jordan-Holder Theorem, one of the most famous theorems of group theory.

Holder, who owed his interest in group theory to his participation in Felix Klein's seminar at Leipzig, made several notable advances in the subject during the last decades of the nineteenth century. The approach to the concept of a normal subgroup through quotient groups appears for the first time in Holder's 1889 paper. He refers to it as an "extraordinary" subgroup; it differs from the "ordinary" subgroup in that only when H is a normal subgroup of G do the cosets aH necessarily form a group under the operation defined by $aH \cdot bH = abH$. Holder called this group of cosets the quotient of the groups G and H, terminology that survives today. He denoted it by the symbol G/H, notation that was introduced by Jordan in 1869 when discussing the existence and properties of the quotient group without explicitly constructing it. Jordan never spoke of a normal subgroup, but rather of a subgroup H being "permutable" with the elements of a group G; that is, H satisfies $aH = Ha$ for all $a \in G$. Jordan's definition of a normal subgroup is, of course, equivalent to the one we use nowadays.

We are now equipped to attack the Jordan-Holder Theorem. To deal conveniently with the theorem, part of the proof is separated as a preliminary lemma.

LEMMA *If H and K are distinct maximal normal subgroups of the group G, then*

a) $H \cap K$ is a maximal normal subgroup of both H and K;
b) $G/H \simeq K/(H \cap K)$ and $G/K \simeq H/(H \cap K)$.

Proof. Since the subgroups H and K are both normal, their product HK is itself a normal subgroup of G (Problem 15, Section 2.6). Plainly, we have the inclusion $H \subseteq HK \subseteq G$. By hypothesis the subgroup H is maximal normal in G, which implies that either $HK = H$ or $HK = G$. Our contention is that $HK = G$. Were this not the case, the equality $HK = H$ would mean $K \subseteq H$ (with $K \neq H$), contradicting the maximality of K; therefore the only possibility left us is that $G = HK$.

A straightforward application of Theorem 2.65 now establishes the isomorphism $G/K \simeq H/(H \cap K)$. But the quotient group G/K is a simple group (Theorem 3.26), so the same must also be true of $H/(H \cap K)$. From this, utilizing Theorem 3.26, we conclude that the group $H \cap K$ is maximal normal in H.

If the letters H and K are interchanged, the symmetry of the hypothesis yields the remaining parts of the lemma.

THEOREM 3.28 (Jordan-Holder) *In a finite group $G \neq \{e\}$, any two composition chains are equivalent.*

Proof. The proof will proceed by induction on the order of the group. If G is simple, the only possible composition chain is the trivial chain $G \supset \{e\}$, so the theorem is certainly true. This establishes the results for groups of order 2, the smallest admissible order. In general, let

$$G = H_0 \supset H_1 \supset \cdots \supset H_{n-1} \supset H_n = \{e\} \tag{1}$$

and

$$G = K_0 \supset K_1 \supset \cdots \supset K_{m-1} \supset K_m = \{e\} \tag{2}$$

be any two composition chains for the group G. Assume further that the theorem holds for all groups having order less than that of G. We distinguish two cases:

Case I: $H_1 = K_1$. In this case, if the set G is deleted from (1) and (2), the resulting chains

$$H_1 \supset H_2 \supset \cdots \supset H_{n-1} \supset H_n = \{e\}$$

$$H_1 \supset K_2 \supset \cdots \supset K_{m-1} \supset K_m = \{e\}$$

both represent composition chains for the subgroup H_1. Since the theorem is assumed to be true for H_1, whose order is less than that of G, these two chains are necessarily equivalent. But $G/H_1 = G/K_1$; hence the given composition chains (1) and (2) must also be equivalent.

Case II: $H_1 \neq K_1$. Either $H_1 \cap K_1 = \{e\}$, or by Theorem 3.28 there exists a composition chain

$$H_1 \cap K_1 \supset L_1 \supset \cdots \supset L_{r-1} \supset L_r = \{e\}$$

for the subgroup $H_1 \cap K_1$. Now, according to the lemma just established, $H_1 \cap K_1$ is a maximal normal subgroup of both H_1 and K_1. It then follows that the two chains

$$G \supset H_1 \supset H_1 \cap K_1 \supset L_1 \supset \cdots \supset L_{r-1} \supset L_r = \{e\} \tag{3}$$

and

$$G \supset K_1 \supset H_1 \cap K_1 \supset L_1 \supset \cdots \supset L_{r-1} \supset L_r = \{e\} \tag{4}$$

are actually composition chains for G. Appealing to the lemma once more, we see that

$$G/H_1 \simeq K_1/(H_1 \cap K_1)$$

and

$$G/K_1 \simeq H_1/(H_1 \cap K_1).$$

Hence the composition quotient groups obtained from (3) are isomorphic in pairs to those obtained from the chain (4); let us agree to abbreviate this situation by writing (3) $\simeq$ (4).

Next, consider the following two composition chains for the group G:

$$G \supset H_1 \supset H_2 \supset \cdots \supset H_{n-1} \supset H_n \supset \{e\}$$

and

$$G \supset H_1 \supset H_1 \cap K_1 \supset L_1 \supset \cdots \supset L_{r-1} \supset L_r = \{e\}.$$

Since these chains have their first two terms in common, we can conclude from Case I that they must be equivalent. In a like manner,

$$G \supset K_1 \supset K_2 \supset \cdots \supset K_{m-1} \supset K_m = \{e\}$$

and

$$G \supset K_1 \supset H_1 \cap K_1 \supset L_1 \supset \cdots \supset L_{r-1} \supset L_r = \{e\}$$

are equivalent chains for G. Combining our results, we observe that

$$(1) \simeq (3) \simeq (4) \simeq (2),$$

from which it follows that the original chains (1) and (2) are equivalent.

Let us quickly review what has just been learned. First, the preceding theorem implies that the composition chains of a given finite group G must all have the same length. Moreover, by means of the quotient groups of any such chain, we are able to associate with G a finite sequence of simple groups. Up to isomorphism these simple groups depend solely on G and are independent of the particular composition chain from which they were originally obtained. The important point is that the composition quotient groups will to some degree mirror the properties of the given group and provide a hint as to its algebraic structure.

Example 3.15

To illustrate the Jordan-Holder Theorem, consider the group Z_{60} of integers modulo 60 and the two composition chains

$$Z_{60} \supset (3) \supset (6) \supset (12) \supset \{0\},$$

$$Z_{60} \supset (2) \supset (6) \supset (30) \supset \{0\}.$$

After suitable rearrangement the isomorphic pairs of composition quotient groups are

$$Z_{60}/(3) \simeq (2)/(6),$$

$$(3)/(6) \simeq Z_{60}/(2),$$

$$(6)/(12) \simeq (30)/\{0\},$$

$$(12)/\{0\} \simeq (6)/(30).$$

All these quotient groups are simple, being cyclic groups of prime order; the first pair of isomorphic groups is of order 3, the last pair is of order 5, and the remaining groups are all of order 2. Note that the product of all these orders is 60.

We now take a brief look at a wide class of groups that contains, among others, all commutative groups.

DEFINITION 3.11 A group G is *solvable* if it has a normal chain

$$G = H_0 \supset H_1 \supset \cdots \supset H_{n-1} \supset H_n = \{e\}$$

in which each of the quotient groups H_{i-1}/H_i is commutative $(i = 1, 2, \ldots, n)$. Such a chain is called a *solvable chain* for G.

Solvable groups are closely connected with the possible solution of polynomial equations by radicals, which explains the unexpected designation. (The name "solvable group" or, as employed by some authors, "soluble group" was introduced by Camille Jordan in 1867.) Groups satisfying Definition 3.11 were first studied by Evariste Galois in the earliest stages of the development of group theory. The famous theorem of Galois (1832) that the general polynomial equation of degree $\geqslant 5$ is not solvable by radicals has as its basis the nonsolvability of a finite group naturally associated with the equation. The phrase "solvable by radicals" means that there is a formula that gives the roots of the polynomial in terms of the usual algebraic operations and the extraction of roots, beginning with the coefficients.

It should be clear at once that any commutative group $G \neq \{e\}$ is solvable, since in this case the trivial chain $G \supset \{e\}$ is a solvable chain. Examples of noncommutative solvable groups are furnished by G_s, the group of symmetries of the square, and the symmetric group S_3. Since $\{e, r_1, r_2, r_3\}$ is a subgroup of index 2 in G_s, the resulting quotient group is commutative. Moreover, $\{e, r_1, r_2, r_3\}$ is itself a commutative group, so the normal chain

$$G_s \supset \{e, r_1, r_2, r_3\} \supset \{e\}$$

has commutative quotient groups, and therefore G_s is a solvable group. As for S_3, a solvable chain is

$$S_3 \supset \left\{\begin{pmatrix}1 & 2 & 3\\ 1 & 2 & 3\end{pmatrix}, \begin{pmatrix}1 & 2 & 3\\ 2 & 3 & 1\end{pmatrix}, \begin{pmatrix}1 & 2 & 3\\ 3 & 1 & 2\end{pmatrix}\right\} \supset \left\{\begin{pmatrix}1 & 2 & 3\\ 1 & 2 & 3\end{pmatrix}\right\}.$$

In more recent times (1963), Feit and Thompson have shown in their paper "Solvability of Groups of Odd Order" that all finite groups of odd order are solvable. Thus if one is interested in nonsolvable groups, these are to be found among the noncommutative groups of even order (note that $o(G_s) = 8$ and $o(S_3) = 6$).

Some basic properties of solvable groups are given in our next theorem.

THEOREM 3.29 *If G is a solvable group, then every subgroup of G is solvable and every homomorphic image of G is solvable.*

Proof. Let

$$H = H_0 \supset H_1 \supset \cdots \supset H_{n-1} \supset H_n = \{e\}$$

be a fixed solvable chain for G. Given a subgroup K of G, define $K_i = K \cap H_i$ $(i = 0, 1, \ldots, n)$. We intend to prove that the chain

$$K = K_0 \supset K_1 \supset \cdots \supset K_{n-1} \supset K_n = \{e\}$$

is a solvable chain for the subgroup K. First, observe that K_i is a normal subgroup of K_{i-1}; in fact, for each $a \in K_{i-1}$,

$$\begin{aligned} aK_ia^{-1} &= (aK_ia^{-1}) \cap K \\ &\subseteq (aH_ia^{-1}) \cap K \subseteq H_i \cap K = K_i. \end{aligned}$$

Also,

$$K_i = K \cap H_i = K \cap H_{i-1} \cap H_i = K_{i-1} \cap H_i,$$

so that

$$K_{i-1}/K_i = K_{i-1}/(K_{i-1} \cap H_i).$$

According to Theorem 2.66, we have the isomorphism

$$K_{i-1}/(K_{i-1} \cap H_i) \simeq K_{i-1}H_i/H_i.$$

But the quotient group $K_{i-1}H_i/H_i$ is commutative, being a subgroup of the commutative group H_{i-1}/H_i. This implies that all the quotient groups K_{i-1}/K_i are commutative, as required.

As for the second part of the theorem, let f be a homomorphism from the group G, onto the group G'. Setting $H'_i = f(H_i)$ for $i = 0, 1, \ldots, n$, we obtain a chain for G':

$$G' = H'_0 \supset H'_1 \supset \cdots \supset H'_{n-1} \supset H'_n = \{e'\}.$$

By the corollary to Theorem 2.54 the subgroup H'_i is normal in H'_{i-1}. What little difficulty there is arises in showing the corresponding quotient group H'_{i-1}/H'_i to be commutative. For this we define mappings $f_i\colon H_{i-1}/H_i \to H'_{i-1}/H'_i$ by taking

$$f_i(aH_i) = f(a)H'_i, \qquad a \in H_{i-1} \qquad (i = 1, 2, \ldots, n).$$

It is easily seen that each f_i is well defined; indeed, if $aH_i = bH_i$, then $a^{-1}b \in H_i$ and hence

$$f(a)^{-1}f(b) = f(a^{-1}b) \in f(H_i) = H'_i,$$

which, in turn, implies that

$$f_i(aH_i) = f(a)H'_i = f(b)H'_i = f_i(bH_i).$$

Next, note that each f_i is itself a homomorphism; if $a, b \in H_{i-1}$,

$$\begin{aligned} f_i((aH_i)(bH_i)) &= f_i(abH_i) = f(ab)H'_i \\ &= f(a)f(b)H'_i = (f(a)H'_i)(f(b)H'_i). \end{aligned}$$

Finally, because $f(H_i) = H'_i$, the function f_i maps onto the set H'_{i-1}/H'_i. From these facts we are able to conclude that

$$H_{i-1}/H_i \simeq H'_{i-1}/H'_i,$$

and since H_{i-1}/H_i is a commutative group, so also is the quotient group H'_{i-1}/H'_i.

COROLLARY *If G is solvable and H is a proper normal subgroup of G, then the quotient group G/H is solvable.*

In the way of a converse to this theorem we prove

THEOREM 3.30 *Let H be a nontrivial normal subgroup of G. If both H and G/H are solvable groups, then G is itself solvable.*

Proof. To begin, let

$$G/H = K'_0 \supset K'_1 \supset \cdots \supset K'_{n-1} = \{eH\}$$

be a solvable chain for the quotient group G/H. Using the Correspondence Theorem, we obtain a sequence of subgroups

$$G = K_0 \supset K_1 \supset \cdots \supset K_{n-1} \supset K_n = H$$

descending from G to H with the property that K_i is a normal subgroup of K_{i-1}; in fact, $K_i = \text{nat}_H^{-1}(K_i')$. Furthermore, by Theorem 2.65,

$$K_{i-1}/K_i \simeq K_{i-1}'/K_i',$$

so the quotient group K_{i-1}/K_i is commutative. Since H is solvable, it has a solvable chain running from H to $\{e\}$, say,

$$H = H_0 \supset H_1 \supset \cdots \supset H_{m-1} \supset H_m = \{e\}.$$

By stringing these sequences of subgroups together we can construct a solvable chain for G:

$$G \supset K_1 \supset \cdots \supset K_{n-1} \supset H \supset H_1 \supset \cdots \supset H_{m-1} \supset \{e\}.$$

Therefore G is a solvable group.

In conjunction, Theorems 3.29 and 3.30 tell us that if H is a nontrivial normal subgroup of G, then G is a solvable group if and only if both H and G/H are solvable groups. The decision regarding solvability is often facilitated by this criterion.

The following theorem gives further insight into the nature of the composition chains for a finite solvable group.

THEOREM 3.31 *Let G be a finite solvable group. Then the quotient groups of any composition chain for G are cyclic groups of prime order.*

Proof. We proceed by induction on the order of G. If $o(G)$ is a prime, there is nothing to prove, since $G \supset \{e\}$ is the only composition chain. (This remark takes care of the basis for the induction when $o(G) = 2$.) In the contrary case, G has a nontrivial normal subgroup H. (Why?) By the induction assumption, H and G/H, being solvable, have composition chains whose quotient groups are of prime order, say, the chains

$$H = H_0 \supset H_1 \supset \cdots \supset H_{n-1} \supset H_n = \{e\}$$

and

$$G/H = K_0' \supset K_1' \supset \cdots \supset K_{m-1}' \supset K_m' = \{eH\}.$$

Taking $K_i = \text{nat}_H^{-1}(K_i')$ for $i = 0, 1, \ldots, m$, it follows from Theorem 2.64 that

$$G = K_0 \supset K_1 \supset \cdots \supset K_{m-1} \supset K_m = H,$$

where K_i is a normal subgroup of K_{i-1}. As before, Theorem 2.65 implies that

$$K_{i-1}/K_i \simeq K'_{i-1}/K'_i,$$

whence K_{i-1}/K_i is (cyclic) of prime order. Linking these subgroups, we obtain (after removing the second occurrence of H) a composition chain for G, all of whose quotient groups are cyclic groups of prime order. By the Jordan-Holder Theorem the same is true of every composition chain.

Let us now use these ideas to establish a result that lies at the root of the existence, for each $n \geqslant 5$, of polynomial equations of degree n that cannot be solved by radicals.

THEOREM 3.32 *For $n \geqslant 5$ the symmetric group S_n is not solvable.*

Proof. Since A_n is a simple group for $n \geqslant 5$ (Theorem 2.50), the normal chain

$$S_n \supset A_n \supset \{\imath\}$$

is actually a composition chain for S_n. However, A_n is not commutative and consequently is not a cyclic group. In light of Theorem 3.31, we deduce that S_n cannot be solvable for $n \geqslant 5$.

At the same time, we have

COROLLARY *For $n \geqslant 5$ the alternating group A_n is not solvable.*

Proof. Assume to the contrary that A_n is solvable ($n \geqslant 5$). Since S_n/A_n is also solvable, being a commutative group, Theorem 3.30 implies that S_n must be solvable, an impossibility.

William Burnside pursued the study of groups, particularly solvable groups, in some 50 of the 150 research papers that he wrote around the turn of the century. In 1904, Burnside established a famous solvability criterion: If p and q are primes, then any group G with $o(G) = p^m q^n$ is solvable. This result represented the final generalization of a large number of special cases that had been proven for particular exponents by Sylow, Frobenius, Jordan, Cole, and Burnside himself.

A group G all of whose elements satisfy the equation $x^n = e$ for some fixed positive integer n is said to be of finite exponent n. In 1902, Burnside posed a question on the relation between the finiteness of the exponent and the finiteness of the group: If G is a finitely generated group with finite exponent, is G necessarily finite? This question, which became known as the

Burnside Problem, had a stimulating effect on the future development of group theory and remained open for years. A negative solution was obtained in 1968 (in a difficult paper of some 330 pages) when it was shown that for every $m \geqslant 2$ and odd $n \geqslant 665$ there exists an infinite group with m generators and exponent n. Before that it was known only that the Burnside Problem's answer was positive for $n = 1, 2, 3, 4$, or 6.

PROBLEMS 3.4

1. Check that each of the following chains represents a composition chain for the indicated group.

a) For the group Z_{36} of integers modulo 36:

$$Z_{36} \supset (3) \supset (9) \supset (18) \supset \{0\}.$$

b) For the group G_s of symmetries of the square:

$$G_s \supset \{e, r_2, d_1, d_2\} \supset \{e, d_1\} \supset \{e\}.$$

c) For a cyclic group (a) of order 30:

$$(a) \supset (a^5) \supset (a^{10}) \supset \{e\}.$$

d) For the symmetric group S_3:

$$S_3 \supset \left\{\begin{pmatrix}1 & 2 & 3\\1 & 2 & 3\end{pmatrix}, \begin{pmatrix}1 & 2 & 3\\2 & 3 & 1\end{pmatrix}, \begin{pmatrix}1 & 2 & 3\\3 & 1 & 2\end{pmatrix}\right\} \supset \left\{\begin{pmatrix}1 & 2 & 3\\1 & 2 & 3\end{pmatrix}\right\}.$$

2. Find a composition chain for the symmetric group S_4. [*Hint:* Consider the set A_4 of even permutations].

3. Show that no infinite cyclic group possesses a composition chain.

4. Prove that if G is a finite cyclic group of order 2^n, then there is exactly one composition chain for G.

5. Derive the following result: If the quotient group G/H is of prime order, then H is a maximal normal subgroup of G. [*Hint*: Use the Correspondence Theorem.]

6. Prove that the cyclic subgroup (n) is a maximal normal subgroup of the additive group Z if and only if n is a prime number.

7. Establish that the following two composition chains for Z_{24} are equivalent:

$$Z_{24} \supset (3) \supset (6) \supset (12) \supset \{0\},$$

$$Z_{24} \supset (2) \supset (4) \supset (12) \supset \{0\}.$$

8. Find all composition chains for (a) the group Z_{30} of integers modulo 30 and (b) the group G_s of symmetries of the square. Verify the Jordan-Holder Theorem in both cases. [*Hint:* There are four composition chains for each group.]

9. Apply the Jordan-Holder Theorem to the group Z_n of integers modulo n to prove that a positive integer n may be factored uniquely (apart from the order of the factors) into a product of primes.

10. Let G be the direct product of three nontrivial groups G_k for $k = 1, 2, 3$; that is to say, $G = G_1 \times G_2 \times G_3$ with the binary operation $\cdot$ defined by

$$(a, b, c) \cdot (a', b', c') = (aa', bb', cc').$$

a) Verify that

$$G \supset \{e_1\} \times G_2 \times G_3 \supset \{e_1\} \times \{e_2\} \times G_3 \supset \{e_1\} \times \{e_2\} \times \{e_3\}$$

is a normal chain for G.

b) If the original groups G_k are simple, prove that the foregoing chain must be a composition chain for G.

c) State and prove an analogous result for n groups.

11. Suppose that H is a proper normal subgroup of the group G and

$$G \supset K_1 \supset \cdots \supset K_r \supset H \supset L_1 \supset \cdots \supset L_n \supset \{e\}$$

is a composition chain for G. Prove that

$$G/H \supset K_1/H \supset \cdots \supset K_r/H \supset \{H\}$$

represents a composition chain for the quotient group G/H.

12. For each of the solvable groups of Problem 1, show that the indicated composition chains have cyclic quotient groups of prime order.

13. Prove that the only solvable simple groups $G \neq \{e\}$ are the groups of prime order.

14. Establish that the symmetric group S_4 is solvable. [*Hint:* The subgroup

$$H = \left\{\begin{pmatrix}1 & 2 & 3 & 4\\ 1 & 2 & 3 & 4\end{pmatrix}, \begin{pmatrix}1 & 2 & 3 & 4\\ 2 & 1 & 4 & 3\end{pmatrix}, \begin{pmatrix}1 & 2 & 3 & 4\\ 3 & 4 & 1 & 2\end{pmatrix}, \begin{pmatrix}1 & 2 & 3 & 4\\ 4 & 3 & 2 & 1\end{pmatrix}\right\}$$

is normal in A_4.]

15. If G_1 and G_2 are solvable groups, show that their direct product $G_1 \times G_2$ is also a solvable group.

16. Let G be the set of all 2×2 matrices of the form $\begin{bmatrix} a & b \\ 0 & c \end{bmatrix}$, where a, b, c are real numbers and $ac \neq 0$. Verify that G is a solvable group. [*Hint:* Consider the subgroup H of all elements of G in which $a = c = 1$.]

17. Let G be a p-group of order p^n, p a prime. Prove that G is a solvable group. [*Hint:* Use induction to show that G has a normal chain $G = H_0 \supset H_1 \supset \cdots \supset H_{n-1} = \{e\}$ with the index $[H_i : H_{i-1}] = p$.]

18. Suppose that H and K are solvable subgroups of the group G, with H normal in G. Prove that the subgroup KH is also solvable.

19. a) Show that any group of order pq, where p and q are distinct primes, is solvable.

b) Let G be a group of order pqr, where p, q, r are distinct primes. If G contains a nontrivial normal subgroup, prove that G is solvable.

4

Ring Theory

. . . mathematical proofs, like diamonds, are hard as well as clear, and will be touched with nothing but strict reasoning.

JOHN LOCKE

4.1 EMMY NOETHER

Amalie Emmy Noether (1882–1935) was one of the foremost of algebraists and probably the greatest female mathematician yet. She was born in the German university town of Erlangen, where her father, Max Noether, was professor of mathematics and a celebrated algebraic geometer. In her early years, Emmy Noether seemed to favor the study of foreign languages over mathematics; in 1900 she passed the test for prospective schoolteachers of English and French.

At the turn of the century it was not an easy matter for a woman to obtain a university education. The University of Erlangen, for instance, declared that the admission of women would "overthrow all academic order." In spite of this policy, Emmy Noether and one other woman were permitted to audit classes at Erlangen, along with a thousand regularly enrolled male students. During the winter semester of 1903–1904, still as an auditor, Noether moved to the University of Göttingen to attend the lectures of Felix Klein and David Hilbert. After one semester, Noether returned to Erlangen, where it had just become possible for female students to enroll with the same status as male students. Three years later (1907), she finished her doctoral dissertation, "On Complete Systems of Invariants for Ternary Biquadratic Forms." Her thesis advisor was her father's friend and colleague Paul Gordan, who, reigning supreme in the field, was called the "King of Invariant Theory" by his admiring contemporaries. Noether later called the thesis "a jungle of formulas"; its style was in sharp contrast to the extreme generality and abstractness of approach that would eventually characterize her work.

During the years between 1908 and 1915, Noether worked in Erlangen without compensation, carrying out her own research and occasionally lecturing for her father when he was ill. Her impressive invariant-theoretic knowledge appeared to be useful to Hilbert in his work on the mathematical aspects of the new general theory of relativity, and in 1915 he invited Noether to come to Göttingen. Hilbert tried repeatedly to secure a lecturership for her but ran afoul of the strong prejudice against women teaching in universities. During a University Senate meeting held to discuss a possible appointment, Hilbert finally exploded, "I do not see that the sex of the candidate is an argument against her admission as a Privatdozent. After all we are a university, not a bathhouse." Hilbert's efforts to arrange the position never did succeed; still, Noether was permitted to lecture at

Göttingen in courses that were announced under Hilbert's name. In 1922 she was designated "unofficial assistant professor," a purely honorary title without accompanying responsibilities or salary. After some urging, the authorities finally provided Noether with a modest but regular remuneration for the courses she taught. When the Nazi regime began in 1933, a decree was issued dismissing all faculty members who were Jewish or had Jewish family members. Deprived of the right to teach in Germany, Noether emigrated to the United States, where she accepted a visiting professorship at Bryn Mawr College in Pennsylvania. Only a year and a half of her life remained to her there; in the spring of 1935, at the height of her creative powers, Noether died unexpectedly after an operation.

Perhaps inspired by Hilbert's (1899) axiomatization of Euclidean geometry, Noether sought a system of abstract axioms from which to deduce the facts of ring theory. Ring theory was in its early stage of development at that time, and an axiomatic approach was novel to the subject. In the period 1920–1926, Noether opened new frontiers in algebra through the publication of five papers focusing on what was then called "the general theory of ideals." Her now classical memoir "Theory of Ideals in Rings" (1921) exerted the greatest influence on mathematics as a whole and was responsible for much of the recognition that she received. This paper demonstrated the fundamental importance of chain conditions on ideals. (A ring satisfies the ascending chain condition if every sequence $I_1 \subseteq I_2 \subseteq \cdots \subseteq I_n \subseteq \ldots$ of ideals in the ring ends after a finite number of terms.) Having formulated the concept of primary ideals, Noether proved that in a commutative ring

Emmy Noether (1882–1935)

satisfying the ascending chain condition, every ideal can be represented as a product of primary ideals. A second article, "Abstract Construction of Ideal Theory in the Domain of Algebraic Number Fields" (1926), provides an abstract characterization of those rings whose ideals are uniquely expressible as a finite product of prime ideals. With this theory, Noether introduced into abstract algebra an analog of Euclid's fundamental theorem of arithmetic.

Göttingen was an international center of mathematics during the time Noether was there, attracting visiting scholars from around the world. In 1924–1925 the circle of pupils and researchers that always surrounded Noether made its most brilliant acquisition, a graduating Amsterdam student named B. L. van der Waerden, at 22 years of age one of the brightest mathematical talents in Europe. Van der Waerden quickly mastered Noether's algebraic theories, enhanced them with new findings of his own, and more than anyone else helped to spread her ideas. In 1931 he published his famous *Moderne Algebra*, a large part of which is a polished and fully detailed treatment of Noether's lectures. Reprinted in numerous editions and translated into many languages, *Moderne Algebra* has been the instrument by which the mathematical community at large became Emmy Noether's students.

4.2 ELEMENTARY PROPERTIES OF RINGS

This chapter is devoted to the study of algebraic structures having two suitably restricted binary operations. Familiar examples are provided by the number systems of arithmetic (the integers, rational numbers, and real numbers) and the algebra of sets. We shall be concerned primarily with the generalized structure known as a ring. The concept of a ring is due to both Richard Dedekind (1831–1916) and Leopold Kronecker (1823–1891). Kronecker called these systems "orders"; the name "ring" or "number ring" passed into mathematics through David Hilbert's 1897 paper "The Theory of Algebraic Number Fields." Abraham Fraenkel (1891–1965) introduced the general notion of an abstract ring in 1914, although his formulation is not the one in common use today. The current definition of a ring seems to have made its appearance in 1917 and gained popularity through its use in Emmy Noether's monumental paper "Theory of Ideals in Rings," published in 1921. Since a ring is basically a combination of a commutative group and a semigroup, our previous experience with groups will prove to be of considerable help. As the reader will see, many of the important concepts of group theory have natural generalizations to systems with two operations.

To enable us to gain more information than could be obtained by studying each operation separately, there must be some relationship between the two operations. The most common requirement is that one of them be distributive over the other: Given an algebraic system $(S, *, \circ)$, the operation $\circ$ is said to be *left distributive* over the operation $*$ if

$$a \circ (b * c) = (a \circ b) * (a \circ c)$$

and *right distributive* if

$$(b * c) \circ a = (b \circ a) * (c \circ a)$$

for all elements $a, b, c \in S$. We speak of the operation $\circ$ as being *distributive* over $*$ whenever both of these conditions hold. If $\circ$ happens to be a commutative operation, then $\circ$ is distributive over $*$ if and only if it is left (right) distributive.

Let us now formally define the principal object of interest in this chapter, the notion of a ring.

DEFINITION 4.1 A *ring* is an ordered triple $(R, +, \cdot)$ consisting of a nonempty set R and two binary operations $+$ and $\cdot$ defined on R such that

1. $(R, +)$ is a commutative group,
2. $(R, \cdot)$ is a semigroup, and
3. the operation $\cdot$ is distributive over the operation $+$.

The reader should understand clearly that $+$ and $\cdot$ represent abstract, unspecified operations and not ordinary addition and multiplication. For convenience, however, one invariably refers to the operation $+$ as *addition* and to the operation $\cdot$ as *multiplication.* In the light of this terminology it is natural then to speak of the commutative group $(R, +)$ as the *additive group* of the ring and of $(R, \cdot)$ as the *multiplicative semigroup* of the ring.

By analogy with the integers the unique identity element for addition is called the *zero element* of the ring and is denoted by the usual symbol 0. The unique additive inverse of an element $a \in R$ will hereafter be written as $-a$.

To minimize the use of parentheses in expressions involving both operations, we shall stipulate that multiplication is to be performed before addition. Accordingly, the expression $a \cdot b + c$ stands for $(a \cdot b) + c$ and not for $a \cdot (b + c)$. Because of the general associative law, parentheses can also be omitted when writing out sums and products of more than two elements.

With these remarks in mind we can now give a more elaborate definition of a ring. A ring $(R, +, \cdot)$ consists of a nonempty set R together with two binary operations $+$ and $\cdot$ of addition and multiplication on R for which the following conditions are satisfied:

1. $a + b = b + a$,
2. $(a + b) + c = a + (b + c)$,

3. there exists an element 0 in R such that $a + 0 = a$ for every $a \in R$,
4. for each $a \in R$ there exists an element $-a \in R$ such that $a + (-a) = 0$,
5. $(a \cdot b) \cdot c = a \cdot (b \cdot c)$, and
6. $a \cdot (b + c) = a \cdot b + a \cdot c$ and $(b + c) \cdot a = b \cdot a + c \cdot a$,

where it is understood that a, b, c represent arbitrary elements of R.

A ring $(R, +, \cdot)$ is said to be a *finite ring* if, naturally enough, the set R of its elements is a finite set. By placing restrictions on the multiplication operation, several other specialized types of rings are obtained.

DEFINITION 4.2

1. A *commutative ring* is a ring $(R, +, \cdot)$ in which multiplication is a commutative operation, $a \cdot b = b \cdot a$ for all $a, b \in R$. (In case $ab = ba$ for a particular pair a, b, we express this fact by saying that a and b *commute*.)
2. A *ring with identity* is a ring $(R, +, \cdot)$ in which there exists an identity element for the operation of multiplication, normally represented by the symbol 1, so that $a \cdot 1 = 1 \cdot a = a$ for all $a \in R$.

Given a ring $(R, +, \cdot)$ with identity 1, an element $a \in R$ is said to be *invertible*, or to be a *unit*, whenever a possesses a (two-sided) inverse with respect to multiplication. The multiplicative inverse of a is unique, whenever it exists, and will be denoted by a^{-1}, so that $a \cdot a^{-1} = a^{-1} \cdot a = 1$. In the future the set of all invertible elements of the ring will be designated by the symbol R^*. It follows easily that the system $(R^*, \cdot)$ forms a group, known as the *group of invertible elements* of R. In this connection, notice that R^* is certainly nonempty, since if nothing else, 1 and -1 belong to R^*. One must not assume, however, that 1 and -1 are necessarily distinct.

A consideration of several concrete examples will serve to bring these ideas into focus.

Example 4.1
If Z, Q, $R^{\#}$ denote the sets of integers, rational, and real numbers, respectively, then the systems

$$(Z, +, \cdot), \qquad (Q, +, \cdot), \qquad (R^{\#}, +, \cdot)$$

are all examples of rings; here, $+$ and $\cdot$ are taken to be ordinary addition and multiplication. In each of these cases the ring is commutative and has the integer 1 for an identity element.

Example 4.2
Let X be a nonempty set and $P(X)$ denote the collection of all subsets of X. Both the systems $(P(X), \cup, \cap)$ and $(P(X), \cap, \cup)$ fail to be rings, since neither $(P(X), \cup)$ nor $(P(X), \cap)$ forms a group. However, in Example

2.20 we showed $(P(X), \Delta)$ to be a commutative group, where Δ indicates the symmetric difference operation

$$A \Delta B = (A - B) \cup (B - A).$$

Since $(P(X), \cap)$ is clearly a commutative semigroup, to establish that the system $(P(X), \Delta, \cap)$ constitutes a ring, it is necessary to verify only the left distributivity of the operation $\cap$ over Δ. For this we require the set identity (Problem 8, Section 1.1)

$$A \cap (B - C) = (A \cap B) - (A \cap C).$$

The argument now proceeds as follows: For all subsets $A, B, C \subseteq X$,

$$\begin{aligned} A \cap (B \Delta C) &= A \cap [(B - C) \cup (C - B)] \\ &= [A \cap (B - C)] \cup [A \cap (C - B)] \\ &= [(A \cap B) - (A \cap C)] \cup [(A \cap C) - (A \cap B)] \\ &= (A \cap B) \Delta (A \cap C). \end{aligned}$$

Example 4.3

The group $(Z_n, +_n)$ of integers modulo n was obtained on defining the sum of two congruence classes by

$$[a] +_n [b] = [a + b].$$

In the same way an operation $\cdot_n$ of multiplication can be introduced in Z_n; for $[a], [b] \in Z_n$, let

$$[a] \cdot_n [b] = [ab].$$

This definition presents a problem similar to that encountered with addition; namely, we must show the product to be well defined and not dependent on the particular representatives of the congruence classes that are chosen. To be more specific, we want to satisfy ourselves that if $[a'] = [a]$ and $[b'] = [b]$, then

$$[a'] \cdot_n [b'] = [a] \cdot_n [b],$$

or equivalently, $[a'b'] = [ab]$. Now $a' \in [a'] = [a]$ and $b' \in [b'] = [b]$, which signifies that

$$a' \equiv a (\text{mod } n) \qquad \text{and} \qquad b' \equiv b (\text{mod } n).$$

Using Theorem 2.14, it may be concluded that $a'b' \equiv ab (\text{mod } n)$ and so the product $a'b' \in [ab]$. But this means that $[a'b'] = [ab]$, as desired (Theorem 2.18 applies).

We omit the detailed verification of the fact that $(Z_n, +_n, \cdot_n)$ is a commutative ring with identity (not unexpectedly, known as the *ring of integers modulo n*), remarking that the various ring axioms hold in Z_n

simply because they hold in Z. The computation required to establish the distributive law, for instance, is

$$\begin{aligned}[a]\cdot_n([b]+_n[c]) &= [a]\cdot_n[b+c]\\ &= [a(b+c)]\\ &= [ab+ac]\\ &= [ab]+_n[ac] = [a]\cdot_n[b]+_n[a]\cdot_n[c].\end{aligned}$$

Observe, incidentally, that the congruence class [1] serves as the multiplicative identity.

Following our usual practice, we shall leave off the brackets from the elements of Z_n, thereby making no genuine notational distinction between a congruence class and its smallest nonnegative representative.

Example 4.4

Consider the set $M_2(R^\#)$ of all 2×2 matrices with entries from $R^\#$. With the usual rules of matrix addition and multiplication, $(M_2(R^\#), +, \cdot)$ becomes a noncommutative ring with identity. Much of the proof is contained in Example 2.23; the distributive laws must both be verified, a straightforward though tedious task.

It is not difficult to see that $M_2(R^\#)$ can be generalized in several ways. One way is to select the entries of our 2×2 matrices from some arbitrary ring $(R, +, \cdot)$, rather than from the real numbers. The resulting set of matrices is denoted by $M_2(R)$. The verification that $(M_2(R), +, \cdot)$ forms a ring proceeds exactly as in the case of $(M_2(R^\#), +, \cdot)$.

We may extend the definitions further to include the set $M_n(R)$ of all $n \times n$ matrices with entries from R ($n \geqslant 2$). An element of $M_n(R)$ is a square array

$$A = \begin{bmatrix} a_{11} & a_{12} & \cdots & a_{1n} \\ \vdots & \vdots & & \vdots \\ a_{n1} & a_{n2} & \cdots & a_{nn} \end{bmatrix} \qquad a_{ij} \in R;\ i, j = 1, 2, \ldots, n.$$

This is sometimes abbreviated by writing $A = (a_{ij})$, in which a_{ij} is the entry appearing in the ith row and jth column. Addition of two $n \times n$ matrices is defined by adding corresponding entries, while multiplication is performed "row-by-column"; in notation,

$$(a_{ij}) + (b_{ij}) = (a_{ij} + b_{ij}),$$

$$(a_{ij}) \cdot (b_{ij}) = (c_{ij}), \qquad \text{where } c_{ij} = \sum_{k=1}^{n} a_{ik}b_{kj}.$$

It is a matter of direct checking to see that $(M_n(R), +, \cdot)$ is also a noncommutative ring.

If $(R, +, \cdot)$ has an identity element 1, then the $n \times n$ identity matrix is that matrix $I_n = (\delta_{ij})$ whose entries are the Kronecker deltas δ_{ij} defined by

$$\delta_{ij} = \begin{cases} 1 & \text{if } i = j, \\ 0 & \text{if } i \neq j. \end{cases}$$

All the entries on the diagonal of I_n are 1, and all the other entries are 0. It is an easy exercise to show that if $A \in M_n(R)$, then

$$I_n \cdot A = A = A \cdot I_n.$$

The consequence of this is that the matrix I_n formed from the identity of R is an identity for multiplication in the ring of matrices $(M_n(R), +, \cdot)$.

Example 4.5

For our last example, let X be a nonempty set and let $(R, +, \cdot)$ be an arbitrary ring. We adopt the notation map (X, R) to indicate the set of all mappings from X into R:

$$\text{map}(X, R) = \{f \mid f : X \to R\}.$$

For ease of notation, let us also agree to write map R in place of map (R, R).

Now the elements of map (X, R) can be combined by performing algebraic operations on their functional values. More specifically, the pointwise sum and product of f and g, denoted by $f + g$ and $f \cdot g$, respectively, are the functions that satisfy

$$(f + g)(x) = f(x) + g(x), \qquad (f \cdot g)(x) = f(x) \cdot g(x) \qquad (x \in X).$$

We shall often indulge in this harmless duplication of symbols whereby $+$ and $\cdot$ are used with two different meanings.

It is readily verified that the above definitions provide map(X, R) with the structure of a ring. We simply point out that the zero element of this ring is the constant function whose sole value is 0, and the additive inverse $-f$ of f is characterized by the rule $(-f)(x) = -f(x)$.

Note that the algebraic properties of map(X, R) are determined by what happens in the ring $(R, +, \cdot)$ (the set X furnishes only the points for the pointwise operations). For instance, if $(R, +, \cdot)$ has a multiplicative identity 1, then the ring (map (X, R), $+$, $\cdot$) likewise possesses an identity element, namely, the constant function defined by $1(x) = 1$ for all $x \in X$.

Although it is logically correct (and often convenient) to speak of a ring as an ordered triple, the notation becomes unwieldy as one progresses farther into the theory. We shall therefore adopt the usual convention of designating a ring, say $(R, +, \cdot)$, simply by the set symbol R and assume that $+$ and $\cdot$ are known. The reader should realize, however, that a given set may perfectly well be the underlying set of several different rings. Let us also agree to abbreviate $a + (-b)$ as $a - b$ and subsequently refer to this expression as the *difference* between a and b. As a final concession to brevity, juxtaposition without a median dot will be used to denote the product of two ring elements.

With these conventions on record, let us begin our formal development of ring theory.

THEOREM 4.1 *If R is a ring, then for any $a, b, c \in R$,*

1. $0a = a0 = 0$,
2. $a(-b) = (-a)b = -(ab)$,
3. $(-a)(-b) = ab$, *and*
4. $a(b - c) = ab - ac$, $(b - c)a = ba - ca$.

Proof. These turn out, in the main, to be simple consequences of the distributive laws. For instance, from $0 + 0 = 0$ it follows that

$$0a = (0 + 0)a = 0a + 0a.$$

Thus by the cancellation law for the additive group $(R, +)$ we have $0a = 0$. In a like manner, one obtains $a0 = 0$. The proof of (2) requires the fact that each element of R has a unique additive inverse (Problem 5 at the end of this section). Since $b + (-b) = 0$,

$$ab + a(-b) = a(b + (-b)) = a0 = 0,$$

which then implies that $-(ab) = a(-b)$. The argument that $(-a)b$ is also the additive inverse of ab proceeds similarly. This leads immediately to (3):

$$(-a)(-b) = -(-a)b = -(-(ab)) = ab.$$

The last assertion is all but obvious, since $-(-x) = x$ in any additive group.

There is one very simple ring that consists only of the additive identity 0, with addition and multiplication given by $0 + 0 = 0$, $0 \cdot 0 = 0$; this ring is usually called the *trivial ring*.

COROLLARY *Let R be a ring with identity 1. If R is not the trivial ring, then the elements 0 and 1 are distinct.*

Proof. Since $R \neq \{0\}$, some nonzero element $a \in R$ exists. If 0 and 1 were equal, it would follow that $a = a1 = a0 = 0$, an obvious contradiction.

Convention Let us assume, once and for all, that any ring with identity contains more than one element. This will rule out the possibility that 0 and 1 coincide.

We now make several remarks about the concept of zero divisors (the term "divisors of zero" is also in common use):

DEFINITION 4.3 If R is a ring and $0 \neq a \in R$, then a is called a *left* (*right*) *zero divisor* in R if there exists some $b \neq 0$ in R such that $ab = 0 (ba = 0)$. A *zero divisor* is any element of R that is either a left or right zero divisor.

According to this definition, 0 is not a zero divisor, and if R contains an identity 1, then 1 is not a zero divisor, nor is any element of R that happens to possess a multiplicative inverse. An obvious example of a ring with zero divisors is Z_n, where the integer $n > 1$ is composite; if $n = n_1 n_2$ in Z $(0 < n_1 \leqslant n_2 < n)$, then the product $n_1 \cdot_n n_2 = 0$ in Z_n.

For the most part we shall be studying rings without zero divisors. In such rings it is possible to conclude from the equation $ab = 0$ that either $a = 0$ or $b = 0$.

One can express the property of being with or without zero divisors in the following useful way.

THEOREM 4.2 *A ring R is without zero divisors if and only if it satisfies the cancellation laws for multiplication; that is, for all $a, b, c \in R$, $ab = ac$ and $ba = ca$, where $a \neq 0$, implies that $b = c$.*

Proof. Suppose that R is without zero divisors and let $ab = ac$, $a \neq 0$. Then the product $a(b - c) = 0$, which means that $b - c = 0$ and $b = c$. The argument is the same for the equation $ba = ca$. Conversely, let R satisfy the cancellation laws and assume that $ab = 0$, with $a \neq 0$. We then have $ab = a0$, whence by cancellation $b = 0$. Similarly, $b \neq 0$ implies that $a = 0$, proving that there are no zero divisors in R.

By an *integral domain* is meant a commutative ring with identity that has no zero divisors. Perhaps the best-known example of an integral domain is the ring of integers, hence the choice of terminology. Theorem 4.2 shows that the cancellation laws for multiplication hold in any integral domain.

The reader should be warned that many authors do not insist on the presence of a multiplicative identity when defining integral domains; in this case the term "integral domain" would merely indicate a commutative ring without zero divisors.

We change direction somewhat to deal with the situation where a subset of a ring again constitutes a ring. Formally speaking,

DEFINITION 4.4 Let $(R, +, \cdot)$ be a ring and $S \subseteq R$ be a nonempty subset of R. If the system $(S, +, \cdot)$ is itself a ring (using the induced operations), then $(S, +, \cdot)$ is said to be a *subring* of $(R, +, \cdot)$.

This definition is adequate but unwieldy, since all the aspects of the definition of a ring must be checked in determining whether a given subset is a subring. In seeking a simpler criterion, notice that $(S, +, \cdot)$ is a subring of $(R, +, \cdot)$ provided that $(S, +)$ is a subgroup of $(R, +)$, $(S, \cdot)$ is a subsemigroup of $(R, \cdot)$, and the two distributive laws are satisfied in S. But the distributive and associative laws hold automatically for elements of S as a consequence of their validity in R. Since these laws are inherited from R, they are not necessarily required in defining a subring.

Taking our cue from these remarks, we could just as well define a subring as follows. The system $(S, +, \cdot)$ forms a subring of the ring $(R, +, \cdot)$ if and only if

1. S is a nonempty subset of R,
2. $(S, +)$ is a subgroup of $(R, +)$, and
3. the set S is closed under multiplication.

To add the final touch, even this definition can be improved upon; the reader will recall that $(S, +)$ is a subgroup of the group $(R, +)$ provided that $a - b \in S$ whenever $a, b \in S$. By these observations we are led to a set of closure conditions that make it somewhat easier to verify that a particular subset is actually a subring.

THEOREM 4.3 *Let R be a ring and $\varnothing \neq S \subseteq R$. Then S is a subring of R if and only if*

1. $a, b \in S$ imply that $a - b \in S$ (closure under differences),
2. $a, b \in S$ imply that $ab \in S$ (closure under multiplication).

If S is a subring of the ring R, then the zero element of S is that of R; moreover, the additive inverse of an element of the subring S is the same as its inverse as a member of R. Verification of these assertions is left as an exercise.

Example 4.6

Every ring R has two obvious subrings, namely, the set $\{0\}$ consisting only of the zero element, and R itself. These two subrings are usually referred to as the *trivial subrings* of R; all other subrings (if any exist) are called *nontrivial*. We shall use the term *proper subring* to mean a subring that is different from R.

Example 4.7
The set Z_e of even integers forms a subring of the ring Z of integers, for

$$2n - 2m = 2(n - m) \in Z_e,$$
$$(2n)(2m) = 2(2nm) \in Z_e.$$

This example also illustrates a fact worth bearing in mind: In a ring with identity a subring need not contain the identity element.

Example 4.8
The ring Z_{12} of integers modulo 12 has several nontrivial subrings, namely, $\{0, 6\}$, $\{0, 4, 8\}$, $\{0, 3, 6, 9\}$, and $\{0, 2, 4, 6, 8, 10\}$. The subring lattice of Z_{12} is

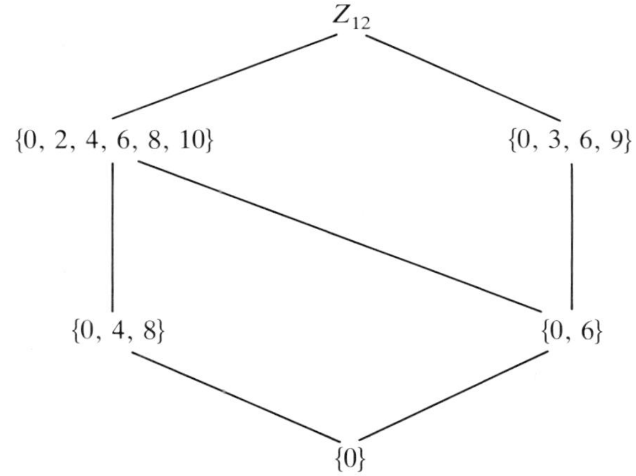

Before stating our next theorem, let us define the *center* of a ring R, denoted by cent R, to be the set

$$\text{cent } R = \{a \in R \mid ar = ra \text{ for all } r \in R\}.$$

Phrased otherwise, cent R consists of those elements that commute with every member of R. It should be apparent that a ring R is commutative if and only if cent $R = R$.

THEOREM 4.4 *For any ring* R, cent R *is a subring of* R.

Proof. To be conscientious about details, first observe that cent R is nonempty; at the very least, the zero element $0 \in \text{cent } R$. Now pick any two elements a, b in cent R. By the definition of center we know that $ar = ra$ and $br = rb$ for every choice of $r \in R$. Thus, for arbitrary $r \in R$,

$$(a - b)r = ar - br = ra - rb = r(a - b),$$

which implies that $a - b \in \text{cent } R$. A similar argument affirms that the product ab also lies in cent R. In the light of Theorem 4.3 these are sufficient conditions for the center to be a subring of R.

It has already been remarked that when a ring has an identity, this need not be true of its subrings. Other interesting situations may arise.

1. Some subring has a multiplicative identity, but the entire ring does not.
2. Both the ring and one of its subrings possess identity elements, but they are distinct.

In each of the cited cases the identity for the subring is necessarily a divisor of zero in the larger ring. To justify this claim, let $1' \neq 0$ denote the identity element of the subring S; we assume further that $1'$ does not act as an identity for the whole ring R. Accordingly, some element $a \in R$ exists for which $a1' \neq a$. It is clear that

$$(a1')1' = a(1'1') = a1',$$

or $(a1' - a)1' = 0$. Since neither $a1' - a$ nor $1'$ is zero, the ring R has zero divisors, and in particular, $1'$ is a zero divisor.

Example 4.9

To present a simple illustration of a ring in which the second of the aforementioned possibilities occurs, consider the set $R = Z \times Z$, consisting of ordered pairs of integers. One converts R into a ring by defining addition and multiplication componentwise:

$$(a, b) + (c, d) = (a + c, b + d),$$

$$(a, b)(c, d) = (ac, bd).$$

A routine calculation will show that $Z \times \{0\} = \{(a, 0) | a \in Z\}$ forms a subring with identity element $(1, 0)$. This obviously differs from the identity of the entire ring R, which turns out to be the ordered pair $(1, 1)$. By our previous remarks, $(1, 0)$ must be a zero divisor in R; in fact, $(1, 0)(0, 1) = (0, 0)$, where $(0, 0)$ serves as the zero element of R.

If R is an arbitrary ring and n a positive integer, then the *nth power* a^n of an element $a \in R$ is defined by the inductive conditions $a^1 = a$ and $a^n = a^{n-1}a$. From this the usual laws of exponents follow at once:

$$a^n a^m = a^{n+m}, \qquad (a^n)^m = a^{nm} \qquad (n, m \in Z_+).$$

To establish these rules, fix m and proceed by induction on n. Observe also that if two elements $a, b \in R$ happen to commute, then so do all powers of a and b, whence $(ab)^n = a^n b^n$ for each positive integer n.

In the event that R possesses an identity element 1 and a^{-1} exists, negative powers of a can be introduced by interpreting a^{-n} as $(a^{-1})^n$, where $n > 0$. With the definition $a^0 = 1$ the symbol a^n now has a well-defined meaning for every integer n—at least when attention is restricted to invertible elements.

Paralleling the exponent notation for powers, there is the notation of integral multiples of an element $a \in R$. For each positive integer n we define the n*th natural multiple* na recursively as follows:

$$1a = a \qquad \text{and} \qquad na = (n-1)a + a, \qquad \text{when } n > 1.$$

If it is also agreed to let $0a = 0$ and $(-n)a = -(na)$, then the definition of na can be extended to all integers. Integral multiples satisfy several identities that are easy to establish:

$$(n + m)a = na + ma,$$

$$(nm)a = n(ma),$$

$$n(a + b) = na + nb,$$

for $a, b \in R$ and arbitrary integers n and m. In addition to these rules there are two further properties resulting from the distributive law, namely,

$$n(ab) = (na)b = a(nb) \qquad \text{and} \qquad (na)(mb) = (nm)(ab).$$

Experience impels us to emphasize that the expression na should not be regarded as a ring product; indeed, the integer n may not even be a member of R. The entire symbol na is just a convenient way of indicating a certain sum of elements of R. However, when there is an identity for multiplication, it is possible to represent na as a product of two ring elements, namely, $na = (n1)a$.

To proceed further with our program, we must first frame a definition.

DEFINITION 4.5 Let R be an arbitrary ring. If there exists a positive integer n such that $na = 0$ for all $a \in R$, then the smallest positive integer with this property is called the *characteristic* of the ring. If no such positive integer exists (that is, $n = 0$ is the only integer for which $na = 0$ for all a in R), then R is said to be of *characteristic zero*. We shall write char R for the characteristic of R.

The rings of integers, rational numbers, and real numbers are all standard examples of systems having characteristic zero (some writers prefer the expression "characteristic infinity"). On the other hand, the ring $P(X)$ of subsets of a fixed set X is of characteristic 2, since

$$2A = A \,\Delta\, A = (A - A) \cup (A - A) = \varnothing$$

for every subset $A \subseteq X$.

Although the definition of characteristic makes an assertion about every element of the ring, in rings with identity the characteristic is completely determined by the identity element. We reach this conclusion below.

THEOREM 4.5 *If R is any ring with identity* 1, *then R has characteristic $n > 0$ if and only if n is the least positive integer for which $n1 = 0$.*

Proof. If char $R = n > 0$, then $na = 0$ for every $a \in R$ and so, in particular, $n1 = 0$. Were $m1 = 0$, where $0 < m < n$, it would necessarily follow that

$$ma = (m1)a = 0a = 0$$

for every element $a \in R$. The implication is that char $R < n$, which is impossible. One establishes the converse in much the same way.

As we have seen, multiplication exerts a strong influence on the additive structure of a ring through the distributive law. The following corollary to Theorem 4.5 shows that by sufficiently restricting the multiplication in a ring R it is possible to reach some interesting conclusions regarding the characteristic of R.

COROLLARY 1 *In an integral domain R, all the nonzero elements have the same additive order; this order is the characteristic of the domain when* char $R > 0$ *and is infinite when* char $R = 0$.

Proof. To verify this assertion, suppose first that char $R = n > 0$. According to the definition of characteristic, each element $0 \neq a \in R$ will then possess a finite additive order m, with $m \leqslant n$. (Recall that for an element $a \neq 0$ of the group $(R, +)$ to have order m means that $ma = 0$ and $ka \neq 0$ if $0 < k < m$.) But the relation $0 = ma = (m1)a$ implies that $m1 = 0$, since R is assumed to be free of zero divisors. We therefore conclude from the theorem that $n \leqslant m$, whence m and n are equal. In consequence every nonzero element of R has additive order n.

A somewhat similar argument can be employed when char $R = 0$. The equation $ma = 0$ would lead, as before, to $m1 = 0$ or $m = 0$. In this case every nonzero element $a \in R$ must be of infinite order.

The last result serves to bring out another useful point, which we place on record as

COROLLARY 2 *An integral domain R has positive characteristic if and only if $na = 0$ for some $0 \neq a \in R$ and some integer $n \in Z_+$.*

Continuing in this vein, let us next show that not any commutative group can serve as the additive group of an integral domain.

THEOREM 4.6 *The characteristic of an integral domain is either zero or a prime number.*

Proof. Let R be of positive characteristic n and assume that n is not a prime. Then n has a nontrivial factorization $n = n_1 n_2$, with $1 < n_1 \leqslant n_2 < n$. It follows that

$$0 = n1 = (n_1 n_2)1 = (n_1 n_2)1^2 = (n_1 1)(n_2 1).$$

By supposition, R is without zero divisors, so either $n_1 1 = 0$ or $n_2 1 = 0$. Since both n_1 and n_2 are less than n, this contradicts the choice of n as the least positive integer for which $n1 = 0$. We therefore conclude that char R must be prime.

COROLLARY *If R is a finite integral domain, then* char $R = p$, *where p is a prime.*

Turning again to the general theory, let R be any ring with identity and consider the set $Z1$ of integral multiples of the identity; stated symbolically,

$$Z1 = \{n1 \mid n \in Z\}.$$

From the relations

$$n1 - m1 = (n - m)1, \qquad (n1)(m1) = (nm)1$$

one can easily infer that $Z1$ itself forms a (commutative) ring with identity. The order of the additive cyclic group $(Z1, +)$ is simply the characteristic of the given ring R.

When R happens to be an integral domain, then $Z1$ is a *subdomain* of R, that is, $Z1$ is also an integral domain with respect to the operations in R. In fact, $Z1$ is the smallest subdomain of R, in the sense that it is contained in every other subdomain of R. If R is a domain of characteristic p, where p is a prime, then we are able to deduce considerably more: Each nonzero element of $Z1$ is invertible. Before establishing this, first observe that the set $Z1$, regarded as an additive cyclic group of order p, consists of p distinct elements, namely, the p sums $n1$, where $n = 0, 1, \ldots, p - 1$. Now let $n1$ be any nonzero element of $Z1 (0 < n < p)$. Since n and p are relatively prime, integers r and s exist for which $rp + sn = 1$. But then

$$1 = (rp + sn)1 = r(p1) + (s1)(n1).$$

As $p1 = 0$, we obtain the equation $1 = (s1)(n1)$, so that $s1$ serves as the multiplicative inverse of $n1$ in $Z1$. The value of these remarks will have to await further developments (in particular, see Section 4.3).

A homomorphism from the ring R into the ring R' is, as one might suspect, a function $f: R \to R'$ that preserves both the ring operations. This amounts to applying the familiar homomorphism concept to the underlying additive group and multiplicative semigroup of R. For a precise definition,

DEFINITION 4.6 Let R and R' be two rings. By a (ring) *homomorphism* from R into R' is meant a function $f: R \to R'$ such that

$$f(a + b) = f(a) + f(b), \qquad f(a \cdot b) = f(a) \cdot f(b)$$

for every pair of elements $a, b \in R$. A homomorphism that is also a one-to-one mapping is called an *isomorphism*.

We emphasize that the $+$ and $\cdot$ occurring on the left-hand sides of the equations in Definition 4.6 are those of R, whereas the $+$ and $\cdot$ occurring on the right-hand sides are those of R'. This use of the same symbols for the operations of addition and multiplication in two different rings need cause no confusion provided that the reader gives careful attention to the context in which the notation is employed.

The usual terminology applies: If f is a homomorphism from R into R', then the image $f(R)$ of R under f is called the *homomorphic image* of R. When $R = R'$, we speak of f as being a homomorphism of R into itself or an *endomorphism* of R. Likewise, an isomorphism of the ring R onto itself is termed an *automorphism* of R.

The knowledge of a few examples should help to establish ideas.

Example 4.10
Let R and R' be arbitrary rings and $f: R \to R'$ be the function that sends each element of R onto the zero element of R'. Then

$$f(a + b) = 0 = 0 + 0 = f(a) + f(b) \qquad (a, b \in R)$$

$$f(ab) = 0 = 00 = f(a)f(b)$$

so that f is a homomorphism. As in the group-theoretic case, this particular mapping is called the *trivial homomorphism*.

Example 4.11
Consider the ring Z of integers and the ring Z_n of integers modulo n. Define $f: Z \to Z_n$ by taking $f(a) = [a]$; that is, map each integer into the congruence class containing it. That f is a homomorphism follows directly from the definition of the operations in Z_n:

$$f(a + b) = [a + b] = [a] +_n [b] = f(a) +_n f(b),$$

$$f(ab) = [ab] = [a] \cdot_n [b] = f(a) \cdot_n f(b).$$

Example 4.12
In the ring map (x, R), define τ_a to be the function that assigns to each $f \in \text{map}(X, R)$ its value at a fixed element $a \in X$; in other words, $\tau_a(f) = f(a)$. Then τ_a is a homomorphism from map (X, R) into R, known as the *evaluation homomorphism* at a. We need only observe that

$$\tau_a(f + g) = (f + g)(a) = f(a) + g(a) = \tau_a(f) + \tau_a(g),$$

$$\tau_a(fg) = (fg)(a) = f(a)g(a) = \tau_a(f)\tau_a(g).$$

We now list some of the structural features that are preserved under homomorphisms.

THEOREM 4.7 *If f is a homomorphism from the ring R into the ring R', the following hold:*

1. $f(0) = 0$ *and*
2. $f(-a) = -f(a)$ *for all* $a \in R$.

If, in addition, R and R' are both rings with identity and $f(R) = R'$, then

3. $f(1) = 1$ *and*
4. $f(a^{-1}) = f(a)^{-1}$ *for each invertible element* $a \in R$.

Two comments regarding part (3) of the above theorem are in order. First, it is evident that

$$f(a)1 = f(a) = f(a1) = f(a)f(1)$$

for any a in R. Knowing this, one might be tempted to appeal (incorrectly) to the cancellation law to conclude that $f(1) = 1$; what is actually required is the fact that multiplicative identities are unique. Second, if the hypothesis that f map onto the set R' is omitted, then it can only be inferred that $f(1)$ is the identity for the homomorphic image $f(R)$. The element $f(1)$ need not serve as an identity for the entire ring R' and, indeed, it may very well happen that $f(1) \neq 1$.

We also observe, in passing, that, by virtue of statement (2),

$$f(a - b) = f(a) + f(-b) = f(a) - f(b).$$

In short, any ring homomorphism preserves differences as well as sums and products.

A further fact that deserves mention is that direct and inverse images of subrings under homomorphisms are again subrings. The proof, while not immediately obvious, is certainly not obscure.

THEOREM 4.8 *Let f be a homomorphism from the ring R into the ring R'. Then*

a) *for each subring S of R, $f(S)$ is a subring of R'; and*
b) *for each subring S' of R', $f^{-1}(S')$ is a subring of R.*

Proof. In compliance with Theorem 2.54, $f(S)$ and $f^{-1}(S')$ are subgroups of the additive groups of R and R', respectively. Thus the argument is reduced to showing that each of these sets is closed under multiplication. Suppose that $f(a)$ and $f(b)$ are arbitrary elements of $f(S)$. Then both a and b belong to the set S, as does their product ab (S being a subring of R); hence $f(a)f(b) = f(ab) \in f(S)$, which entails that $f(S)$ is closed under multiplication. The assertion involving $f^{-1}(S')$ can be dealt with similarly.

In agreement with our previous use of the term, two rings R and R' are said to be *isomorphic* if there exists an isomorphism from R onto R'; we indicate this by writing $R \simeq R'$. Isomorphic rings are algebraically indistinguishable, although they may be quite different in the nature of their elements.

Before we announce our next result, another definition is necessary.

DEFINITION 4.7 A ring R is said to be *imbedded* in a ring R' if there exists some subring S' of R' such that $R \simeq S'$.

In general, if a ring R is imbedded in a ring R', then R' is referred to as an *extension* of R, and we say that R can be *extended* to R'. The most important cases are those in which one passes from a given ring R to an extension possessing some property not present in R. As a simple application, let us prove that an arbitrary ring can be imbedded in an extension ring with identity.

THEOREM 4.9 (Dorroh Extension Theorem) *Any ring R can be imbedded in a ring with identity.*

Proof. Consider the Cartesian product $R \times Z$, where

$$R \times Z = \{(r, n) | r \in R;\ n \in Z\}.$$

If addition and multiplication are defined by

$$(a, n) + (b, m) = (a + b, n + m),$$

$$(a, n)(b, m) = (ab + ma + nb, nm),$$

then it is a simple matter to verify that $R \times Z$ forms a ring; we leave the actual details as an exercise. Note that this system has a multiplicative identity, namely, the pair $(0, 1)$, since

$$(a, n)(0, 1) = (a0 + 1a + n0, n1) = (a, n)$$

and, similarly,

$$(0, 1)(a, n) = (a, n).$$

Next, consider the subset $R \times \{0\}$ of $R \times Z$ consisting of all pairs of the form $(a, 0)$. Since

$$(a, 0) - (b, 0) = (a - b, 0),$$
$$(a, 0)(b, 0) = (ab, 0),$$

it is evident that $R \times \{0\}$ constitutes a subring of $R \times Z$. A straightforward calculation, which we omit, shows that $R \times \{0\}$ is isomorphic to the given ring R under the mapping $f : R \to R \times \{0\}$ defined by $f(a) = (a, 0)$. This process of extension therefore imbeds R in $R \times Z$, a ring with identity.

A point to be made in connection with the preceding theorem is that the imbedding process may be carried out even if the given ring has an identity to start with. Of course, in this case the construction has no particular merit; indeed, the original identity element serves only to introduce divisors of zero into the extended ring.

Although Theorem 4.9 shows that we could confine our study to rings with identity, it is nonetheless desirable to develop as much of the theory as possible without the assumption of such an element. Thus unless an explicit statement is made to the contrary, the subsequent discussions will not presuppose the existence of a multiplicative identity.

PROBLEMS 4.2

1. Let $(R, +)$ be an arbitrary commutative group with identity element 0. If an operation of multiplication is defined by taking $a \cdot b = 0$ for all $a, b \in R$, show that the system $(R, +, \cdot)$ forms a commutative ring.

2. Consider the set Z of integers with the usual operations. If a new addition and multiplication are defined by

$$a * b = a + b + 1, \qquad a \circ b = a + b + ab,$$

prove that the system $(Z, *, \circ)$ is a commutative ring with identity.

3. Establish the following statements:

a) The group Z_{12}^* of invertible elements of the ring Z_{12} is isomorphic to the Klein 4-group.

b) The group Z_5^* of invertible elements of the ring Z_5 is isomorphic to the cyclic group Z_4.

c) The group Z_9^* of invertible elements of the ring Z_9 is cyclic with generators 2 and 5.

4. Prove that the only ring R in which the two operations are equal (that is, $a + b = ab$ for all $a, b \in R$) is the trivial ring $R = \{0\}$.

5. In a ring R with identity, prove that each of the following is true:
   a) The identity element for multiplication is unique.
   b) If $a \in R$ has a multiplicative inverse, then a^{-1} is unique.
   c) If an element $a \in R$ is invertible, then so also is $-a$.
   d) No zero divisor of R can possess a multiplicative inverse.
6. a) If the set X contains more than one element, show that every nonempty proper subset of X is a zero divisor in the ring $P(X)$.
   b) For the ring map $R^{\#}$, verify that the functions $f(x) = x + |x|$ and $g(x) = x - |x|$ are both zero divisors.
7. Assume that R is a ring with identity having no zero divisors. For $a, b \in R$, confirm that
   a) $ab = 1$ if and only if $ba = 1$;
   b) if $a^2 = b^2$, then either $a = b$ or $a = -b$;
   c) if a is invertible and $a = a^{-1}$, then either $a = 1$ or $a = -1$.
8. Let a, b be commuting elements of the ring R. Use induction on $n \in Z_+$ to derive the binomial expansion

$$(a + b)^n = a^n + \binom{n}{1} a^{n-1}b + \cdots + \binom{n}{k} a^{n-k}b^k + \cdots + \binom{n}{n-1} ab^{n-1} + b^n, \qquad \text{where } \binom{n}{k} = \frac{n!}{k!(n-k)!}$$

   is the usual binomial coefficient.
9. An element a of a ring R is said to be *idempotent* if $a^2 = a$ and *nilpotent* if $a^n = 0$ for some $n \in Z_+$. Show that
   a) a nonzero idempotent element cannot be nilpotent;
   b) every nonzero nilpotent element is a zero divisor;
   c) if R is an integral domain, the only nilpotent element is the zero element;
   d) if R is an integral domain, the identity element is the only nonzero idempotent element.
10. If a is a nilpotent element of R, a ring with identity, establish that $1 + a$ and $1 - a$ are both invertible in R. [*Hint:* If $a^n = 0$, show that $(1 + a)^{-1} = 1 - a + a^2 + \cdots + (-1)^{n-1}a^{n-1}$.]
11. For the ring $M_2(R^{\#})$ of 2×2 real matrices, verify that
   a) $\begin{bmatrix} 1 & 1 \\ 0 & 0 \end{bmatrix}$ is a zero divisor;
   b) $\begin{bmatrix} 1 & 0 \\ 1 & 0 \end{bmatrix}$ is an idempotent element;
   c) $\begin{bmatrix} 1 & -1 \\ 1 & -1 \end{bmatrix}$ is a nilpotent element.

12. In the ring Z_{10} of integers modulo 10, prove that $\{0, 2, 4, 6, 8\}$ forms a commutative subring with identity element 6. Also show that 6 is a zero divisor in Z_{10}.

13. Prove that the function ring map $R^{\#}$ has no idempotent elements other than 0 and 1.

14. Prove that the set $S = \{a + b\sqrt{2} | a, b \in Z\}$ is a subring of the ring $R^{\#}$ of real numbers.

15. Using the sets S below, determine whether S is a subring of the ring $M_2(Z)$ of 2×2 matrices with integral entries:

a) $S = \left\{\begin{bmatrix} a & a \\ b & b \end{bmatrix} \middle| a, b \in Z\right\}$;

b) $S = \left\{\begin{bmatrix} a & 0 \\ a & 0 \end{bmatrix} \middle| a \in Z\right\}$;

c) $S = \left\{\begin{bmatrix} a & 0 \\ 0 & a \end{bmatrix} \middle| a \in Z\right\}$;

d) $S = \left\{\begin{bmatrix} a & 0 \\ 0 & 2a \end{bmatrix} \middle| a \in Z\right\}$;

e) $S = \left\{\begin{bmatrix} a & 0 \\ 0 & 0 \end{bmatrix} \middle| a \in Z\right\}$;

f) $S = \left\{\begin{bmatrix} a & b \\ 0 & 0 \end{bmatrix} \middle| a, b \in Z\right\}$.

16. a) Let a be an idempotent element of the ring R. Show that the set

$$S = \{ara | r \in R\}$$

forms a subring of R with identity element a.

b) Illustrate part (a) using the matrix ring $M_2(Z)$ and idempotent $\begin{bmatrix} 1 & 0 \\ 1 & 0 \end{bmatrix}$.

17. In the ring $M_2(R^{\#})$, deduce that

$$\text{cent } M_2(R^{\#}) = \left\{\begin{bmatrix} a & 0 \\ 0 & a \end{bmatrix} \middle| a \in R^{\#}\right\}.$$

18. Let R be an arbitrary ring and $n \in Z_+$. If the set S_n is defined by

$$S_n = \{a \in R | n^k a = 0 \text{ for some } k > 0\},$$

determine whether S_n is a subring of R.

19. Establish the following assertions concerning the characteristic of a ring R:

a) if there exists an integer k such that $ka = 0$ for all $a \in R$, then k is divisible by char R;

b) if char $R > 0$, then char $S \leqslant$ char R for any subring S of R;

c) if R is an integral domain and S is a subdomain of R, then char S = char R.

20. Let R be a ring with a finite number of elements, say $a_1, a_2, \ldots, a_n$, and let n_i be the order of a_i regarded as a member of the additive group of R. Prove that the characteristic of R is the least common multiple of the integers $n_i (i = 1, 2, \ldots, n)$.

21. Suppose that R is a ring with identity such that char $R = n > 0$. If n is not prime, show that R has divisors of zero.

22. If R is a ring that has no nonzero nilpotent elements, deduce that all the idempotent elements of R belong to cent R. [*Hint:* If $a^2 = a$, then $(ara - ar)^2 = (ara - ra)^2 = 0$ for all $r \in R$.]

23. Assume that R is a ring with the property that $a^2 + a \in \text{cent } R$ for every element a in R. Prove that R is necessarily a commutative ring. [*Hint:* Utilize the expression $(a + b)^2 + (a + b)$ to show first that $ab + ba$ lies in the center for all $a, b \in R$.]

24. Let $(G, +)$ be a commutative group and R be the set of all (group) homomorphisms of G into itself. The pointwise sum $f + g$ and composition $f \circ g$ of two functions $f, g \in R$ are defined by the usual rules

$$(f + g)(x) = f(x) + g(x), \qquad (f \circ g)(x) = f(g(x)) \qquad (x \in G).$$

Show that the resulting system $(R, +, \circ)$ forms a ring. At the same time, determine the invertible elements of R.

25. Given that f is a homomorphism from the ring R into the ring R', prove that

a) $f(\text{cent } R) \subseteq \text{cent } f(R)$;

b) if char $R > 0$, then char $S \leqslant$ char R for any subring S of R;

c) if a is a zero divisor of R, then $f(a)$ need not be a zero divisor of R';

d) if R has a positive characteristic, then char $f(R) \leqslant$ char R.

26. Consider the rings $R = \{a + b\sqrt{2} | a, b \in Z\}$ and $R' = \{a + b\sqrt{3} | a, b \in Z\}$. Show that the mapping $f: R \to R'$ given by $f(a + b\sqrt{2}) = a + b\sqrt{3}$ is not a homomorphism.

27. Determine which of the following mappings are homomorphisms between the indicated rings:

a) $f: Z \to Z_e$ defined by $f(a) = 2a$;

b) $f: Z \times Z \to Z$ defined by $f(a, b) = a$;

c) $f: R^\# \to R^\#$ defined by $f(a) = |a|$;

d) $f: R \to R$ defined by $f(a + b\sqrt{2}) = a - b\sqrt{2}$, where $R = \{a + b\sqrt{2} | a, b \in Z\}$;

e) $f: Z \times Z \to M_2(Z)$ defined by $f(a, b) = \begin{bmatrix} a & 0 \\ 0 & b \end{bmatrix}$;

f) $f: M_2(R^\#) \to R^\#$ defined by $f(a) = \det a$.

28. Let R be a ring with identity. For any invertible element $a \in R$, show that the function $f_a: R \to R$ defined by $f_a(x) = axa^{-1}$ is an automorphism of R.

29. Suppose that R consists of all matrices in $M_2(Z)$ of the form $\begin{bmatrix} n & 0 \\ 2n & 0 \end{bmatrix}$. Prove that R is a ring that is isomorphic to the ring Z of integers.

30. Let f be a homomorphism from the ring R into itself and let S be the set of elements that are left fixed by f; that is,

$$S = \{a \in R \mid f(a) = a\}.$$

Establish that S forms a subring of R.

31. Given that $f: Z \to Z$ is a homomorphism, show that f is either the identity mapping or the trivial homomorphism. [*Hint:* Since $f(1)$ is an idempotent element of Z, Problem 9(d) applies.]

32. Suppose that R is a ring of characteristic n. If addition and multiplication are defined in $R \times Z_n$ by

$$(x, a) + (y, b) = (x + y, a +_n b),$$

$$(x, a)(y, b) = (xy + ay + bx, a \cdot_n b),$$

prove that $R \times Z_n$ is an extension ring of R of characteristic n. Show also that $R \times Z_n$ has an identity element.

33. A *Boolean ring* is a ring with identity in which every element is idempotent. Show that the following are Boolean rings:

a) the ring Z_2 of integers modulo 2;

b) the ring $P(X)$ of subsets of some nonempty set X;

c) the function ring $R = \{f \mid f: R^\# \to Z_2\}$, using pointwise operations.

34. Prove that any Boolean ring R is commutative. [*Hint:* First establish that $a = -a$ for all $a \in R$.]

35. Show that any ring R (not necessarily with identity) in which each element is idempotent can be imbedded in a Boolean ring. [*Hint:* Let $R' = R \times Z_2$ and mimic the argument of Theorem 4.9.]

36. a) Let R be a commutative ring with identity and S the set of all idempotent elements of R. Define a new sum of a and b in S by taking

$$a +' b = a + b - 2ab.$$

With + as the addition, prove that S constitutes a Boolean ring, known as the *idempotent Boolean ring of R*.

b) In particular, obtain the idempotent Boolean ring of Z_{12}.

c) Show that the idempotent Boolean ring of any integral domain is isomorphic to Z_2.

4.3 IDEALS

In this section we introduce an important class of subrings, known as *ideals*, whose role in ring theory is similar to that of the normal subgroups in the study of groups. As the reader will see, ideals lead to the construction of quotient rings that are the appropriate analogs of quotient groups.

Ideals arose out of the attempt of nineteenth century algebraists to "widen the field of arithmetic" to more general sorts of numbers. Ernst Kummer (1810–1893) was the first to observe explicitly that the elements of some such number systems, unlike the ordinary integers, did not admit a unique factorization into primes. A simple example of this phenomenon occurs in the ring of complex numbers built up from a root of the equation $x^2 = -5$, where 9 can be decomposed into "prime" factors in two ways:

$$9 = 3 \cdot 3 = (2 + \sqrt{-5})(2 - \sqrt{-5}).$$

To restore the highly desirable unique factorization property in situations such as this, Kummer was led to invent additional prime numbers (1847), which he adjoined to the ring under consideration. Kummer called these new objects "ideal complex numbers," never saying what they actually were but rather describing their behavior. In an extremely questionable analogy he compared the new numbers with the element fluorine, which chemists had not yet isolated and which they described only as it behaves in its compounds. The unique factorization law is reestablished when the factorization is expressed in the presence of ideal complex numbers.

The ideal complex numbers, which seemed mysterious and artificial, served as a starting point for a totally different theory of ideals by two of Kummer's followers, Leopold Kronecker and Richard Dedekind. Kronecker allegedly achieved this by 1859 but published nothing on the subject until 1882; whereas Dedekind's conception of ideals appeared in his supplement to the second edition (1871) of Dirichlet's *Number Theory*. Dedekind took an ideal to be an infinite set I of algebraic integers such that if $\alpha, \beta \in I$, then $\alpha \pm \beta \in I$ and $\gamma\alpha \in I$, where γ is an arbitrary algebraic integer; in essence, he identified Kummer's "ideal complex number" with the set of those numbers that it divides. Dedekind's innovative idea of defining his main object of study, the ideal, to be a set (or "system") rather than a number was a major departure from the established practice of the time; it can be regarded as the first appearance of the set-theoretic approach to modern algebra, which was to become so influential.

A similar notion of ideal occurred independently to Kronecker, who was Kummer's favorite pupil (from Kummer's days as a young schoolteacher) and who succeeded Kummer as professor of mathematics at the University of Berlin. Rather than publishing immediately, Kronecker made his results on ideal theory known in conversations with a small circle of friends and pupils. After 20 years of refinement and improvement he finally brought them out in a paper, "Main Features of an Arithmetic Theory of Algebraic Magnitudes," in 1882 in honor of the fiftieth anniversary of Kummer's doctorate. Because Kronecker's cramped expository style was difficult to follow, his version of ideal theory, or "divisor theory" as he called it, attracted little attention. By contrast, the marvelous clarity with which Dedekind presented the basic concepts made his formulation the one that achieved universal acceptance. For both Dedekind and Kronecker the central theorem was that in the ring of integers of an algebraic number field, every ideal can be represented uniquely, except for the ordering of the factors, as a product of prime ideals. (The ideals that play the role of the ordinary prime numbers are called *prime ideals*.)

While ideals were introduced initially in the context of algebraic numbers, they were soon freed from this restriction and given a definition in terms of abstract rings. Any subring is closed under multiplication, but we require ideals to possess a stronger type of multiplicative closure; namely, they must be closed under multiplication by an arbitrary ring element.

DEFINITION 4.8 A subring I of the ring R is said to be a *two-sided ideal* of R if and only if $r \in R$ and $a \in I$ imply that both $ra \in I$ and $ar \in I$.

Viewed otherwise, Definition 4.8 asserts that whenever one of the factors in a product belongs to I, then the product itself must be in I. This may be roughly summarized by saying that the set I "absorbs" products.

Taking stock of Theorem 4.3, which gives a minimal set of conditions to be a subring, our current definition of a two-sided ideal may be reformulated as follows.

DEFINITION 4.9 Let I be a nonempty subset of a ring R. Then I is a two-sided ideal of R if and only if

1. $a, b \in I$ imply that $a - b \in I$ and
2. $r \in R$ and $a \in I$ imply that both products $ra, ar \in I$.

If condition (2) of Definition 4.9 is weakened so as to require only that the product ra belongs to I for every choice of $r \in R$ and $a \in I$, we are led to the notion of a *left ideal; right ideals* are defined in a symmetric way. Needless to say, if the ring R happens to be commutative—the most important case as far as we shall be concerned—then there is no distinction between left, right, and two-sided ideals.

Convention In what follows, let us agree that the term "ideal," unmodified, will always mean two-sided ideal.

Before proceeding further, we pause to examine this concept by means of three specific examples.

Example 4.13
With respect to the usual matrix operators, the set

$$T_2(Z) = \left\{ \begin{bmatrix} a & b \\ 0 & c \end{bmatrix} \middle| a, b, c \in Z \right\}$$

forms a noncommutative ring with identity (in fact, a subring of the ring $M_2(Z)$ of all 2×2 matrices with integer entries). We propose to show that in this ring the subset

$$I = \left\{ \begin{bmatrix} a & b \\ 0 & 0 \end{bmatrix} \middle| a, b \in Z \right\}$$

is an ideal. In the first place,

$$\begin{bmatrix} a & b \\ 0 & 0 \end{bmatrix} - \begin{bmatrix} a' & b' \\ 0 & 0 \end{bmatrix} = \begin{bmatrix} a - a' & b - b' \\ 0 & 0 \end{bmatrix},$$

whence I is closed under differences. Given any element $\begin{bmatrix} r & s \\ 0 & t \end{bmatrix}$ of the ring $T_2(Z)$, we have

$$\begin{bmatrix} r & s \\ 0 & t \end{bmatrix}\begin{bmatrix} a & b \\ 0 & 0 \end{bmatrix} = \begin{bmatrix} ra & rb \\ 0 & 0 \end{bmatrix} \quad \text{and} \quad \begin{bmatrix} a & b \\ 0 & 0 \end{bmatrix}\begin{bmatrix} r & s \\ 0 & t \end{bmatrix} = \begin{bmatrix} ar & as + bt \\ 0 & 0 \end{bmatrix}.$$

Both of these products belong to I, allowing us to conclude that I is an ideal of $T_2(Z)$.

A further comment is in order. Although I happens to be a subring of the ring $M_2(Z)$, it fails to be an ideal, since I does not "swallow up" the product

$$\begin{bmatrix} 1 & 0 \\ 1 & 0 \end{bmatrix}\begin{bmatrix} 1 & 1 \\ 0 & 0 \end{bmatrix} = \begin{bmatrix} 1 & 1 \\ 1 & 1 \end{bmatrix}.$$

But on the positive side of the ledger, I turns out to be a right ideal of $M_2(Z)$. The required multiplication goes as follows:

$$\begin{bmatrix} a & b \\ 0 & 0 \end{bmatrix}\begin{bmatrix} r & s \\ u & t \end{bmatrix} = \begin{bmatrix} ar + bu & as + bt \\ 0 & 0 \end{bmatrix} \in I.$$

Example 4.14
For each integer $a \in Z$, let (a) represent the set consisting of all integral multiples of a; that is,

$$(a) = \{na \mid n \in Z\}.$$

The following relations confirm (a) to be an ideal of the ring of integers:

$$na - ma = (n - m)a,$$

$$m(na) = (mn)a \qquad (n, n \in Z).$$

In particular, since $(2) = Z_e$, the ring of even integers forms an ideal of Z. Notice, too, that $(0) = \{0\}$ and $(1) = Z$.

Example 4.15

Another illustration is furnished by $\text{map}(X, R)$, the ring of mappings from the set X into the ring R (see Example 4.4). For a fixed element $x \in X$ we denote by I_x the set of all mappings which take on the value 0 at x:

$$I_x = \{f \in \text{map}(X, R) | f(x) = 0\}.$$

Now choose $f, g \in I_x$ and $h \in \text{map}(X, R)$. From the definition of the ring operations in $\text{map}(X, R)$,

$$(f - g)(x) = f(x) - g(x) = 0 - 0 = 0,$$

while

$$(fh)(x) = f(x)h(x) = 0h(x) = 0,$$

and, in a similar manner, $(hf)(x) = 0$. Thus $f - g$, fh and hf all belong to I_x, which implies that I_x is an ideal.

More generally, if S is any nonempty subset of X, then

$$I = \{f \in \text{map}(X, R) | f(x) = 0 \text{ for all } x \in S\}$$

constitutes an ideal of $\text{map}(X, R)$. Since $I = \bigcap_{x \in S} I_x$, we have a situation in which the intersection of ideals is once again an ideal. Theorem 4.11 shows that this is no accident.

The following theorem presents a fact that, despite its apparent simplicity, will be frequently applied in the sequel.

THEOREM 4.10 *If I is a proper (right, left, two-sided) ideal of a ring R with identity, then no element of I possesses a multiplicative inverse; that is, $I \cap R^* = \varnothing$.*

Proof. Let I be an ideal of R and suppose that there is some member $a \neq 0$ of I such that a^{-1} exists in R. (The theorem is trivial when $I = \{0\}$.) Since I is closed under multiplication by arbitrary ring elements, it follows that $1 = a^{-1}a \in I$. By the same reasoning, I contains $r = r1$ for every r in R; that is, $R \subseteq I$, whence the equality $I = R$. This contradicts the hypothesis that I is a proper subset of R.

Note that, en route, we have also established

COROLLARY *In a ring with identity, no proper (right, left, two-sided) ideal contains the identity element.*

We now take up some of the standard methods for constructing new ideals from given ones. To begin with simpler things:

THEOREM 4.11 *Let $\{I_i\}$ be an arbitrary collection of (right, left, two-sided) ideals of the ring R, where i ranges over some index set. Then $\cap I_i$ is also a (right, left, two-sided) ideal of R.*

Proof. We give the proof for the case in which the ideals are two-sided. First, observe that the intersection $\cap I_i$ is nonempty, since each of the ideals I_i must contain the zero element of the ring. Suppose that the elements $a, b \in \cap I_i$ and $r \in R$. Then a and b are members of I_i, where i varies over the indexing set. Inasmuch as I_i is assumed to be an ideal of R, it follows that $a - b$, ar and ra all lie in the set I_i. But this is true for every value of i, whence the elements $a - b$, ar and ra belong to $\cap I_i$, making $\cap I_i$ an ideal of R.

Consider, for the moment, an arbitrary ring R and a nonempty subset S of R. By the symbol (S) we shall mean the set

$$(S) = \cap\{I \mid S \subseteq I;\ I \text{ is an ideal of } R\}.$$

The collection of all ideals that contain S is not empty, since the entire ring itself is an ideal containing any subset of R; thus the set (S) exists and satisfies the inclusion $S \subseteq (S)$. By virtue of Theorem 4.11, (S) forms an ideal of R, known as the *ideal generated by the set* S. It is noteworthy that whenever I is any ideal of R with $S \subseteq I$, then necessarily $(S) \subseteq I$. For this reason, one often speaks of (S) as being the smallest ideal of R to contain the set S. It should be apparent that corresponding remarks apply to the one-sided ideals generated by S.

If S consists of a finite number of elements, say $a_1, a_2, \ldots, a_n$, then the ideal that they generate is customarily denoted by $(a_1, a_2, \ldots, a_n)$. Such an ideal is said to be *finitely generated* with the given elements a_i as its generators. An ideal (a) generated by just one ring element is termed a *principal ideal.*

A natural undertaking is to determine the precise form of the members of the various ideals (right, left, two-sided) generated by a single element, say a, of an arbitrary ring R. The right ideal generated by a is called a *principal right ideal* and is denoted by $(a)_r$. Being closed with respect to multiplication on the right, $(a)_r$ necessarily contains all products ar $(r \in R)$, as well as the elements na (n an integer), and hence includes their sum $ar + na$.

As usual, the notation na represents the n-fold sum of a. It is a fairly simple matter to check that the set of elements of the form $ar + na$ constitutes a right ideal of R. Observe, too, that the element a is a member of the ideal, since $a = a0 + 1a$. These remarks make it clear that

$$(a)_r = \{ar + na \mid r \in R;\ n \in Z\}.$$

When there is an identity element present, the term na becomes superfluous, since in this setting we may write the expression $ar + na$ more simply as

$$ar + na = ar + a(n1) = a(r + n1) = ar',$$

where $r' = r + n1$ is some ring element. Thus the set $(a)_r$ consists of all right multiples of a by elements of R. If R is a ring with identity, we shall frequently employ the more suggestive notation aR in place of $(a)_r$; that is,

$$(a)_r = aR = \{ar \mid r \in R\}.$$

Similar remarks apply, of course, to the principal left ideal $(a)_l$ generated by a.

As a general comment, observe that the products ar $(r \in R)$ form the set of elements of a right ideal of R even when the ring does not possess an identity. The difficulty, however, is that this ideal need not contain a itself.

With regard to the two-sided ideal (a) generated by a, the situation is more complicated. Certainly the elements ras, ra, as and na must all belong to the ideal (a) for every choice of $r, s \in R$ and $n \in Z$. In general, the sum of two elements ras and $r'as'$ is no longer of the same form, so in order to have closure under addition, any finite sum $\sum r_i a s_i$, where $r_i, s_i \in R$, is also required to be in (a). The reader will experience no difficulty in showing that the principal ideal generated by a is given by

$$(a) = \left\{ na + ra + as + \sum_{\text{finite}} r_i a s_i \mid r, s, r_i, s_i \in R;\ n \in Z \right\}.$$

In case R happens to have an identity, this description of (a) reduces to the set of all finite sums $\sum_{\text{finite}} r_i a s_i$.

A particularly important type of ring is a principal ideal ring, which we now define.

DEFINITION 4.10 A ring R is said to be a *principal ideal ring* if every ideal I of R is of the form $I = (a)$ for some $a \in R$.

The following theorem furnishes an example of such rings.

THEOREM 4.12 *The ring Z of integers is a principal ideal ring; in fact, if I is an ideal of Z, then $I = (n)$ for some nonnegative integer n.*

Proof. If $I = \{0\}$, the theorem is trivially true, since the zero ideal $\{0\}$ is the principal ideal generated by 0. Suppose then that I does not consist of the zero element alone. Now, if $m \in I$, $-m$ also lies in I, so the set I contains positive integers. Let n designate the least positive integer in I. As I forms an ideal of Z, each integral multiple of n must belong to I, whence $(n) \subseteq I$.

To establish the inclusion $I \subseteq (n)$, let k be an arbitrary element of I. By the Division Algorithm, integers q and r exist for which $k = qn + r$, with $0 \leqslant r < n$. Since k and qn are both members of I, it follows that $r = k - qn \in I$. If $r > 0$, we would have a contradiction to the assumption that n is the smallest positive integer in I. Accordingly, $r = 0$ and $k = qn \in (n)$. Thus only multiples of n belong to I, implying that $I \subseteq (n)$. The two inclusions show that $I = (n)$ and the argument is complete.

Let us now describe certain binary operations on the set of all ideals of R. Similar considerations apply to the sets of right and left ideals, but for economy of effort we concentrate on two-sided ideals. Given a finite number of ideals $I_1, I_2, \ldots, I_n$ of the ring R, one defines their sum in the natural way:

$$I_1 + I_2 + \cdots + I_n = \{a_1 + a_2 + \cdots + a_n | a_i \in I_i\}.$$

Then $I_1 + I_2 + \cdots + I_n$ is likewise an ideal of R and is the smallest ideal of R that contains every I_i; phrased in another way, $I_1 + I_2 + \cdots + I_n$ is the ideal generated by the union $I_1 \cup I_2 \cup \cdots \cup I_n$. In the special case of two ideals I and J our definition reduces to

$$I + J = \{a + b | a \in I;\ b \in J\}.$$

More generally, let $\{I_i\}$ be an arbitrary indexed collection of ideals of R. The *sum* of this collection may be denoted by $\sum I_i$ and is the ideal of R whose members are all possible finite sums of elements from the various ideals I_i:

$$\sum I_i = \left\{ \sum_{\text{finite}} a_i | a_i \in I_i \right\}.$$

The reader will take care to remember that, although $\{I_i\}$ may be an infinite family of ideals, only finite sums of elements of R are involved in the definition above. An alternative description of $\sum I_i$ could be given by

$$\sum I_i = \{\sum a_i | a_i \in I_i;\ \text{all but a finite number of the } a_i \text{ are } 0\},$$

where it is understood that $\sum$ represents an arbitrary sum with one or more terms. Just as $\cap I_i$ can be interpreted as the largest ideal of R contained in every I_i, the sum $\sum I_i$ supplies the dual notion of the smallest ideal containing every I_i.

We now come to a less elementary but extremely useful notion, namely, the *product* of ideals. Once again, assume that I and J are two ideals of the ring R. To be consistent with our earlier definition of the sum $I + J$, we should define the product IJ to be the collection of all simple products ab, where $a \in I$ and $b \in J$. Unfortunately, the resulting set fails to form an ideal. (Why?) To counter this difficulty, we instead take the elements of IJ to be all possible finite sums of simple products; stated explicitly,

$$IJ = \left\{ \sum_{\text{finite}} a_i b_i | a_i \in I;\ b_i \in J \right\}.$$

With this definition, IJ indeed becomes an ideal of R. Suppose that $x, y \in IJ$ and $r \in R$; then,

$$x = a_1 b_1 + a_2 b_2 + \cdots + a_n b_n, \qquad y = a'_1 b'_1 + a'_2 b'_2 + \cdots + a'_m b'_m,$$

where the a_i and a'_i are in I, and the b_i and b'_i are in J. From this we obtain

$$x - y = a_1 b_1 + \cdots + a_n b_n + (-a'_1) b'_1 + \cdots + (-a'_m) b'_m,$$
$$rx = (ra_1) b_1 + (ra_2) b_2 + \cdots + (ra_n) b_n.$$

Now the elements $-a'_i$ and ra_i necessarily lie in I, so $x - y$ and $rx \in IJ$; likewise, $xr \in IJ$, making IJ an ideal of R. In point of fact, IJ is just the ideal generated by the set of all products ab, $a \in I$, $b \in J$.

There is no difficulty in extending the above remarks to any finite number of ideals $I_1, I_2, \ldots, I_n$ of the ring R. A moment's thought shows that the product $I_1 I_2 \cdots I_n$ is the ideal consisting of finite sums of terms of the form $a_1 a_2 \cdots a_n$, with a_i in I_i. It is perhaps appropriate to point out that, because of the associative law for multiplication in R, the notation $I_1 I_2 \cdots I_n$ is unambiguous. A special case immediately presents itself, namely, the situation in which all the ideals are alike, say equal to the ideal I. Here, we see that I^n is the set of finite sums of products of n elements from I:

$$I^n = \left\{ \sum_{\text{finite}} (a_{i_1} a_{i_2} \cdots a_{i_n}) | a_{i_k} \in I \right\}.$$

An ideal I of the ring R is said to be a *nil ideal* if each element $a \in I$ is nilpotent; that is, if there exists a positive integer n for which $a^n = 0$, where n depends on the particular element a. Naturally enough, the ideal I will be termed *nilpotent* provided that $I^n = \{0\}$ for some positive integer n. The condition $I^n = \{0\}$ is equivalent to requiring that for every choice of n elements $a_1, a_2, \ldots, a_n \in I$ (distinct or not), the product $a_1 a_2 \cdots a_n = 0$. In particular, $a^n = 0$ for all a in I, whence every nilpotent ideal is automatically a nil ideal.

To furnish an example, $I = \{0, 2, 4\}$ is a nilpotent ideal of the ring Z_8 of integers modulo 8; here, it will be found that $I^3 = \{0\}$. More generally, the principal ideal $(p) = pZ_{p^n}$ is nilpotent in Z_{p^n}, for any prime p and $n > 1$.

The next item on the program is to analyze the behavior of ideals under homomorphisms.

THEOREM 4.13 *Let f be a homomorphism from the ring R onto the ring R'. Then*

a) *for each ideal I of R, $f(I)$ is an ideal of R'; and*
b) *for each ideal I' or R', $f^{-1}(I')$ is an ideal of R.*

Proof. Theorem 4.8 shows us that $f(I)$ is a subring of the ring R'. To see that it is actually an ideal, pick $r' \in R'$ and $f(a) \in f(I)$, where $a \in I$. Since f carries R onto R', we may write $r' = f(r)$ for some r in R. Then the product $ra \in I$, I being an ideal of R, and so

$$r'f(a) = f(r)f(a) = f(ra) \in f(I).$$

Likewise, $f(a)r' \in f(I)$, which makes $f(I)$ an ideal of R'.

Given an ideal I' of R', we know that $f^{-1}(I')$ is a subring of R (Theorem 4.8 enters again). Suppose that $a \in f^{-1}(I')$, so that $f(a) \in I'$, and let r be a typical element of R. Then $f(ra) = f(r)f(a) \in I'$; in other words, the product ra lies in $f^{-1}(I')$. A similar argument reveals that $ar \in f^{-1}(I')$ and hence $f^{-1}(I')$ forms an ideal of R'.

If f is a homomorphism from the ring R into the ring R', then the *kernel* of f is the set

$$\ker f = \{a \in R \mid f(a) = 0\}.$$

Ignoring multiplication in the rings, this is just the definition of the kernel of f when regarded as a homomorphism between the underlying additive groups of R and R'. As in the case of groups, the homomorphism f is a one-to-one function if and only if $\ker f = \{0\}$.

So that our analogy between ideals and normal subgroups may make sense, we should anticipate the theorem below:

THEOREM 4.14 *The kernel* $\ker f$ *of a homomorphism f from a ring R into a ring R' is an ideal of R.*

Proof. We have previously observed that the trivial subring $\{0\}$ constitutes an ideal of R'. Inasmuch as $\ker f = f^{-1}(0)$, the stated conclusion follows from Theorem 4.13.

Once these results are available, it is a straightforward matter to derive the Correspondence Theorem for Ideals. Mimicking the argument of Theorem 2.64, we obtain the result below; its proof is left as an exercise.

THEOREM 4.15 (Correspondence Theorem) *Let f be a homomorphism from the ring R onto the ring R'. Then there is a one-to-one correspondence between those ideals I of R such that* $\ker f \subseteq I$ *and the set of all ideals I' of R'; specifically, I' is given by $I' = f(I)$.*

This seems to be a natural place to insert an example.

Example 4.16
Consider an arbitrary ring R with identity and the mapping $f: Z \to R$ given by $f(n) = n1$. At the risk of being repetitious, let us again emphasize that $n1$ means the n-fold sum of 1. A simple computation shows that f, so defined, is a homomorphism from the ring Z of integers into the ring R:

$$f(n + m) = (n + m)1 = n1 + m1 = f(n) + f(m)$$

and

$$f(nm) = (nm)1 = n(m1) = (n1)(m1) = f(n)f(m).$$

Since $\ker f$ constitutes an ideal of Z, a principal ideal ring, it follows that

$$\ker f = \{n \in Z | n1 = 0\} = (p)$$

for some nonnegative integer p. A moment's reflection should convince the reader that the integer p is just the characteristic of R. In particular, any ring R with identity that is of characteristic zero will contain a subring isomorphic to the integers; more specifically, $Z \simeq Z1$, where 1 is the identity of R.

We now turn our attention to the matter of cosets in a ring. Given an ideal I of the ring R, the system $(I, +)$ is a normal subgroup of the additive group $(R, +)$. Thus it is possible to construct the quotient group of R by I. In our present notation the cosets assume the form $a + I = \{a + i | i \in I\}$. Two cosets $a + I$ and $b + I$ are equal if and only if $a - b \in I$ (Theorem 2.39).

As before, we shall employ the symbol R/I to denote the collection of all cosets of I in R; that is,

$$R/I = \{a + I | a \in R\}.$$

It follows from Theorem 2.46 and Problem 22, Section 2.6, that if addition of cosets is defined by the rule

$$(a + I) + (b + I) = (a + b) + I,$$

then R/I becomes a commutative (additive) group. The set R/I can be endowed with the structure of a ring by introducing multiplication in the obvious way; just put

$$(a + I)(b + I) = ab + I.$$

One is faced with the usual problem of showing that this operation is well defined, so that the product of two cosets $a + I$ and $b + I$ does not depend on the particular representatives a and b. To this end, suppose that

$$a + I = a' + I \qquad \text{and} \qquad b + I = b' + I.$$

Then $a - a' = i_1$ and $b - b' = i_2$ for some $i_1, i_2 \in I$. From this we conclude that

$$\begin{aligned} ab - a'b' &= a(b - b') + (a - a')b' \\ &= ai_2 + i_1 b' \in I, \end{aligned}$$

since both the products ai_2 and $i_1 b'$ must lie in I (remember that I is an ideal of R). The implication, of course, is that $ab + I = a'b' + I$, or in equivalent terms that $(a + I)(b + I) = (a' + I)(b' + I)$; hence our definition of multiplication in R/I is meaningful.

The verification that R/I, under the operations indicated above, constitutes a ring is easy, and the details are safely left to the reader's care. To assure completeness, we simply state

THEOREM 4.16 *If I is an ideal of the ring R, then R/I is also a ring, known as the quotient ring of R by I.*

Example 4.17

Consider Z_8, the ring of integers modulo 8, and the ideal $I = \{0, 4\}$. It is easily seen that the quotient ring Z_8/I has four elements, namely, the cosets

$$0 + I = \{0, 4\}, \qquad 2 + I = \{2, 6\},$$
$$1 + I = \{1, 5\}, \qquad 3 + I = \{3, 7\}.$$

Constructing the operation tables for this ring, we obtain:

+	{0, 4}	{1, 5}	{2, 6}	{3, 7}
{0, 4}	{0, 4}	{1, 5}	{2, 6}	{3, 7}
{1, 5}	{1, 5}	{2, 6}	{3, 7}	{0, 4}
{2, 6}	{2, 6}	{3, 7}	{0, 4}	{1, 5}
{3, 7}	{3, 7}	{0, 4}	{1, 5}	{2, 6}

	{0, 4}	{1, 5}	{2, 6}	{3, 7}
{0, 4}	{0, 4}	{0, 4}	{0, 4}	{0, 4}
{1, 5}	{0, 4}	{1, 5}	{2, 6}	{3, 7}
{2, 6}	{0, 4}	{2, 6}	{0, 4}	{2, 6}
{3, 7}	{0, 4}	{3, 7}	{2, 6}	{1, 5}

The reader cannot fail to notice the isomorphism between the quotient ring Z_8/I and the ring Z_4.

Example 4.18

In the ring Z of integers, consider the principal ideal (n), where n is a nonnegative integer. The cosets of (n) in Z take the form

$$a + (n) = \{a + kn \mid k \in Z\},$$

from which it is clear that the cosets of (n) are precisely the congruence classes modulo n. What we earlier described as the operations for congruence classes in Z_n can now be viewed as coset operations in Z_n:

$$(a + (n)) + (b + (n)) = a + b + (n);$$

$$(a + (n))(b + (n)) = ab + (n).$$

In short, the ring Z_n of integers modulo n could just as well be interpreted as the quotient ring of Z by (n).

At an earlier stage we saw that to any normal subgroup H of a group G there corresponds a homomorphism having H as its kernel; one need only consider the natural mapping onto the quotient group G/H. It is not surprising that an entirely similar result carries over to ring theory. Indeed, each ideal I of a ring R determines a quotient ring R/I and, in consequence, a natural mapping $\text{nat}_I: R \to R/I$ that sends each $a \in R$ to its coset $a + I$; explicitly,

$$\text{nat}_I(a) = a + I.$$

Paralleling our study of groups, we get

THEOREM 4.17 *Let I be an ideal of the ring R. Then the natural mapping nat_I is a homomorphism from R onto the quotient ring R/I with kernel I.*

Proof. By Theorem 2.57, nat_I is known to be a homomorphism between the additive groups of R and R/I, with $\ker(\text{nat}_I) = I$. All that is required here is to show that multiplication is preserved by nat_I. But this presents no problem:

$$\text{nat}_I(ab) = ab + I = (a + I)(b + I) = \text{nat}_I(a)\,\text{nat}_I(b),$$

making nat_I a ring homomorphism.

Knowing this, we may take the homomorphism stipulated in the Correspondence Theorem to be the natural mapping nat_I. The statement then reads: If I is an ideal of the ring R, then there is a one-to-one correspondence between those ideals J of R such that $I \subseteq J$ and the set of all ideals of the quotient ring R/I.

Without further delay, let us derive the Fundamental Homomorphism Theorem for Rings.

THEOREM 4.18 (Fundamental Homomorphism Theorem) *If f is a homomorphism from the ring R onto the ring R', then $R/\ker f \simeq R'$.*

Proof. As in the group-theoretic version, we define a function $\bar{f}: R/\ker f \to R'$, the induced mapping, by taking $\bar{f}(a + \ker f) = f(a)$; it is this function that establishes the announced isomorphism. From

Theorem 2.61 we already know that, via $\bar{f}$, $R/\ker f$ and R' are isomorphic as groups. It remains to settle the question of whether $\bar{f}$ preserves the multiplication operation in $R/\ker f$. Since f is a ring homomorphism,

$$\begin{aligned}\bar{f}((a + \ker f)(b + \ker f)) &= \bar{f}(ab + \ker f)\\ &= f(ab)\\ &= f(a)f(b)\\ &= \bar{f}(a + \ker)\bar{f}(b + \ker f),\end{aligned}$$

establishing the result claimed.

We draw attention to the fact that, by virtue of the definition of the induced mapping, any ring homomorphism f admits the factorization

$$f = \bar{f} \circ \mathrm{nat}_{\ker f}.$$

Example 4.19
For a simple illustration of these ideas we turn to the rings Z_4 and Z_2. Define the function $f: Z_4 \to Z_2$ by

$$f(0) = f(2) = 0, \qquad f(1) = f(3) = 1.$$

It is a tedious process to check that f is a homomorphism, and we pass over the details. The reader should verify several cases to be satisfied that this property actually holds.

In the present case the kernel of f turns out to be the two-element set $\{0, 2\}$. Hence

$$Z/\ker f = \{\{0, 2\}, \{1, 3\}\}.$$

The operation tables for the quotient ring $Z_4/\ker f$ are shown below:

$+$	$\{0, 2\}$	$\{1, 3\}$
$\{0, 2\}$	$\{0, 2\}$	$\{1, 3\}$
$\{1, 3\}$	$\{1, 3\}$	$\{0, 2\}$

$\cdot$	$\{0, 2\}$	$\{1, 3\}$
$\{0, 2\}$	$\{0, 2\}$	$\{0, 2\}$
$\{1, 3\}$	$\{0, 2\}$	$\{1, 3\}$

Theorem 4.18 asserts that $Z_4/\ker f \simeq Z_2$; indeed, this is fairly evident from an inspection of the displayed tables.

As an application of the Fundamental Theorem, we propose to prove that any (ring) homomorphism onto the ring of integers is uniquely determined by its kernel. Contrast this with the situation for groups; both i_Z and $-i_Z$ are homomorphisms from the additive group Z of integers into itself. The kernel of each of these mappings is $\{0\}$, but clearly $i_Z \neq -i_Z$.

We begin by establishing a lemma that is of independent interest.

LEMMA *The only nontrivial homomorphism from the ring Z of integers into itself is the identity map i_Z.*

Proof. Because each positive integer n may be written as $n = 1 + 1 + \cdots + 1$ (n summands), the operation-preserving nature of f implies that $f(n) = nf(1)$. On the other hand, if n is an arbitrary negative integer, then $-n \in Z_+$ and so

$$f(n) = f(-(-n)) = -f(-n) = -(-n)f(1) = nf(1).$$

Plainly, $f(0) = 0 = 0f(1)$. The upshot is that $f(n) = nf(1)$ for every n in Z. Because f is not identically zero, we must have $f(1) = 1$; to see that this is so, simply apply the cancellation law to the relation $f(m) = f(m1) = f(m)f(1)$, where $f(m) \neq 0$. One finds in this way that $f(n) = n = i_Z(n)$ for all $n \in Z$, making f the identity map on Z.

COROLLARY *There is at most one homomorphism under which an arbitrary ring R is isomorphic to the ring Z.*

Proof. Suppose that the rings R and Z are isomorphic under two functions $f, g: R \to Z$. Then, the composition $f \circ g^{-1}$ is a homomorphic mapping from the ring Z onto itself. Since this is the case, the lemma just proved implies that $f \circ g^{-1} = i_Z$, or $f = g$.

We now have the necessary information to prove the following result.

THEOREM 4.19 *Any homomorphism from an arbitrary ring R onto the ring Z of integers is uniquely determined by its kernel.*

Proof. Let f and g be two homomorphisms from the ring R onto Z with the property that $\ker f = \ker g$. Our aim, of course, is to show that f and g must be the same function. Now by Theorem 4.18 the quotient rings $R/\ker f$ and $R/\ker g$ are both isomorphic to the ring of integers via the induced mappings $\bar{f}$ and $\bar{g}$, respectively. The assumption that f and g have a common kernel, when combined with the preceding corollary, forces $\bar{f} = \bar{g}$. It follows at once from the factorizations

$$f = \bar{f} \circ \mathrm{nat}_{\ker f}, \qquad g = \bar{g} \circ \mathrm{nat}_{\ker g}$$

that the functions f and g are themselves identical.

PROBLEMS 4.3

1. If I is a right ideal and J a left ideal of the ring R such that $I \cap J = \{0\}$, prove that $ab = 0$ for all $a \in I$, $b \in J$.
2. Given an ideal I of the ring R, define the set $C(I)$ by

$$C(I) = \{r \in R | ra - ar \in I \text{ for all } a \in R\}.$$

Verify that $C(I)$ forms a subring of R.
3. a) Show by example that if I and J are both ideals of the ring R, then $I \cup J$ need not be an ideal.
   b) If $\{I_i\}(i = 1, 2, \ldots)$ is a collection of ideals of the ring R such that

$$I_1 \subseteq I_2 \subseteq \cdots \subseteq I_n \subseteq \cdots,$$

prove that $\cup I_i$ is also an ideal of R.
4. Let R be an arbitrary ring and $R \times Z$ be the extension ring constructed in Theorem 4.9. Establish that
   a) $R \times \{0\}$ is an ideal of $R \times Z$;
   b) $Z \simeq \{0\} \times Z$;
   c) if a is an idempotent element of R, then the pair $(-a, 1)$ is idempotent in $R \times Z$, while $(a, 0)$ is a zero divisor.
5. Let I be an ideal of R, a commutative ring with identity. For an element $a \in R$ the ideal generated by the set $I \cup \{a\}$ is denoted by (I, a). Assuming that $a \notin I$, show that

$$(I, a) = \{i + ra | i \in I, r \in R\}.$$

6. Establish each of the assertions below:
   a) $M_2(Z_e)$ is an ideal of the matrix ring $M_2(Z)$.
   b) The set

$$I = \left\{ \begin{bmatrix} 0 & a \\ 0 & 0 \end{bmatrix} \middle| a \in R^\# \right\}$$

is an ideal of the ring $T_2(R^\#)$ of 2×2 triangular matrices.
   c) The matrix ring $M_2(R^\#)$ has no nontrivial ideals. [*Hint:* If $\begin{bmatrix} a & b \\ c & d \end{bmatrix} \in I$ with, say, $d \neq 0$, then

$$\begin{bmatrix} 0 & 0 \\ 0 & 1 \end{bmatrix} \begin{bmatrix} a & b \\ c & d \end{bmatrix} \begin{bmatrix} 0 & 0 \\ 0 & 1 \end{bmatrix} = \begin{bmatrix} d & 0 \\ 0 & 0 \end{bmatrix} \in I;$$

this, together with similar conditions, puts the identity matrix into I.]
7. Show that Z is a subring of the ring Q of rational numbers but is not an ideal of Q.

8. Prove that every subring of Z forms an ideal. Do the same for the ring Z_n.
9. Consider the set $R = \{a + bi | a, b \in Z\}$, where $i^2 = -1$. Verify that R forms a ring with respect to ordinary addition and multiplication and that $I = \{a + 2bi | a, b \in Z\}$ is an ideal of this ring.
10. a) Determine whether $Z \times Z_e$ is an ideal of the ring $Z \times Z$.
    b) In the ring $P(X)$ of all subsets of a set X, show that the collection of all finite subsets of X forms an ideal.
11. a) Let I be an ideal of the ring R and define

$$\text{ann } I = \{r \in R | ra = 0 \text{ for all } a \in I\}.$$

    Prove that ann I forms an ideal of R, called the *annihilator* of I.
    b) In the ring Z_{24}, find the annihilator of the ideal $\{0, 8, 16\}$.
12. If I and J are ideals of the ring R, define their product to be the set

$$IJ = \{\sum a_j b_i | a_i \in I,\ b_j \in J\}$$

where $\sum$ denotes a finite sum with one or more terms. Establish that IJ forms an ideal of R with $IJ \subseteq I \cap J$.
13. Prove that the product of ideals is distributive over the sum of ideals; that is,

$$I(J + K) = IJ + IK$$

for any three ideals I, J, and K of a ring R.
14. Let R be an arbitrary ring. For each fixed element $a \in R$, define the left-multiplication function $T_a : R \to R$ by taking $T_a(x) = ax$. If T_R denotes the set of all such functions, prove the following:
    a) T_a is a (group) homomorphism of the additive group of R into itself;
    b) T_R forms a ring, where multiplication is taken to be functional composition;
    c) the mapping $f(a) = T_a$ determines a homomorphism of R onto the ring T_R;
    d) the kernel of f is the ideal ann R;
    e) if for each $0 \neq a \in R$, some $b \in R$ exists such that $ab \neq 0$, then $R \simeq T_R$. (In particular, this holds whenever R has an identity element.)
15. Let I be an ideal of the ring R. Establish each of the following:
    a) R/I has no divisors of zero if and only if $ab \in I$ implies that either a or b belongs to I.
    b) R/I is commutative if and only if $ab - ba \in I$ for all a, b in R.
    c) R/I has an identity element if and only if there is some $e \in R$ such that $ae - a \in I$ and $ea - a \in I$ for all a in R.
    d) Whenever R is a commutative ring with identity, then so is the quotient ring R/I.

16. Let R be a commutative ring with identity and let N denote the set of all nilpotent elements in R. Verify that

a) The set N forms an ideal of R. [*Hint:* If $a^n = b^m = 0$ for integers n and m, consider $(a - b)^{n+m}$.]

b) The quotient ring R/N has no nonzero nilpotent elements.

17. If R is an arbitrary ring and $n \in Z_+$, prove that

a) the sets $I_n = \{na | a \in R\}$ and $J_n = \{a \in R | na = 0\}$ are both ideals of R;

b) $\text{char}(R/I_n)$ divides n;

c) if char $R \neq 0$, then char R divides $n \cdot \text{char}(R/J_n)$.

18. Prove that any homomorphic image of a principal ideal ring is again a principal ideal ring. [*Hint:* If f is a homomorphism from the principal ideal ring R onto the ring R', then $f((a)) = (f(a))$ for each $a \in R$.]

19. If I is an ideal of the ring R, prove that

a) R/I is a simple ring if and only if there is no ideal J of R with $I \subset J \subset R$ (a ring is said to be *simple* provided that it has no nontrivial ideals);

b) if R is a principal ideal ring, then so is the quotient ring R/I; in particular, Z_n is a principal ideal ring for each $n \in Z_+$.

20. Verify that (up to isomorphism) the only homomorphic images of the ring Z of integers are the rings Z_n, $\{0\}$ and Z itself.

21. Suppose that S is a subring and I an ideal of the ring R. If $S \cap I = \{0\}$, prove that S is isomorphic to a subring of the quotient ring R/I. [*Hint:* Utilize the mapping $f(a) = a + I$, where $a \in S$.]

22. If $R = Z \times Z$ is the ring of Example 4.8, show that $R/Z \simeq Z$. [*Hint:* Consider the homomorphism $f(a, b) = a$.]

23. A ring homomorphism $f: Z_9 \to Z_3$ from Z_9 onto Z_3 is defined by taking

$$f(0) = f(3) = f(6) = 0, \qquad f(1) = f(4) = f(7) = 1,$$
$$f(2) = f(5) = f(8) = 3.$$

Describe the cosets in $Z_9/\ker f$ and set up an isomorphism so that $Z_9/\ker f \simeq Z_3$.

24. Let $R = \text{map}\, R^{\#}$ be the ring of real-valued functions on $R^{\#}$. If

$$I = \{f \in R | f(0) = 0\},$$

prove that the quotient ring R/I is isomorphic to $R^{\#}$.

25. Consider the ring $T_2(Z)$ of 2×2 triangular matrices with integral entries.

a) Prove that the mapping $f: T_2(Z) \to Z$ given by

$$f\left(\begin{bmatrix} a & b \\ 0 & c \end{bmatrix}\right) = c$$

is a homomorphism from $T_2(Z)$ onto Z.

b) Describe the ideals $f^{-1}(Z_e)$ and ker f.

c) Set up an isomorphism so that $T_2(Z)/\ker f \simeq Z$.

26. If $m, n \in Z_+$, with $m|n$, confirm that the mapping $f([a]) = [a]$ is a homomorphism from the ring Z_n onto the ring Z_m. Describe ker f.

27. Assuming that f is a homomorphism from the ring R onto the ring R', prove the following:

a) The mapping $\bar{f}: M_n(R) \to M_n(R')$ defined by $\bar{f}((a_{ij})) = (f(a_{ij}))$ is a ring homomorphism.

b) $\ker \bar{f} = \{(a_{ij}) \in M_n(R)\, a_{ij} \in \ker f\}$.

c) $\bar{f}$ is an isomorphism if and only if f is an isomorphism.

28. Let f be a homomorphism from the commutative ring R onto the ring R'. If I and J are ideals of R, verify each of the assertions below:

a) $f(I + J) = f(I) + f(J)$;

b) $f(IJ) = f(I)f(J)$;

c) $f(I \cap J) \subseteq f(I) \cap f(J)$, with equality when either $I \supseteq \ker f$ or $J \supseteq \ker f$.

29. If I is an ideal of the ring R, prove that

a) $M_n(I)$ is an ideal of the matrix ring $M_n(R)$;

b) the quotient ring $M_n(R)/M_n(I)$ is isomorphic to $M_n(R/I)$. [*Hint:* Consider the mapping $f: M_n(R) \to M_n(R/I)$ given by $f((a_{ij})) = (a_{ij} + I)$.]

30. A *commutator* in a ring R is any element of the form $[a, b] = ab - ba$, where $a, b \in R$. Let $[R, R]$ denote the ideal of R generated by the set of all commutators. Prove that

a) R is a commutative ring if and only if $[R, R] = \{0\}$;

b) for an ideal I of R the quotient ring R/I is commutative if and only if $[R, R] \subseteq I$; in particular, $R/[R, R]$ is commutative.

31. Establish the following assertions:

a) The set

$$S = \left\{ \begin{bmatrix} a & b \\ 0 & a \end{bmatrix} \middle| a, b \in Z \right\}$$

is a subring of the ring $T_2(Z)$ of 2×2 triangular matrices with integral entries.

b) The set

$$I = \left\{ \begin{bmatrix} 0 & c \\ 0 & 0 \end{bmatrix} \middle| c \in Z \right\}$$

forms an ideal of S with $S/I \simeq Z$.

32. Prove that the rings

$$R_1 = \{a + b\sqrt{2} | a, b \in Z\}$$

and

$$R_2 = \left\{\begin{bmatrix} a & 2b \\ b & a \end{bmatrix} \middle| a, b \in Z\right\}$$

are isomorphic.

33. Define the function $f : Z_{15} \to Z_{15}$ by the rule $f(a) = 6a$. Show that f is a homomorphism and use the Fundamental Homomorphism Theorem to obtain $Z_{15}/\{0, 5, 10\} \simeq (0, 3, 6, 9, 12\}$.

4.4 INTEGRAL DOMAINS AND FIELDS

In the preceding section a hierarchy of special rings was established by imposing more and more restrictions on the multiplicative semigroup of a ring. At first glance, one might be tempted to require that the multiplicative semigroup actually be a group; such an assumption would be far too demanding in that this situation can take place only in the trivial ring consisting of zero alone. A less stringent condition would be the following: The nonzero elements comprise a group under multiplication. This leads to the notion of a field.

DEFINITION 4.11 A ring F is said to be a *field* provided that the set $F - \{0\}$ is a commutative group under the multiplication of F (the identity of this group will be written as 1).

Definition 4.11 implicitly assumes that any field F contains at least one element that is different from zero; $F - \{0\}$ must be nonempty, serving as the set of elements of a group. It is also to be remarked that, since $a0 = 0 = 0a$ for any $a \in F$, all the members of F commute under multiplication and not merely the nonzero elements. Similarly, the relation $1 \cdot 0 = 0 = 0 \cdot 1$ implies that 1 is the identity for the entire ring F. A field is a commutative ring with identity in which each nonzero element possesses an inverse under multiplication.

Occasionally, we shall find it convenient to drop the requirement of commutativity in the consideration of a field, in which case the resulting system is called a *division ring* or *skew field*; that is to say, a ring is a division ring if its nonzero elements form a group (not necessarily commutative) with respect to multiplication.

The concept of a field evolved during the last half of the nineteenth century through the research of Richard Dedekind and Leopold Kronecker on algebraic numbers and algebraic functions. The first systematic treatment of the theory of algebraic numbers (that is, numbers that are roots of polynomials $a_0x^n + a_1x^{n-1} + \cdots + a_{n-1}x + a_n$ with integers $a_0, a_1, \ldots, a_n$ as coefficients) was published by Dedekind in a supplement to the second edition of Dirichlet's *Number Theory*, which he prepared in 1871. With this work the notion of a number field became firmly established in mathematics. Dedekind used the German word "korper" (a "body" of numbers, equivalent to the English "field") to mean any set F of nonzero algebraic numbers such that if α and β belong to F, so also do $\alpha \pm \beta$, $\alpha\beta$, and α/β; that is, F has the property that the rational operations of algebra can be performed within F. The fields that Dedekind considered were therefore restricted to real and complex numbers. Kronecker followed a short time later (1882) with his own theory of fields, or "domains of rationality" as he called them. Kronecker's definition was more general than Dedekind's, since it included fields of rational functions with any number of variables, but it still did not attain complete generality. The earliest exposition of the theory from the abstract point of view was given by Heinrich Weber, who had early espoused the axiomatic approach to groups. In 1893, Weber published an article on the foundations of Galois theory in which he presented the definition of a (commutative) field, substantially as we know it today. Weber emphasized the primacy of the group concept as the fundamental structure of algebra: "A group becomes a field if it can have two kinds of composition, of which the first is called addition and the second multiplication." The axiomatization process was carried to its completion by E. V. Huntington and Leonard Dickson, who in 1903 gave the earliest sets of independent postulates for abstract fields.

After this preamble, let us look at several standard examples of fields.

Example 4.20
The rational numbers Q and the real numbers $R^{\#}$, with respect to ordinary addition and multiplication, are the most obvious illustrations of fields. More interesting is the following subset of $R^{\#}$:

$$F = \{a + b\sqrt{2} \mid a, b \in Q\}.$$

It is easily checked that F is a commutative ring with identity; the additive and multiplicative identity elements are

$$0 = 0 + 0\sqrt{2}, \qquad 1 = 1 + 0\sqrt{2}.$$

As usual, it is the multiplicative inverses that cause trouble. Pick a nonzero element of F; that is, let $a + b\sqrt{2} \in F$ with a and b not both zero. Then $a^2 - 2b^2 \neq 0$ (otherwise, $\sqrt{2}$ would be rational) so that division by $a^2 - 2b^2$ is permissible. By using the customary rationalization process,

$$\begin{aligned}(a + b\sqrt{2})^{-1} &= \frac{1}{a + b\sqrt{2}} = \frac{1}{a + b\sqrt{2}}\,\frac{a - b\sqrt{2}}{a - b\sqrt{2}} \\ &= \frac{a}{a^2 - 2b^2} + \frac{-b}{a^2 - 2b^2}\sqrt{2}.\end{aligned}$$

Since a and b are rational numbers, $a/(a^2 - 2b^2)$ and $-b/(a^2 - 2b^2)$ are both rational. Thus $(a + b\sqrt{2})^{-1}$ has the required form to be a member of F. Notice that if a and b were restricted to the set of integers, then F would not be a field but merely an integral domain.

Example 4.21
Consider the set $C = R^{\#} \times R^{\#}$ of ordered pairs of real numbers. To turn C into a field, we define addition and multiplication by

$$(a, b) + (c, d) = (a + c, b + d),$$
$$(a, b)(c, d) = (ac - bd, ad + bc).$$

The reader may verify without difficulty that C, together with these operations, is a commutative ring with identity. In this setting, the pair $(1, 0)$ serves as the multiplicative identity, and $(0, 0)$ is the zero element of the ring. Given any nonzero element (a, b) of C, either $a \neq 0$ or $b \neq 0$, so that $a^2 + b^2 > 0$; thus

$$\left(\frac{a}{a^2 + b^2}, \frac{-b}{a^2 + b^2}\right)$$

exists in C and has the property that

$$(a, b)\left(\frac{a}{a^2 + b^2}, \frac{-b}{a^2 + b^2}\right) = \left(\frac{a^2 + b^2}{a^2 + b^2}, \frac{a(-b) + ab}{a^2 + b^2}\right) = (1, 0).$$

This shows that each nonzero member of C has an inverse under multiplication, thereby proving the system C to be a field.

It is worth pointing out that the field C contains a subring that is isomorphic to the field of real numbers. If

$$R^{\#} \times \{0\} = \{(a, 0) | a \in R^{\#}\},$$

it follows that $R^{\#} \simeq R^{\#} \times \{0\}$ via the mapping f defined by $f(a) = (a, 0)$. Inasmuch as the distinction between these systems is only one of notation, we customarily identify the real number a with the corresponding ordered pair $(a, 0)$; in this sense, $R^{\#}$ may be regarded as a subring of C.

Now the definition of the operations in C enables us to express an arbitrary element $(a, b) \in C$ as

$$(a, b) = (a, 0) + (b, 0)(0, 1),$$

where the pair $(0, 1)$ is such that $(0, 1)^2 = (0, 1)(0, 1) = (-1, 0)$. Introducing the symbol i as an abbreviation for $(0, 1)$, we have

$$(a, b) = (a, 0) + (b, 0)i.$$

Finally, if it is agreed to replace pairs of the form $(a, 0)$ by the first component a, the displayed representation becomes

$$(a, b) = a + bi, \qquad \text{with} \qquad i^2 = -1.$$

In other words, the field C as defined initially is nothing more than a disguised version of the familiar complex number system.

Example 4.22

For an illustration of a division ring that is not a field, we turn to the ring of (Hamilton's) *real quaternions*. To introduce this ring, let the set H consist of all ordered 4-tuples of real numbers:

$$H = \{(a, b, c, d) | a, b, c, d \in R^{\#}\}.$$

Addition and multiplication of the elements of H are defined by the rules

$$(a, b, c, d) + (a', b', c', d') = (a + a', b + b', c + c', d + d'),$$

$$\begin{aligned}(a, b, c, d)(a', b', c', d') = (&aa' - bb' - cc' - dd',\\ &ab' + ba' + cd' - dc',\\ &ac' - bd' + ca' + db',\\ &ad' + bc' - cb' + da').\end{aligned}$$

A certain amount of tedious, but nonetheless straightforward, calculation shows that the resulting system is a ring (known as the *ring of real quaternions*) in which $(0, 0, 0, 0)$ and $(1, 0, 0, 0)$ act as the zero and identity elements, respectively.

Let us next introduce some special symbols by putting

$$1 = (1, 0, 0, 0), \qquad i = (0, 1, 0, 0), \qquad j = (0, 0, 1, 0), \qquad k = (0, 0, 0, 1).$$

The elements $1, i, j, k$ have a number of distinctive properties; specifically, 1 is the multiplicative identity of H and

$$i^2 = j^2 = k^2 = -1,$$

$$ij = k, \qquad jk = i, \qquad ki = j, \qquad ji = -k, \qquad kj = -i, \qquad ik = -j.$$

These relations demonstrate that the commutative law for multiplication fails to hold in H, so H definitely falls short of being a field.

As in Example 4.21, the definition of the algebraic operations in H permits us to write each quaternion in the form

$$(a, b, c, d) = (a, 0, 0, 0)1 + (b, 0, 0, 0)i + (c, 0, 0, 0)j + (d, 0, 0, 0)k.$$

Since the subring $\{(r, 0, 0, 0) | r \in R^{\#}\}$ is isomorphic to $R^{\#}$, the notation can be further simplified on replacing $(r, 0, 0, 0)$ by the element r itself;

by adopting these conventions the real quaternions may henceforth be regarded as the set

$$H = \{a + bi + cj + dk \mid a, b, c, d \in R^{\#}\},$$

with addition and multiplication performed as for polynomials (subject to the rules of the last paragraph).

The main point in our investigation is that any nonzero quaternion $q = a + bi + cj + dk$ (in other words, one of a, b, c, d must be different from zero) is a multiplicatively invertible element. By analogy with the complex numbers, each quaternion has a *conjugate*, defined as follows:

$$\bar{q} = a - bi - cj - dk.$$

It is easily verified that the product

$$q\bar{q} = \bar{q}q = a^2 - (bi + cj + dk)^2 = a^2 + b^2 + c^2 + d^2 \neq 0,$$

thus exhibiting that q has the multiplicative inverse

$$q^{-1} = (a^2 + b^2 + c^2 + d^2)^{-1}\bar{q}.$$

Incidentally, the totality of all members of H of the form $(a, b, 0, 0) = a + bi$, the *special quaternions*, constitute a subring isomorphic to C; as substitutes, one might also consider the set of all elements $(a, 0, b, 0)$ or all elements $(a, 0, 0, b)$. In this light the real quaternions may be viewed as a suitable generalization of the complex numbers.

The following theorem shows that any field is without divisors of zero, and consequently a system in which the cancellation law for multiplication holds.

THEOREM 4.20 *Every field F is an integral domain.*

Proof. Since every field is a commutative ring with identity, we need only prove that F contains no zero divisors. To this purpose, suppose that $a, b \in F$, with $ab = 0$. If the element $a \neq 0$, then it must possess a multiplicative inverse $a^{-1} \in F$. Then the hypothesis that $ab = 0$ yields

$$0 = a^{-1}0 = a^{-1}(ab) = 1b = b,$$

as desired.

There obviously exist integral domains that are not fields; a prime example is the ring Z of integers. However, an integral domain having only a finite number of elements must necessarily be a field.

THEOREM 4.21 *Any integral domain R with only a finite number of ideals is a field.*

Proof. Let a be any nonzero element of R. Consider the set of principal ideals (a^n), where $n \in Z_+$:

$$(a^n) = \{ra^n | r \in R\}.$$

Since R has only a finite number of distinct ideals, it follows that $(a^m) = (a^n)$ for certain positive integers m, n with $m < n$. Now a^m, as an element of (a^m), must lie in (a^n). This being so, some $\bar{r} \in R$ exists for which $a^m = \bar{r}a^n$. By use of the cancellation law,

$$1 = \bar{r}a^{n-m} = (\bar{r}a^{n-m-1})a.$$

Because multiplication is commutative, we therefore have $a^{-1} = \bar{r}a^{n-m-1}$. This argument shows that every nonzero element of R is invertible; hence R forms a field.

COROLLARY *Any finite integral domain is a field.*

Because the aforementioned corollary is so basic a result, we offer a second proof. The "counting argument" involved in this latter proof adapts to a variety of situations in which the underlying ring is finite.

The reasoning proceeds as follows. Suppose that $a_1, a_2, \ldots, a_n$ are the members of the integral domain R. For a fixed nonzero element $a \in R$ we consider the n products $aa_1, aa_2, \ldots, aa_n$. These products are all distinct, since if $aa_i = aa_j$, the cancellation law (valid in any integral domain) would yield $a_i = a_j$. It follows that each element of R must be of the form aa_i for some choice of i. In particular, some $\bar{a}_i \in R$ exists such that $a\bar{a}_i = 1$. From the commutativity of multiplication we infer that $a^{-1} = \bar{a}_i$, whence every nonzero element of R possesses a multiplicative inverse.

There are no finite division rings that are not fields. To put it another way, in a finite system in which all the field properties except the commutativity of multiplication are assumed, the multiplication must also be commutative. Proving this renowned result is far from being as elementary as the case of a finite integral domain and is omitted (see Reference 49 at the end of the book for details).

Finite division rings do exist, since the ring Z_p of integers modulo a prime p furnishes an example of such. A finite division ring $R = \{0, 1, a, b\}$ in which the number of elements is not a prime can be described by the following operation tables:

$+$	0	1	a	b
0	0	1	a	b
1	1	0	b	a
a	a	b	0	1
b	b	a	1	0

$\cdot$	0	1	a	b
0	0	0	0	0
1	0	1	a	b
a	0	a	b	1
b	0	b	1	a

The proof that every finite division ring must be a field was given virtually at the same time in 1905 by Joseph H. M. Wedderburn (1882–1948) and Leonard Eugene Dickson (1874–1954) in papers entitled "A Theorem on Finite Algebras" and "On Finite Algebra," respectively. Wedderburn's article contains three reasonably different demonstrations of this celebrated result, the last two of which had been laid out by Dickson. Owing to priority of publication, it has gone into the mathematical literature as Wedderburn's Theorem.

L. E. Dickson was prominent among a small circle of those who greatly influenced the rapid development of American mathematics at the turn of the century. He received the first doctorate in mathematics from the newly founded University of Chicago in 1896, became an assistant professor there in 1900, and remained at Chicago until his retirement in 1939.

Reflecting the abstract interests of his thesis advisor, the distinguished E. H. Moore, Dickson initially pursued the study of finite simple groups. Whether there exist two nonisomorphic simple groups with the same order had been a longstanding problem; in 1900, Dickson showed that there are infinitely many such examples (no others have subsequently been found). His classic treatise *Linear Groups* (1901) lists all the isomorphisms between the simple groups then known. Around 1905, Dickson discovered several further infinite families of finite simple groups; and until 1955 no additional finite simple groups were produced.

By 1906, Dickson's prodigious output had already reached 126 papers. He would jokingly remark that, while his honeymoon was a success, he managed to get only two research articles written. His monumental *History of the Theory of Numbers* (1919), which appeared in three volumes totaling more than 1600 pages, took nine years to complete; by itself this would have been a life's work for an ordinary man. One of the century's most prolific mathematicians, Dickson wrote 267 papers and 18 books covering a broad range of topics in his field. An enduring bit of legend is his barb against applicable mathematics: "Thank God that number theory is unsullied by applications." In recognition for his work, Dickson was the first recipient of the Cole Prize in algebra, awarded in 1928 by the American Mathematical Society. (The prize is named after Frank Nelson Cole who in 1892–1893 had determined the three simple groups with orders between 201 and 660.) Without scholars such as Dickson, the United States could not have attained its worthy place in the mathematical world.

After this aside, let us take a closer look at the algebraic structure of the ring Z_n of integers modulo n.

It has been previously shown that, for each positive integer n, Z_n forms a commutative ring with identity. A reasonable question is: For precisely what values of n, if any at all, will this ring turn out to be a field? For a quick answer: n must be a prime number. What could be simpler or more natural? This fact is brought out by the following theorem.

THEOREM 4.22 *A nonzero element* $[a] \in Z_n$ *is invertible in the ring* Z_n *if and only if* a *and* n *are relatively prime integers.*

Proof. If a and n are relatively prime, then there exist integers r and s such that $ar + ns = 1$. This implies that

$$\begin{aligned}[1] = [ar + ns] &= [ar] +_n [ns] \\ &= [ar] +_n [0] = [a] \cdot_n [r],\end{aligned}$$

showing the congruence class $[r]$ to be the multiplicative inverse of $[a]$.

Now to the "only if" part. Assume $[a]$ to be a multiplicatively invertible element of Z_n, say with inverse $[b]$. We thus have $[ab] = [a] \cdot_n [b] = [1]$, so there exists an integer k for which $ab - 1 = kn$. But then $ab + n(-k) = 1$; hence a and n are relatively prime integers.

COROLLARY *The zero divisors of* Z_n *are precisely the nonzero elements of* Z_n *that are not invertible.*

Proof. Naturally, no zero divisor of Z_n can possess a multiplicative inverse. On the other hand, suppose that $[a] \neq [0]$ is not invertible in Z_n, so that $\gcd(a, n) = d$, where $1 < d < n$. Then, $a = rd$ and $n = sd$ for suitable nonzero integers r and s. This leads to

$$[a] \cdot_n [s] = [as] = [rds] = [rn] = [0].$$

Since the defining properties of s rule out the possibility that $[s] = [0]$, it follows that $[a]$ is a zero divisor of Z_n.

These results may be conveniently summarized in the following statement.

THEOREM 4.23 *The ring* Z_n *of integers modulo* n *is a field if and only if* n *is a prime number. If* n *is composite, then* Z_n *is not an integral domain, and the zero divisors of* Z_n *are those nonzero elements* $[a]$ *for which* $\gcd(a, n) \neq 1$.

Every field necessarily has at least two elements, 1 being different from 0. Theorem 4.23 indicates that there is a field having this minimum number as its number of elements, namely, Z_2.

As an interesting application of these ideas, consider the following assertion: If there exists a homomorphism $f: Z \to F$ of the ring Z of integers onto a field F, then F is necessarily a finite field with a prime number of elements. By the Fundamental Homomorphism Theorem, $Z/\ker f \simeq F$. But $\ker f = (n)$ for some positive integer n, since Z is a principal ideal domain. In this connection, observe that $n \neq 0$; otherwise, Z would be isomorphic to a field, an impossibility. Taking stock of the fact that $Z/(n) = Z_n$, we are thus able to conclude that $Z_n \simeq F$, in consequence of which F has n elements. At this point, Theorem 4.23 comes to our aid; since F, and in turn its isomorphic image Z_n, forms a field, n must be a prime number.

A useful counting function is the so-called *Euler phi-function* (*totient*), defined as follows: $\phi(1) = 1$ and, for each integer $n > 1$, $\phi(n)$ is the number of invertible elements in the ring Z_n. By virtue of Theorem 4.22, $\phi(n)$ may also be characterized as the number of positive integers $< n$ that are relatively prime to n. For instance, $\phi(6) = 2$, $\phi(9) = 6$, and $\phi(12) = 4$; it should be equally clear that whenever p is a prime number, then $\phi(p) = p - 1$.

LEMMA *If G_n is the subset of Z_n defined by*

$$G_n = \{[a] \in Z_n \mid a \text{ is relatively prime to } n\},$$

then $(G_n, \cdot_n)$ forms a finite group of order $\phi(n)$.

Proof. In the light of the preceding remarks, $(G_n, \cdot_n)$ is simply the group of invertible elements of Z_n.

This leads at once to a classical result of Euler concerning the phi-function; the simplicity of the argument illustrates the advantage of the algebraic approach to number theory.

THEOREM 4.24 (Euler-Fermat) *If n is a positive integer and a is relatively prime to n, then $a^{\phi(n)} \equiv 1 \pmod{n}$.*

Proof. The congruence class $[a]$ can be viewed as an element of the multiplicative group $(G_n, \cdot_n)$. Since this group has order $\phi(n)$, it follows that $[a]^{\phi(n)} = [1]$ or, equivalently, $a^{\phi(n)} \equiv 1 \pmod{n}$. (Recall that if G is a finite group of order k, then $x^k = 1$ for all $x \in G$.)

There is an interesting relationship between fields and the lack of ideals; what we shall show is that fields have as trivial an ideal structure as possible.

THEOREM 4.25 *Let R be a commutative ring with identity. Then R is a field if and only if R has no nontrivial ideals.*

Proof. Assume first that R is a field. We wish to show that the trivial ideals $\{0\}$ and R are its only ideals. Let us suppose to the contrary that there exists some nontrivial ideal I of R. By our assumption the subset I is such that $I \neq \{0\}$ and $I \neq R$. This means that there exists some nonzero element $a \in I$. Since R is taken to be a field, a has a multiplicative inverse a^{-1} present in R. By the definition of ideal we thus obtain $1 = a^{-1}a \in I$, which in turn implies that $I = R$, contradicting our choice of I.

Conversely, suppose that the ring R has no nontrivial ideals. Given a nonzero element $a \in R$, consider the principal ideal (a) generated by a:

$$(a) = \{ra \mid r \in R\}.$$

Now (a) cannot be the zero ideal, inasmuch as $a = 1a \in (a)$, with $a \neq 0$. It follows from the hypothesis that the only other possibility is that $(a) = R$. In particular, since $1 \in (a)$, an element $\bar{r} \in R$ exists for which $\bar{r}a = 1$. Multiplication is commutative, so $\bar{r} = a^{-1}$. Therefore each nonzero element of R is multiplicatively invertible, and we are finished.

In view of this last result the ring Z of integers fails to be a field, since it contains the nontrivial ideal Z_e.

Theorem 4.25 is useful in revealing the nature of homomorphisms between fields. We exploit it to prove the following theorem.

THEOREM 4.26 *Let f be a homomorphism from the field F into the field F'. Then either f is the trivial homomorphism or else f is one-to-one.*

Proof. The proof consists of noticing that since $\ker f$ is an ideal of the field F, either $\ker f = \{0\}$ or else $\ker f = F$. The condition $\ker f = \{0\}$ implies that f is a one-to-one function. On the other hand, if it happens that $\ker f = F$, then each element of F is carried onto 0; that is to say, f is the trivial homomorphism.

COROLLARY *Any homomorphism of a field F onto itself is an automorphism of F.*

Any ring with identity that is a subring of a field must of necessity be an integral domain. Turning the situation around, one might ask whether each integral domain can be considered apart from isomorphism as a subring of some field. More formally: Can a given integral domain be imbedded in a field? In the finite case there is clearly no difficulty, since any finite integral domain already forms a field.

Our concern with this question arises from the natural desire to solve the linear equation $ax = b$, where $a \neq 0$. A major drawback to the notion of an integral domain is that it does not always furnish a solution within the system. Of course, any such solution would have to be unique, since $ax_1 = b = ax_2$ implies that $x_1 = x_2$ by the cancellation law. It hardly seems necessary to point out that when the integral domain happens to be a field, the equation $ax = b$ $(a \neq 0)$ is always solvable, since one need only take $x = a^{-1}b$.

We begin our discussion of this problem with an obvious definition.

DEFINITION 4.12 By a *subfield* of a field F is meant any subring F' of F that is itself a field.

For example, the ring Q of rational numbers is a subfield of the real field $R^{\#}$; the same is true of the field $F = \{a + b\sqrt{2} | a, b \in Q\}$.

Surely, the set F' will be a subfield of the field F provided that (1) F' is a subgroup of the additive group of F and (2) $F' - \{0\}$ is a subgroup of the multiplicative group $F - \{0\}$. Recalling our minimal set of conditions for determining subgroups (see Theorem 2.25), it follows that F' will be a subfield of F if and only if the following requirements are met:

1. F' is a nonempty subset of F containing at least one nonzero element,
2. $a, b \in F'$ imply that $a - b \in F'$, and
3. $a, b \in F'$, with $b \neq 0$, imply that $ab^{-1} \in F'$.

The next theorem furnishes a clue as to the nature of the field in which we wish to imbed a given integral domain.

THEOREM 4.27 *Let the integral domain R be a subring of the field F. If the set F' is defined by*

$$F' = \{ab^{-1} | a, b \in R;\ b \neq 0\},$$

then F' forms a subfield of F with $R \subseteq F'$; in fact, F' is the smallest (in the sense of inclusion) subfield of F containing R.

Proof. Note first that $1 = 1 \cdot 1^{-1} \in F'$, so that $F' \neq \{0\}$. Now consider two arbitrary elements x, y of F'. With reference to the definition of F' we then have

$$x = ab^{-1}, \qquad y = cd^{-1}$$

for a suitable choice of $a, b, c, d \in R$, where $b \neq 0$, $d \neq 0$. A simple computation shows that

$$x - y = (ad - bc)(bd)^{-1} \in F'.$$

Also, if y is nonzero (that is, whenever $c \neq 0$), we conclude that

$$xy^{-1} = (ad)(cb)^{-1} \in F'.$$

By virtue of the remarks following Definition 4.12 this is sufficient to establish that the set F' is a subfield of F. Furthermore,

$$a = a1 = a1^{-1} \in F'$$

for each a in R, implying that $R \subseteq F'$. Finally, any subfield of F that contains R necessarily includes all products ab^{-1}, with a, $0 \neq b \in R$, and hence contains F'.

Theorem 4.27 began with an integral domain already imbedded in a field. In the general case it becomes necessary to construct the imbedding field. Since the expression ab^{-1} might not always exist, one must now work with ordered pairs (a, b), where $b \neq 0$. Our thinking is that (a, b) will play a role analogous to ab^{-1} in the foregoing theorem.

We first frame a definition that will help us to discuss this situation.

DEFINITION 4.13 Let R be an integral domain. A (*classical*) *field of quotients of* R is any field $Q_{cl}(R)$ such that

1. $R \subseteq Q_{cl}(R)$ and
2. each element of $Q_{cl}(R)$ is of the form ab^{-1}, where $a, b \in R$ and $b \neq 0$.

Our intention is to prove that an arbitrary integral domain R, or at least an isomorphic copy of R, has a field of quotients. As a starting point, consider the set of ordered pairs

$$S = \{(a, b) | a, b \in R;\ b \neq 0\}.$$

A relation $\sim$ may be introduced in S by taking

$$(a, b) \sim (c, d) \qquad \text{if and only if} \qquad ad = bc.$$

(We have in mind the previous theorem, in which $ab^{-1} = cd^{-1}$ if and only if $ad = bc$.)

It is not difficult to verify that the relation $\sim$ is an equivalence relation in S. The transitive property is the least obvious. To see that it holds, assume that $(a, b) \sim (c, d)$ and $(c, d) \sim (e, f)$, so that

$$ad = bc, \qquad cf = de.$$

Now multiply the first equation by f and the second by b to get

$$adf = bcf, \qquad bcf = bde.$$

Putting these together, we obtain $afd = bed$. Since $d \neq 0$, the cancellation law gives us $af = be$, which is exactly the condition that $(a, b) \sim (e, f)$.

Next we label those elements of S that are equivalent to the pair (a, b) by the symbol a/b; in other words,

$$\begin{aligned} a/b &= \{(c, d) | (a, b) \sim (c, d)\} \\ &= \{(c, d) | ad = bc\}. \end{aligned}$$

The collection of all equivalence classes a/b relative to $\sim$ will be denoted by the symbol $Q_{cl}(R)$:

$$Q_{cl}(R) = \{a/b \mid a,\ b \in R;\ b \neq 0\}.$$

From Theorem 1.8 we know that the elements of $Q_{cl}(R)$ constitute a partition of the set S. That is, the ordered pairs in S fall into disjoint classes, called *formal fractions*, each class consisting of equivalent pairs and non-equivalent pairs belonging to different classes. Further, two such classes a/b and c/d are identical if and only if $ad = bc$.

With these remarks in mind, let us proceed to define the operations of addition and multiplication required to make $Q_{cl}(R)$ into a ring. We do this in the same way as in elementary arithmetic, that is, by means of the formulas

$$a/b + c/d = (ad + bc)/bd$$

and

$$(a/b)(c/d) = ac/bd.$$

Note, incidentally, that since $b \neq 0$ and $d \neq 0$ imply that $bd \neq 0$, the right-hand sides of these equations do indeed represent equivalence classes of the type under consideration.

As usual, our first task is to justify that these operations are well defined; in other words, it is necessary to show that the sum and product are independent of the particular elements of R used in their definition. Let us present the argument for addition in detail. Suppose, then, that $a/b = a'/b'$ and $c/d = c'/d'$; we must show that

$$(ad + bc)/bd = (a'd' + b'c')/b'd'.$$

From what is given, it follows at once that

$$ab' = ba', \qquad cd' = dc'.$$

These equations yield

$$\begin{aligned}(ad + bc)(b'd') - (a'd' + b'c')(bd) &= (ab' - ba')(dd') + (cd' - dc')(bb') \\ &= 0(dd') + 0(bb') = 0.\end{aligned}$$

By the definition of the equality of equivalence classes this amounts to saying that

$$(ad + bc)/bd = (a'd' + b'c')/b'd',$$

which proves addition to be well defined. In much the same way, one can establish that

$$ac/bd = a'c'/b'd'.$$

The next lemma reveals the algebraic nature of $Q_{cl}(R)$ under these operations.

LEMMA *For any integral domain R the system $Q_{cl}(R)$ forms a field.*

Proof. It is an entirely straightforward matter to show that $Q_{cl}(R)$ is a commutative ring. We leave the reader to make the necessary verifications and merely point out that $0/b$ serves as the zero element, while $(-a)/b$ is the negative of a/b.

That the equivalence class c/c, where c is any fixed nonzero element of R, constitutes the multiplicative identity is evidenced by the following calculation:

$$(a/b)(c/c) = ac/bc = a/b.$$

The last of these equalities holds for arbitrary a/b in $Q_{cl}(R)$, since $(ac)b = (bc)a$.

To confirm that every nonzero element of $Q_{cl}(R)$ is invertible, suppose that a/b is not the zero of $Q_{cl}(R)$; that is to say, $a \neq 0$. Then the equivalence class b/a is also a member of $Q_{cl}(R)$ and has the property that

$$(a/b)(b/a) = ab/ba = ab/ab.$$

Inasmuch as ab/ab plays the role of the identity element for $Q_{cl}(R)$, we see at once that $(a/b)^{-1} = b/a$. The net result of this is that $Q_{cl}(R)$ is a field.

This lemma paves the way for

THEOREM 4.28 *Any integral domain R possesses a classical field of quotients.*

Proof. We begin by establishing that the field $Q_{cl}(R)$ contains a subring isomorphic to R. For this, consider the subset K of $Q_{cl}(R)$ consisting of all elements of the form $a/1$, where 1 is the multiplicative identity of R:

$$K = \{a/1 \in Q_{cl}(R) | a \in R\}.$$

The reader may easily check that K is a subring of $Q_{cl}(R)$. An obvious (onto) mapping $f: R \to K$ is defined by taking $f(a) = a/1$. Since the condition $a/1 = b/1$ implies that $a1 = b1$, or $a = b$, we see that f is a one-to-one function. Furthermore, it has the property of preserving both addition and multiplication:

$$f(a + b) = (a + b)/1 = a/1 + b/1 = f(a) + f(b),$$

and

$$f(ab) = ab/1 = (a/1)(b/1) = f(a)f(b).$$

By this function, R can be isomorphically imbedded in $Q_{cl}(R)$.

By identifying R with K we may henceforth regard R as actually being contained in $Q_{cl}(R)$. In practice, one simply replaces the fraction $a/1 \in K$ by the corresponding element $a \in R$.

All that remains to complete the proof is to verify that each member a/b of $Q_{cl}(R)$ can be written as ab^{-1}. It should be clear that

$$a/b = (a/1)(1/b) = (a/1)(b/1)^{-1}.$$

Replacing $a/1$ by a and $b/1$ by b, the displayed equation assumes the more familiar form $a/b = ab^{-1}$. The point is this: The set $Q_{cl}(R)$ may now be interpreted as consisting of all quotients ab^{-1}, where $a, b \in R$ and $b \neq 0$.

Thus $Q_{cl}(R)$ satisfies the conditions of Definition 4.13, thereby becoming a classical ring of quotients of R.

Let us record an immediate consequence of Theorem 4.28.

THEOREM 4.29 *A ring is an integral domain if and only if it is a subring of a field.*

Proof. We have just seen that any integral domain R is (isomorphic to) a subring of the field of quotients $Q_{cl}(R)$. On the other hand, a field is an integral domain, and so all its subrings are integral domains.

At this point we leave the theory of quotients and turn to prime fields. Clearly, any field F has at least one subfield, namely, F itself; a field that does not possess any proper subfields is called a *prime field.*

Example 4.23

The field Q of rational numbers is the simplest example of a prime field. To demonstrate this, suppose that F is any subfield of Q and let $a \in F$ be any nonzero element. Since F is a subfield of Q, it must contain the product $aa^{-1} = 1$. In turn, $n = n1 \in F$ for any n in Z; in other words, F contains all the integers. It then follows that every rational number $n/m = nm^{-1}(m \neq 0)$ also lies in F, so $F = Q$.

Example 4.24

For each prime p the field Z_p of integers modulo p is a prime field. The reasoning here depends on the fact that the additive group $(Z_p, +_p)$ is a finite group of prime order and therefore by Lagrange's Theorem has no nontrivial subgroups.

An observation that will not detain us long is that each field F contains a unique prime subfield. To make things more specific, let $\{F_i\}$ be the collection of all subfields of F. Then the intersection $\cap F_i$ is also a subfield of F. Now if F' is any subfield of the field $\cap F_i$, then $F' \in \{F_i\}$, whence $\cap F_i \subseteq F'$; the implication is that $F' = \cap F_i$, forcing $\cap F_i$ to be a prime field. As regards the uniqueness assertion, suppose that K_1 and K_2 are both prime subfields of F. Then, $K_1 \cap K_2$ is a subfield of F as well as K_1, with $K_1 \supseteq K_1 \cap K_2$. But K_1 can possess no proper subfields, which signifies that $K_1 = K_1 \cap K_2$. Likewise, $K_2 = K_1 \cap K_2$, whence $K_1 = K_2$.

We conclude this section by showing that, to within isomorphism, the rational number field and the fields Z_p are the only prime fields.

THEOREM 4.30 *Any prime field F is isomorphic either to Q, the field of rational numbers, or to one of the fields Z_p of integers modulo a prime p.*

Proof. To begin, let 1 be the multiplicative identity of F and define the mapping $f: Z \to F$ by $f(n) = n1$ for any integer n. Then, f is a homomorphism from Z onto the subring $Z1$ of integral multiples of 1. In compliance with Theorem 4.18 we therefore have $Z/\ker f \simeq Z1$. But $\ker f$ is an ideal of Z, a principal ideal domain, whence $\ker f = (n)$ for some nonnegative integer n. The possibility that $n = 1$ can be ruled out; otherwise, $1 = f(1) = 0$ or, what amounts to the same thing, $F = \{0\}$.

Note further that if $n \neq 0$, then n must in fact be a prime number. Suppose to the contrary that $n = n_1 n_2$, where $1 < n_i < n$. Since $n \in \ker f$, it follows that

$$(n_1 1)(n_2 1) = (n_1 n_2)1 = n1 = 0,$$

yielding the contradiction that the field F has divisors of zero. (This result is not entirely unexpected because the integer n is the characteristic of F and as such must be a prime, whenever $n \neq 0$.)

The preceding discussion indicates that two possibilities arise: either

1. $Z1 \simeq Z/(p) = Z_p$ for some prime p, or
2. $Z1 \simeq Z/(0) = Z$.

Turning to a closer analysis of these cases, assume first that $Z1 \simeq Z_p$, with p prime. Inasmuch as the ring of integers modulo a prime forms a field, the subring $Z1$ must itself be a field. But F, being a prime field, contains no proper subfields. Accordingly, $Z1 = F$, which leads to the isomorphism $F \simeq Z_p$.

For the final stage of the proof, consider the situation in which $Z1 \simeq Z$. Under these circumstances the subring $Z1$ is an integral domain, but not a field. Taking stock of Theorem 4.27, as well as the hypothesis that F is a prime field, we conclude that

$$F = \{ab^{-1} \mid a, b \in Z1;\ b \neq 0\}$$
$$= \{(n1)(m1)^{-1} \mid n, m \in Z;\ m \neq 0\}.$$

It is now a purely routine matter to verify that the fields F and Q are isomorphic under the mapping $g(n/m) = (n1)(m1)^{-1}$; we leave the details as an exercise.

Since every field contains a unique prime subfield, the following subsidiary result is of interest.

COROLLARY *Every field contains a subfield that is isomorphic either to the field Q or to one of the fields Z_p.*

PROBLEMS 4.4

1. Let F consist of all matrices in $M_2(Q)$ of the form

$$\begin{bmatrix} a & 0 \\ 0 & a \end{bmatrix}.$$

Prove that F is a field.

2. If r is a root of the equation $x^2 + x + 1 = 0$, show that the set

$$F = \{a + br \mid a, b \in Q\}$$

forms a field under ordinary addition and multiplication.

3. a) Assuming that R is a division ring, prove that cent R forms a field.

b) Show that, in a field, every subring with identity is an integral domain.

4. For any field F, establish that the additive group is not isomorphic to the multiplicative group; that is, $(F, +) \not\simeq (F - \{0\}, \cdot)$. [*Hint:* If $f: F - \{0\} \to F$ is an isomorphism, consider $f((-1)^2)$.]

5. In the field C of complex numbers, define the mapping $f: C \to C$ by sending each number to its conjugate; in other words, $f(a + bi) = a - bi$. Verify that f is an automorphism of C.

6. Prove that the product $F_1 \times F_2$ of two fields F_1 and F_2 is not a field.

7. Consider the following subset of $M_2(R^\#)$:

$$K = \left\{ \begin{bmatrix} a & b \\ -b & a \end{bmatrix} \middle| \, a, b \in R^\# \right\}$$

Prove that K forms a field that is isomorphic to the field C of complex numbers. [*Hint:* Define the mapping $f: C \to K$ by

$$f(a + bi) = \begin{bmatrix} a & b \\ -b & a \end{bmatrix}.]$$

8. Prove that there are infinitely many real quaternions q for which $q^2 = -1$. [*Hint:* Choose $q = bi + cj + dk$, where $b^2 + c^2 + d^2 = 1$.]

9. In the ring H of real quaternions, define the norm of an element q to be $N(q) = q\bar{q}$, where $\bar{q}$ denotes the conjugate of q. Establish the following properties of the norm:

a) $N(q) = 0$ if and only if $q = 0$.

b) $N(\bar{q}) = N(q)$.

c) $N(q_1 q_2) = N(q_1)N(q_2)$.

d) If $q \neq 0$, then $q^{-1} = N(q)^{-1}q$.

e) If $q \neq 0$, $N(q^{-1}) = N(q)^{-1}$.

10. Give an example of a field F and two nonzero elements $a, b \in F$ such that $a^2 + b^2 = 0$.

11. Prove that the center of the ring H of real quaternions is $R^{\#}$. [*Hint:* If $q = a + bi + cj + dk$, then $qi = iq$ implies that $c = d = 0$; similarly, $b = 0$.]

12. Let K consist of all matrices in $M_2(C)$ of the form

$$\begin{bmatrix} \alpha & \beta \\ -\bar{\beta} & \bar{\alpha} \end{bmatrix}$$

where $\bar{\alpha}$, $\bar{\beta}$ are the conjugates of α, $\beta \in C$. Verify that

a) K forms a noncommutative subring of $M_2(C)$;

b) The ring H of real quaternions is isomorphic to K; [*Hint:* For $q = a + bi + cj + dk \in H$, write

$$q = (a + bi) + (c + di)j = \alpha + \beta j$$

and define the mapping $f: H \to K$ by $f(q) = \begin{bmatrix} \alpha & \beta \\ -\bar{\beta} & \bar{\alpha} \end{bmatrix}$.]

c) If $q \in H$, the norm $N(q) = \det f(q)$.

13. For the ring H of real quaternions, prove that

a) the set $V = \{a + bi + cj + dk \mid a, b, c, d \in Z\}$ forms a subring of H;

b) the group of invertible elements of V is

$$V^* = \{1, -1, i, -i, j, -j, k, -k\};$$

[*Hint:* Take the norm of $q_1 q_2 = 1$.]

c) V^* is isomorphic to the multiplicative group of eight matrices

$$\begin{bmatrix} 1 & 0 \\ 0 & 1 \end{bmatrix}, \begin{bmatrix} -1 & 0 \\ 0 & -1 \end{bmatrix}, \begin{bmatrix} 0 & 1 \\ -1 & 0 \end{bmatrix}, \begin{bmatrix} 0 & -1 \\ 1 & 0 \end{bmatrix},$$

$$\begin{bmatrix} 0 & i \\ i & 0 \end{bmatrix}, \begin{bmatrix} 0 & -i \\ -i & 0 \end{bmatrix}, \begin{bmatrix} i & 0 \\ 0 & -i \end{bmatrix}, \begin{bmatrix} -i & 0 \\ 0 & i \end{bmatrix}.$$

This matrix group is called the *quaternion group* and is denoted by Q_8.

[*Hint:* Problem 12(b) suggests that 1, i, j, k correspond to the matrices

$$\begin{bmatrix} 1 & 0 \\ 0 & 1 \end{bmatrix}, \begin{bmatrix} i & 0 \\ 0 & i \end{bmatrix}, \begin{bmatrix} 0 & 1 \\ -1 & 0 \end{bmatrix}, \begin{bmatrix} 0 & i \\ i & 0 \end{bmatrix}, \text{ respectively.]}$$

14. a) Give an example to show that a nonzero homomorphic image of an integral domain need not be an integral domain.

b) Prove that any nonzero homomorphic image of a field is a field.

15. a) If F is a finite field, say $F = \{0, 1, a_1, \ldots, a_n\}$, prove that the product $a_1 a_2 \cdots a_n = -1$. [*Hint:* Each of the elements of the set $\{a_1, \ldots, a_n\}$ has an inverse in this set.]

b) Apply part (a) to the field Z_p to show that

$$(p - 1)! \equiv -1 (\text{mod } p)$$

for any prime number p. This result, published in 1770, is named *Wilson's Theorem* after John Wilson (1741–1793).

16. Knowing that Z_{13} is a field, obtain the multiplicative inverse of each nonzero element.

17. Deduce the following facts concerning the Euler phi-function:

a) If $\gcd(n, m) = 1$, then $\phi(nm) = \phi(n)\phi(m)$. [*Hint:* $\gcd(a, nm) = 1$ if and only if $\gcd(a, n) = 1$ and $\gcd(a, m) = 1$.]

b) For any prime p and $n > 0$, $\phi(p^n) = p^n - p^{n-1} = p^n(1 - 1/p)$. [*Hint:* The integers k such that $1 < k < p^n$ and $\gcd(k, p) \neq 1$ are $p, 2p, \ldots, p^{n-1}p$.]

c) If $n = p_1^{k_1} p_2^{k_2} \cdots p_r^{k_r}$ is the prime factorization of an integer $n > 1$, then

$$\phi(n) = n(1 - 1/p_1)(1 - 1/p_2) \cdots (1 - 1/p_r).$$

d) $\phi(720) = 192$ and $\phi(36{,}000) = 9600$.

18. a) Let R be a finite commutative ring with identity. Prove that every nonzero element of R is either a zero divisor or invertible.

b) Find an example of a nonzero element in a commutative ring with identity that is neither a zero divisor nor invertible.

19. Given the set

$$H_n = \{[a] \in Z_n \mid [a] \text{ is not a zero divisor of } Z_n\},$$

show that $(H_n, \cdot_n)$ forms a group of order $\phi(n)$. [*Hint:* See Problem 18(a).]

20. a) Derive Fermat's Little Theorem: If p is a prime number and $a \not\equiv 0 (\text{mod } p)$, then $a^{p-1} \equiv 1 (\text{mod } p)$.

b) If $\gcd(a, n) = 1$, show that the unique solution of the linear congruence $ax \equiv b (\text{mod } n)$ is $x \equiv ba^{\phi(n)-1} (\text{mod } n)$.

c) Solve the linear congruences $5x \equiv 3 (\text{mod } 9)$ and $2x \equiv 11 (\text{mod } 15)$.

21. a) Prove that every field is a principal ideal domain.
b) Show that the ring $R = \{a + b\sqrt{2} | a, b \in Z\}$ is not a field by exhibiting a nontrivial ideal of R.

22. Let f be a homomorphism from the ring R into the ring R' and suppose that R has a subring F that is a field. Establish that either $F \subseteq \ker f$ or else R' contains a subring isomorphic to F.

23. Derive the following results:
a) The identity element of a subfield is the same as that of the field.
b) If $\{F_i\}$ is an index collection of subfields of the field F, then $\cap F_i$ is also a subfield of F.
c) A subring F' of a field F is a subfield of F if and only if F' contains at least one nonzero element and $a^{-1} \in F'$ for every nonzero $a \in F'$.
d) A subset F' of a finite field F is a subfield of F if and only if F' contains more than one element and is closed under addition and multiplication.

24. a) For a fixed prime p, consider the subset S of $R^{\#}$ defined by

$$S = \{a + b\sqrt{p} | a, b \in Q\}.$$

Show that S is a subfield of $R^{\#}$.
b) Prove that any subfield of the field $R^{\#}$ must contain the rational numbers.

25. Prove that if the field F is of characteristic $p > 0$, then every subfield of F must have characteristic p.

26. Let F be a field of characteristic $p > 0$. Show that for a given $n \in Z_+$,

$$F' = \{a \in F | a^{p^n} = a\}$$

is a subfield of F.

27. Let F be a field, F' a subfield of F, and f an automorphism of F. We say that f *fixes* an element $a \in F$ in case $f(a) = a$. Prove the following assertions:
a) The set of all automorphisms of F forms a group in which the binary operation is composition of functions.
b) The set of all automorphisms of F that fix each element of F' constitute a group.
c) If G is a group of automorphisms of F, then the set of all elements of F that are fixed by G (that is, the set $F(G) = \{a \in F | f(a) = a$ for all $f \in G\}$) is a subfield of F, known as the *fixed field of* G.

28. For a fixed prime p, show that the set of rational numbers

$$R_p = \{a/b \in Q | p \nmid b\}$$

forms an integral domain and find its field of quotients.

29. Prove that any field is (isomorphic to) its own field of quotients; in particular, $R = Q_{cl}(R)$ for every finite integral domain R.

30. Establish that any field containing the integral domain R as a subring contains the field of quotients $Q_{cl}(R)$. In this sense, $Q_{cl}(R)$ is the smallest field containing R.

31. The sets $R_1 = \{a + b\sqrt{2} \mid a, b \in Z\}$ and $R_2 = \{a + bi \mid a, b \in Z\}$ both form integral domains under the obvious operations, but not fields. Describe the field of quotients of each.

4.5 WILLIAM ROWAN HAMILTON

William Rowan Hamilton (1805–1865) was a child prodigy whose maturity was all that his early precocity promised. The fourth of nine children of a practicing attorney, he was born in Dublin, Ireland, and except for short visits elsewhere spent his whole life in the same city. Hamilton's preliminary education, which was carried on at home, was mainly in languages, classics, and mathematics. The young Hamilton was proficient in Oriental as well as European tongues. By the time he was 7, he could read Latin, Greek, Hebrew, French, and Italian; at the age of 13 he had a working knowledge of as many languages as the years he had lived. Hamilton's artistic inclination was just as strong; often he said that, although he made his living as a mathematician, at heart he was a poet.

The young Hamilton's thoughts were turned to mathematics by a meeting in 1818 with Zerah Colburn, an American youngster who gave a demonstration in Dublin as a "lightning calculator." The meeting was in the nature of a public test of arithmetic skills, and Hamilton came out second best, which only induced this competitive youth to begin a furious study of the original texts of Euclid (in Greek), Newton (in Latin), and Laplace (in French). At 17 he detected an error in Laplace's *Mécanique Céleste*, which was communicated to the President of the Royal Irish Academy. After reading Hamilton's article on the subject the President is supposed to have remarked, "This young man, I do not say will be, but is, the first mathematician of his age."

In 1823, Hamilton took the entrance examination for Trinity College, Dublin, and came out first in a field of 100 candidates. Once he was there, his progress continued to be brilliant. He achieved the previously unheard-of distinction of winning the highest possible marks in both mathematics and English verse. While still an undergraduate (1827), he was appointed professor of astronomy and superintendent of the observatory; it was understood that, except for twelve annual lectures, Hamilton would be left free to pursue his own lines of interest. Soon afterwards, he was elected Royal Astronomer of Ireland, a position that he retained until his death.

Hamilton devoted himself entirely to theoretical studies and did hardly anything as a practical astronomer, perhaps because the research equipment was poor and inadequate and he himself lacked the technical training. He was knighted in 1835 as an honor for his work in optics, particularly the result that light refracts in certain biaxial crystals in a conical configuration of rays.

While Hamilton's earliest papers were occupied with geometrical optics and mechanics, his landmark discovery came in algebra with the quaternions; these objects provided the first example of a logically consistent number system in which the commutative law for multiplication fails. In 1833 he read his paper "Theory of Algebraic Couples" to the Royal Irish Academy, interpreting the complex numbers as an algebra of ordered pairs of real numbers with relevant operations. Hamilton subsequently tried to extend his ideas to ordered triples, or "hypercomplex numbers" of the form $a + bi + cj$, where j is a new imaginary unit much like the complex unit $i = \sqrt{-1}$. For ten years he was baffled by the problem of defining multiplication of such numbers so as to preserve the familiar properties of the real and complex number systems. Each morning at breakfast, one of his little boys would ask, "Well, Papa, can you multiply the triplets?" Each morning, Hamilton was obliged to reply sadly, "No, I can only add and subtract them."

Hamilton's difficulty with multiplication is evident from the simple product

$$(a + bi + cj)(a - bi - cj) = a^2 + b^2 + c^2 - 2bcij.$$

In analogy with the complex numbers the product ought to equal $a^2 + b^2 + c^2$. This would happen if $ij = 0$, which Hamilton was "tempted for a moment to fancy." He was entirely correct to refrain from the assumption that $ij = 0$, however, since it would imply that $0 = (ij)^2 = i^2j^2 = (-1)(-1) = 1$. For a number of years, Hamilton was at an impasse over what to do with the product ij.

In a moment of insight, which occurred while he was strolling with his wife by Brougham Bridge in Dublin in 1843, Hamilton suddenly realized that his difficulties would vanish if he made two compromises. First, he should use quadruplets $a + bi + cj + dk$, or "quaternions" as he called them, rather than triplets for his new numbers. Second, the commutative law for multiplication had to be abandoned. Hamilton was so excited by the unorthodox nature of his inspiration that he stopped in his walk and carved the basic multiplication laws for the quaternion units i, j, k in the bridge's stonework. From this time until his death 22 years later, Hamilton's scientific life was devoted to an elaboration of the theory of quaternions. These investigations were published in two treatises, the *Lectures on Quaternions* in 1853 and the massive *Elements of Quaternions* (762 closely printed pages), posthumously edited by Hamilton's son in 1866. Though the quaternions have fallen short of the expectations to which they gave rise, they nonetheless represent a creative effort of the first order.

William Rowan Hamilton (1805–1865)

Today one can still find the commemorative tablet on Brougham Bridge which reads, "Here as he walked by on the 16th of October 1843 Sir William Rowan Hamilton in a flash of genius discovered the fundamental formula for quaternion multiplication $i^2 = j^2 = k^2 = ijk = -1$ and cut it in a stone of this bridge."

4.6 CERTAIN SPECIAL IDEALS

This section is devoted to a study of certain special types of ideals, most notably maximal and prime ideals. On the whole, our hypotheses will restrict us to commutative rings with identity. The requirement is motivated to some extent by the fact that many of the standard examples of ring theory have this property. Another reason, which is perhaps more important from the conceptual point of view, is that the most satisfactory and complete results occur here. We begin our discussion with the following definition.

DEFINITION 4.14 An ideal I of the ring R is said to be a *maximal ideal* provided that $I \neq R$ and whenever J is an ideal of R with $I \subset J \subseteq R$, then $J = R$.

Expressed somewhat loosely, an ideal is maximal if it is not the whole ring and is not properly contained in any larger proper ideal; the only ideal to contain a maximal ideal properly is the ring itself.

It is usually quite awkward to prove that an ideal is maximal directly from Definition 4.14. We therefore need several theorems that will help to determine whether or not a given ideal is actually maximal, but that are, in general, easier to apply than Definition 4.14. One such result is presented below.

THEOREM 4.31 *Let I be a proper ideal of the ring R. Then I is a maximal ideal if and only if $(I, a) = R$ for any element $a \notin I$. Here (I, a) denotes the ideal generated by $I \cup \{a\}$.*

Proof. First, notice that (I, a) satisfies $I \subset (I, a) \subseteq R$. These inclusions imply that if I were a maximal ideal of R, we would necessarily have $(I, a) = R$. On the other hand, assume that J is an ideal of the ring R with the property that $I \subset J \subseteq R$. If a is any element of J that does not lie in I, then $I \subset (I, a) \subseteq J$. The requirement that $(I, a) = R$ would thus force J to be all of R, and we could conclude that I is a maximal ideal.

A knowledge of several moderately simple examples will provide some basis for understanding these ideas.

Example 4.25
We propose to show that in the ring Z of integers the maximal ideals correspond to the prime numbers. More precisely, the principal ideal (n), where $n > 1$, is maximal if and only if n is a prime.

Suppose that (n) is a maximal ideal of Z but that the integer n is not prime. Then we can write $n = n_1 n_2$, with $1 < n_1 \leqslant n_2 < n$. This implies that the ideals (n_1) and (n_2) are such that

$$(n) \subset (n_1) \subset Z, \qquad (n) \subset (n_2) \subset Z,$$

contrary to the maximality of (n).

For the opposite direction, let us now assume n to be a prime. Pick any integer a not in the principal ideal (n), so that $n \nmid a$. Combined with the fact that n is a prime number, this immediately gives $\gcd(n, a) = 1$. Thus by Theorem 1.14, integers r and s exist for which $rn + sa = 1$. Inasmuch as n and a are both members of the ideal $((n), a)$, so is the sum $rn + sa$, resulting in $1 \in ((n), a)$; consequently, $((n), a) = Z$. Since this is true for every $a \notin (n)$, we may conclude from Theorem 4.31 that (n) is a maximal ideal of Z.

Example 4.26
For another illustration of the practicality of Theorem 4.31 we take $R = \text{map}\, R^{\#}$, a commutative ring with identity. Consider the set M of

all functions in R that vanish at 0:

$$M = \{f \in R \mid f(0) = 0\}.$$

Evidently, M forms an ideal of R; we contend that it is actually a maximal ideal. To see this, let f be an arbitrary element of R that is not in M. If i denotes the identity map on $R^{\#}$ (that is, $i(x) = x$), a quick check reveals that $(i^2 + f^2)(x) \neq 0$ for all $x \in R^{\#}$. Hence the function $i^2 + f^2$ will possess a multiplicative inverse in the ring R. Since $i \in M$, it follows that

$$i^2 + f^2 \in (M, f) = \{h + gf \mid h \in M, g \in R\}.$$

But the presence of an invertible element in the ideal (M, f) forces $(M, f) = R$, from which our contention is apparent.

Our immediate goal is to obtain a general result assuring the existence of suitably many maximal ideals. As will be seen presently, the crucial step in the proof depends on Zorn's Lemma (see the Appendix at the end of the book), an exceedingly powerful tool that is almost indispensable in modern mathematics. Zorn's Lemma (traditionally called a lemma but in fact an equivalent form of the Axiom of Choice) asserts:

Zorn's Lemma *If $(S, \leqslant)$ is a partially ordered set with the property that every chain in S has an upper bound in S, then S possesses at least one maximal element.*

Clearly, some partial orderings are more useful than others in applications of Zorn's Lemma. In our later investigations we shall frequently take S to be a family of subsets of a given set and the partial ordering to be the usual inclusion relation; an upper bound of any chain of elements would simply be their set-theoretic union. For this particular setting, Zorn's Lemma may be formulated as follows:

Let $\mathscr{A}$ be a nonempty family of subsets of some fixed nonempty set with the property that for each chain $\mathscr{C}$ in $\mathscr{A}$, the union $\cup\mathscr{C}$ also belongs to $\mathscr{A}$. Then $\mathscr{A}$ contains a set that is maximal in the sense that it is not properly contained in any member of $\mathscr{A}$.

Before Max Zorn (1906–) put forward his maximal principle, algebraists used the Axiom of Choice chiefly in the guise of Zermelo's Well-Ordering Theorem, with the uncomfortable feeling that it was a tool that was inappropriate to algebra. Attempts were made to replace the Axiom by some other device that would make existence proofs shorter and more algebraic. Zorn, then at Hamburg, seems to have formulated his alternative to the Axiom in 1933. It was quickly disseminated and, by late 1934, found users in the United States, who dubbed it "Zorn's Lemma." (An inaccurate label, of course, since a lemma is a little theorem, usually proved on the way to a big theorem.) When Zorn lectured on his principle to the American Mathematical Society that October, he was urged to publish his result; it appeared the following year in the Society's *Bulletin* as "A Remark on Method in Transfinite Algebra."

We owe the following theorem to Wolfgang Krull (1898–1971), a leading developer of commutative ring theory in the period 1920–1940. Since Krull spent the academic year 1921–1922 at Göttingen, following his doctorate from Freiburg in 1921, he is often listed as one of Emmy Noether's "pupils." No doubt her earlier researches had a considerable influence on his noteworthy *Idealtheorie* (1935). Krull established his theorem on maximal ideals in 1929 by means of the Well-Ordering Theorem. The proof that we present is taken from Zorn's 1935 paper, where it appears rather unobtrusively as an application of his new "axiom."

Because this may be the reader's first contact with Zorn's Lemma, we proceed in somewhat leisurely fashion to establish

THEOREM 4.32 (Krull-Zorn) *In a commutative ring with identity, each proper ideal is contained in a maximal ideal.*

Proof. Let I be any proper ideal of R, a commutative ring with identity. We define a family of ideals of R by taking

$$\mathscr{A} = \{J \mid I \subseteq J;\ J \text{ is a proper ideal of } R\}.$$

This family is obviously nonempty, since I itself belongs to $\mathscr{A}$.

Now consider an arbitrary chain $\{J_i\}$ of ideals in $\mathscr{A}$. Our aim, of course, is to establish that $\cup J_i$ is again a member of $\mathscr{A}$. To this purpose, let the elements $a, b \in \cup J_i$ and $r \in R$. Then there exist indices j and k for which $a \in J_j$, $b \in J_k$. As the collection $\{J_i\}$ forms a chain, either $J_j \subseteq J_k$ or else $J_k \subseteq J_j$. For definiteness, suppose that $J_j \subseteq J_k$ so that both $a, b \in J_k$. But J_k is an ideal of R; hence the difference $a - b \in J_k \subseteq \cup J_i$. Also the products ar and ra lie in $J_k \subseteq \cup J_i$. All this shows $\cup J_i$ to be an ideal of R. Note that $\cup J_i \neq R$, since $1 \notin J_i$ for all i. Finally, $I \subseteq \cup J_i$, whence the union $\cup J_i \in \mathscr{A}$.

Therefore on the basis of Zorn's Lemma the family $\mathscr{A}$ contains a maximal element M. It follows directly from the definition of $\mathscr{A}$ that M is a proper ideal of the ring R with $I \subseteq M$. We assert that M is in fact a maximal ideal. To see this, suppose that J is any ideal of R for which $M \subset J \subseteq R$. Since M is a maximal element of the family $\mathscr{A}$, J cannot belong to $\mathscr{A}$. Accordingly, the ideal J must be improper, which is to say that $J = R$. We thus conclude that M is a maximal ideal of R, completing the proof.

Remark. Theorem 4.32 can be extended to rings without identity and, more specifically, to any finitely generated ring; this latter hypothesis is sufficient to ensure that the union of any chain in $\mathscr{A}$ is still a proper subset of R. The point to keep in mind is that all these results assert the existence of certain maximal ideals but provide no clue as to how to find them.

From Theorem 4.32 it is a short step to:

THEOREM 4.33 *An element of a commutative ring R with identity is invertible if and only if it belongs to no maximal ideal of R.*

Proof. If an element of R is invertible, then it belongs to no proper ideal of R (Theorem 4.10) and, in particular, to no maximal ideal. Conversely, suppose that the element $a \in R$ lies in no maximal ideal of R. Then the principal ideal (a) must be improper; otherwise, $a \in (a) \subseteq M$ for some maximal ideal M. But the condition $(a) = R$ is equivalent to the invertibility of a in R.

Although maximal ideals were defined for arbitrary rings, we shall abandon a degree of generality and for the time being limit our discussion almost exclusively to commutative rings with identity. A ring of this kind is, of course, much easier to handle than one that is not commutative. Another advantage stems from the fact that each ideal, other than the ring itself, will be contained in a maximal ideal. Thus until further notice we shall assume that *all given rings are commutative with identity*, even when this is not explicitly mentioned. To be sure, a good deal of the subsequent material could be presented without this additional restriction.

The Krull-Zorn Theorem has many important applications throughout ideal theory. For the moment we content ourselves with giving an elementary proof of a somewhat special result; although the fact involved is rather interesting, there will be no occasion to make use of it.

THEOREM 4.34 *In a ring R having exactly one maximal ideal M, the only idempotents are 0 and 1.*

Proof. Assume that the theorem is false; that is, suppose that there exists an idempotent $a \in R$, with $a \neq 0, 1$. The relation $a^2 = a$ implies that $a(1 - a) = 0$, so that a and $1 - a$ are both zero divisors. Hence by Problem 5(d), Section 4.2, neither the element a nor $1 - a$ is invertible in R. But this means that the principal ideals (a) and $(1 - a)$ are both proper ideals of the ring R. As such, they must be contained in M, the sole maximal of R. Accordingly, the elements a and $1 - a$ lie in M, whence

$$1 = a + (1 - a) \in M.$$

This leads at once to the contradiction that $M = R$.

Although more elementary proofs are possible, Theorem 4.34 can be used to show that a field has no idempotents except 0 and 1. A full justification of this statement consists of first establishing that the zero ideal is the only maximal ideal in a field.

We now come to a characterization of maximal ideals in terms of their quotient rings.

THEOREM 4.35 *Let I be a proper ideal of the ring R. Then I is a maximal ideal if and only if the quotient ring R/I is a field.*

Proof. To begin, let I be a maximal ideal of R. Since R is a commutative ring with identity, the quotient ring R/I also has these properties. Thus to prove that R/I is a field, it suffices to show that each nonzero element of R/I has a multiplicative inverse. Now if the coset $a + I \neq I$, then $a \notin I$. By virtue of the fact that I is a maximal ideal, the ideal (I, a) generated by I and a must be the whole ring R:

$$R = (I, a) = \{i + ra \mid i \in I,\ r \in R\};$$

that is to say, every element of R is expressible in the form $i + ra$, where $i \in I$ and $r \in R$. The identity element 1, in particular, may be written as $1 = \bar{i} + \bar{r}a$ for suitable choice of $\bar{i} \in I$, $\bar{r} \in R$. But then, the difference $1 - \bar{r}a \in I$. This obviously implies that

$$1 + I = \bar{r}a + I = (\bar{r} + I)(a + I),$$

which asserts that $\bar{r} + I = (a + I)^{-1}$. Hence R/I is a field.

For the opposite direction we suppose that R/I is a field and J is any ideal of R for which $I \subset J \subseteq R$. The argument consists of showing that $J = R$, for then I will be a maximal ideal. Since I is a proper subset of J, there exists an element $a \in J$ with $a \notin I$. Consequently, the coset $a + I \neq I$, the zero element of R/I. Since R/I is assumed to be a field, $a + I$ must have an inverse under multiplication,

$$(a + I)(b + I) = ab + I = 1 + I,$$

for some coset $b + I \in R/I$. It then follows that $1 - ab \in I \subset J$. But the product ab also lies in J—recall that a is an element of the ideal J—implying that the identity element $1 = (1 - ab) + ab \in J$. This in turn yields $J = R$, as desired.

Example 4.27

Consider the ring Z_e of even integers, a commutative ring without identity. In this ring the principal ideal (4) generated by the integer 4 is a maximal ideal, where

$$(4) = \{4(2j) + 4k \mid j, k \in Z\} = 4Z.$$

The argument might be expressed as follows. If n is any element not in (4), then n is an even integer not divisible by 4; consequently, n can be expressed in the form $n = 4m + 2$ for some integer m. We then have

$$2 = 4(-m) + n \in ((4), n),$$

so that $Z_e = (2) = ((4), n)$. By virtue of Theorem 4.31 this is sufficient to demonstrate the maximality of the ideal (4).

Now note that in the quotient ring $Z_e/(4)$,

$$(2 + (4))(2 + (4)) = 4 + (4) = (4).$$

The ring $Z_e/(4)$ therefore possesses divisors of zero and cannot be a field. The point that we wish to make is that the assumption of a multiplicative identity is essential to Theorem 4.35.

We now shift our attention from maximal ideals to prime ideals. Before formally defining this notion, let us turn to the ring Z of integers for motivation. Specifically, consider the principal ideal (p) generated by a prime number p. If the product $ab \in (p)$, where $a, b \in Z$, then p divides ab. But if a prime divides a product, it necessarily divides one of the factors. This being the case, either $a \in (p)$ or $b \in (p)$. The ideal (p) thus has the interesting property that, whenever (p) contains a product, at least one of the factors must belong to (p). This observation serves to suggest and partly to illustrate the next definition.

DEFINITION 4.15 An ideal I of the ring R is a *prime* ideal if, for all a, b in R, $ab \in I$ implies that either $a \in I$ or $b \in I$.

By induction, Definition 4.15 can easily be extended to finitely many elements: An ideal I of R is prime if, whenever a product $a_1 a_2 \cdots a_n$ of elements of R belongs to I, then at least one of the $a_i \in I$. In this connection we should caution the reader that many authors insist that the term "prime ideal" always means a proper ideal.

Example 4.28

A commutative ring R with identity is an integral domain if and only if the zero ideal $\{0\}$ is a prime ideal of R.

Example 4.29

The prime ideals of the ring Z are precisely the ideals (n), where n is a prime number, together with the two trivial ideals $\{0\}$ and Z. From the foregoing we already know that if n is a prime, then the principal ideal (n) is a prime ideal of Z. On the other hand, consider any ideal (n) with n composite $(n \neq 0, 1)$, say $n = n_1 n_2$, where $1 < n_1 \leqslant n_2 < n$. Certainly the product $n_1 n_2 = n \in (n)$. However, since neither n_1 nor n_2 is an integral multiple of n, $n_1 \notin (n)$ and $n_2 \notin (n)$. Hence when n is composite, the ideal (n) cannot be prime. Notice also that although $\{0\}$ is prime, it is not a maximal ideal of Z.

Example 4.30

For an illustration of a ring possessing a nontrivial prime ideal that is not maximal, take $R = Z \times Z$, where the operations are performed componentwise. One may readily verify that $Z \times \{0\}$ is a prime ideal of R. Since

$$Z \times \{0\} \subset Z \times Z_e \subset R,$$

with $Z \times Z_e$ an ideal of R, $Z \times \{0\}$ fails to be maximal.

By analogy with Theorem 4.35 the prime ideals of a ring may be characterized in the following manner.

THEOREM 4.36 *Let I be a proper ideal of the ring R. Then I is a prime ideal if and only if the quotient ring R/I is an integral domain.*

Proof. First, take I to be a prime ideal of R. Since R is a commutative ring with identity, so is the quotient ring R/I. It remains therefore only to verify that R/I is free of zero divisors. For this, assume that

$$(a + I)(b + I) = I.$$

In other words, the product of these two cosets is the zero element of the ring R/I. The foregoing equation is plainly equivalent to requiring that $ab + I = I$ or, what amounts to the same thing, $ab \in I$. Since I is assumed to be a prime ideal, one of the factors a or b must be in I. But this means that either the coset $a + I = I$ or else $b + I = I$; hence R/I is without zero divisors.

To prove the converse, we simply reverse the argument. Accordingly, suppose that R/I is an integral domain and the product $ab \in I$. In terms of cosets this means that

$$(a + I)(b + I) = ab + I = I.$$

By hypothesis, R/I contains no divisors of zero, so $a + I = I$ or $b + I = I$. In any event, one of a or b belongs to I, forcing I to be a prime ideal of R.

There is an important class of ideals that are always prime, namely, the maximal ideals. From the several ways of proving this result we choose the argument given below; another approach involves the use of Theorems 4.35 and 4.36.

THEOREM 4.37 *In a commutative ring with identity, every maximal ideal is a prime ideal.*

Proof. Assume that I is a maximal ideal of the ring R, a commutative ring with identity, and the product $ab \in I$ with $a \notin I$. We propose to show that $b \in I$. The maximality of I implies that the ideal generated by I and a must be the whole ring: $R = (I, a)$. Hence there exist elements $i \in I$, $r \in R$ such that $1 = i + ra$. Since both ab and i belong to I, we conclude that

$$b = 1b = (i + ra)b = ib + r(ab) \in I,$$

from which it follows that I is a prime ideal of R.

We should point out that without the assumption of an identity element this last result does not remain valid; a specific illustration is the ring Z_e of even integers, where (4) forms a maximal ideal which is not prime.

Of course, the converse of Theorem 4.37 does not hold. Example 4.30 shows that nontrivial prime ideals exist that fail to be maximal ideals. The special properties of Boolean rings and principal ideal domains guarantee that the notions of primeness and maximality are equivalent for these important classes of rings. A Boolean ring is a ring in which every element is idempotent.

THEOREM 4.38 *Let R be a Boolean ring. A nontrivial ideal I of R is prime if and only if it is a maximal ideal.*

Proof. It is sufficient to show that if the ideal I is prime, then I is also maximal. To see this, suppose that J is an ideal of R with the property that $I \subset J \subseteq R$; what we must prove is that $J = R$. If a is any element of J not in I, then $a(1 - a) = 0 \in I$. Using the fact that I is a prime ideal with $a \notin I$, we infer that $1 - a \in I \subset J$. As both the elements a and $1 - a$ lie in J, it follows that

$$1 = a + (1 - a) \in J.$$

The ideal J thus contains the identity and, consequently, $J = R$. Since no proper ideal lies between I and the whole ring R, we conclude that I is a maximal ideal.

Remark. Since every integral domain contains the two trivial prime ideals, the use of the term "prime ideal" in a principal ideal domain customarily excludes these from consideration.

THEOREM 4.39 *Let R be a principal ideal domain. A nontrivial ideal (a) of R is prime if and only if it is a maximal ideal.*

Proof. Assume that (a) is a prime ideal and let I be any ideal of R satisfying $(a) \subset I \subseteq R$. Because R is a principal ideal ring, an element $b \in R$ exists for which $I = (b)$. Now $a \in (a) \subset (b)$; hence $a = rb$ for some choice of r in R. By supposition, (a) is a prime ideal, so either $r \in (a)$ or $b \in (a)$. The possibility that $b \in (a)$ leads immediately to the contradiction $(b) \subseteq (a)$. Therefore the element $r \in (a)$, which implies that $r = sa$ for suitable choice of s in R, or $a = rb = (sa)b$. Since $a \neq 0$ and R is an integral domain, we must have $1 = sb$. This, of course, means that the identity element $1 \in (b) = I$, whence $I = R$, making (a) a maximal ideal of R. Theorem 4.37 takes care of the converse.

COROLLARY *A nontrivial ideal of the ring Z is prime if and only if it is maximal.*

Let us detour briefly to introduce a concept that plays an important role in many aspects of ideal theory.

DEFINITION 4.16 Let I be an ideal of the ring R. The *nil radical of I*, designated by $\sqrt{I}$, is the set

$$\sqrt{I} = \{r \in R \mid r^n \in I \text{ for some } n \in Z_+ \ (n \text{ varies with } r)\}.$$

We observe that the nil radical of I may equally well be characterized as the set of elements $r \in R$ whose image $r + I$ in the quotient ring R/I is nilpotent. The nil radical of the zero ideal is sometimes referred to as the nil radical of the ring R; this set consists of all nilpotent elements of R and accounts for the use of the term.

Although it is not obvious from the definition, $\sqrt{I}$ is actually an ideal of the ring R that contains I. In the first place, if a and b are elements of $\sqrt{I}$, then suitably chosen integers $n, m \in Z_+$ exist such that $a^n \in I$ and $b^m \in I$. Now every term in the binomial expansion of $(a - b)^{n+m}$ contains either a^n or b^m as a factor. This implies that $(a - b)^{n+m}$ lies in I and therefore the difference $a - b \in \sqrt{I}$. Next, if r is any element of R, then $(ra)^n = r^n a^n \in I$, so $ra \in \sqrt{I}$; thus $\sqrt{I}$ is indeed an ideal of R. That $I \subseteq \sqrt{I}$ should be clear from the definition.

Example 4.31
In the ring Z, let us show that if $n = p_1^{k_1} p_2^{k_2} \cdots p_r^{k_r}$ is a factorization of the positive integer $n \neq 1$ into distinct primes p_j, then

$$\sqrt{(n)} = (p_1 p_2 \cdots p_r).$$

Indeed, if the integer $a = p_1 p_2 \cdots p_r$ and $k = \max\{k_1, k_2, \ldots, k_r\}$, then we have $a^k \in (n)$; this makes it clear that $(p_1 p_2 \cdots p_r) \subseteq \sqrt{(n)}$. On the other hand, if some positive integral power of the integer m is divisible by n (that is, if $m \in \sqrt{(n)}$), then m itself must be divisible by each of the primes $p_1, p_2, \ldots, p_r$, and hence a member of the ideal

$$(p_1) \cap (p_2) \cap \cdots \cap (p_r) = (p_1 p_2 \cdots p_r).$$

As concrete illustrations of this situation, observe that $\sqrt{(12)} = \sqrt{(2^2 3)} = (6)$ and $\sqrt{(8)} = (2)$.

Some of the basic properties of the nil radical of an ideal are assembled in the theorem below.

THEOREM 4.40 *If I and J are two ideals of the ring R, then*

1. $\sqrt{IJ} = \sqrt{I \cap J} = \sqrt{I} \cap \sqrt{J}$,
2. $\sqrt{I + J} = \sqrt{\sqrt{I} + \sqrt{J}} \supseteq \sqrt{I} + \sqrt{J}$,
3. $I^k \subseteq J$ *for some* $k \in Z_+$ *implies that* $\sqrt{I} \subseteq \sqrt{J}$, *and*
4. $\sqrt{\sqrt{I}} = \sqrt{I}$.

Proof. Since property (1) is the only fact that will be explicitly required in the body of the text, we shall content ourselves with its derivation; the proofs of the remaining assertions are quite elementary and are left as an exercise.

Now if $a^n \in IJ$, then $a^n \in I \cap J$, and so $a^n \in I$, $a^n \in J$. We thus conclude that $\sqrt{IJ} \subseteq \sqrt{I \cap J} \subseteq \sqrt{I} \cap \sqrt{J}$. On the other hand, if it happens that $a \in \sqrt{I} \cap \sqrt{J}$, positive integers n, m must exist for which $a^n \in I$ and $a^m \in J$. This implies that the element $a^{n+m} = a^n a^m \in IJ$; hence $a \in \sqrt{IJ}$. Accordingly, $\sqrt{I} \cap \sqrt{J} \subseteq \sqrt{IJ}$, and the desired equality follows.

DEFINITION 4.17 An ideal I of the ring R is said to be a *semiprime ideal* if and only if $I = \sqrt{I}$.

In effect, Definition 4.17 states that an ideal I is semiprime if and only if $a^n \in I$ for some $n \in Z_+$ implies that a itself lies in I. Our next result characterizes semiprime ideals by the quotient rings which they determine.

THEOREM 4.41 *An ideal I of the ring R is a semiprime ideal if and only if the quotient ring R/I has no nonzero nilpotent elements.*

Proof. Suppose that $a + \sqrt{I}$ is a nilpotent element of $R/\sqrt{I}$. Then there exists some $n \in Z_+$ such that $(a + \sqrt{I})^n = a^n + \sqrt{I} = \sqrt{I}$; that is to say, the element $a^n \in \sqrt{I}$. Hence $(a^n)^m = a^{nm} \in I$ for some positive integer m. This implies that $a \in \sqrt{I}$ and, consequently, that $a + \sqrt{I} = \sqrt{I}$, the zero element of $R/\sqrt{I}$.

As regards the converse, assume that R/I has no nonzero nilpotent elements and let $a \in \sqrt{I}$. Then, for some positive integer n, $a^n \in I$. Passing to the quotient ring R/I, this means simply that $(a + I)^n = I$; in other words, $a + I$ is nilpotent in R/I. By supposition we must have $a + I = I$ and, in consequence, the element $a \in I$. Our argument shows that $\sqrt{I} \subseteq I$; since the reversed inclusion always holds, $I = \sqrt{I}$, so that I is semiprime.

This being proved, it is not hard to establish

COROLLARY *If P is a prime ideal of the ring R, then P is semiprime.*

Proof. Because P is prime, R/P possesses no zero divisors and, in particular, no nonzero nilpotent elements.

This corollary provides another good reason why a semiprime ideal was named as it was; being a semiprime ideal in a ring is a bit weaker than being prime.

In addition to the nil radical a number of other radicals are in circulation; one of the more prominent ones is considered over the course of the next few pages. The various formulations are not, in general, equivalent to one another, and this has given rise to a certain confusion and ambiguity in the use of the term. Indeed, whenever the word "radical" is encountered by itself, the reader should take pains to discover precisely what is meant by it. By way of removing some of this confusion we shall give a qualifying adjective to the type of radical which appears here.

DEFINITION 4.18 The *Jacobson radical* of a ring R, denoted by rad R, is the set

$$\text{rad } R = \cap\{M \mid M \text{ is a maximal ideal of } R\}.$$

If rad $R = \{0\}$, then R is said to be a ring without Jacobson radical or, more briefly, R is a *semisimple ring.*

The Jacobson radical always exists, since we know by Theorem 4.32 that any commutative ring with identity contains at least one maximal ideal. It is also immediately obvious from the definition that rad R forms an ideal of R that is contained in each maximal ideal.

To fix ideas, let us determine the Jacobson radical in two specific rings.

Example 4.32
The ring Z of integers is a semisimple ring. According to Example 4.25, the maximal ideals of Z are precisely the principal ideals generated by the prime numbers; thus

$$\text{rad } R = \cap\{(p) | p \text{ is a prime number}\}.$$

Since no nonzero integer is divisible by every prime, we see at once that the radical rad $R = \{0\}$.

Example 4.33
A more penetrating illustration is furnished by the ring $R = \text{map}(X, F)$, where X is an arbitrary set and F a field. For any element $x \in X$, consider the function $\tau_x f = f(x)$ that assigns to each function in R its value at x. It is easily checked that τ_x is a homomorphism of R into F; since R contains all the constant functions, this homomorphism actually maps onto the field F. Thus by Problem 10 at the end of this section its kernel is the maximal ideal

$$M_x = \{f \in R | f(x) = 0\}.$$

Because rad $R \subseteq \cap M_x = \{f \in R | f(x) = 0 \text{ for all } x \in X\} = \{0\}$, it follows that R must be a semisimple ring.

Our next theorem establishes a basic connection between the Jacobson radical and invertibility of ring elements.

THEOREM 4.42 *Let I be an ideal of the ring R. Then $I \subseteq$ rad R if and only if each element of the coset $1 + I$ has an inverse in R.*

Proof. We begin by assuming that $I \subseteq$ rad R and that there is some element $a \in I$ for which $1 + a$ is not invertible. Our object, of course, is to derive a contradiction. By Theorem 4.33 the element $1 + a$ must belong to some maximal ideal M of the ring R. Since $a \in$ rad R, a is also contained in M, and therefore $1 = (1 + a) - a$ lies in M. But this means that $M = R$, which is clearly impossible.

To prove the converse, suppose that each member of $1 + I$ has a multiplicative inverse in R, but $I \nsubseteq$ rad R. By definition of the Jacobson radical there will exist a maximal ideal M of R with $I \nsubseteq M$. Now if a is any element of I that is not in M, the maximality of M implies that the ideal $(M, a) = R$. Knowing this, we can express the identity element 1 in the form $1 = m + ra$ for suitable choice of $m \in M$ and $r \in R$. But then $m = 1 - ra \in 1 + I$, so m possesses an inverse. The conclusion is untenable, since no proper ideal contains an invertible element.

The form that this result takes when I is the principal ideal generated by $a \in \operatorname{rad} R$ furnishes a characterization of the Jacobson radical in terms of elements rather than ideals. Although actually a corollary to the theorem just proved, it is important enough to be singled out as a theorem.

THEOREM 4.43 *In any ring R an element $a \in \operatorname{rad} R$ if and only if $1 - ra$ is invertible for each $r \in R$.*

This theorem adapts itself to many uses. Three fairly short and instructive applications are presented below.

COROLLARY 1 *An element a is invertible in the ring R if and only if the coset $a + \operatorname{rad} R$ is invertible in the quotient ring $R/\operatorname{rad} R$.*

Proof. Assume that the coset $a + \operatorname{rad} R$ has an inverse in $R/\operatorname{rad} R$, so that

$$(a + \operatorname{rad} R)(b + \operatorname{rad} R) = 1 + \operatorname{rad} R$$

for some $b \in R$. Then $1 - ab$ lies in $\operatorname{rad} R$. We now appeal to Theorem 4.43, with $r = 1$, to conclude that the product $ab = 1 - 1(1 - ab)$ is invertible; this in turn forces the element a to have an inverse in R. The other direction of the corollary is all but obvious.

COROLLARY 2 *The only idempotent element in* $\operatorname{rad} R$ is 0.

Proof. Let the element $a \in \operatorname{rad} R$ with $a^2 = a$. Taking $r = 1$ in the preceding theorem, we see that $1 - a$ has an inverse in R, say $(1 - a)b = 1$, where $b \in R$. This leads immediately to

$$a = a(1 - a)b = (a - a^2)b = 0,$$

which completes the proof.

COROLLARY 3 *Every nil ideal of R is contained in* $\operatorname{rad} R$.

Proof. Let N be a nil ideal of R and suppose that $a \in N$. For every $r \in R$ we then have $ra \in N$, so the product ra is nilpotent. Problem 10, Section 4.2, therefore implies that $1 - ra$ is invertible in R. This shows that the element a lies in $\operatorname{rad} R$, from which it follows that $N \subseteq \operatorname{rad} R$.

The result below provides a convenient method for manufacturing semisimple rings; its proof utilizes both implications of the last theorem.

THEOREM 4.44 *For any ring R the quotient ring $R/\operatorname{rad} R$ is semisimple; that is,* $\operatorname{rad}(R/\operatorname{rad} R) = \{0\}$.

Proof. Before becoming involved in details, let us remark that since rad R constitutes an ideal of R, we may certainly form the quotient ring $R/\text{rad } R$. To simplify notation somewhat, we will temporarily denote rad R by I.

Suppose that the coset $a + I \in \text{rad}(R/I)$. Our strategy is to show that the element $a \in I$, since then $a + I = I$, which would imply that $\text{rad}(R/I)$ consists of only the zero element of R/I. Since $a + I$ is a member of $\text{rad}(R/I)$, Theorem 4.43 asserts that

$$(1 + I) - (r + I)(a + I) = 1 - ra + I$$

is invertible in R/I for each choice of $r \in R$. Accordingly, there exists a coset $b + I$ (depending, of course, on both r and a) such that

$$(1 - ra + I)(b + I) = 1 + I.$$

This is plainly equivalent to requiring that

$$1 - (b - rab) \in I = \text{rad } R.$$

Again appealing to Theorem 4.43, we conclude that the element

$$b - rab = 1 - 1(1 - b + rab)$$

has an inverse c in R. But then

$$(1 - ra)(bc) = (b - rab)c = 1,$$

so that $1 - ra$ possesses a multiplicative inverse in R. Since this argument holds for every $r \in R$, it follows that $a \in \text{rad } R = I$, as desired.

PROBLEMS 4.6

In the following set of problems, all rings are assumed to be commutative with identity.

1. a) Prove that $Z \times Z_e$ is a maximal ideal of the ring $Z \times Z$.
   b) Show that the ring R is a field if and only if $\{0\}$ is a maximal ideal of R.

2. Prove that a proper ideal M of the ring R is maximal if and only if for every element $r \notin M$ there exists some $a \in R$ such that $1 + ra \in M$.

3. Let f be a homomorphism from the ring R onto the ring R'. Prove that
   a) if M is a maximal ideal of R with $M \supseteq \ker f$, then $f(M)$ is a maximal ideal of R',
   b) if M' is a maximal ideal of R', then $f^{-1}(M')$ is a maximal ideal of R,
   c) the mapping $M \to f(M)$ defines a one-to-one correspondence between the set of maximal ideals of R that contain $\ker f$ and the set of all maximal ideals of R'.

4. If M_1 and M_2 are distinct maximal ideals of the ring R, establish the equality $M_1M_2 = M_1 \cap M_2$.

5. Let M be a proper ideal of the ring R. Prove that M is a maximal ideal if and only if, for each ideal I of R, either $I \subseteq M$ or else $I + M = R$.

6. An ideal I of the ring R is said to be *minimal* if $I \neq \{0\}$ and no ideal J of R exists such that $\{0\} \subset J \subset I$.

a) Prove that a nonzero ideal I of R is a minimal ideal if and only if $(a) = I$ for each nonzero element $a \in I$.

b) Verify that the ring Z of integers has no minimal ideals.

7. Let I be a proper ideal of the ring R. Show that I is a prime ideal if and only if the complement of I in R is closed under multiplication.

8. In the ring $R = \text{map}\, R^{\#}$, define the set I by

$$I = \{f \in R \mid f(1) = f(-1) = 0\}.$$

Establish that I is an ideal of R, but not a prime ideal.

9. a) With the aid of Theorem 4.35 and Example 4.25, obtain another proof of the fact that Z_p is a field if and only if p is a prime number.

b) Prove that in Z_n the maximal ideals are the principal ideals $(p) = pZ_n$, where p is a prime dividing n.

10. Given that f is a homomorphism from the ring R onto the ring R', verify that

a) R' is a field if and only if $\ker f$ is a maximal ideal of R,

b) R' is an integral domain if and only if $\ker f$ is a prime ideal of R.

11. Show that in the ring $R = Z \times Z$, where the operations are performed componentwise, $Z_e \times Z_e$ is a prime ideal.

12. Consider the commutative ring R consisting of all matrices in $M_2(R^{\#})$ of the form $\begin{bmatrix} a & b \\ 0 & 0 \end{bmatrix}$. Prove that the set

$$M = \left\{ \begin{bmatrix} 0 & b \\ 0 & 0 \end{bmatrix} \middle| b \in R^{\#} \right\}$$

forms a maximal ideal of R. [*Hint:* The mapping

$$f\left(\begin{bmatrix} a & b \\ 0 & 0 \end{bmatrix}\right) = a$$

is a homomorphism from R onto $R^{\#}$.]

13. a) Find all maximal and prime ideals of the ring Z_{12}.

b) In the ring Z_{36}, obtain $\sqrt{I}$ for $I = (4)$ and $I = (6)$.

14. a) Show that if P_1 and P_2 are two ideals of the ring R such that $P_1 \not\subseteq P_2$ and $P_2 \not\subseteq P_1$, then the ideal $P_1 \cap P_2$ is not prime.

b) Let $\{P_i\}$ be a chain of prime ideals of the ring R. Prove that $\cup P_i$ and $\cap P_i$ are both prime ideals of R.

15. Prove that if I is an ideal of the ring R and P is a prime ideal of I, then P is an ideal of the whole ring R.

16. Assume that P is a proper prime ideal of the ring R with the property that the quotient ring R/P is finite. Show that P must be a maximal ideal of R.

17. An ideal I of the ring R is called *primary* if the conditions $ab \in I$ and $a \notin I$ imply $b^n \in I$ for some positive integer n. Establish the following facts concerning primary ideals:

a) Every prime ideal is a primary ideal.

b) If I is a primary ideal of R, then its nil radical $\sqrt{I}$ is the smallest (in the sense of inclusion) prime ideal of R containing I.

c) An ideal I of the ring R is primary if and only if every zero divisor of the quotient ring R/I is nilpotent.

d) The primary ideals of the ring Z are precisely the ideals (p^n), where p is a prime number and $n \geqslant 1$, together with the two trivial ideals.

18. Let f be a homomorphism from the ring R onto the ring R'. Prove that

a) if P is a prime (primary) ideal of R with $P \supseteq \ker f$, then $f(P)$ is a prime (primary) ideal of R';

b) if P' is a prime (primary) ideal of R', then $f^{-1}(P')$ is a prime (primary) ideal of R;

c) the mapping $P \to f(P)$ defines a one-to-one correspondence between the set of prime (primary) ideals of R that contain $\ker f$ and the set of all prime (primary) ideals of R'.

19. a) For any ideal I of R, prove that I and $\sqrt{I}$ are contained in precisely the same prime ideals of R.

b) Using part (a), deduce that whenever I is a prime ideal of R, then $I = \sqrt{I}$.

20. Let f be a homomorphism from the ring R onto the ring R'. Prove that

a) if I is an ideal of R with $I \supseteq \ker f$, then $\sqrt{f(I)} = f(\sqrt{I})$,

b) if I' is an ideal of R', then $\sqrt{f^{-1}(I')} = f^{-1}(\sqrt{I'})$.

21. Verify that the intersection of semiprime ideals of the ring R is again a semiprime ideal of R.

22. If I is an ideal of the ring R, prove that $\sqrt{I}$ is the smallest (in the set-theoretic sense) semiprime ideal of R that contains I.

23. a) Let $I_1, I_2, \ldots, I_n$ be arbitrary ideals of the ring R and P be a prime ideal of R. If $I_1 I_2 \cdots I_n \subseteq P$, establish that $I_i \subseteq P$ for at least one value of i. [*Hint:* If $I_i \nsubseteq P$ for all i, choose $a_i \in I_i - P$ and consider the element $a = a_1 a_2 \cdots a_n$.]

b) Assume that M is a maximal ideal of R. Prove that for each integer $n \in Z_+$ the only prime ideal containing M^n is M.

24. A ring R is said to be a *local ring* if it has exactly one maximal ideal. Establish that the conditions below are equivalent:

a) R is a local ring, with maximal ideal M;

b) the set of noninvertible elements of R coincides with M;

c) the set of noninvertible elements of R forms an ideal;

d) the sum of two noninvertible elements of R is again noninvertible;

e) for each element $r \in R$, either r or $1 - r$ is invertible.

25. Describe the Jacobson radical of the ring Z_n of integers modulo n. In particular, show that Z_n is semisimple if and only if n is a square-free integer. [*Hint:* See Problem 9(b).]

26. Let f be a homomorphism from the ring R onto the ring R'. Show that $f(\text{rad } R) \subseteq \text{rad } R'$ and, whenever $\ker f \subseteq \text{rad } R$, then $\text{rad } R = f^{-1}(\text{rad } R')$.

27. Let R be a ring in which every ideal is prime. Show that R is a field. [*Hint:* For $0 \neq a \in R$, $a \in (a^2)$.]

28. Prove the following assertions:

a) A Boolean ring R is a field if and only if $R \simeq Z_2$.

b) A proper ideal I of a Boolean ring R is a maximal ideal if and only if $R/I \simeq Z_2$.

4.7 DIVISIBILITY THEORY IN INTEGRAL DOMAINS

As the section title suggests, this section is concerned with the problem of factoring elements of an integral domain. The particular impetus is furnished by the ring of integers, where the Fundamental Theorem of Arithmetic states that every integer $n > 1$ can be written, in an essentially unique way, as a product of prime numbers; for example, the integer $360 = 2 \cdot 2 \cdot 2 \cdot 3 \cdot 3 \cdot 5$. We are interested here in the possibility of extending the factorization theory of the ring Z and, in particular, the aforementioned Fundamental Theorem of Arithmetic to a more general setting. Needless to say, any reasonable abstraction of these number-theoretic ideas depends on a suitable interpretation of prime elements (the building blocks for the study of divisibility questions in Z) in integral domains. Except for certain definitions, which we prefer to have available for arbitrary rings, our hypothesis will, for the most part, restrict us to integral domains. The plan is to proceed from the most general results about divisibility, prime elements, and uniqueness of factorization to stronger results concerning specific classes of integral domains.

Throughout this section the rings considered are assumed to be commutative, and it is supposed that each possesses an identity element.

One of the first mathematicians to appreciate the importance of unique factorization for more general sorts of numbers was Ernst Kummer (1810–1893). Kummer, who turned from the study of Protestant theology to mathematics, earned his doctorate from the University of Halle at the age of 23. He went on to hold professorships at the universities of Breslau and Berlin. Shortly after taking his degree, Kummer became interested in Fermat's famous "Last Theorem," that is, the assertion that the equation $x^p + y^p = z^p$, where p is an odd prime, has no solution in integers x, y, z all different from zero. Kummer's idea was to introduce a complex number λ satisfying $\lambda^p = 1$ in order to factor the left-hand side of the equation $x^p + y^p = z^p$ as

$$(x + y)(x + \lambda y)\cdots(x + \lambda^{p-1}y) = z^p.$$

He then addressed the proof of the insolubility of Fermat's equation. Unfortunately, Kummer's purported proof had tacitly assumed that the complex numbers $x + \lambda^k y$ with which he was dealing were representable in essentially one way as a product of "primes." Dirichlet, who had made similar attempts himself, pointed out (1843) that this assumption holds only for certain values of p. To overcome this lack of unique factorization, Kummer invented new numbers, the "ideal prime factors"; their theory was expounded in complete detail in a series of articles beginning in 1847. Although he failed to solve Fermat's Last Theorem (except for a large number of cases of p), Kummer's attempt to do so helped to open up an entirely new area of mathematics, the theory of ideals. Though important advances have been made since Kummer's time, the Last Theorem has not been proved to this day.

We begin our generalization of the classical theory of prime factors by imitating, for arbitrary rings, the definition of divisibility of integers:

DEFINITION 4.19 If $a \neq 0$ and b are elements of the ring R, then a is said to *divide* b, in symbols $a|b$, provided that some $c \in R$ exists such that $b = ac$. In case a does not divide b, we shall write $a \nmid b$.

Other language for the divisibility property $a|b$ is that a is a *factor* of b, that b is *divisible* by a, and that b is a *multiple* of a. Whenever the notation $a|b$ is employed, it is to be understood that the element $a \neq 0$; on the other hand, not only may $b = 0$, but in such instances we always have divisibility.

Some immediate consequences of this definition are listed below; the reader should take the time to become convinced of each of them.

THEOREM 4.45 *Let the elements $a, b, c \in R$. Then*

a) $a|0$, $1|a$, $a|a$;

b) $a|1$ *if and only if a is invertible;*

c) *if* $a|b$, *then* $ac|bc$;
d) *if* $a|b$ *and* $b|c$, *then* $a|c$;
e) *if* $c|a$ *and* $c|b$, *then* $c|(ax + by)$ *for every* $x, y \in R$.

Division of elements in a ring R is closely related to ideal inclusion:

$$a|b \text{ if and only if } (b) \subseteq (a).$$

Indeed, $a|b$ means that $b = ac$ for some $c \in R$; thus $b \in (a)$, so that $(b) \subseteq (a)$. Conversely, if $(b) \subseteq (a)$, then there exists an element c in R for which $b = ac$, implying that $a|b$.

Questions concerning divisibility are complicated somewhat by the presence of invertible elements. For if u has a multiplicative inverse, any element of $a \in R$ can be expressed in the form $a = a(uu^{-1})$, so that both $u|a$ and $u^{-1}|a$. An extreme situation occurs in the case of fields, where every nonzero element divides every other element. On the other hand, in the ring Z_e of even integers the element 2 has no divisors at all.

To overcome the difficulty that is produced by invertible elements, we introduce the following definition.

DEFINITION 4.20 Two elements $a, b \in R$ are said to be *associated elements* or simply *associates* if $a = bu$, where u is an invertible element of R.

A simple argument shows that the relation $\sim$, defined on R by taking $a \sim b$ if and only if a is an associate of b, is an equivalence relation with equivalence classes that are sets of associated elements. The associates of the identity are just the invertible elements of R.

Example 4.34
In the case of the ring Z the only associates of an integer $n \in Z$ are $\pm n$, since ± 1 are the only invertible elements.

Example 4.35
Consider the domain $Z(i)$ of *Gaussian integers*, a subdomain of the complex number field, whose elements form the set

$$Z(i) = \{a + bi | a, b \in Z;\ i^2 = -1\}.$$

Here the only invertible elements are ± 1 and $\pm i$. Suppose that $a + bi \in Z(i)$ has a multiplicative inverse $c + di$. Then we must have $(a + bi)(c + di) = 1$, so $(a - bi)(c - di) = 1$. Therefore

$$\begin{aligned} 1 &= (a + bi)(c + di)(a - bi)(c - di) \\ &= (a^2 + b^2)(c^2 + d^2). \end{aligned}$$

From the fact that a, b, c, d are all integers, it follows that $a^2 + b^2 = 1$. The only solutions of this last equation are $a = \pm 1$, $b = 0$ or $a = 0$, $b = \pm 1$. This leads to the four invertible elements ± 1, $\pm i$. In consequence the class of associates determined by any Gaussian integer $a + bi$ consists of exactly four members:

$$a + bi, \quad -a - bi, \quad -b + ai, \; b - ai.$$

Since associated elements are rather closely related, it is not surprising that they have similar properties; for instance:

THEOREM 4.46 *Let a, b be nonzero elements of an integral domain R. Then the following statements are equivalent:*

1. *a and b are associates,*
2. *both $a|b$ and $b|a$,*
3. *$(a) = (b)$.*

Proof. To prove the equivalence of (1) and (2), suppose that $a = bu$, where u is an invertible element; then, also, $b = au^{-1}$, so both $a|b$ and $b|a$. Going in the opposite direction, if $a|b$, we can write $b = ax$ for some $x \in R$, while from $b|a$ it follows that $a = by$ with $y \in R$. Therefore $b = (by)x = b(yx)$. Since $b \neq 0$, the cancellation law implies that $1 = yx$. Hence y is an invertible element of R, with $a = by$, proving that a and b must be associates. The equivalence of (2) and (3) stems from our earlier remarks relating division of ring elements to ideal inclusion.

We now have a certain amount of information about divisibility in integral domains, but the basic question remains unanswered: When does a ring possess a factorization theory in which the analog of the Fundamental Theorem of Arithmetic holds? To this end, let us introduce two new classes of elements, prime and irreducible elements; when the ring is specialized to the ring of integers, these concepts are equivalent and yield the usual notion of a prime number.

DEFINITION 4.21

1. A nonzero element $p \in R$ is called a *prime* if and only if p is not invertible and $p|ab$ implies either $p|a$ or else $p|b$.
2. A nonzero element $q \in R$ is said to be *irreducible* (or *nonfactorizable*) if and only if q is not invertible and in every factorization $q = bc$ with b, $c \in R$, either b or c is invertible.

Briefly, an irreducible element q is an element that cannot be factored in R in a nontrivial way; the only factors of q are its associates and the invertible elements of R. In such rings as division rings and fields, where each nonzero element possesses a multiplicative inverse, the concept of an irreducible element is of no significance.

Observe also that every element that is an associate of an irreducible (prime) element is itself irreducible (prime). It follows by an easy induction argument that if a product $a_1 a_2 \dots a_n$ is divisible by a prime p, then p must divide at least one of the factors $a_i (i = 1, 2, \dots, n)$.

LEMMA *In an integral domain R, any prime element p is irreducible.*

Proof. Suppose that $p = ab$ for some $a, b \in R$. Since p is prime, either $p|a$ or $p|b$; say p divides b, so that there exists some element c in R for which $b = pc$. We then have $abc = pc = b$. It follows from the cancellation law that $ac = 1$; hence a is invertible. This allows us to conclude that p must be an irreducible element of R.

Although prime elements are irreducible in integral domains, the converse is not always true, as we shall see later on. In the context of principal ideal domains—our primary interest in this section—the notions of an irreducible element and a prime element coincide. This is brought out in the theorem below.

THEOREM 4.47 *Let R be a principal ideal domain. A nonzero element $p \in R$ is irreducible if and only if it is prime.*

Proof. By what we have just proved, p prime always implies that p is irreducible. So assume that p is an irreducible element and that p divides the product ab, say $pc = ab$, with $c \in R$. As R is a principal ideal ring, the ideal generated by p and a,

$$(p, a) = (d)$$

for some choice of d in R; hence $p = rd$ for suitable $r \in R$. But p is irreducible by hypothesis, so either r or d must be an invertible element.

If d happened to possess an inverse, we would have $(p, a) = R$. Thus there would exist elements $s, t \in R$ for which $1 = sp + ta$. Then

$$b = b1 = bsp + bta = bsp + pct = p(bs + ct),$$

which implies that $p|b$.

On the other hand, if r is invertible in R, then $d = r^{-1}p \in (p)$, whence $(d) \subseteq (p)$. It follows that the element $a \in (p)$ and, in consequence, $p|a$. At any rate, if $p|ab$, then p must divide one of the factors, making p a prime element of R.

We next take up two theorems having to do with the ideal structure of a principal ideal domain; the first result has considerable theoretical importance and will, in particular, serve as our basic tool for this section.

THEOREM 4.48 *Let R be a principal ideal domain. If* $\{I_n\}$, $n \in Z_+$, *is any infinite sequence of ideals of R satisfying*

$$I_1 \subseteq I_2 \subseteq \cdots \subseteq I_n \subseteq I_{n+1} \subseteq \cdots,$$

then an integer m exists such that $I_n = I_m$ *for all* $n > m$.

Proof. It is an easy matter to verify that $I = \cup I_n$ is an ideal of R (see the argument of Theorem 4.32). Being an ideal of a principal ideal ring, $I = (a)$ for suitable choice of $a \in R$. Now the element a must lie in one of the ideals of the union, say the ideal I_m. For $n > m$ it then follows that

$$I = (a) \subseteq I_m \subseteq I_n \subseteq I;$$

hence $I_n = I_m$, as required.

In asserting the equivalence of maximal and prime ideals in principal ideal domains, Theorem 4.39 failed to identify these ideals; this situation is taken care of by our next theorem. First, let us define a principal ideal of the ring R to be a *maximal principal ideal* if it is maximal (with respect to inclusion) in the set of proper principal ideals of R.

LEMMA *Let R be an integral domain. For a nonzero element* $p \in R$ *the following hold:*

a) *p is an irreducible element of R if and only if* (p) *is a maximal principal ideal;*

b) *p is a prime element of R if and only if the principal ideal* $(p) \neq R$ *is prime.*

Proof. To begin, we suppose that p is an irreducible element of R and that (a) is any principal ideal for which $(p) \subset (a) \subseteq R$. As $p \in (a)$, we must have $p = ra$ for some $r \in R$. The fact that p is an irreducible element implies that either r or a is invertible. Were r allowed to possess a multiplicative inverse, then $a = r^{-1}p \in (p)$, from which it follows that $(a) \subseteq (p)$, an obvious contradiction. Accordingly, the element a is invertible, whence $(a) = R$. This argument shows that no principal ideal lies between (p) and the whole ring R, so (p) is a maximal principal ideal.

On the other hand, let (p) be a maximal principal ideal of R. For a proof by contradiction, assumes that p is not an irreducible element. Then p admits a factorization $p = ab$ where $a, b \in R$ and neither

a nor b is invertible (the alternative possibility that p has an inverse implies that $(p) = R$, so it may be ruled out). Now if the element a were in (p), then $a = rp$ for some choice of $r \in R$. Hence $p = ab = (rp)b$. Using the cancellation law, we could deduce that $1 = rb$; but this results in the contradiction that b is invertible. Therefore $a \notin (p)$, yielding the proper inclusion $(p) \subset (a)$. Next, observe that if $(a) = R$, then a will possess an inverse, contrary to assumption. We thus conclude that $(p) \subset (a) \subset R$, which denies that (p) is a maximal principal ideal. Our original supposition is false, and a must be an irreducible element of R.

With regard to the second assertion of the lemma, suppose that p is any prime element of R. To see that the principal ideal (p) is in fact a prime ideal, we let the product $ab \in (p)$. Then there exists an element $r \in R$ for which $ab = rp$; hence $p|ab$. By hypothesis, p is a prime element, so either $p|a$ or $p|b$. Translating this into ideals, either $a \in (p)$ or $b \in (p)$; in consequence, (p) is a prime ideal of R.

The converse is proved in much the same way. Let (p) be a prime ideal and $p|ab$. Then $ab \in (p)$. Using the fact that (p) is a prime ideal, it follows that one of a or b lies in (p). This means that either $p|a$ or else $p|b$ and makes p a prime element of R.

For principal ideal domains, all this may be summarized by the following theorem.

THEOREM 4.49 *Let R be a principal ideal domain. The nontrivial ideal (p) is a maximal (prime) ideal of R if and only if p is an irreducible (prime) element of R.*

An immediate consequence of this theorem is that every nonzero noninvertible element of R is divisible by some prime.

COROLLARY *Let $a \neq 0$ be a noninvertible element of the principal ideal domain R. Then there exists a prime $p \in R$ such that $p|a$.*

Proof. Since a is not invertible, the principal ideal $(a) \neq R$. Thus by Theorem 4.32 a maximal ideal M of R exists such that $(a) \subseteq M$. But the preceding result tells us that every maximal ideal is of the form $M = (p)$, where p is a prime element of R (in this setting, there is no distinction between prime and irreducible elements). Thus $(a) \subseteq (p)$, which is to say that $p|a$.

Let us sketch a second proof of the foregoing corollary that avoids the use of Theorem 4.32. First, put $(a) = I_1$. If I_1 is not already a maximal ideal, then an ideal I_2 of R exists such that $I_1 \subset I_2$. By the same reasoning, if I_2 is not maximal, then $I_1 \subset I_2 \subset I_3$ for some ideal I_3. Appealing to Theorem

4.48, this process must terminate after a finite number of steps; in other words, we can eventually find a maximal ideal M of R containing $I_1 = (a)$. As before, $M = (p)$, with p a prime in R; the remainder of the proof is like that above.

If R is an integral domain with the property that every noninvertible element of R can be expressed uniquely (up to invertible elements as factors and the order of factors) as a product of irreducible elements, then we say that R is a unique factorization domain. For a more formal definition:

DEFINITION 4.22 An integral domain R is a *unique factorization domain* in case the following two conditions hold:

1. every element $a \in R$ that is neither zero nor invertible can be factored into a finite product of irreducible elements;
2. if $a = p_1 p_2 \cdots p_n = q_1 q_2 \cdots q_m$ are two factorizations of a into irreducible elements, then $n = m$, and there is a permutation π of the indices such that p_i and $q_{\pi(i)}$ are associates $(i = 1, 2, \ldots, n)$.

In short, an integral domain is a unique factorization domain if it possesses a factorization theory in which the analog of the Fundamental Theorem of Arithmetic holds. We intend to show that any principal ideal domain is a unique factorization domain; toward this goal, let us first prove:

THEOREM 4.50 *If R is a principal ideal domain, then every element of R that is neither zero nor invertible has a factorization into a finite product of primes.*

Proof. Consider any nonzero noninvertible element $a \in R$. By the last corollary a prime p_1 in R exists with $p_1 | a$. Then $a = p_1 a_1$ for some (nonzero) $a_1 \in R$, whence $(a) \subseteq (a_1)$. Were $(a) = (a_1)$, we would have $a_1 = ra$ for suitable $r \in R$; it would follow that $a = p_1 a_1 = p_1 r a$, or $1 = p_1 r$, resulting in the contradiction that p_1 is invertible. Consequently, we have the proper inclusion $(a) \subset (a_1)$.

Repeat the procedure, now starting with a_1, to obtain an increasing chain of principal ideals

$$(a) \subset (a_1) \subset (a_2) \subset \cdots \subset (a_n) \subset \cdots,$$

with $a_{n-1} = p_n a_n$ for some prime $p_n \in R$. This process goes on as long as a_n is not an invertible element of R. But Theorem 4.48 asserts that the displayed chain of ideals eventually terminates; in other words, a_n must possess an inverse for some n, and

$$(a) \subset (a_1) \subset (a_2) \subset \cdots \subset (a_n) = R.$$

We conclude from this that the element a is expressible as the finite product of primes

$$a = p_1 p_2 \cdots p_{n-1} p'_n,$$

where $p'_n = p_n a_n$, being an associate of a prime, is itself prime.

COROLLARY *In a principal ideal domain R, every nontrivial ideal is the product of a finite number of prime (maximal) ideals.*

Proof. If the element $0 \neq a \in R$ is not invertible, then a has a representation as a finite product of primes, say $a = p_1 p_2 \cdots p_n$, where each p_i is a prime element of R. It then follows that

$$(a) = (p_1 p_2 \cdots p_n) = (p_1)(p_2) \cdots (p_n),$$

with (p_i) a prime ideal of R.

Having proved the existence of a prime factorization in principal ideal domains, one is naturally led to the question of uniqueness. Our next theorem is the focal point of this section.

THEOREM 4.51 *Every principal ideal domain R is a unique factorization domain.*

Proof. Theorem 4.50 shows that each noninvertible element $0 \neq a \in R$ has a prime factorization. To establish uniqueness, let us suppose that a can be represented as a product of primes in two ways, say

$$a = p_1 p_2 \cdots p_n = q_1 q_2 \cdots q_m, \qquad n \leqslant m,$$

where the p_i and q_i are all primes. Since $p_1 | (q_1 q_2 \cdots q_m)$, it follows that p_1 divides some $q_i (1 \leqslant i \leqslant m)$; renumbering, if necessary, we may suppose that $p_1 | q_1$. Now p_1 and q_1 are both prime elements of R, with $p_1 | q_1$, so they must be associates: $q_1 = p_1 u_1$ for some invertible element $u_1 \in R$. Canceling the common factor p_1, we are left with

$$p_2 \cdots p_n = u_1 q_2 \cdots q_m.$$

Continuing this argument, we arrive (after n steps) at

$$1 = u_1 u_2 \cdots u_n q_{n+1} \cdots q_m.$$

Since the q_i are not invertible, this forces $m = n$. It has also been shown that every p_i has some q_j as an associate, and conversely. Thus the two prime factorizations are identical, apart from the order in which the factors appear and from replacement of factors by associates.

Attention is called to the fact that the converse of Theorem 4.51 is not true; in the next section we shall give an example of a unique factorization domain that is not a principal ideal domain.

A useful fact to bear in mind is that in a unique factorization domain R, any irreducible element $p \in R$ is necessarily prime. Suppose that p divides the product ab, say $pc = ab$. Let

$$a = p_1 \cdots p_n, \qquad b = q_1 \cdots q_m, \qquad \text{and} \qquad c = t_1 \cdots t_s$$

be the unique factorizations of a, b, and c into irreducible factors. We then have

$$p_1 \cdots p_n q_1 \cdots q_m = ab = pt_1 \cdots t_s.$$

Since the factorization of ab into irreducibles is unique, the element p must be an associate of one of the p_i or q_i, and, consequently, p divides either a or b.

Another interesting class of integral domains, which we propose to look at now, is provided by the so-called Euclidean domains; these arose out of attempts to generalize the familiar Division Algorithm for ordinary integers to arbitrary rings. The precise definition follows.

Definition 4.23 An integral domain R is said to be *Euclidean* if a function $\delta : R \to Z_+ \cup \{0\}$ (the *Euclidean valuation*) exists such that the following conditions are satisfied:

1. $\delta(a) = 0$ if and only if $a = 0$;
2. for all $a, b \in R$, $\delta(ab) = \delta(a)\delta(b)$;
3. given $a, b \in R$, with $b \neq 0$, there exist elements $q, r \in R$ (the *quotient* and *remainder*) such that $a = qb + r$, where $\delta(r) < \delta(b)$.

As simple examples of Euclidean domains, we may take

1. any field F, with valuation defined by $\delta(a) = 1$ for all nonzero $a \in F$ and $\delta(0) = 0$;
2. the ring Z of integers, with valuation defined by $\delta(a) = |a|$ for all $a \in Z$;
3. the Gaussian integers $Z(i)$, with valuation defined by $\delta(a + bi) = a^2 + b^2$ for all $a + bi \in Z(i)$ (the details appear in Theorem 4.55).

Several elementary properties of Euclidean domains appear in the lemma below.

Lemma *Let R be a Euclidean domain with valuation δ. Then*

1. *$\delta(1) = 1$;*
2. *an element $0 \neq a \in R$ is invertible if and only if $\delta(a) = 1$;*
3. *if two elements $a, b \in R$ are associates, then $\delta(a) = \delta(b)$.*

Proof. Assertion (1) follows from the fact that $a = a1$ for any nonzero $a \in R$; hence

$$\delta(a) = \delta(a1) = \delta(a)\delta(1).$$

Since $\delta(a)$ is a positive integer, we may cancel it to obtain $\delta(1) = 1$.

To prove (2), suppose that $a \neq 0$ has an inverse in R, so that $ab = 1$ for some choice of $b \in R$. Then, using (1), we obtain

$$\delta(a)\delta(b) = \delta(ab) = \delta(1) = 1.$$

Inasmuch as $\delta(a)$ and $\delta(b)$ are both positive integers, this equation holds if and only if $\delta(a) = \delta(b) = 1$. Conversely, assume that the element $0 \neq a \in R$ is such that $\delta(a) = 1$. Applying the "division algorithm" to 1 and a, $q, r \in R$ exist for which $1 = qa + r$, where $\delta(r) < \delta(a) = 1$. The implication is that $\delta(r) = 0$, or $r = 0$; in other words, $1 = qa$, making a an invertible element of R.

If a and b are associates, then $a = ub$, with u invertible in R. This means that

$$\delta(a) = \delta(ub) = \delta(u)\delta(b) = 1\delta(b) = \delta(b),$$

by (2).

In general, the elements q and r in the "division algorithm" are not uniquely determined. We shall prove the following.

THEOREM 4.52 *The quotient and remainder in condition (3) of the definition of a Euclidean domain R are unique if and only if*

$$\delta(a + b) \leqslant \max\{\delta(a), \delta(b)\}$$

for all $a, b \in R$.

Proof. Suppose that there exist $a, b \in R$ such that $\delta(a + b) > \max\{\delta(a), \delta(b)\}$. Then we would have

$$b = 0(a + b) + b = 1(a + b) - a,$$

with both $\delta(-a) = \delta(a) < \delta(a + b)$ and $\delta(b) < \delta(a + b)$. This exhibits a lack of uniqueness of quotient and remainder in condition (3).

On the other hand, assume that the stated inequality holds and that for $a, b \neq 0$ in R there are two representations:

$$a = qb + r \qquad \text{with} \qquad \delta(r) < \delta(b);$$

$$a = q'b + r' \qquad \text{with} \qquad \delta(r') < \delta(b).$$

We conclude that

$$\begin{aligned}\delta(q - q')\delta(b) &= \delta((q - q')b) \\ &= \delta(r - r') \leqslant \max\{\delta(r), \delta(-r')\} < \delta(b),\end{aligned}$$

whence $\delta(q - q') < 1$. But then $\delta(q - q') = 0$ or, equivalently, $q - q' = 0$. This in its turn gives $r - r' = 0$, and uniqueness follows.

The Euclidean ring Z with the usual absolute value as the valuation δ does not satisfy the property that $\delta(a + b) \leqslant \max\{\delta(a), \delta(b)\}$ for all $a, b \in Z$; hence there is no uniqueness in the "division algorithm" as described in condition (3). We have, for instance, the two representations

$$-7 = 4(-2) + 1 \quad \text{with} \quad |1| < |-2|,$$

$$-7 = 3(-2) - 1 \quad \text{with} \quad |-1| < |-2|.$$

We could remedy this drawback by requiring that remainders always be positive.

Unique factorization in Z follows ultimately from the Division Algorithm. It is not surprising that in rings where there is an analog of division with remainder we can obtain a similar factorization theory. The main line of argument consists of showing that every Euclidean domain is a principal ideal domain; the converse is false, although it is admittedly difficult to find examples of principal ideal domains that are not Euclidean.

THEOREM 4.53 *Every Euclidean domain is a principal ideal domain.*

Proof. Let R be a Euclidean domain with valuation δ and I be an ideal of R; ignoring the trivial case, we may suppose that $I \neq \{0\}$. Consider the set S defined by

$$S = \{\delta(a) | a \in I;\ a \neq 0\}.$$

Since S is a nonempty subset of positive integers, by the Well-Ordering Principle it has a least element. Pick $b \in I$ so that $\delta(b)$ is minimal in S. Our contention is that $I = (b)$.

Let a be an arbitrary member of I. By the definition of Euclidean domain, elements $q, r \in R$ exist for which $a = qb + r$, where $\delta(r) < \delta(b)$. Now $r = a - qb$ lies in I, since I is an ideal containing both a and b. If $r \neq 0$, the inequality $\delta(r) < \delta(b)$ would contradict our choice of b as the element in I with smallest positive value of δ. Consequently, we must have $r = 0$, and so $a = qb \in (b)$; this implies that $I \subseteq (b)$. The reverse inclusion clearly holds, since $b \in I$, thereby completing the proof.

COROLLARY *Every Euclidean domain is a unique factorization domain.*

It would seem inappropriate to conclude this section without some mention of the *quadratic number fields*. The elements of these domains form the sets

$$Q(\sqrt{n}) = \{a + b\sqrt{n} | a, b \in Q\},$$

with $n \neq 1$ a *square-free integer*, that is, an integer that is not divisible by the square of any positive integer >1. When $n < 0$, we may view

$Q(\sqrt{n})$ as a subdomain of the complex number system C and represent its elements in the standard form $a + b\sqrt{n}i$. It is not difficult to show that if n_1, n_2 are square-free integers, then $Q(\sqrt{n_1}) = Q(\sqrt{n_2})$ if and only if $n_1 = n_2$.

Each element $\alpha = a + b\sqrt{n}$ of the quadratic field $Q(\sqrt{n})$ gives rise to another element $\bar{\alpha} = a - b\sqrt{n}$ of $Q(\sqrt{n})$, which we shall call the *conjugate* of α (for $n < 0$, $\bar{\alpha}$ is the usual complex conjugate of α). A simple argument will establish that the mapping $f: Q(\sqrt{n}) \to Q(\sqrt{n})$ defined by $f(\alpha) = \bar{\alpha}$ is an isomorphism.

To study divisibility properties of $Q(\sqrt{n})$, it is convenient to make use of the concept of the norm of an element (an analog of the absolute value notion in Z):

DEFINITION 4.24 For each element $\alpha = a + b\sqrt{n}$ in $Q(\sqrt{n})$ the *norm* $N(\alpha)$ of α is simply the product of α and its conjugate $\bar{\alpha}$:

$$N(\alpha) = \alpha\bar{\alpha} = (a + b\sqrt{n})(a - b\sqrt{n}) = a^2 - b^2 n.$$

Some properties of the norm function that follow easily from the definition are listed below.

LEMMA *For all $\alpha, \beta \in Q(\sqrt{n})$ the following hold:*

1. *$N(\alpha) = 0$ if and only if $\alpha = 0$;*
2. *$N(\alpha\beta) = N(\alpha)N(\beta)$;*
3. *$N(1) = 1$.*

Proof. Given $\alpha = a + b\sqrt{n}$ in $Q(\sqrt{n})$, $N(\alpha) = a^2 - b^2 n = 0$ if and only if both $a = b = 0$ (that is, $\alpha = 0$); otherwise, we would contradict the choice of n as a square-free integer.

Since the mapping $f(\alpha) = \bar{\alpha}$ is an isomorphism, N is a multiplicative function in the sense that

$$N(\alpha\beta) = \alpha\beta\overline{\alpha\beta} = \alpha\beta\bar{\alpha}\bar{\beta} = \alpha\bar{\alpha}\beta\bar{\beta} = N(\alpha)N(\beta)$$

for all $\alpha, \beta \in Q(\sqrt{n})$.

The proof of assertion (3) follows from the fact that

$$N(1) = N(1^2) = N(1)N(1) = N(1)^2,$$

whence $N(1) = 1$.

Although $Q(\sqrt{n})$ has been labeled as a field, we actually have not proved this to be the case; it is high time to remedy this situation.

THEOREM 4.54 *For each square-free integer n the system* $Q(\sqrt{n})$ *forms a field; in fact,* $Q(\sqrt{n})$ *is a subfield of C.*

Proof. The reader may easily verify that $Q(\sqrt{n})$ is a commutative ring with identity. It remains only to establish that each nonzero element of $Q(\sqrt{n})$ has a multiplicative inverse in $Q(\sqrt{n})$. Now, if $0 \neq \alpha \in Q(\sqrt{n})$, then the element $\beta = \bar{\alpha}/N(\alpha)$ evidently lies in $Q(\sqrt{n})$; furthermore, the product

$$\alpha\beta = \alpha(\bar{\alpha}/N(\alpha)) = N(\alpha)/N(\alpha) = 1,$$

so that β serves as the inverse of α.

Contained in each quadratic field $Q(\sqrt{n})$ is the integral domain

$$Z(\sqrt{n}) = \{a + b\sqrt{n} \mid a, b \in Z\}.$$

Since $Z(\sqrt{n})$ is closed under conjugation, the norm function enables us to get a clear view of the sets of invertible and irreducible elements in these domains. The multiplicative property of the norm, for instance, transfers any factorization $\alpha = \beta\gamma$ of an element $\alpha \in Z(\sqrt{n})$ into a factorization $N(\alpha) = N(\beta)N(\gamma)$ of the integer $N(\alpha)$. This is particularly helpful in proving the lemma below.

LEMMA *For any* $\alpha \in Z(\sqrt{n})$ *the following hold:*

1. $N(\alpha) = \pm 1$ *if and only if* α *is invertible in* $Z(\sqrt{n})$;
2. *if* $N(\alpha) = \pm p$, *where p is a prime number, then* α *is an irreducible element of* $Z(\sqrt{n})$.

Proof. As regards (1), observe that if $N(\alpha) = \pm 1$, then $\alpha\bar{\alpha} = \pm 1$; thus $\alpha \mid 1$, which is to say that α is invertible. To prove the converse, let α be an invertible element of $Z(\sqrt{n})$, so that $\alpha\beta = 1$ for some β in $Z(\sqrt{n})$. Then

$$N(\alpha)N(\beta) = N(\alpha\beta) = N(1) = 1.$$

Since $N(\alpha)$ and $N(\beta)$ are both integers, this implies that $N(\alpha) = \pm 1$.

Next suppose that α has the property that $N(\alpha) = \pm p$, where p is a prime number. As $N(\alpha) \neq 0, \pm 1$, the element α is neither 0 nor invertible in $Z(\sqrt{n})$. If $\alpha = \beta\gamma$ is a factorization of α in $Z(\sqrt{n})$, then

$$N(\beta)N(\gamma) = N(\alpha) = \pm p,$$

from which it follows that one of $N(\beta)$ or $N(\gamma)$ must have the value ± 1. From the first part of the lemma we may thus conclude that either β or γ is invertible in $Z(\sqrt{n})$, while the other is an associate of α. Accordingly, α is an irreducible element of $Z(\sqrt{n})$.

Example 4.36
Let us find all invertible elements in $Z(i) = Z(\sqrt{-1})$, the domain of Gaussian integers, by finding those members α of $Z(i)$ for which $N(\alpha) = 1$ (in this setting, the norm assumes nonnegative values). If $\alpha = a + bi \in Z(i)$ and $N(\alpha) = 1$, then $a^2 + b^2 = 1$, with $a, b \in Z$. This equation is possible only if $a = \pm 1$ and $b = 0$, or $a = 0$ and $b = \pm 1$. Hence the only choice for invertible elements in $Z(i)$ are ± 1 and $\pm i$.

Perhaps the most obvious approach to the question of unique factorization in the quadratic domains $Z(\sqrt{n})$ is to try to show that they are Euclidean domains (a natural candidate for the Euclidean valuation is $\delta(\alpha) = |N(\alpha)|$). In the following theorem we shall do precisely this for the domains $Z(\sqrt{-1})$, $Z(\sqrt{-2})$, $Z(\sqrt{2})$, and $Z(\sqrt{3})$. Although there are other Euclidean quadratic domains, our attention is restricted to these few for which the "division algorithm" is easily established.

THEOREM 4.55 *Each of the domains $Z(\sqrt{n})$, where $n = -1, -2, 2, 3$, is Euclidean; hence they are unique factorization domains.*

Proof. The strategy employed in the proof is to show that the function δ defined on $Z(\sqrt{n})$ by $\delta(\alpha) = |N(\alpha)|$ is a Euclidean valuation for $n = -1, -2, 2, 3$. It is clear that $\delta(\alpha) = 0$ if and only if $\alpha = 0$. Since both the norm and the absolute value function are multiplicative, we also have

$$\delta(\alpha\beta) = |N(\alpha\beta)| = |N(\alpha)|\,|N(\beta)| = \delta(\alpha)\delta(\beta)$$

whenever α, $\beta \in Z(\sqrt{n})$. Thus our major effort is concentrated on establishing condition (3) of Definition 4.23.

If $\beta \neq 0$, the product $\alpha\beta^{-1} \in Q(\sqrt{n})$ and so may be written in the form $\alpha\beta^{-1} = a + b\sqrt{n}$, with $a, b \in Q$. Select integers x and y (the nearest integers to a and b) such that

$$|a - x| \leqslant 1/2, \qquad |b - y| \leqslant 1/2.$$

Now set $\sigma = x + y\sqrt{n}$. Then $\sigma \in Z(\sqrt{n})$, and norm formula (valid also in $Q(\sqrt{n})$) shows that

$$\begin{aligned} |N(\alpha\beta^{-1} - \sigma)| &= |N((a - x) + (b - y)\sqrt{n})| \\ &= |(a - x)^2 - n(b - y)^2|. \end{aligned}$$

But the manner in which x and y were chosen implies that

$$-n/4 \leqslant (a - x)^2 - n(b - y)^2 \leqslant 1/4, \qquad \text{if } n > 0,$$

$$0 \leqslant (a - x)^2 - n(b - y)^2 \leqslant 1/4 + (-n)1/4, \qquad \text{if } n < 0.$$

In terms of the function δ this means that

$$\delta(\alpha\beta^{-1} - \sigma) = |(a - x)^2 - n(b - y)^2| < 1$$

for $n = -1, -2, 2, 3$. Putting $\rho = \beta(\alpha\beta^{-1} - \sigma)$, we have $\alpha = \sigma\beta + \rho$. Since α and $\sigma\beta$ are in $Z(\sqrt{n})$, this equation implies that ρ is in $Z(\sqrt{n})$ also. Moreover, for the previously indicated values of n,

$$\delta(\rho) = \delta(\beta(\alpha\beta^{-1} - \sigma)) = \delta(\beta)\delta(\alpha\beta^{-1} - \sigma) < \delta(\beta).$$

Definition 4.23 is therefore satisfied in its entirety when $n = -1$, -2, 2, 3 and the corresponding quadratic domains $Z(\sqrt{n})$ are Euclidean.

COROLLARY *The domain $Z(i)$ of Gaussian integers is a unique factorization domain.*

In proving our next result we shall need the following preliminary fact:

LEMMA *If $Z(\sqrt{n})$ is a unique factorization domain, then 2 is not an irreducible element of $Z(\sqrt{n})$.*

Proof. One of the integers n or $n - 1$ is even, so $2|n(n - 1)$. Since

$$(n + \sqrt{n})(n - \sqrt{n}) = n^2 - n = n(n - 1),$$

it follows that $2|(n + \sqrt{n})(n - \sqrt{n})$. But $2 \nmid (n + \sqrt{n})$ and $2 \nmid (n - \sqrt{n})$, since neither $(n/2) + \frac{1}{2}\sqrt{n}$ nor $(n/2) - \frac{1}{2}\sqrt{n}$ lies in $Z(\sqrt{n})$. We have an instance in which 2 divides a product of two members of $Z(\sqrt{n})$ but divides neither factor individually. Hence 2 is not prime in $Z(\sqrt{n})$ and consequently cannot be irreducible. (By the remarks on page 333 the two concepts are equivalent in unique factorization domains.)

As a companion to Theorem 4.55, let us now determine those negative values of n for which $Z(\sqrt{n})$ is a unique factorization domain.

THEOREM 4.56 *For $n < 0$, $Z(\sqrt{n})$ is a unique factorization domain only when $n = -1$ or $n = -2$.*

Proof. Since $Z(\sqrt{-1})$ and $Z(\sqrt{-2})$ are already known to be unique factorization domains, attention may be restricted to the case in which $n \leqslant -3$. We shall show that, for these values of n, 2 is an irreducible element of $Z(\sqrt{n})$; an appeal to the lemma is then in order.

Suppose that 2 were not irreducible in $Z(\sqrt{n})$. Then there would exist α, $\beta \in Z(\sqrt{n})$ for which $2 = \alpha\beta$, where neither α nor β is invertible; as a result, $|N(\alpha)| \geqslant 2$ and $|N(\beta)| \geqslant 2$. Because

$$|N(\alpha)|\,|N(\beta)| = |N(\alpha\beta)| = |N(2)| = 4,$$

with $|N(\alpha)|$ and $|N(\beta)|$ both positive integers, we would necessarily have $|N(\alpha)| = |N(\beta)| = 2$. To prove that 2 is irreducible in $Z(\sqrt{n})$, it therefore suffices to show that there are no elements of $Z(\sqrt{n})$ of norm ± 2.

Let $\gamma = a + b\sqrt{n}$ be a typical member of $Z(\sqrt{n})$. If $n \leqslant -3$ and $b \neq 0$, then

$$N(\gamma) = a^2 - b^2 n = a^2 + b^2(-n) \geqslant 0 + 1 \cdot 3 > \pm 2,$$

while if $b = 0$, then

$$N(\gamma) = a^2 - b^2 n = a^2 \neq \pm 2.$$

In short, when $n \leqslant -3$, it is not possible to find an element in $Z(\sqrt{n})$ having norm ± 2; hence 2 is irreducible, and, by the lemma, $Z(\sqrt{n})$ is not a unique factorization domain.

Example 4.37

To round out the picture with an example of the failure of unique factorization, let us consider the quadratic domain $Z(\sqrt{-5})$. Observe that the element 9 has two factorizations in $Z(\sqrt{-5})$, namely,

$$9 = 3 \cdot 3 = (2 + \sqrt{-5})(2 - \sqrt{-5}).$$

Clearly, no two of the factors $3, 2 + \sqrt{-5}$, and $2 - \sqrt{-5}$ are associates of each other—the only invertible elements in $Z(\sqrt{-5})$ are ± 1—and it can easily be shown that they are all irreducible. Indeed, were any one of

$$\alpha_1 = 3, \qquad \alpha_2 = 2 + \sqrt{-5}, \qquad \alpha_3 = 2 - \sqrt{-5}$$

reducible, we would write $\alpha_i = \beta\gamma$, where neither of the elements β, $\gamma \in Z(\sqrt{-5})$ is invertible. Taking norms, it follows that

$$9 = N(\alpha_i) = N(\beta)N(\gamma), \qquad N(\beta),\ N(\gamma) \in Z_+,$$

which in turn yields $N(\beta) = N(\gamma) = 3$. Hence if $\beta = a + b\sqrt{-5}$, we find that we must solve the equation $a^2 + 5b^2 = 3$ for integers a and b; but this equation obviously has no solutions in Z ($b \neq 0$ implies that $a^2 + 5b^2 \geqslant 5$, and if $b = 0$, then $a^2 = 3$). Thus we have exhibited two genuinely different factorizations of the element 9 into irreducibles, so unique factorization does not hold in $Z(\sqrt{-5})$.

This example has the additional feature of showing that the concepts of an irreducible element and of a nonzero prime do not always coincide in an arbitrary integral domain. Specifically, we have $(2 + \sqrt{-5})|3 \cdot 3$, but $(2 + \sqrt{-5}) \nmid 3$, so $2 + \sqrt{-5}$ cannot be a prime element of $Z(\sqrt{-5})$.

PROBLEMS 4.7

1. If u is a nonzero element of the integral domain R, prove that the following statements are equivalent:
   a) u and 1 are associates;
   b) u is an invertible element of R;
   c) the principal ideal $(u) = R$.
2. Given that I, J, and K are ideals of a principal ideal domain R, derive the following relations:
   a) If $I = (a)$ and $J = (b)$, then $IJ = (ab)$; in particular, $I^n = (a^n)$.
   b) $I(J \cap K) = IJ \cap IK$.
   c) $I + (J \cap K) = (I + J) \cap (I + K)$.
   d) $I \cap (J + K) = (I \cap J) + (I \cap K)$.
   e) If $I + J = R$, then $IJ = I \cap J$.
3. Verify that any two nonzero elements of a unique factorization domain possess a greatest common divisor. [*Hint:* If $a = p_1^{k_1} p_2^{k_2} \cdots p_r^{k_r}$ and $b = p_1^{l_1} p_2^{l_2} \cdots p_r^{l_r}$ (p_i irreducible), then $\gcd(a, b) = p_1^{j_1} p_2^{j_2} \cdots p_r^{j_r}$, where $j_i = \min(k_i, l_i)$.]
4. Let R be an integral domain. Prove that R is a unique factorization domain if and only if every nontrivial principal ideal of R is the product of a finite number of maximal principal ideals and these ideals are unique up to a permutation of order.
5. Show that $\delta(a) = |a|$ is not a Euclidean valuation on the domain Q.
6. Assuming that R is a Euclidean domain with valuation δ, prove the statements below:
   a) For nonzero $a, b \in R$, if $a|b$ and $\delta(a) = \delta(b)$, then a and b are associates. [*Hint:* Show that $b|a$.]
   b) For nonzero a, $b \in R$, $\delta(ab) > \delta(a)$ if and only if b is not an invertible element. (*Hint:* Use the "division algorithm" to write $a = q(ab) + r$.]
   c) If n is any integer such that $\delta(1) + n \geqslant 0$, then the function $\delta' : R - \{0\} \to Z$ defined by $\delta'(a) = \delta(a) + n$ is also a Euclidean valuation on R.

7. In the domain $Z(i)$, let $a = -4 + i$ and $b = 5 + 3i$. Show that:

a) the Euclidean valuation δ defined by $\delta(r + si) = r^2 + s^2$ does not satisfy $\delta(a + b) \leqslant \max\{\delta(a), \delta(b)\}$;

b) the element a has the two representations

$$a = (-1 + i)b + (4 - i) = (-1)b + (1 + 4i)$$

and hence uniqueness fails in the "division algorithm" for $Z(i)$.

8. If $\alpha \in Q(\sqrt{d})$, show that some $n \in Z$ exists for which $N(n\alpha)$ is an integer.

9. Verify that the group $Z(i)^*$ of invertible elements in $Z(i)$ is isomorphic to K_4, the Klein 4-group.

10. a) Let R be a principal ideal domain and $p \in R$ be an irreducible element. Prove that the quotient ring $R/(p)$ is a field.

b) Use part (a) to show that $Z(i)/(1 - i)$ and $Z(i)/(2 + i)$ are both fields.

11. Verify that there are infinitely many irreducible elements in $Z(i)$. [*Hint:* Any prime $p \equiv 1 (\text{mod } 4)$ can be expressed as the sum of two squares of integers.]

12. Confirm the following.

a) The set $I = \{a + b\sqrt{-3} \mid a, b \in Z;\ a - b \text{ is even}\}$ is an ideal of the quadratic domain $Z(\sqrt{-3})$.

b) The set $S = \{a + 2bi \mid a, b \in Z;\ i^2 = -1\}$ is a subring, but not an ideal, of the domain $Z(i)$.

13. In $Z(\sqrt{2})$, show that $\pm 1 \pm \sqrt{2}$ and $\pm 3 \pm 2\sqrt{2}$ are invertible elements.

14. For each ideal I in $Z(i)$, the domain of Gaussian integers, establish that the quotient ring $Z(i)/I$ is finite. [*Hint:* Write $I = (\alpha)$ and use the "division algorithm" on α and any $\beta \in Z(i)$.]

15. Verify that in a Euclidean domain an element is prime if and only if it is irreducible.

16. Establish the following assertions (where n is a square-free integer):

a) For $n < -1$ the only invertible elements of the quadratic domain $Z(\sqrt{n})$ are ± 1.

b) For $n > 1$, $Z(\sqrt{n})$ has infinitely many invertible elements. [*Hint:* If a_1, b_1 is a solution of the equation $a^2 - nb^2 = \pm 1$, conclude that a_k, b_k is also a solution, where $a_k + b_k\sqrt{n} = (a_1 + b_1\sqrt{n})^k$, $k \in Z_+$.]

c) The invertible elements of $Z(\sqrt{2})$ are precisely the elements of the form $\pm(1 + \sqrt{2})^n$, $n \in Z_+$. [*Hint:* If u is any positive invertible element of $Z(\sqrt{2})$, then $(1 + \sqrt{2})^n \leqslant u < (1 + \sqrt{2})^{n+1}$ for some $n \in Z_+$; hence

$$1 \leqslant u(1 + \sqrt{2})^{-n} < 1 + \sqrt{2}.$$

Assuming that $u(1 + \sqrt{2})^{-n} = a + b\sqrt{2}$, show that $a = 1$, $b = 0$.]

17. a) Factor each of the following elements into primes: $11 + 7i$ in $Z(i)$; $4 + 7\sqrt{2}$ in $Z(\sqrt{2})$; $4 - \sqrt{-3}$ in $Z(\sqrt{-3})$.

b) Show that in the quadratic domain $Z(\sqrt{6})$ the relation $6 = (\sqrt{6})^2 = 3 \cdot 2$ does not violate unique factorization.

18. Prove that the domain $Z(\sqrt{-6})$ is not a unique factorization domain by discovering two distinct factorizations of the element 10. Do the same for element 9 in the domain $Z(\sqrt{-7})$.

19. Show that the quadratic domain $Z(\sqrt{-5})$ is not a principal ideal domain. [*Hint:* Consider the ideal $(3, 2 + \sqrt{-5})$.]

20. Describe the field of quotients of the quadratic domain $Z(\sqrt{n})$.

21. Verify that the principal ideal $(\sqrt{2})$ is a maximal ideal of the quadratic domain $Z(\sqrt{2})$.

22. Given that I is an ideal of the domain $Z(\sqrt{n})$, n square-free, prove that

a) if $\bar{I}$ denotes the set of all conjugates of elements of I, then $\bar{I}$ is also an ideal of $Z(\sqrt{n})$;

b) if I is a principal ideal, say $I = (\alpha)$, then $\bar{I} = (\bar{\alpha})$; hence $I\bar{I} = (N(\alpha))$.

23. Let n be a square-free integer with $n \equiv 1 (\text{mod } 4)$. Establish that $Z(\sqrt{n})$ is not a unique factorization domain. [*Hint:* Show that 2 is irreducible in $Z(\sqrt{n})$.]

24. Suppose that R is an integral domain with the property that whenever $I_1 \supseteq I_2 \supseteq \cdots \supseteq I_n \supseteq \cdots$ is a descending chain of ideals of R, then there exists an integer m such that $I_m = I_n$ for all $m > n$. Prove that R must be a field. [*Hint:* For $0 \neq a \in R$, consider the chain $(a) \supseteq (a^2) \supseteq (a^3) \supseteq \cdots$.]

4.8 POLYNOMIAL RINGS

The next step in our program is to apply some of the previously developed theory to a particular class of rings, the so-called polynomial rings. For the moment we shall merely remark that these are rings whose elements consist of "polynomials" with coefficients from a fixed, but otherwise arbitrary, ring. The most interesting results occur when the coefficients are taken from a field.

As a first order of business, we seek to formalize the intuitive idea of what is meant by a polynomial. Out of the veritable multitude of results concerning polynomials we will then attempt to assemble those facets of the theory whose discussion reinforces the concepts and theorems expounded earlier; we hope thereby to convey a rough idea of how the classical arithmetic of polynomials fits into ring theory.

To begin with simpler things, given an arbitrary ring R, let poly R denote the totality of all infinite sequences

$$f = (a_0, a_1, a_2, \ldots, a_k, \ldots)$$

of elements of R in which at most a finite number of the a_k are nonzero; in other words, $(a_0, a_1, a_2, \ldots) \in \text{poly } R$ if and only if there is some nonnegative integer n such that $a_k = 0$ for all $k > n$, where the particular integer n varies from sequence to sequence. Such sequences are called *polynomials over the ring R*, or just polynomials over R. (Our choice of terminology will be justified shortly.)

At times it will be convenient to use the notation

$$(a_0, a_1, \ldots, a_n, 0, 0, \ldots)$$

for the polynomial with last nonzero term a_n; when $n = 0$, we allow the possibility $a_0 = 0$ in order to include the zero polynomial $(0, 0, 0, \ldots)$, each of whose terms is zero. With this agreement the set of polynomials over R may be regarded as the set

$$\text{poly } R = \{(a_0, a_1, \ldots, a_n, 0, 0, \ldots) | a_k \in R;\ n \geqslant 0\}.$$

Thus the sequence $(1, 1, 1, 0, 0, \ldots)$ would be a polynomial over Z_2, while $(1, 0, 1, 0, \ldots, 1, 0, \ldots)$ would not.

We intend to introduce suitable operations in the set poly R so that the resulting system forms a ring containing R as a subring. At the outset it should be made perfectly clear that two polynomials

$$f = (a_0, a_1, a_2, \ldots) \quad \text{and} \quad g = (b_0, b_1, b_2, \ldots)$$

are considered to be equal if and only if they are equal term by term:

$$f = g \text{ if and only if } a_k = b_k \text{ for all } k \geqslant 0.$$

Now polynomials may themselves be added and multiplied as follows:

$$f + g = (a_0 + b_0, a_1 + b_1, \ldots)$$

and

$$fg = (c_0, c_1, c_2, \ldots),$$

where, for each $k \geqslant 0$, c_k is given by

$$c_k = \sum_{i+j=k} a_i b_j = a_0 b_k + a_1 b_{k-1} + \cdots + a_{k-1} b_1 + a_k b_0.$$

It is understood that the above summation runs over all integers $i, j \geqslant 0$ subject to the restriction that $i + j = k$. Note that if $a_k = 0$ for $k > n$ and $b_k = 0$ for $k > m$, then $a_k + b_k = 0$ for $k > \max\{m, n\}$ and $\sum_{i+j=k} a_i b_j = 0$ for $k > m + n$; thus both the sum $f + g$ and product fg belong to poly R.

A routine check establishes that with these two definitions, poly R becomes a ring. To verify a distributive law, for instance, take

$$f = (a_0, a_1, \ldots), \qquad g = (b_0, b_1, \ldots), \qquad h = (c_0, c_1, \ldots).$$

One quickly finds that

$$f(g + h) = (a_0, a_1, \ldots)(b_0 + c_0, b_1 + c_1, \ldots) = (d_0, d_1, \ldots),$$

where

$$\begin{aligned} d_k &= \sum_{i+j=k} a_i(b_j + c_j) \\ &= \sum_{i+j=k} (a_i b_j + a_i c_j) \\ &= \sum_{i+j=k} a_i b_j + \sum_{i+j=k} a_i c_j. \end{aligned}$$

A similar calculation of $fg + fh$ leads to the same general term, so $f(g + h) = fg + fh$. The rest of the details are left to the reader's care. We simply point out that the zero polynomial $(0, 0, 0, \ldots)$ serves as the zero element of this ring, while the additive inverse of an arbitrary member $(a_0, a_1, a_2, \ldots)$ of poly R is, of course, $(-a_0, -a_1, -a_2, \ldots)$.

To summarize what we know so far:

THEOREM 4.57 *The system* poly R *forms a ring, known as the ring of polynomials over* R. *Furthermore, the ring* poly R *is commutative with identity if and only if the given ring* R *has these properties.*

If S represents the subset of all sequences having 0 for every term beyond the first, that is, the set

$$S = \{(a, 0, 0, \ldots) | a \in R\},$$

then it is not particularly difficult to show that S constitutes a subring of poly R that is isomorphic to R; one need only consider the mapping that sends the polynomial $(a, 0, 0, \ldots)$ to the element a. In this sense, poly R contains the original ring R as a subring.

As a result of the aforementioned imbedding, we shall no longer distinguish between an element $a \in R$ and the special polynomial $(a, 0, 0, \ldots)$ of poly R. The elements of R, regarded as polynomials, are hereafter called *constant polynomials*, or just *constants.*

With the aid of some additional notation it is possible to represent polynomials the way we would like them to look. As a first step in this direction, we let ax designate the sequence

$$(0, a, 0, 0, \ldots).$$

That is, ax is the specific member of poly R that has the element a for its second term and 0 for all other terms. More generally, the symbol ax^n, $n \geqslant 1$, denotes the sequence

$$(0, \ldots, 0, a, 0, \ldots),$$

where the element a appears as the $(n+1)$st term in this sequence; for example, we have

$$ax^2 = (0, 0, a, 0, \ldots)$$

and

$$ax^3 = (0, 0, 0, a, 0, \ldots).$$

By use of these definitions, each polynomial

$$f = (a_0, a_1, a_2, \ldots, a_n, 0, 0, \ldots)$$

may be uniquely expressed in the form

$$\begin{aligned} f &= (a_0, 0, 0, \ldots) + (0, a_1, 0, 0, \ldots) + \cdots + (0, \ldots, a_n, 0, \ldots) \\ &= a_0 + a_1 x + a_2 x^2 + \cdots + a_n x^n, \end{aligned}$$

where the elements $a_0, a_1, \ldots, a_n$, (the *coefficients* of f) lie in R. As a notational device, we shall often write this as $f = \sum_{k=1}^{n} a_k x^k$ or just $f = \sum a_k x^k$.

Using sigma notation, the definitions of addition and multiplication of polynomials assume the form

$$\sum a_k x^k + \sum b_k x^k = \sum (a_k + b_k) x^k,$$

and

$$(\sum a_k x^k)(\sum b_k x^k) = \sum c_k x^k,$$

where

$$c_k = \sum_{i+j=k} a_i b_j = \sum_{i=0}^{k} a_i b_{k-i}.$$

We should emphasize that, according to our definition, x is simply a new symbol, or indeterminant, that is totally unrelated to the ring R and in no sense represents an element of R. To indicate the indeterminant x, it is common practice to write $R[x]$ for the set poly R and $f(x)$ for any member of the same. From now on we shall make exclusive use of this notation.

Remark. If the ring R happens to have a multiplicative identity 1, many authors identify the polynomial $0 + 1x + 0x^2 + 0x^3 + \cdots$ with x, thereby treating x itself as a special member of $R[x]$, namely, the polynomial $x = (0, 1, 0, 0, \ldots)$. From this view, ax becomes an actual product of members of $R[x]$:

$$ax = (a, 0, 0, \ldots)(0, 1, 0, 0, \ldots).$$

Concerning the notation of polynomials, it is customary to omit terms with zero coefficients and to replace $(-a_k)x^k$ by $-a_kx^k$. Although x is not to be considered as an element of $R[x]$, we shall nonetheless take the liberty of writing the term $1x^k$ as $x^k (k \geqslant 1)$. With these conventions, one should view, for example, the polynomial

$$1 - x^2 - 4x^4 \in Z[x]$$

as representing the sequence $(1, 0, -1, 0, -4, 0, 0, \ldots)$.

An important definition in connection with polynomials is that of degree, given below.

Definition 4.25 Given a nonzero polynomial

$$f(x) = a_0 + a_1x + \cdots + a_nx^n, \qquad a_n \neq 0,$$

in $R[x]$, we call a_n the *leading coefficient* of $f(x)$; and the integer n the *degree* of the polynomial.

The degree of any nonzero polynomial is therefore a nonnegative integer; no degree is assigned to the zero polynomial. Note that the polynomials of degree 0 are precisely the nonzero constant polynomials. If R is a ring with identity, a polynomial whose leading coefficient is 1 is said to be *monic polynomial.*

As a matter of notation, we shall hereafter write $\deg f(x)$ for the degree of any nonzero polynomial $f(x) \in R[x]$.

Suppose then that $f(x)$, $g(x) \in R[x]$, with $\deg f(x) = n$ and $\deg g(x) = m$, so that

$$f(x) = a_0 + a_1x + \cdots + a_nx^n, \qquad a_n \neq 0,$$

$$g(x) = b_0 + b_1x + \cdots + b_mx^m, \qquad b_m \neq 0.$$

From the definition of multiplication in $R[x]$ the reader may easily check that all the coefficients of $f(x)g(x)$ beyond the $(n + m)$th are zero, whence

$$f(x)g(x) = a_0b_0 + (a_0b_1 + a_1b_0)x + \cdots + a_nb_mx^{n+m}.$$

If we assume that one of a_n or b_m is not a divisor of zero in R, then $a_nb_m \neq 0$ and

$$\deg(f(x)g(x)) = n + m = \deg f(x) + \deg g(x).$$

This certainly holds if R is taken to be an integral domain, or again if R has an identity and one of a_n or b_m is the identity element or invertible.

The foregoing argument serves to establish the first part of the next theorem; the proof of the second assertion is left as an exercise.

THEOREM 4.58 *If $f(x)$ and $g(x)$ are nonzero polynomials in $R[x]$, then*

a) *either* $f(x)g(x) = 0$ *or* $\deg(f(x)g(x)) \leqslant \deg f(x) + \deg g(x)$, *with equality whenever R is an integral domain;*
b) *either* $f(x) + g(x) = 0$ *or* $\deg(f(x) + g(x)) \leqslant \max\{\deg f(x), \deg g(x)\}$.

The notion of degree may also be used to prove the following corollary.

COROLLARY *If the ring R is an integral domain, then so is its polynomial ring $R[x]$.*

Proof. We observed earlier that whenever R is a commutative ring with identity, these properties carry over to $R[x]$. To see that $R[x]$ has no zero divisors, select $f(x) \neq 0$, $g(x) \neq 0$ in $R[x]$. Then

$$\deg(f(x)g(x)) = \deg f(x) + \deg g(x) \geqslant 0;$$

hence the product cannot be the zero polynomial. (The zero polynomial has no degree assigned to it.)

Example 4.38
As an illustration of what might happen if R had zero divisors, consider Z_8, the ring of integers modulo 8. Taking

$$f(x) = 1 + 2x, \qquad g(x) = 4 + x + 4x^2,$$

we obtain $f(x)g(x) = 4 + x + 6x^2$, so that

$$\deg(f(x)g(x)) = 2 < 1 + 2 = \deg f(x) + \deg g(x).$$

Although many properties of the ring R carry over to the associated polynomial ring $R[x]$, it should be pointed out that for no ring R does $R[x]$ form a field. In fact, when R is a field or, for that matter, an integral domain, no element of $R[x]$ that has positive degree can possess a multiplicative inverse. Suppose that $f(x) \in R[x]$, with $\deg f(x) > 0$; if $f(x)g(x) = 1$ for some $g(x)$ in $R[x]$, we could obtain the contradiction

$$0 = \deg 1 = \deg(f(x)g(x)) = \deg f(x) + \deg g(x) \neq 0.$$

The degree of a polynomial is used in the factorization theory of $R[x]$ in much the same way as the absolute value is employed in Z. It is through the degree concept that induction can be utilized in $R[x]$ to develop a polynomial counterpart of the familiar division algorithm. One can subsequently establish that the ring $F[x]$ with coefficients in a field forms a Euclidean domain in which the degree function is used to define the required Euclidean valuation.

Before embarking on this program we wish to introduce several new ideas. To this purpose, let R be a ring with identity; assume further that R' is any ring containing R as a subring (that is, R' is an extension of R) and let r be an arbitrary element of R'. For each polynomial

$$f(x) = a_0 + a_1x + \cdots + a_nx^n$$

in $R[x]$ we may define $f(r) \in R'$ by taking

$$f(r) = a_0 + a_1r + \cdots + a_nr^n.$$

The element $f(r)$ is said to be the result of *substituting* r for x in $f(x)$. Suffice it to say that the addition and multiplication used in defining $f(r)$ are those of the ring R', not those of $R[x]$.

Now suppose that $f(x)$, $g(x)$ are polynomials in $R[x]$ and $r \in \operatorname{cent} R'$. We leave the reader to prove that if

$$h(x) = f(x) + g(x), \qquad k(x) = f(x)g(x),$$

then

$$h(r) = f(r) + g(r), \qquad k(r) = f(r)g(r).$$

This being so, it may be concluded that the mapping $\phi_r : R[x] \to R'$, which sends $f(x)$ to $f(r)$, is a homomorphism of $R[x]$ into R'. Such a homomorphism will be called the *substitution homomorphism determined by* r and its range denoted by the symbol $R[r]$:

$$\begin{aligned} R[r] &= \{f(r) \mid f(x) \in R[x]\} \\ &= \{a_0 + a_1r + \cdots + a_nr^n \mid a_k \in R;\ n \geqslant 0\}. \end{aligned}$$

It is a simple matter to show that $R[r]$ constitutes a subring of R'; in fact, $R[r]$ is the subring of R' generated by the set $R \cup \{r\}$. Since R has an identity element 1, $1x = x \in R[x]$, and so $r \in R[r]$. Note also that $R[r] = R$ if and only if $r \in R$.

The foregoing remarks justify part of the next theorem.

THEOREM 4.59 *Let R be a ring with identity, R' an extension ring of R, and the element $r \in \operatorname{cent} R'$. Then there is a unique homomorphism $\phi_r : R[x] \to R'$ such that $\phi_r(x) = r$, $\phi_r(a) = a$ for all $a \in R$.*

Proof. We need to verify that ϕ_r is unique. Suppose, then, that there is another homomorphism $\tau : R[x] \to R$ satisfying the indicated conditions and consider any polynomial $f(x) = a_0 + a_1x + \cdots + a_nx^n \in R[x]$. By assumption, $\tau(a_k) = a_k$ for each coefficient a_k, while $\tau(x^k) = \tau(x)^k = r^k$. Taking stock of the fact that τ is a homomorphism,

$$\begin{aligned}\tau(f(x)) &= \tau(a_0) + \tau(a_1)\tau(x) + \cdots + \tau(a_n)\tau(x)^n \\ &= a_0 + a_1 r + \cdots + a_n r^n = f(r) = \phi_r(f(x)).\end{aligned}$$

This proves that $\tau = \phi_r$, yielding the uniqueness conclusion.

Without some commutativity assumption the above remarks need not hold. If we let

$$h(x) = (x - a)(x - b) = x^2 - (a + b)x + ab,$$

then

$$h(r) = r^2 - (a + b)r + ab.$$

Lacking the hypothesis that $r \in \text{cent } R'$, we cannot conclude that

$$(r - a)(r - b) = r^2 - ar - rb + ab$$

will equal $h(r)$; in other words, $h(x) = f(x)g(x)$ does not always imply that $h(r) = f(r)g(r)$.

Whenever $f(r) = 0$, we call the element r a *root* or *zero* of the polynomial $f(x)$. Of course, a given polynomial $f(x) \in R[x]$ might not have a root in R; we shall see later that when R is a field, an extension field R' of R always exists in which $f(x)$ possesses a root. It is perhaps appropriate to point out at this time that the problem of obtaining all roots of a polynomial $f(x) \in R[x]$ is equivalent to that of finding all elements $r \in R'$ for which $f(x) \in \ker \phi_r$.

After this brief digression, let us now state and prove the Division Algorithm for polynomials.

THEOREM 4.60 (Division Algorithm) *Let R be a commutative ring with identity and $f(x)$, $g(x) \neq 0$ be polynomials in $R[x]$, with the leading coefficient of $g(x)$ an invertible element. Then unique polynomials $q(x)$, $r(x) \in R[x]$ exist such that*

$$f(x) = q(x)g(x) + r(x),$$

where either $r(x) = 0$ or $\deg r(x) < \deg g(x)$.

Proof. The proof is by induction on the degree of $f(x)$. First, note that if $f(x) = 0$ or $f(x) \neq 0$ and $\deg f(x) < \deg g(x)$, a representation meeting the requirements of the theorem exists on taking $q(x) = 0$ and $r(x) = f(x)$. Furthermore, if $\deg g(x) = 0$, then $g(x)$ is an invertible element of the ring R, and it suffices to let $q(x) = f(x)g(x)^{-1}$ and $r(x) = 0$.

This being so, we may hereafter assume that $\deg f(x) = n$, $\deg g(x) = m$, where $n \geqslant m \geqslant 1$; that is,

$$f(x) = a_0 + a_1 x + \cdots + a_n x^n, \qquad a_n \neq 0,$$
$$g(x) = b_0 + b_1 x + \cdots + b_m x^m, \qquad b_m \neq 0, \qquad (n \geqslant m).$$

In the case in which $n = 1$, $f(x) = a_0 + a_1 x$ and $g(x) = b_0 + b_1 x$ are both linear polynomials. The requirements of the algorithm are then fulfilled by writing

$$f(x) = a_1 b_1^{-1}(b_0 + b_1 x) + (a_0 - a_1 b_1^{-1} b_0).$$

Now suppose that the theorem is true for all polynomials of degree less than n (the induction hypothesis). The polynomial

$$f_1(x) = f(x) - (a_n b_m^{-1})x^{n-m} g(x)$$

lies in $R[x]$ and, since the coefficient of x^n is $a_n - (a_n b_m^{-1})b_m = 0$, has degree less than n. By supposition there are polynomials $q_1(x)$, $r(x) \in R[x]$ such that

$$f_1(x) = q_1(x)g(x) + r(x),$$

where $r(x) = 0$ or $\deg r(x) < \deg g(x)$. Substituting, we obtain the equation

$$\begin{aligned} f(x) &= (q_1(x) + (a_n b_m^{-1})x^{n-m})g(x) + r(x) \\ &= q(x)g(x) + r(x), \end{aligned}$$

which shows that the desired representation also exists when $\deg f(x) = n$.

As for uniqueness, suppose that

$$f(x) = q(x)g(x) + r(x) = q'(x)g(x) + r'(x),$$

where $r(x)$ and $r'(x)$ satisfy the requirements of the theorem. Subtracting, we get

$$r(x) - r'(x) = (q'(x) - q(x))g(x).$$

Since the leading coefficient of $g(x)$ is invertible, it follows that $q'(x) - q(x) = 0$ if and only if $r(x) - r'(x) = 0$. With this in mind, let $q'(x) - q(x) \neq 0$. Knowing that b_m is not a zero divisor of R, we must have

$$\begin{aligned} \deg(q'(x) - q(x))g(x) &= \deg(q'(x) - q(x)) + \deg g(x) \\ &\geqslant \deg g(x) > \deg(r(x) - r'(x)), \end{aligned}$$

a contradiction; the last inequality relies on the fact that the degrees of $r(x)$ and $r'(x)$ are both less than the degree of $g(x)$. Thus $q'(x) = q(x)$, which in turn implies that $r'(x) = r(x)$.

The polynomials $q(x)$ and $r(x)$ appearing in the Division Algorithm are called, respectively, the *quotient* and *remainder* on dividing $f(x)$ by $g(x)$. In this connection it is important to observe that if $g(x)$ is a monic polynomial, or if R is taken to be a field, one need not assume that the leading coefficient of $g(x)$ is invertible.

We now come to a series of theorems concerning the factorization properties of $R[x]$.

THEOREM 4.61 (Remainder Theorem) *Let R be a commutative ring with identity. If $f(x) \in R[x]$ and $a \in R$, then there exists a unique polynomial $q(x)$ in $R[x]$ such that $f(x) = (x - a)q(x) + f(a)$.*

Proof. All this is scarcely more than an application of the Division Algorithm to the polynomials $f(x)$ and $x - a$. We then obtain

$$f(x) = (x - a)q(x) + r(x),$$

where $r(x) = 0$ or $\deg r(x) < \deg(x - a) = 1$. It follows in either case that $r(x)$ is a constant polynomial, say $r(x) = r \in R$. Substitution of a for x leads to

$$f(a) = (a - a)q(a) + r = r,$$

as desired.

COROLLARY *The polynomial $f(x) \in R[x]$ is divisible by $x - a$ if and only if a is a root of $f(x)$.*

As a member of $Z_6[x]$, the linear polynomial $3x$ has three roots, namely 0, 2, and 4; that is, it has more roots than its degree. This situation cannot happen when the ring of coefficients is an integral domain.

THEOREM 4.62 *Let R be an integral domain and $f(x) \in R[x]$ be a nonzero polynomial of degree n. Then $f(x)$ can have at most n distinct roots in R.*

Proof. The argument proceeds by induction on the degree of $f(x)$. When $\deg f(x) = 0$, the result is trivial, since $f(x)$ cannot have any roots. If $\deg f(x) = 1$, for instance, $f(x) = ax + b$ $(a \neq 0)$, then $f(x)$ has at most one root; indeed, if a is an invertible element, it follows that $-a^{-1}b$ is the only root of $f(x)$. Now suppose that the theorem is true for all polynomials of degree $n - 1$ and let $\deg f(x) = n$. If $f(x)$ has a root a, the preceding corollary gives $f(x) = (x - a)q(x)$, where the polynomial $q(x)$ has degree $n - 1$. Any root a' of $f(x)$ distinct from a must necessarily be a root of $q(x)$ because by substitution,

$$0 = f(a') = (a - a')q(a')$$

and, since R has no zero divisors, $q(a') = 0$. From our induction hypothesis, $q(x)$ has at most $n - 1$ distinct roots. Since the only roots of $f(x)$ are a and those of $q(x)$, $f(x)$ cannot possess more than n distinct roots in R.

With this step forward we can establish

COROLLARY 1 *Let $f(x)$ and $g(x)$ be two nonzero polynomials of degree n over the integral domain R. If there exist $n + 1$ distinct elements $a_k \in R$ ($k = 1, 2, \ldots, n + 1$) such that $f(a_k) = g(a_k)$, then $f(x) = g(x)$.*

Proof. The polynomial $h(x) = f(x) - g(x)$ is such that $\deg h(x) \leqslant n$ and, by supposition, has at least $n + 1$ distinct roots in R. This is impossible unless $h(x) = 0$, whence $f(x) = g(x)$.

COROLLARY 2 *Let $f(x) \in R[x]$, where R is an integral domain, and let S be any infinite subset of R. If $f(a) = 0$ for all $a \in S$, then $f(x)$ is the zero polynomial.*

Example 4.39
Consider the polynomial $x^p - x \in Z_p[x]$, where p is a prime number. Now the nonzero elements of Z_p form a commutative group under multiplication of order $p - 1$. Hence we have $a^{p-1} = 1$, or $a^p = a$ for every $a \neq 0$. This is equally true if $a = 0$. Our example shows that it may very well happen that every element of the underlying ring is a root of a polynomial, yet the polynomial is not zero.

With the Division Algorithm at our disposal we can prove that the ring $F[x]$ is rich in structure.

THEOREM 4.63 *The polynomial ring $F[x]$, where F is a field, forms a Euclidean domain.*

Proof. As has been noted, $F[x]$ is an integral domain. Moreover, the function δ defined by $\delta(0) = 0$ and

$$\delta(f(x)) = 2^{\deg f(x)} \qquad \text{if } f(x) \neq 0$$

is a suitable Euclidean valuation. Only condition (3) of Definition 4.23 fails to be immediate. But if $f(x)$ and $g(x) \neq 0$ are two polynomials in $F[x]$, then there exist $q(x), r(x) \in F[x]$ such that $f(x) = q(x)g(x) + r(x)$, where either $r(x) = 0$ or $\deg r(x) < \deg g(x)$. This means that $\delta(r(x)) = 0$ or $2^{\deg r(x)} < 2^{\deg g(x)}$, whence $\delta(r(x)) < \delta(g(x))$. Thus the function δ satisfies the requisite properties for a Euclidean valuation on $F[x]$.

COROLLARY *$F[x]$ is a principal ideal domain and hence a unique factorization domain.*

For a less existential proof of the fact that $F[x]$ is a principal ideal domain and a considerably more precise description of its ideals, one can repeat—with appropriate modifications—the pedestrian argument used in Theorem 4.12. It will appear that any nontrivial ideal I of $F[x]$ is of the form $I = (f(x))$, where $f(x)$ is a nonzero polynomial of minimal degree in I.

Since a field is trivially a unique factorization domain, part of the last corollary could be regarded as a special case of the next theorem.

THEOREM 4.64 *If R is a unique factorization domain, then so is $R[x]$.*

Proof. Suppose that $R[x]$ is not a unique factorization domain and let S be the set of all nonconstant polynomials in $R[x]$ that do not have a unique factorization into irreducible elements. Select $f(x) \in S$ to be of minimal degree. We may assume that

$$f(x) = p_1(x)p_2(x)\cdots p_r(x) = q_1(x)q_2(x)\cdots q_s(x),$$

where the $p_i(x)$ and $q_j(x)$ are all irreducible and

$$m = \deg p_1(x) \geqslant \deg p_2(x) \geqslant \cdots \geqslant \deg p_r(x),$$

$$n = \deg q_1(x) \geqslant \deg q_2(x) \geqslant \cdots \geqslant \deg q_s(x),$$

with $n \geqslant m > 0$. It is further evident that no $p_i(x) = uq_j(x)$ for any invertible element u. Otherwise, the polynomial obtained on dividing $f(x)$ by $q_j(x)$ will have unique factorization; this implies that $f(x)$ can also be factored uniquely. Let a, b be the leading coefficients of $p_1(x)$, $q_1(x)$, respectively, and define

$$g(x) = af(x) - bp_1(x)x^{n-m}q_2(x)\cdots q_s(x).$$

On one hand, we have

$$\begin{aligned} g(x) &= ap_1(x)p_2(x)\cdots p_r(x) - bp_1(x)x^{n-m}q_2(x)\cdots q_s(x) \\ &= p_1(x)(ap_2(x)\cdots p_r(x) - bx^{n-m}q_2(x)\cdots q_s(x)), \end{aligned}$$

and, on the other hand,

$$\begin{aligned} g(x) &= aq_1(x)q_2(x)\cdots q_s(x) - bp_1(x)x^{n-m}q_2(x)\cdots q_s(x) \\ &= (aq_1(x) - bp_1(x)x^{n-m})q_2(x)\cdots q_s(x). \end{aligned}$$

Now either $g(x) = 0$, which forces $aq_1(x) = bp_1(x)x^{n-m}$, or else $\deg g(x) < \deg f(x)$. In the latter event, $g(x)$ must possess a unique factorization into irreducibles, some of which are $q_2(x), \ldots, q_s(x)$ and

$p_1(x)$. The net result of this is that $p_1(x)|g(x)$, but $p_1(x) \nmid q_i(x)$ for $i > 1$, so that

$$p_1(x)|(aq_1(x) - bp_1(x)x^{n-m}),$$

and therefore $p_1(x)|aq_1(x)$. In either of the two cases considered, we are able to conclude that $p_1(x)$ divides the product $aq_1(x)$; this being so, $aq_1(x) = p_1(x)h(x)$ for some polynomial $h(x) \in R[x]$. Since R is taken to be a unique factorization domain, a has a unique factorization as a product of irreducible elements of R—hence of $R[x]$—say, $a = c_1c_2 \cdots c_k$, where each c_i is irreducible in $R[x]$. (The only factorizations of a as an element of $R[x]$ are those it had as an element of R.) Arguing from the representation

$$c_1c_2 \cdots c_kq_1(x) = p_1(x)h(x),$$

with $p_1(x)$ an irreducible, it follows that each c_i and, in consequence, the element a divides $h(x)$. But then

$$aq_1(x) = p_1(x)ah_1(x)$$

for some $h_1(x)$ in $R[x]$ or, upon canceling, $q_1(x) = p_1(x)h_1(x)$; in other words, $p_1(x)|q_1(x)$. Using the irreducibility of $q_1(x)$ as a member of $R[x]$, $p_1(x)$ must be an associate of $q_1(x)$. However, this conflicts with our original assumptions. Thus we see that $R[x]$ is indeed a unique factorization domain.

Coming back to the corollary to Theorem 4.63, there is an interesting converse that deserves mention, namely, that if R is an integral domain such that the polynomial ring $R[x]$ forms a principal ideal domain, then R is necessarily a field. In verifying this the main point to be proved is that any nonzero element $a \in R$ is invertible in R. By virtue of our hypothesis the ideal generated by x and a must be principal; for instance,

$$(x, a) = (f(x)), \qquad 0 \neq f(x) \in R[x].$$

Since both $x, a \in (f(x))$, it follows that

$$a = g(x)f(x), \qquad \text{and} \qquad x = h(x)f(x)$$

for suitable $g(x)$, $h(x)$ in $R[x]$. The first of these relations signifies that $\deg f(x) = 0$, say $f(x) = a_0$, and as a result, $\deg h(x) = 1$, say $h(x) = b_0 + b_1x$ ($b_1 \neq 0$). We thus obtain $x = a_0(b_0 + b_1x)$. But this means that the product $a_0b_1 = 1$, thereby making a_0 (or, equivalently, $f(x)$) an invertible element of R. The implication is that the ideal (x, a) is the entire ring $R[x]$. It is therefore possible to write the identity element in the form

$$1 = xk_1(x) + ak_2(x),$$

with the two polynomials $k_1(x), k_2(x) \in R[x]$. This can happen only if $ac_0 = 1$, where $c_0 \neq 0$ is the constant term of $k_2(x)$. In consequence the element a has a multiplicative inverse in R, which settles the whole affair.

At the heart of all the interesting questions on factorization in $R[x]$ lies the idea of an irreducible polynomial, which we formulate in a rather general way as follows:

DEFINITION 4.26 Let R be an integral domain. A nonconstant polynomial $f(x) \in R[x]$ is said to be *irreducible over* R, or is an *irreducible polynomial* in $R[x]$, if $f(x)$ cannot be expressed as the product of two polynomials (in $R[x]$) of positive degree. Otherwise, $f(x)$ is termed *reducible* in $R[x]$.

In the case of the principal ideal domain $F[x]$, where F is a field, the irreducible polynomials are precisely the irreducible elements of $F[x]$ (recall that the invertible elements of the polynomial ring $F[x]$ are just the nonzero constant polynomials); by Theorem 4.47 these coincide with the prime elements of $F[x]$. Of the equivalent notions, irreducible polynomial, irreducible element, and prime element, the term "irreducible polynomial" is the one customarily preferred for $F[x]$.

Perhaps we should emphasize that Definition 4.26 applies only to polynomials of positive degree; the constant polynomials are neither reducible nor irreducible. Thus the factorization theory of $F[x]$ concerns only polynomials of degree ≥ 1.

The dependence of Definition 4.26 on the polynomial domain $R[x]$ is essential. It may very well happen that a given polynomial is irreducible when viewed as an element of one domain, yet is reducible in another. One such example is the polynomial $x^2 + 1$; it is irreducible in $R^{\#}[x]$ but reducible in both $C[x]$, where we have $x^2 + 1 = (x + i)(x - i)$, and $Z_2[x]$, where $x^2 + 1 = (x + 1)(x + 1)$. Thus to ask merely whether a polynomial is irreducible, without specifying the coefficient ring involved, is incomplete and meaningless.

More often than not, it is a formidable task to decide when a given polynomial is irreducible over a specific ring. If F is a finite field, say one of the fields of integers modulo a prime, we may actually examine all of the possible roots. To cite a simple illustration, the polynomial $f(x) = x^3 + x + 1$ is irreducible in $Z_2[x]$. If there are any factors of this polynomial, at least one must be linear. But the only possible roots for $f(x)$ are 0 and 1, yet $f(0) = f(1) = 1 \neq 0$, showing that no roots exist in Z_2. However, one should proceed cautiously; the polynomial $x^4 + x^2 + 1$ is reducible in $Z_2[x]$, even though it has no root in Z_2; indeed, $x^4 + x^2 + 1 = (x^2 + x + 1)^2$.

Example 4.40
Any *linear* polynomial $ax + b$, $a \neq 0$, is irreducible in $R[x]$, where R is an integral domain. Indeed, since the degree of a product of two polynomials is the sum of the degree of the factors, it follows that a representation

$$ax + b = g(x)h(x), \qquad g(x), h(x) \in R[x],$$

with $1 \leqslant \deg g(x)$, $1 \leqslant \deg h(x)$ is impossible. This signifies that every reducible polynomial has degree 2 at least.

Example 4.41
The polynomial $x^2 - 2$ is irreducible in $Q[x]$, where Q as usual is the field of rational numbers. In the contrary case we would have

$$\begin{aligned} x^2 - 2 &= (ax + b)(cx + d) \\ &= (ac)x^2 + (ad + bc)x + bd, \end{aligned}$$

where the coefficients $a, b, c, d \in Q$. Accordingly,

$$ac = 1, \qquad ad + bc = 0, \qquad bd = -2,$$

whence $c = 1/a$, $d = -2/b$. Substituting in the relation $ad + bc = 0$, we obtain

$$0 = -2a/b + b/a = (-2a^2 + b^2)/ab.$$

Thus $-2a^2 + b^2 = 0$, or $(b/a)^2 = 2$, which is impossible because $\sqrt{2}$ is not a rational number. Although irreducible in $Q[x]$, the polynomial $x^2 - 2$ is nonetheless reducible in $R^{\#}[x]$; in this case, $x^2 - 2 = (x - \sqrt{2})(x + \sqrt{2})$, and both factors are in $R^{\#}[x]$.

For ease of reference more than to present new concepts, let us summarize in the next theorem some of the results of previous sections (specifically, Theorems 4.35 and 4.49) as applied to the principal ideal domain $F[x]$.

THEOREM 4.65 *If F is a field, the following statements are equivalent:*

a) $f(x)$ is an irreducible polynomial in $F[x]$;
b) the principal ideal $(f(x))$ is a maximal (prime) ideal of $F[x]$;
c) the quotient ring $F[x]/(f(x))$ forms a field.

Since part (c) of Theorem 4.65 is crucial to our later development, it is perhaps best to give an independent proof that $F[x]/(f(x))$ is a field for any irreducible polynomial $f(x) \in F[x]$. We already know that $F[x]/(f(x))$ is a commutative ring with identity, so all that remains is to show that every nonzero element has a multiplicative inverse. For this, let $g(x) + (f(x)) \neq 0 + (f(x))$; then $g(x) \notin (f(x))$. It is easily verified that the set

$$I = \{g(x)r(x) + f(x)s(x) | r(x), s(x) \in F[x]\}$$

forms a nonzero ideal of $F[x]$ containing both $g(x)$ and $f(x)$. By the corollary to Theorem 4.63, I is generated by a unique monic polynomial $h(x)$.

Noting that $f(x) \in I$, we have $f(x) = h(x)d(x)$ for a suitable $d(x) \in F[x]$. The irreducibility of $f(x)$ implies that either $h(x)$ or $d(x)$ is a constant. If $d(x) = d \in F$, then $h(x) = d^{-1}f(x)$, whence $h(x) \in (f(x))$, which is to say that $I \subseteq (f(x))$. But this is impossible, since it leads to the conclusion that $g(x) \in (f(x))$. Accordingly, $h(x)$ is a nonzero constant and therefore invertible. In light of Theorem 4.10 the ideal $I = F[x]$.

Because $1 \in F[x] = I$, we can obtain polynomials $r(x)$ and $s(x)$ in $F[x]$ satisfying $1 = g(x)r(x) + f(x)s(x)$. Now $1 - g(x)r(x) \in (f(x))$ and so

$$\begin{aligned} 1 + (f(x)) &= g(x)r(x) + (f(x)) \\ &= (g(x) + (f(x)))(r(x) + (f(x))). \end{aligned}$$

As a result, $r(x) + (f(x)) = (g(x) + (f(x)))^{-1}$, and we have proved that $F[x]/(f(x))$ is a field.

The theorem on prime factorization of polynomials is stated now.

THEOREM 4.66 (Unique Factorization in $F[x]$) *Each polynomial $f(x) \in F[x]$ of positive degree is the product of a nonzero element of F and irreducible monic polynomials of $F[x]$. Apart from the order of the factors, this factorization is unique.*

Suffice it to say that Theorem 4.66 can be made more explicit for particular polynomial domains. When we deal with polynomials over the complex numbers, the crucial tool is a result known as the *Fundamental Theorem of Algebra*. The name appears to have originated with Gauss at a time when algebra was seen as the theory of equations.

THEOREM 4.67 (Fundamental Theorem of Algebra) *Let C be the field of complex numbers. If $f(x) \in C[x]$ is a polynomial of positive degree, then $f(x)$ has at least one root in C.*

This theorem was surmised as early as 1629 by Albert Girard (1590–1633), who, having examined certain equations of low degree, wrote that "every equation in algebra has as many solutions as the denomination exponent of its highest term indicates." Proofs of Girard's conjecture were attempted in the eighteenth century. The noted French mathematician Jean D'Alembert (1717–1783) published the first proof in 1746, but his argument was far from convincing. It was from this effort that the theorem became known, especially in France, as D'Alembert's Theorem. D'Alembert did not even establish the existence of a root, but only showed the form a root must take; this necessitated a lengthy discussion of complex numbers, something fairly new at the time. Other defective proofs were put forward by Euler (1749), Lagrange (1772), and Laplace (in lectures at the Ecole Polytechnique in 1795).

The first rigorous proof of the Fundamental Theorem was given by a young man in his doctoral dissertation. Carl Friedrich Gauss was only 22 years old when he published his thesis from Helmstadt under the title "A New Proof That Every Rational Integral Function of One Variable Can Be Resolved into Real Factors of the First or Second Degree" (1799). Perhaps the only thing wrong with this landmark of mathematics is the word "new" in the title; his was the first satisfactory demonstration, not an additional proof added to those already known. Gauss devised three other proofs, in the years 1815, 1816, and 1849, the last at the time of the celebration of the golden jubilee of his doctorate. In the first three of these proofs the assumption is made that the coefficients of the polynomial are real numbers; in his fourth demonstration, Gauss at last operated freely with complex numbers, since as he then stated, they "are nowadays apparent to all."

Although many proofs of the result are available, none are strictly algebraic in nature; thus we shall assume the validity of Theorem 4.67 without proof. The reader will experience little difficulty, however, in establishing the following corollary.

COROLLARY 1 *If $f(x) \in C[x]$ is a polynomial of degree $n > 0$, then $f(x)$ can be expressed in $C[x]$ as a product of n (not necessarily distinct) linear factors.*

Another way of stating the above corollary is that the only irreducible polynomials in $C[x]$ are the linear polynomials. Directing our attention now to the real field, we can obtain the form of the prime factorization in $R^\#[x]$ (bear in mind that polynomials with coefficients from $R^\#$ are polynomials in $C[x]$ and therefore have roots in C).

COROLLARY 2 *If $f(x) \in R^\#[x]$ is of positive degree, then $f(x)$ can be factored into linear and irreducible quadratic factors.*

Proof. Since $f(x)$ also belongs to $C[x]$, $f(x)$ factors in $C[x]$ into a product of linear polynomials $x - c_k$, $c_k \in C$. If $c_k \in R^\#$, then $x - c_k \in R^\#[x]$. Otherwise, $c_k = a + bi$, where $a, b \in R^\#$ and $b \neq 0$. But the complex roots of real polynomials occur in conjugate pairs (see Problem 18 at the end of this section), so $\bar{c}_k = a - bi$ is also a root of $f(x)$. Thus

$$(x - c_k)(x - \bar{c}_k) = x^2 - 2ax + (a^2 + b^2) \in R^\#[x]$$

is a factor of $f(x)$. The quadratic polynomial $x^2 - 2ax + (a^2 + b^2)$ is irreducible in $R^{\#}[x]$, since any factorization in $R^{\#}[x]$ is also valid in $C[x]$ and $(x - c_k)(x - \bar{c}_k)$ is its unique factorization in $C[x]$.

An interesting remark, to be recorded without proof, is that if F is a finite field, the polynomial ring $F[x]$ contains irreducible polynomials of every degree (see Theorem 5.35).

This may be a convenient place to introduce the notion of a primitive polynomial.

DEFINITION 4.27 Let R be a unique factorization domain. The *content* of a nonconstant polynomial $f(x) = a_0 + a_1x + \cdots + a_nx^n \in R[x]$, denoted by the symbol cont $f(x)$, is defined to be the greatest common divisor of its coefficients:

$$\text{cont } f(x) = \gcd(a_0, a_1, \ldots, a_n).$$

We call $f(x)$ a *primitive polynomial* if cont $f(x) = 1$.

Viewed otherwise, Definition 4.27 asserts that a polynomial $f(x) \in R[x]$ is primitive if and only if there is no prime element of R that divides all of its coefficients. In this connection it may be noted that in the domain $F[x]$ of polynomials with coefficients from a field F, every nonconstant polynomial is primitive; indeed, there are no primes in F. The reader should also take care to remember that the notion of greatest common divisor and, in consequence, the content of a polynomial are not determined uniquely, but determined only to within associates.

Given a polynomial $f(x) \in R[x]$ of positive degree, it is possible to write $f(x) = cf_1(x)$, where $c \in R$ and $f_1(x)$ is primitive; simply let $c = \text{cont } f(x)$. To a certain extent this reduces the question of factorization in $R[x]$—at least when R is a unique factorization domain—to that of primitive polynomials. By way of specific illustrations we observe that $f(x) = 3x^3 - 4x + 35$ is a primitive polynomial in $Z[x]$, while $g(x) = 12x^2 + 6x - 3 = 3(4x^2 + 2x - 1)$ is not a primitive element of the same, since $g(x)$ has content 3.

Here is another new concept: Suppose that I is a (proper) ideal of R, a commutative ring with identity. There is an obvious mapping $v: R[x] \to (R/I)[x]$; for any polynomial $f(x) \in R[x]$, simply apply nat_I to the coefficients of $f(x)$ so that

$$v(f(x)) = (a_0 + I) + (a_1 + I)x + \cdots + (a_n + I)x^n,$$

or, more briefly, $v(f(x)) = \sum (\text{nat}_I a_k)x^k$. The reader will encounter no difficulty in verifying that v, defined in this way, is a homomorphism of $R[x]$ onto $(R/I)[x]$, the so-called *reduction homomorphism modulo I*. The polynomial $v(f(x))$ is said to be the *reduction of $f(x)$ modulo I*.

Although it might seem to be rather special, the reduction homomorphism will serve us in good stead on several occasions. We make immediate use of it to characterize primitive polynomials.

THEOREM 4.68 *Let R be a unique factorization domain and let $f(x) = a_0 + a_1x + \cdots + a_nx^n \in R[x]$, with $\deg f(x) > 0$. Then $f(x)$ is a primitive polynomial in $R[x]$ if and only if, for each prime element $p \in R$, the reduction of $f(x)$ modulo the principal ideal (p) is nonzero.*

Proof. By definition the reduction of $f(x)$ modulo (p) is

$$v(f(x)) = (a_0 + (p)) + (a_1 + (p))x + \cdots + (a_n + (p))x^n.$$

Thus to say that $v(f(x)) = 0$ for some prime $p \in R$ is equivalent to asserting that $a_k \in (p)$, or rather, $p|a_k$ for all k. But the latter condition signifies that cont $f(x) \neq 1$; hence $f(x)$ is not primitive.

One of the most crucial facts concerning primitive polynomials is Gauss' Lemma, which we prove next.

THEOREM 4.69 (Gauss' Lemma) *Let R be a unique factorization domain. If $f(x)$, $g(x)$ are both primitive polynomials in $R[x]$, then their product $f(x)g(x)$ is also primitive in $R[x]$.*

Proof. Given a prime element $p \in R$, (p) is a prime ideal of R, whence the quotient ring $R' = R/(p)$ forms an integral domain. We next consider the reduction homomorphism v modulo the principal ideal (p). Since $R'[x]$ is an integral domain, it follows that the reduction of $f(x)g(x)$ cannot be the zero polynomial:

$$v(f(x)g(x)) = v(f(x))v(g(x)) \neq 0.$$

The assertion of the theorem is now a direct consequence of our last result.

COROLLARY 1 *If R is a unique factorization domain and $f(x)$, $g(x) \in R[x]$, then*

$$\text{cont}(f(x)g(x)) = \text{cont } f(x) \text{ cont } g(x).$$

Proof. As was noted earlier, we can write $f(x) = af_1(x)$, $g(x) = bg_1(x)$, where $a = \text{cont } f(x)$, $b = \text{cont } g(x)$ and where $f_1(x)$, $g_1(x)$ are primitive in $R[x]$. Therefore $f(x)g(x) = abf_1(x)g_1(x)$. According to the theorem, the product $f_1(x)g_1(x)$ is a primitive polynomial of $R[x]$. This means that the content of $f(x)g(x)$ is simply ab or, what amounts to the same thing, cont $f(x)$ cont $g(x)$.

The next corollary, aside from its importance in Galois theory, has an interest all its own.

COROLLARY 2 *If a polynomial $f(x) \in Z[x]$ is reducible in $Q[x]$, then it is reducible in $Z[x]$.*

Proof. Suppose that $f(x) = u(x)v(x)$, where $u(x)$ and $v(x)$ are in $Q[x]$. Let a and b denote the least common multiples of the denominators of the rational coefficients of $u(x)$ and $v(x)$, respectively. Then $g(x) = au(x)$, $h(x) = bv(x)$ have integral coefficients and satisfy $abf(x) = g(x)h(x)$. If $a_1 = \text{cont } g(x)$ and $b_1 = \text{cont } h(x)$, Corollary 1 gives us $ab \text{ cont } f(x) = a_1 b_1$. It follows that $c = (ab)^{-1}a_1 b_1$ is an integer, being the content of a polynomial in $Z[x]$. Now write $g(x) = a_1 g_1(x)$ and $h(x) = b_1 h_1(x)$, with $g_1(x)$ and $h_1(x)$ primitive polynomials in $Z[x]$. Putting these relations together, we arrive at

$$f(x) = (ab)^{-1}g(x)h(x) = ((ab)^{-1}a_1 b_1 g_1(x))h_1(x),$$

which is the required factorization of $f(x)$.

Any unique factorization domain R, being an integral domain, possesses a field of quotients $K = Q_{cl}(R)$, and we may consider the ring of polynomials $R[x]$ as imbedded in the polynomial ring $K[x]$. The next theorem deals with the relation between the irreducibility of a polynomial in $R[x]$ and its irreducibility when considered as an element of the larger ring $K[x]$. The classic example of this situation is, of course, the polynomial domain $Z[x] \subseteq Q[x]$. Before concentrating our efforts on this relationship, we require a preliminary lemma.

LEMMA *Let R be a unique factorization domain, with field of quotients K. Given a nonconstant polynomial $f(x) \in K[x]$, (nonzero) elements $a, b \in R$ and a primitive polynomial $f_1(x)$ in $R[x]$ exist such that*

$$f(x) = ab^{-1}f_1(x).$$

Furthermore, $f_1(x)$ is unique up to invertible elements of R as factors.

Proof. Inasmuch as K is the quotient field of R, $f(x)$ can be written in the form

$$f(x) = (a_0 b_0^{-1}) + (a_1 b_1^{-1})x + \cdots + (a_n b_n^{-1})x^n,$$

where $a_i, b_i \in R$ and $b_i \neq 0$. Take b to be any common multiple of the b_i; for instance, $b = b_0 b_1 \cdots b_n$. Then $b \neq 0$, and, since the coefficients of $bf(x)$ all lie in R, we have $bf(x) = g(x) \in R[x]$. Accordingly,

$$f(x) = b^{-1}g(x) = ab^{-1}f_1(x),$$

where $f_1(x) \in R[x]$ is a primitive polynomial and $a = \text{cont}\, g(x)$. We emphasize that $f_1(x)$ is of the same degree as $f(x)$, so it cannot be invertible in $R[x]$.

As for uniqueness, suppose that $f(x) = ab^{-1}f_1(x) = cd^{-1}f_2(x)$ are two representations that satisfy the conditions of the theorem. Then

$$adf_1(x) = bcf_2(x).$$

Since $f_1(x)$ and $f_2(x)$ are both primitive, the corollary to Gauss' Lemma implies that we must have $ad = ubc$ for some invertible element $u \in R$. In consequence, $f_1(x) = uf_2(x)$, showing that $f_1(x)$ is unique to within invertible factors in R.

THEOREM 4.70 *Let R be a unique factorization domain, with field of quotients K. If $f(x) \in R[x]$ is an irreducible primitive polynomial, then it is also irreducible as an element of $K[x]$.*

Proof. Assume to the contrary that $f(x)$ is reducible over K. Then $f(x) = g(x)h(x)$, where the polynomials $g(x)$, $h(x)$ are in $K[x]$ and are of positive degree. By virtue of the lemma just proved,

$$g(x) = ab^{-1}g_1(x), \qquad h(x) = cd^{-1}h_1(x),$$

with $a, b, c, d \in R$ and $g_1(x)$, $h_1(x)$ primitive in $R[x]$. Thus

$$bdf(x) = acg_1(x)h_1(x).$$

Now Gauss' Lemma asserts that the product $g_1(x)h_1(x)$ is a primitive polynomial in $R[x]$, whence $f(x)$ and $g_1(x)h_1(x)$ differ by an invertible element of R:

$$f(x) = ug_1(x)h_1(x).$$

Since $\deg g_1(x) = \deg g(x) > 0$, $\deg h_1(x) = \deg h(x) > 0$, the outcome is a nontrivial factorization of $f(x)$ in $R[x]$, contrary to hypothesis.

There is an obvious converse to Theorem 4.70, namely: If the primitive polynomial $f(x) \in R[x]$ is irreducible as an element of $K[x]$, it is also irreducible in $R[x]$. This is justified by the fact that $R[x]$ (or an isomorphic copy thereof) appears naturally as a subring of $K[x]$; thus if $f(x)$ were reducible in $R[x]$, it would obviously be reducible in the larger ring $K[x]$. Our remarks lead to the following conclusion: Given a primitive polynomial $f(x) \in R[x]$, R a unique factorization domain, $f(x)$ is irreducible in $R[x]$ if and only if $f(x)$ is irreducible in $K[x]$.

Our next concern is a generalization of a famous theorem of Eisenstein dealing with the problem of irreducibility. This result is of fundamental importance in the classical theory of polynomials with integral coefficients. The generalization that we have in mind is formulated below.

THEOREM 4.71 *Let R be an integral domain and the nonconstant polynomial $f(x) = a_0 + a_1x + \cdots + a_nx^n \in R[x]$. Suppose that there exists a prime ideal P of R such that*

a) $a_n \notin P$,
b) $a_k \in P$ *for* $0 \leqslant k < n$,
c) $a_0 \notin P^2$.

Then $f(x)$ *is irreducible in* $R[x]$.

Proof. Assume that, contrary to assertion, $f(x)$ is reducible in $R[x]$, say $f(x) = g(x)h(x)$ for polynomials $g(x), h(x) \in R[x]$, where

$$g(x) = b_0 + b_1x + \cdots + b_rx^r,$$

$$h(x) = c_0 + c_1x + \cdots + c_sx^s \qquad (r + s = n;\ r, s > 0).$$

Now consider the reduction of $f(x)$ modulo the ideal P. Invoking hypothesis (b), it can be inferred that

$$\nu(g(x))\nu(h(x)) = \nu(f(x)) = (a_n + P)x^n.$$

Since the polynomial ring $(R/P)[x]$ comprises an integral domain, the only possible factorizations of $(a_n + P)x^n$ are into linear factors. This being so, a moment's reflection shows that

$$\nu(g(x)) = (b_r + P)x^r, \qquad \nu(h(x)) = (c_s + P)x^s.$$

This means that the constant terms of these reductions are zero; that is,

$$b_0 + P = c_0 + P = P.$$

Altogether we have proved that both $b_0, c_0 \in P$, revealing at the same time that $a_0 = b_0c_0 \in P^2$, which is untenable by hypothesis (c). Accordingly, no such factorization of $f(x)$ can occur, and $f(x)$ is indeed irreducible in $R[x]$.

Theorem 4.71 leads almost immediately to the Eisenstein test for irreducibility.

COROLLARY (The Eisenstein Criterion) *Let R be a unique factorization domain and K be its field of quotients. Let $f(x) = a_0 + a_1x + \cdots + a_nx^n$ be a nonconstant polynomial in $R[x]$. Suppose further that for some prime $p \in R$, $p \nmid a_n$, $p | a_k$ for $0 \leqslant k < n$, and $p^2 \nmid a_0$. Then $f(x)$ is irreducible in $K[x]$.*

Proof. We already know that (p) is a prime ideal of R. Taking stock of the theorem, $f(x)$ is an irreducible polynomial of $R[x]$ and hence of $K[x]$ (at this point a direct appeal is made to Theorem 4.70).

This is probably a good time at which to look at some examples.

Example 4.42
Consider the monic polynomial

$$f(x) = x^n + a \in Z[x] \qquad (n > 1),$$

where $a \neq \pm 1$ is a nonzero square-free integer. For any prime p dividing a, p is certainly a factor of all the coefficients except the leading one, and our hypothesis ensures that $p^2 \nmid a$. Thus $f(x)$ fulfills Eisenstein's criterion and so is irreducible over Q. Incidentally, this example shows that there are irreducible polynomials in $Q[x]$ of every degree.

Note also that $x^4 + 4 = (x^2 + 2x + 2)(x^2 - 2x + 2)$; one should not expect Theorem 4.71 to lead to a decision in this case, since, of course, 4 fails to be a square-free integer.

Example 4.43
Eisenstein's test is not directly applicable to the *cyclotomic polynomial*

$$\Phi(x) = \frac{x^p - 1}{x - 1} = x^{p-1} + x^{p-2} + \cdots + x + 1 \in Z[x] \qquad (p \text{ prime}),$$

because no suitable prime is available. This problem is resolved by the observation that $\Phi(x)$ is irreducible in $Z[x]$ if and only if $\Phi(x + 1)$ is irreducible. A simple computation yields

$$\Phi(x + 1) = \frac{(x + 1)^p - 1}{(x + 1) - 1} = (x^p + \tbinom{p}{1}x^{p-1} + \cdots + px)/x$$

$$= x^{p-1} + \tbinom{p}{1}x^{p-2} + \cdots + p \in Z[x].$$

If the Eisenstein criterion is now applied, it is easy to see that all the requirements for the irreducibility of $\Phi(x + 1)$ in $Z[x]$ are satisfied. In the binomial coefficient $\binom{p}{k} = p!/k!(p - k)!$ the numerator is divisible by p for $k < p$ but not the denominator. One finds in this way that the original cyclotomic polynomial $\Phi(x)$ must be irreducible in $Z[x]$ and hence also as a polynomial in $Q[x]$.

PROBLEMS 4.8

1. In the polynomial ring $Z_{12}[x]$, verify that
   a) $4x^2 + 2x + 6$ is a zero divisor;
   b) $4x + 9$ is an idempotent element;
   c) $6x^2$ is a nilpotent element;
   d) $6x + 5$ is invertible.

2. Show that there are invertible elements of every degree in $Z_4[x]$.

3. For any ring R, prove that the set of polynomials in $R[x]$ with the property that all odd powers of x have zero coefficients forms a subring, but not an ideal, of $R[x]$.

4. If R is a commutative ring with identity, establish the following:

a) no monic polynomial in $R[x]$ is a zero divisor;

b) $x - 1$ is a factor of a polynomial

$$f(x) = a_0 + a_1x + \cdots + a_nx^n \in R[x]$$

if and only if $a_0 + a_1 + \cdots + a_n = 0$;

c) the polynomial $1 + ax$ is invertible in $R[x]$ if and only if the element a is nilpotent in R. [*Hint:* See Problem 10, Section 4.2.]

5. Prove that the polynomial ring $Z_n[x]$ is infinite but has positive characteristic.

6. a) Suppose that p is a fixed prime number. In the ring $Z[x]$, define the set P by

$$P = \{a_0 + a_1x + \cdots + a_nx^n \in Z[x] | p \text{ divides } a_0\}.$$

Show that P is a prime ideal of $Z[x]$.

b) Determine whether $Z_e[x]$ forms an ideal of $Z[x]$.

7. Show that the polynomial $x^3 + 5x \in Z_6[x]$ has more roots than its degree.

8. a) Derive the Rational Root Theorem: If s/t, where $\gcd(s, t) = 1$, is a rational root of the polynomial

$$f(x) = a_0 + a_1x + \cdots + a_nx^n \in Z[x],$$

then $s | a_0$ and $t | a_n$.

b) Obtain the rational roots of the following polynomials in $Z[x]$:

$$f(x) = 2x^3 - 5x^2 - x + 6,$$

$$g(x) = x^3 - 4x^2 + x + 6,$$

$$h(x) = x^4 - 3x^2 + 2x + 1.$$

9. Prove the Integral Root Theorem: Any rational root of a monic polynomial $f(x) \in Z[x]$ must be an integral divisor of the constant term of $f(x)$.

10. Consider the substitution homomorphism $\phi_{\sqrt{2}}: Z[x] \to R^{\#}$ defined by $\phi_{\sqrt{2}}(f(x)) = f(\sqrt{2})$. Show that the kernel and image of $\phi_{\sqrt{2}}$ are $(x^2 - 2)$ and $Z(\sqrt{2})$, respectively; hence

$$Z[x]/(x^2 - 2) \simeq Z(\sqrt{2}).$$

11. a) In $Z_p[x]$, where p is a prime, prove that

$$x^{p-1} - 1 = (x-1)(x-2)\cdots(x-(p-1)).$$

[*Hint:* The polynomial $f(x) = x^p - x$ has p distinct roots in Z_p.]

b) Derive Wilson's Theorem: If p is a prime, then

$$(p-1)! \equiv -1 (\text{mod } p).$$

12. Given that R is an integral domain, establish that the only invertible elements of $R[x]$ are the constant polynomials determined by the invertible elements of R; in particular, the only invertible elements of $Z[x]$ are ± 1.

13. For an arbitrary ring R, prove that:

a) If I is an ideal of R, then $I[x]$ forms an ideal of the polynomial ring $R[x]$.

b) If R and R' are isomorphic rings, then $R[x]$ is isomorphic to $R'[x]$.

c) char R = char $R[x]$.

14. Establish the following assertions concerning the polynomial ring $Z[x]$:

a) The ideal

$$(x) = \{a_1 x + a_2 x^2 + \cdots + a_n x^n | a_k \in Z;\ n \geqslant 1\}$$

is a prime ideal of $Z[x]$, but not a maximal ideal. Incidentally, (x) is maximal in $F[x]$, where F is a field.

b) $Z[x]$ is not a principal ideal domain. [*Hint:* Consider $(x, 2)$, the (maximal) ideal of polynomials with even constant terms.]

15. Let P be a prime of R, a commutative ring with identity. Prove that $P[x]$ is a prime ideal of the polynomial ring $R[x]$. If M is a maximal ideal of R, is $M[x]$ a maximal ideal of $R[x]$?

16. Consider the polynomial domain $F[x]$, where F is a field, and a fixed element $r \in F$.

a) Show that the set of all polynomials having r as a root,

$$M_r = \{f(x) \in F[x] | f(r) = 0\},$$

forms a maximal ideal of $F[x]$, with $F[x]/M_r \simeq F$. [*Hint:* $M_r = \ker \phi_r$, where $\phi_r: F[x] \to F$ is the substitution homomorphism induced by r.]

b) Given that F is infinite, establish that $\bigcap_{r \in F} M_r = \{0\}$.

17. Regarding the ring of Example 4.9, show that the polynomial $(a, 0)x^2 \in R[x]$ has infinitely many roots in R.

18. Given $f(x) = a_0 + a_1 x + \cdots + a_n x^n \in C[x]$, define the polynomial $\bar{f}(x)$ by

$$\bar{f}(x) = \bar{a}_0 + \bar{a}_1 x + \cdots + \bar{a}_n x^n,$$

where $\bar{a}_k$ denotes the usual complex conjugate of a_k. Verify that

a) $r \in C$ is a root of $f(x)$ if and only if $\bar{r}$ is a root of $\bar{f}(x)$. [*Hint:* $\overline{f(r)} = \bar{f}(\bar{r})$.]

b) If $f(x) \in R^{\#}[x] \subseteq C[x]$ and r is a complex root of $f(x)$, then $\bar{r}$ is also a root of $f(x)$.

19. Let R be a commutative ring with identity and let $f(x) \in R[x]$. The function $\bar{f}: R \to R$ defined by taking $\bar{f}(r) = f(r)$ for every $r \in R$ is called the *polynomial function induced by* $f(x)$. Assuming that P_R denotes the set of all polynomial functions induced by elements of $R[x]$, prove that:

a) P_R forms a subring of map R, known as the ring of polynomial functions on R.

b) The mapping $\alpha: R[x] \to P_R$ given by $\alpha(f(x)) = \bar{f}$ is a homomorphism of $R[x]$ onto P_R.

c) If the element $r \in R$ is fixed and $I_r = \{\bar{f} \in P_R \mid \bar{f}(r) = 0\}$, then I_r is an ideal of P_R.

20. a) When R is an integral domain, show that distinct polynomials in $R[x]$ induce distinct polynomial functions (in other words, the mapping $\alpha: R[x] \to P_R$ is one-to-one) if and only if R has an infinite number of elements.

b) Give an example of two distinct polynomials that induce the same polynomial function.

21. Let R be a commutative ring with identity and define the function $\delta: R[x] \to R[x]$, the so-called *derivative function*, as follows. If

$$f(x) = a_0 + a_1 x + \cdots + a_n x^n \in R[x],$$

then

$$\delta f(x) = a_1 + 2a_2 x + \cdots + n a_n x^{n-1}.$$

For any $f(x), g(x) \in R[x]$ and any $r \in R$, establish that

a) $\delta(f(x) + g(x)) = \delta f(x) + \delta g(x)$;

b) $\delta(rf(x)) = r\delta f(x)$;

c) $\delta(f(x)g(x)) = \delta f(x) \cdot g(x) + f(x) \cdot \delta g(x)$. [*Hint:* Induct on the number of terms of $f(x)$.]

22. Suppose that R is a commutative ring with identity and let $r \in R$ be a root of the nonzero polynomial $f(x) \in R[x]$. We call r a *multiple root* of $f(x)$ provided that

$$f(x) = (x - r)^n g(x) \qquad (n > 1),$$

where $g(x) \in R[x]$ is a polynomial such that $g(r) \neq 0$. Prove that an element $r \in R$ is a multiple root of $f(x)$ if and only if r is a root of both $f(x)$ and $\delta f(x)$.

23. Let F be a field and $f(x) \in F[x]$ be a polynomial of degree 2 or 3. Deduce that $f(x)$ is irreducible in $F[x]$ if and only if it has no root in F. Give an example that shows that this result need not hold if $\deg f(x) \geqslant 4$.

24. Prove that if a polynomial $f(x) \in F[x]$ (where F is a field of characteristic 0) is irreducible, then all of its roots in any field containing F must be distinct. [*Hint:* First show that $\gcd(f(x), \delta f(x)) = 1$.]

25. Given that I is a proper ideal of R, a commutative ring with identity, establish the assertions below:

a) If $v: R[x] \to (R/I)[x]$ is the reduction homomorphism modulo the ideal I, then $\ker v = I[x]$; hence $R[x]/I[x] \simeq (R/I)[x]$.

b) If the polynomial $f(x) \in R[x]$ is such that $v(f(x))$ is irreducible in $(R/I)[x]$, then $f(x)$ is irreducible in $R[x]$.

c) The polynomial $f(x) = x^3 - x^2 + 1$ is irreducible in $Z[x]$. [*Hint:* Reduce the coefficients modulo 2.]

d) The two polynomials $g(x) = x^3 + (2a + 1)x + (2b + 1)$ and $h(x) = x^3 + (3a - 1)x + (3b + 1)$ are irreducible in $Z[x]$ for all $a, b \in Z$.

26. Prove that if $f(x)$ is an irreducible polynomial in $Z[x]$, then $f(x)$ is also irreducible when regarded as an element of $Q[x]$.

27. Factor $f(x) = 2x^5 - x^4 - 2x^3 + x^2 - 4x + 2$ into irreducible factors in each of the polynomial rings $Q[x]$, $R^{\#}[x]$, and $C[x]$. [*Hint:* Begin by using the Rational Root Theorem.]

28. For the polynomial ring $Z_2[x]$, establish that

a) a polynomial $f(x) \in Z_2[x]$ has $x + 1$ as a factor if and only if $f(x)$ has an even number of nonzero coefficients;

b) the only irreducible polynomials of degree 4 are $x^4 + x + 1$, $x^4 + x^3 + 1$, and $x^4 + x^3 + x^2 + x + 1$.

29. Apply Eisenstein's Criterion to confirm that the following polynomials are irreducible in $Q[x]$:

$$f(x) = x^2 + 1,$$

$$g(x) = x^2 - x + 1,$$

$$h(x) = 2x^5 - 6x^3 + 9x^2 - 15.$$

[*Hint:* Consider $f(x + 1)$ and $g(-x)$.]

30. Show that any polynomial $f(x) \in R^{\#}[x]$, with $\deg f(x) \geqslant 3$, is reducible.

31. Assume that F is a field and $\tau: F[x] \to F[x]$ is the mapping that sends each polynomial to the polynomial with coefficients reversed; that is,

$$\tau(a_0 + a_1 x + \cdots + a_n x^n) = a_n + a_{n-1} x + \cdots + a_0 x^n.$$

Prove that

a) τ is an automorphism of $F[x]$;

b) $r \in F - \{0\}$ is a root of $f(x) \in F[x]$ if and only if r^{-1} is a root of $\tau(f(x))$;

c) $f(x) \in F[x]$ is irreducible if and only if $\tau(f(x))$ is irreducible.

32. Show that the quotient rings $Q[x]/(x^2 + 1)$ and $Q[x]/(x^2 - 1)$ are not isomorphic.

33. Let R be an integral domain and $\mu: R[x] \to R$ be the mapping that sends each polynomial to the sum of its coefficients; that is,

$$\mu(a_0 + a_1 x + \cdots + a_n x^n) = a_0 + a_1 + \cdots + a_n.$$

Verify the assertions below:

a) μ is a homomorphism from $R[x]$ onto R.

b) The set

$$P = \{a_0 + a_1 x + \cdots + a_n x^n \in R[x] | a_0 + a_1 + \cdots + a_n = 0\}$$

forms a prime ideal of $R[x]$.

c) $P = (x - 1)$. [*Hint:* By the Remainder Theorem, any $f(x) \in R[x]$ can be written $f(x) = (x - 1)g(x) + f(1)$.]

d) $R[x]/(x - 1) \simeq R$.

34. For a ring R, show that the ring $M_n(R[x])$ of $n \times n$ matrices with entries from $R[x]$ is isomorphic to the polynomial ring $M_n(R)[x]$.

35. If F is a field, show that the polynomial ring $F[x]$ is semisimple. [*Hint:* Use Theorem 4.47.]

5

Field Theory

The mind of the mathematician goes forth in the universe of logic in quest of things that are there.

C. J. KEYSER

5.1 EVARISTE GALOIS

No episode in the history of mathematics is more moving than the life of Evariste Galois (1811–1832), the tormented French genius who died at the age of 20 from wounds received in a mysterious duel. During the short years of his life he left a body of work—for the most part published posthumously—that earned him a place among the leading creative mathematicians of all time.

The son of a village mayor, Galois was educated solely by his mother until he was twelve years old. In 1823 he entered the College Louis-le-Grand, a famous preparatory school in Paris. Galois was a good student during his first two years, but then his boredom with the uninspiring curriculum became apparent. Forced to repeat his third year because of poor work, he took advantage of the chance to enroll in a mathematics course. This initial contact with the subject was a revelation to Galois. Fascinated by the rigor of Legendre's *Géométrie*, he read the text with full understanding in one sitting, as one might a novel. The elementary textbooks on algebra did not satisfy him, so Galois learned his algebraic analysis directly from the great treatises of Lagrange; and then he turned to the original papers of Abel. A year early, he took the entrance examination for the Ecole Polytechnique, which was the premier institution of higher learning in mathematics. But he was rejected as being deficient in some of the basics. This was but a first embittering experience for him. In his last year at the preparatory school, 1828–1829, Galois attended a course of L. P. E. Richard, a gifted teacher of mathematics and one perceptive enough to recognize Galois's exceptional talent. Encouraged by Richard, the 17-year-old Galois published his first paper (1829), a small article on continued fractions; and he submitted some of his research on the relationship between algebraic equations and their associated groups to the Academy of Sciences.

The year 1829 was disastrous for Galois. Augustin-Louis Cauchy, a member of the Academy and a professor at the Ecole Polytechnique, was appointed to review Galois's paper; he either forgot or lost the communication, as well as another presented a week later. Then came the suicide of Galois's father after a series of violent political attacks directed toward him. Beyond this, Galois failed the competitive examination for entry into the Ecole Polytechnique for the second and final time.

Frustrated in every attempt to enter the Ecole Polytechnique, Galois enrolled in the less prestigious Ecole Normale, which trained future secondary school teachers. He submitted a new version of his investigations to the Academy, hoping to enter it in the competition for the Grand Prize in Mathematics, the pinnacle of mathematical honor. This time it was entrusted to the permanent secretary, Joseph Fourier (1768–1830), who died shortly thereafter without examining the manuscript. It was never retrieved from among his papers. A further disappointment was reserved for Galois. In January 1831 he submitted his paper for the third time under the title "*Une Mémoire sur les Conditions de Résolubilité des Equations par Radicaux.*" After a delay of some six months, during which Galois wrote to the president of the Academy asking what had happened, the paper was rejected by the referee Siméon Denis Poisson (1781–1840). At the conclusion of his report, Poisson remarked:

> His arguments are not sufficiently clear, nor developed enough for us to judge their correctness.... It is hoped that the author would publish his work in its entirety so that we can form a definite opinion.

In the wake of the July Revolution of 1830, which toppled Charles X from the throne and installed Louis Philippe, Galois threw himself into the political turmoil with his habitual fury. He published a blistering letter criticizing the director of the Ecole Normale for failing to support the Revolution; for this he was promptly expelled. No longer a student, Galois continued his mathematical research to a certain extent but devoted most of his effort to political activism. At a republican banquet he sarcastically proposed a toast to the monarchy of Louis Philippe while holding an open pocket-knife in his hand; he was arrested the next morning for threatening the life of the king. Though he was acquitted six weeks later when the jury rendered a verdict of "unproven," he was arrested again in less than a month. This time his arrest was a preventive measure; he was sentenced to six months' imprisonment on the technical charge of illegally wearing a uniform. Thus it happened that Galois spent ten months of his last year of life as a political prisoner.

Put on parole upon the announcement of a cholera epidemic in Paris, Galois was provoked to a duel in murky circumstances. (The theory has been advanced that the challenger was hired by the police, who arranged the confrontation to eliminate a person they considered a dangerous radical.) On the eve of the duel, apparently sure that he would die, Galois wrote a letter to a friend describing the contents of the memoir Poisson had rejected. Its seven pages, hastily written, contain a summary of the discoveries he had been unable to develop. The letter ends with the plea:

> Eventually there will be, I hope, some people who will find it profitable to decipher this mess.

Evariste Galois (1812–1832)

Galois spent the rest of the night annotating and making corrections to some of his papers; next to a theorem, he scrawled:

> There are a few things to be completed in this proof. I do not have time.

The duel took place early in the morning of May 30, 1832. Galois was grievously wounded by a shot in the abdomen and left to lie where he had fallen. A passing peasant found him and took him to the hospital. He died the next morning of peritonitis, attended by his younger brother. Galois tried to console his brother, saying, "Do not cry. I need all my courage to die at twenty." He was buried in a common ditch at the cemetery of Montparnasse; the exact location is unknown.

5.2 VECTOR SPACES, BASES, AND DIMENSION

For almost 200 years after the invention of analytic geometry by René Descartes (1596–1650), analytic methods dominated geometry. Toward the middle of the nineteenth century, more and more geometers came to see that the method of coordinates, based on a set of arbitrary axes, expressed geometrical facts in a complicated and roundabout manner; intuitive content became dependent on an awkward process of calculation. This dissatisfaction with Cartesian techniques led to attempts to find a symbolic calculus that would deal more directly and simply with geometrical objects, without the intervention of an irrelevant frame of reference. One of the pioneers "to free geometry from the hieroglyphics of analysis" was Hermann Günther Grassmann (1809–1877), a distinguished mathematician whose own generation passed him by.

The son of a schoolmaster in Stettin, Germany, Grassmann began his career in 1827 at the University of Berlin, where he studied theology and classical languages for six semesters. About 1832 he began intensive independent study of mathematics; and in 1834 he was hired to teach the subject at a Berlin industrial school, where he remained for slightly more than a year. At the beginning of 1836, Grassmann returned to Stettin, where he spent the remainder of his life teaching mathematics, the sciences, and religion at various secondary schools. He had no university training in mathematics and was never to hold a position at a university, although he repeatedly sought one. In 1844 he published his chief work, a book known briefly as the *Ausdehnungslehre* (Calculus of Extensions). He set forth therein a theory that covered much the same ground as Hamilton's quaternions, then being independently elaborated; but where Hamilton was still thinking in terms of a three-dimensional analog of the complex numbers, Grassmann's system was presented in a form applicable to n-dimensional Euclidean space. Hamilton's work enjoyed a slight priority of publication, while Grassmann was the first to create his system and first to publish a coherent exposition of the new theory. As for terminology, it was Hamilton who introduced (1844) the words "vector" and "scalar" in their precise mathematical sense.

Grassmann's *Ausdehnungslehre* was a remarkable achievement for a schoolteacher in his early thirties who had never attended a single university lecture in mathematics. Grassmann's basic algebraic entity, which he called an "extensive quantity," was a formal sum $\sum \alpha_k e_k$, where the α_k are real numbers and the e_k are a (finite) "system of units." He defined the addition of two quantities and their multiplication by a real number by setting

$$\sum \alpha_k e_k + \sum \beta_k e_k = \sum (\alpha_k + \beta_k) e_k,$$

$$\alpha(\sum \alpha_k e_k) = \sum (\alpha \alpha_k) e_k,$$

and immediately proved the algebraic properties of these operations. (His extensive quantities, in current terms, form an n-dimensional vector space over $R^{\#}$ with basis $e_1, e_2, \ldots$.) Guided by geometric intuition, Grassmann proceeded to develop, essentially as is done today, the notions of subspace, linear span, independence, and basis. Thus the birth of linear algebra, which is so often associated with Cayley's famous 1858 "Memoir on the Theory of Matrices," may arguably be said to have occurred in 1844.

The *Ausdehnungslehre* is very difficult to understand. Endowing his "science of extended quantities" with the utmost generality that it could support, Grassmann shrouded it in philosophical abstraction. The result was a forbidding treatise that was accessible only with extreme difficulty (indeed, it is almost unreadable) and virtually ignored by the mathematical community. The fate of this original work was tragic; it remained unsold, and its publisher was forced to dispose of the whole edition as waste paper. In 1862, Grassmann brought out a greatly revised version of the *Ausdehnungslehre*, having the 300 copies printed at his own expense in his brother's shop. The second edition followed its predecessor into temporary oblivion. Disappointed, Grassmann gradually abandoned mathematics and turned his talents to the study of Sanskrit and the Indo-European languages. In this sphere he won public recognition and was awarded, in the last year of his life, an honorary doctorate from the University of Tübingen.

In 1888 the mathematical logician Guiseppe Peano (1858–1932) published *The Geometrical Calculus According to the Ausdehnungslehre of H. Grassmann*, a short book that elaborated the geometric ideas that Grassmann established as the basis of his research. This treatment contained an axiomatic definition of the concept of an abstract vector space, almost in its present form. Whereas Grassmann's publications were characterized by their abstruseness, nothing could be clearer than Peano's exposition. Nevertheless, his book had as little influence on the mathematics of the time as did Grassmann's; the definition of an abstract vector space became widely known only in the early 1920's, when various writers restated the axioms as if the concept were quite new.

Roughly speaking, a vector space is an additive commutative group whose members can be suitably multiplied by the elements of some field. The precise requirements are set out in the definition below.

DEFINITION 5.1 A *vector space*, or *linear space*, over a field is an ordered triple $(V, F, \cdot)$ consisting of

1. a commutative group V (the operation written as $+$) whose elements are called *vectors*,
2. a field F whose elements are called *scalars*, and
3. an operation $\cdot$ of *scalar multiplication* connecting the group and field such that for each $c \in F$ and $x \in V$ there is a uniquely defined element $c \cdot x \in V$; this operation is required to satisfy the following properties:

a) $(c_1 + c_2) \cdot x = (c_1 \cdot x) + (c_2 \cdot x)$,

b) $(c_1 c_2) \cdot x = c_1 \cdot (c_2 \cdot x)$,

c) $c \cdot (x + y) = (c \cdot x) + (c \cdot y)$,

d) $1 \cdot x = x$, where 1 is the field identity element.

Although the addition sign has been used in two contexts in the above definition—to designate the operation of the group and to designate one of the operations of the field—no confusion should arise from this practice. It will always be clear in any given situation whether vectors or scalars are being added. When both vector addition and scalar multiplication are involved in an expression, we follow our usual understanding in omitting parentheses: Multiplication takes precedence over addition.

In the sequel a vector space over the field F will be denoted by $V(F)$ or, if there can be no doubt about the field, simply by V. For further convenience we shall drop the symbol $\cdot$ and write cx in place of $c \cdot x$.

The reader should realize that a vector space is markedly different from the algebraic systems studied earlier in that the "products" of scalar multiplication employ elements from both F and V (technically speaking, scalar multiplication is a function $\alpha: F \times V \to V$, where $\alpha(c, x)$ is represented by cx). This is typical of certain situations in which we step outside a set to introduce additional structure into it. The scalar field F, although it is an essential part of Definition 5.1, may be thought to lurk in the background, with V being the primary object of interest; of course, the operations in F influence the structure of V through their appearance in the identities of part (3) of the definition. In passing, let us emphasize that the property $1x = x$ is quite essential; without it, every field and commutative group would produce a vector space under the trivial scalar multiplication $cx = 0$ for all $c \in F$ and $x \in V$.

Note too that a given field K is a vector space over any of its subfields F and, in particular, over itself. Just take V to be the additive group of K and define cx to be the usual multiplication in the field K of the element $x \in K$ by $c \in F$. In this way, for instance, the field C of complex numbers may be regarded as a vector space over the field $R^{\#}$ of real numbers.

Before we discuss the implications of the axioms of a vector space, it may be well to give a selection of examples. The formal verification that each example described actually constitutes a vector space is left as an exercise.

Example 5.1
To begin with the prototype of a vector space, let $V_n(F)$ denote the set of all ordered n-tuples of elements from an arbitrary field F; that is, the members of $V_n(F)$ are of the form $(a_1, a_2, \ldots, a_n)$, where $a_i \in F$ for $i = 1, 2, \ldots, n$. To make $V_n(F)$ into a commutative group, addition of two n-tuples is defined to be componentwise addition:

$$(a_1, a_2, \ldots, a_n) + (b_1, b_2, \ldots, b_n) = (a_1 + b_1, a_2 + b_2, \ldots, a_n + b_n).$$

In particular, the n-tuple whose every component is zero serves as the additive identity, while

$$-(a_1, a_2, \ldots, a_n) = (-a_1, -a_2, \ldots, -a_n).$$

This system becomes a vector space if multiplication by a scalar is also expressed in terms of components as follows:

$$c(a_1, a_2, \ldots, a_n) = (ca_1, ca_2, \ldots, ca_n), \qquad c \in F.$$

Since the set $V_n(F)$ can be viewed as consisting of all $1 \times n$ matrices over F, this example has a natural generalization, which we take up next.

Example 5.2
Consider the additive group $M_n(F)$ of $n \times n$ matrices with entries from a field F. A vector space is obtained if scalar multiplication is introduced in the obvious way:

$$c(a_{ij}) = (ca_{ij}), \qquad c \in F, \qquad (a_{ij}) \in M_n(F).$$

Example 5.3
The set $F[x]$ of polynomials with coefficients from a field F admits a vector space structure. Our definitions of the operations are the expected ones: If

$$p(x) = a_0 + a_1x + a_2x^2 + \cdots + a_nx^n$$

and

$$q(x) = b_0 + b_1x + b_2x^2 + \cdots + b_mx^m,$$

where $m \geqslant n$, then

$$p(x) + q(x) = (a_0 + b_0) + (a_1 + b_1)x + \cdots + (a_m + b_m)x^m,$$

$$cp(x) = (ca_0) + (ca_1)x + \cdots + (ca_n)x^n.$$

Needless to say, in the above sum $p(x) + q(x)$ we set $a_k = 0$ for $n + 1 \leqslant k \leqslant m$.

Example 5.4
For any field F, Example 5.1 may be generalized by using infinitely many elements of F. Just take $V_\infty(F)$ to be the totality of all infinite sequences from F. The members of $V_\infty(F)$ may be conveniently identified with row vectors

$$(a_1, a_2, a_3, \ldots), \qquad a_k \in F,$$

having infinitely many components. In this case, $(a_1, a_2, a_3, \ldots) = (b_1, b_2, b_3, \ldots)$ if and only if $a_k = b_k$ for all $k \in Z_+$. Vector addition and multiplication by a scalar $c \in F$ are performed componentwise, so

$$(a_1, a_2, a_3, \ldots) + (b_1, b_2, b_3, \ldots) = (a_1 + b_1, a_2 + b_2, a_3 + b_3, \ldots),$$

$$c(a_1, a_2, a_3, \ldots) = (ca_1, ca_2, ca_3, \ldots).$$

Using these operations, we find $V_\infty(F)$ to be a vector space over F.

Example 5.5
Given a field F, consider map(X, F), the collection of all functions from an arbitrary set X into F. For $f, g \in \text{map}(X, F)$ and $c \in F$, the natural way of defining $f + g$ and cf is pointwise:

$$(f + g)(x) = f(x) + g(x),$$

$$(cf)(x) = cf(x), \qquad x \in X.$$

With vector addition and scalar multiplication so understood, map(X, F) becomes a vector space.

Example 5.6
As a final, and not quite so simple, example of a vector space, we start with an additively written commutative group G in which every nonzero element has order p (p a prime); that is to say, $px = 0$ for all $x \in G$. For $x \in G$ and $[n] \in Z_p$, take the product $[n]x$ to mean

$$[n]x = \begin{cases} nx & \text{if } n > 0 \\ 0 & \text{if } n = 0 \\ (-n)(-x) & \text{if } n < 0 \end{cases}$$

With this definition of scalar multiplication, G may be regarded as a vector space over the field Z_p of integers modulo p.

A further comment on notation: Rather than introduce a new symbol, we shall use 0 to designate the zero elements both of V and of F. The additive inverse of a scalar $c \in F$ shall be denoted by $-c$, while the inverse of a vector $x \in V$ is also represented by its negative, $-x$. These conventions should not lead to ambiguity provided that the reader notes carefully the context in which the notation is employed.

Some immediate consequences of Definition 5.1 are embodied in our first theorem.

THEOREM 5.1 *If V is a vector space over the field F and $x \in V$, $c \in F$, then*

1. $0x = 0$,
2. $c0 = 0$,
3. $-(cx) = (-c)x = c(-x)$.

Proof. To establish (1), we use the field result $0 + 1 = 1$. Then

$$0x + x = 0x + 1x = (0 + 1)x = 1x = x = 0 + x.$$

Since V is a group under vector addition, the cancellation law yields $0x = 0$.

The proof of the second part of the theorem follows from the group result $0 + x = x$. We have

$$c0 + cx = c(0 + x) = cx = 0 + cx.$$

Again the cancellation law gives the desired conclusion.

Finally, to obtain (3), observe that

$$0 = 0x = [c + (-c)]x = cx + (-c)x.$$

This implies that $(-c)x = -(cx)$. Similarly,

$$0 = c0 = c[x + (-x)] = cx + c(-x),$$

which forces $c(-x) = -(cx)$.

As the reader should expect by now, a formal investigation of vector spaces involves consideration of such notions as subsystems, operation-preserving functions, quotient structures, and so on. Following our standard pattern of presentation, we begin the study with the question of subsystems. In the case of vector spaces, the sub-vector spaces are customarily referred to as subspaces.

DEFINITION 5.2 Let V be a vector space over the field F and $W \subseteq V$, with $W \neq \varnothing$. Then W is a *subspace* of V if, under the same operations as for V, W is itself a vector space.

Since $W \subseteq V$, much of the algebraic structure of W is inherited from V. The minimum conditions that W must satisfy to be a subspace are:

1. $(W, +)$ is a subgroup of $(V, +)$;
2. W is closed under scalar multiplication.

The usual criterion for deciding whether $(W, +)$ is a subgroup of $(V, +)$ is to see if W is closed under differences. The second of the above conditions implies that $-x = (-1)x$ will belong to W whenever x is an element of W. Because $x - y = x + (-y)$, condition (2), together with the closure of W under addition, is sufficient to guarantee that W be closed under differences. This observation allows us to reformulate Definition 5.2 as follows.

DEFINITION 5.3 A nonempty subset W is a subspace of the vector space V over F if it has the properties

1. $x, y \in W$ imply that $x + y \in W$,
2. $x \in W$ and $c \in F$ imply that $cx \in W$.

According to the next theorem, the two conditions of this definition can be combined into a single easily applied criterion.

THEOREM 5.2 *Let V be a vector space over the field F and W be a nonempty subset of V. Then W is a subspace of V if and only if $cx + dy \in W$ whenever $x, y \in W$ and $c, d \in F$.*

Proof. If W is a subspace of V, then in view of its closure properties $cx + dy \in W$ for all $x, y \in W$ and $c, d \in F$. Conversely, if this condition holds, W must contain $1x + 1y = x + y$ and $cx + 0y = cx$ for each choice of $x, y \in W$ and $c \in F$. Thus W is closed with respect to both vector space operations, making it a subspace of V.

Example 5.7

Every vector space V has two *trivial* subspaces, namely, V itself and the zero subspace $\{0\}$. A subspace other than V is said to be a *proper subspace.*

Example 5.8

In $V_n(F)$, consider the set W of vectors whose components add up to zero:

$$W = \left\{(a_1, a_2, \ldots, a_n) \,\middle|\, \sum_{k=1}^{n} a_k = 0\right\}.$$

If $(a_1, a_2, \ldots, a_n)$ and $(b_1, b_2, \ldots, b_n)$ are arbitrary elements of W, then the sum

$$c(a_1, a_2, \ldots, a_n) + d(b_1, b_2, \ldots, b_n) = (ca_1 + db_1, \ldots, ca_n + db_n)$$

is such that

$$\sum_{k=1}^{n} (ca_k + db_k) = c \sum_{k=1}^{n} a_k + d \sum_{k=1}^{n} b_k = c0 + d0 = 0.$$

The implication is that W is a subspace of $V_n(F)$.

Example 5.9

Given a field F, let $F_n[x]$ symbolize the set of all polynomials over F of degree less than or equal to n, together with the zero polynomial; that is,

$$F_n[x] = \{f(x) \in F[x] | f(x) = 0 \text{ or } \deg f(x) \leqslant n\}.$$

Whenever $cf(x) + dg(x) \neq 0$, we have

$$\deg (cf(x) + dg(x)) \leqslant \max\{\deg f(x), \deg g(x)\},$$

which ensures that $F_n[x]$ is a subspace of the vector space $F[x]$. It will be slightly inaccurate, but convenient, to refer to $F_n[x]$ as "the subspace of polynomials of degree n or less." (Note that $F[x]$ is also a ring, but $F_n[x]$ does not form a subring of it.)

We next consider operations on the subspaces of a vector space that produce other subspaces, the most important of which are the sum and intersection. If $W_1, W_2, \ldots, W_n$ are nonempty subsets of the vector space V, then their *(linear) sum* is defined to be the set

$$W_1 + W_2 + \cdots + W_n = \{x_1 + x_2 + \cdots + x_n | x_i \in W_i \text{ for all } i\}.$$

The following simple, but quite useful, fact will be needed on a variety of occasions.

THEOREM 5.3 *If V is a vector space over the field F and $W_i (i = 1, 2, \ldots, n)$ are all subspaces of V, then so is their sum $W_1 + W_2 + \cdots + W_n$.*

Proof. The sum $W_1 + W_2 + \cdots + W_n$ is certainly not empty, since each W_i contains the zero vector and hence $0 = 0 + 0 + \cdots + 0 \in W_1 + W_2 + \cdots + W_n$. Suppose that x and y are typical elements of the sum. Then $x = x_1 + x_2 + \cdots + x_n$ and $y = y_1 + y_2 + \cdots + y_n$, where x_i, $y_i \in W_i$ for $i = 1, 2, \ldots, n$. Given scalars c, $d \in F$,

$$\begin{aligned} cx + dy &= c(x_1 + x_2 + \cdots + x_n) + d(y_1 + y_2 + \cdots + y_n) \\ &= (cx_1 + dy_1) + \cdots + (cx_n + dy_n). \end{aligned}$$

Inasmuch as W_i is a subspace of V, it follows that $cx_i + dy_i \in W_i$ for each i. This puts the vector $cx + dy$ in $W_1 + W_2 + \cdots + W_n$, in consequence of which the sum forms a subspace of V.

From our earlier work with other subsystems we would expect the intersection of subspaces of V to be a subspace again; indeed, this is the content of the next theorem. As it would be repetitious to present the details again, the proof is left to the reader.

THEOREM 5.4 *If V is a vector space over the field F and if W_i is an indexed collection of subspaces of V, then their intersection $\cap W_i$ is also a subspace of V.*

In analogy with the corresponding ideas for groups and rings, this result provides an effective means for generating subspaces: Starting with a nonempty subset S of the vector space V over F, we define

$$[S] = \cap\{W | S \subseteq W;\ W \text{ is a subspace of } V\}.$$

There is at least one subspace containing S, namely, the entire space V, so that $[S]$ exists and is unique. We infer at once from Theorem 5.4 that $[S]$ is a subspace of V, called the *subspace generated* or *spanned by the set S*. More important still, $[S]$ is the smallest subspace containing S, in the sense of being included in every subspace that contains S.

Remark. To cover the case in which S is the empty set, many authors adopt the convention that $[\varnothing] = \{0\}$.

The following theorem gives an alternative and more constructive description of $[S]$; before obtaining this we need one more definition.

DEFINITION 5.4 Suppose that V is a vector space over the field F and $x_1, x_2, \ldots, x_n \in V$. Any finite sum of the form

$$c_1x_1 + c_2x_2 + \cdots + c_nx_n = \sum_{i=1}^{n} c_ix_i,$$

where each $c_i \in F$ is said to be a *linear combination* of the vectors x_1, $x_2, \ldots, x_n$. A linear combination is called *trivial* if all its coefficients $c_i = 0$ and *nontrivial* if at least one coefficient is different from zero.

Using this new terminology, Theorem 5.2 may be rephrased in another way: W is a subspace of the vector space V if and only if W is a nonempty subset of V that is closed under the formation of linear combinations.

THEOREM 5.5 *If S is a nonempty subset of the vector space V over F, then $[S]$, the subspace generated by S, consists of all (finite) linear combinations of elements of S:*

$$[S] = \left\{\sum_{k=1}^{n} c_kx_k | c_k \in F,\ x_k \in S,\ n \in Z_+\right\}.$$

Proof. As a temporary device, let us denote the set on the right-hand side of the displayed equation by span S. Evidently, a linear combination of linear combinations of elements from S can again be written as a linear combination of members of S, so that span S is itself a subspace of V. By definition, $x = 1x \in \text{span}\, S$ for each vector x in S and hence $S \subseteq \text{span}\, S$. Since $[S]$ is the smallest subspace of V to contain the set S, it follows that $[S] \subseteq \text{span}\, S$. On the other hand, being a subspace that contains S, $[S]$ must of necessity contain all linear combinations of elements of S. Accordingly, the inclusion span $S \subseteq [S]$ also holds, whence the announced equality.

By taking $S = \{x\}$ we arrive at the following subsidiary result:

COROLLARY *The subspace generated by a single vector $x \in V$ is given by $[x] = \{cx \mid c \in F\}$.*

Observe that S may be an infinite set, but $[S]$ consists only of finite sums; it does not even make sense here to talk about infinite sums.

Example 5.10

Let S be the subset of $V_n(F)$ whose elements are the n vectors $e_1 = (1, 0, \ldots, 0)$, $e_2 = (0, 1, \ldots, 0), \ldots, e_n = (0, 0, \ldots, 1)$; in general, e_k is the vector with 1 in the kth component position and 0 elsewhere:

$$e_k = (\delta_{1k}, \delta_{2k}, \ldots, \delta_{nk}), \qquad k = 1, 2, \ldots, n.$$

Here $[S]$, the subspace spanned by these vectors, is all of $V_n(F)$. Indeed, for any n-tuple $(a_1, a_2, \ldots, a_n)$ over F we have

$$\begin{aligned}(a_1, a_2, \ldots, a_n) &= a_1(1, 0, \ldots, 0) + a_2(0, 1, \ldots, 0) + \cdots + a_n(0, 0, \ldots, 1)\\ &= a_1e_1 + a_2e_2 + \cdots + a_ne_n \in [S].\end{aligned}$$

Let us see what happens when we generalize this situation to the sequence space $V_\infty(F)$. In this context, e_k now denotes the infinite sequence whose kth term is 1, while all other terms are 0:

$$e_k = (\delta_{1k}, \delta_{2k}, \delta_{3k}, \ldots), \qquad k = 1, 2, \ldots.$$

As before, let S be the set of all e_k. Recall that in forming the linear span $[S]$, only finite linear combinations of elements of S are utilized. It follows therefore that $[S]$ is not $V_\infty(F)$, but rather the (proper) subspace consisting of those vectors having only a finite number of nonzero entries:

$$[S] = \{(a_1, a_2, \ldots, a_n, 0, 0, \ldots) \mid a_k \in F,\ n \in Z_+\}.$$

The first part of this example serves to suggest, as well as to illustrate, the next definition.

DEFINITION 5.5 A vector space V is *finitely generated* when it contains a finite subset that spans V; if V is spanned by the vectors $x_1, x_2, \ldots, x_n$, this will be indicated by writing $V = [x_1, x_2, \ldots, x_n]$.

The union of subspaces, unlike their intersection, is not necessarily a subspace. For instance, in the vector space $M_2(F)$, take U to be the subset consisting of all scalar matrices

$$\begin{bmatrix} a & 0 \\ 0 & a \end{bmatrix}, \qquad a \in F,$$

and W to be the set of all matrices of the form

$$\begin{bmatrix} 0 & b \\ -b & 0 \end{bmatrix}, \qquad b \in F.$$

That U and W are both subspaces of $M_2(F)$ is easily verified. Note, however, that while the matrices

$$\begin{bmatrix} 1 & 0 \\ 0 & 1 \end{bmatrix} \quad \text{and} \quad \begin{bmatrix} 0 & 1 \\ -1 & 0 \end{bmatrix}$$

belong to $U \cup W$, their sum

$$\begin{bmatrix} 1 & 0 \\ 0 & 1 \end{bmatrix} + \begin{bmatrix} 0 & 1 \\ -1 & 0 \end{bmatrix} = \begin{bmatrix} 1 & 1 \\ -1 & 1 \end{bmatrix}$$

fails to be a member of the union. From this we conclude that $U \cup W$ is not a subspace of $M_2(F)$.

While the union of subspaces need not be a subspace, the subspace generated by the union always exists and equals the sum of the given subspaces. We now establish this fact.

THEOREM 5.6 *If U and W are subspaces of the vector space V, then $U + W$ is the smallest subspace containing both U and W; in symbols,*

$$U + W = [U \cup W].$$

Proof. Since $u + w = 1u + 1w$, any vector in $U + W$ can be expressed as a linear combination of elements from $U \cup W$. But Theorem 5.5 asserts that $[U \cup W]$ consists of all such linear combinations; hence $U + W \subseteq [U \cup W]$. For the opposite inclusion, observe that both $U \subseteq U + W$ and $W \subseteq U + W$, so their union $U \cup W \subseteq U + W$. In other words, $U + W$ is a subspace of V that contains $U \cup W$. As $[U \cup W]$ is, by definition, the smallest subspace with this property, we conclude that $[U \cup W] \subseteq U + W$.

This characterization of the linear span of $U \cup W$ is usually much easier to apply in specific cases than the definition itself. In the space $M_2(F)$, for example, where U and W consist of matrices of the form

$$\begin{bmatrix} a & 0 \\ 0 & a \end{bmatrix} \quad \text{and} \quad \begin{bmatrix} 0 & b \\ -b & 0 \end{bmatrix},$$

respectively, we have

$$[U \cup W] = \left\{ \begin{bmatrix} a & b \\ -b & a \end{bmatrix} \middle| a, b \in F \right\}.$$

To see this, one need only observe that each of the matrices on the right can be written as

$$\begin{bmatrix} a & b \\ -b & a \end{bmatrix} = \begin{bmatrix} a & 0 \\ 0 & a \end{bmatrix} + \begin{bmatrix} 0 & b \\ -b & 0 \end{bmatrix} \in U + W.$$

Although each vector x belonging to the sum $U + W$ admits a representation $x = u + w$ with $u \in U$ and $w \in W$, this expression is generally not unique. The next theorem provides a necessary and sufficient condition for u and w to be uniquely determined by the vector x.

THEOREM 5.7 *Let U and W be two subspaces of the vector space V. Then the following conditions are equivalent:*

1. $U \cap W = \{0\}$.
2. *Every vector x of the sum $U + W$ is uniquely representable in the form $x = u + w$, where $u \in U$, $w \in W$.*

Proof. We begin by assuming that $U \cap W = \{0\}$. Suppose further that a vector $x \in U + W$ has two representations:

$$x = u_1 + w_1 = u_2 + w_2, \qquad u_i \in U, \qquad w_i \in W.$$

Then $u_1 - u_2 = w_2 - w_1$. But the left side is in U and the right side is in W, so both sides belong to $U \cap W$. It follows therefore that

$$u_1 - u_2 = w_2 - w_1 = 0 \qquad \text{or} \qquad u_1 = u_2, \qquad w_1 = w_2.$$

In other words, x is uniquely expressible in the form $u + w$.

Conversely, assume that statement (2) holds and the vector $z \in U \cap W$. We may then express z in two different ways as the sum of a vector in U and a vector in W, namely, $z = z + 0$ (here $z \in U$ and $0 \in W$) and $z = 0 + z$ (here $0 \in U$ and $z \in W$). The uniqueness assumption of condition (2) then implies that $z = 0$ or, rather, $U \cap W = \{0\}$.

It may have occurred to the reader that different sets of vectors can generate the same space. A simple illustration is furnished by $V_2(F)$, where

$$[(-1, 0), (0, -1), (1, 1)] = [(1, 0), (0, 1)] = V_2(F).$$

The question that we wish to raise is whether, given a vector space V, there is a smallest number of vectors that generate V. A related question: If $[S] = V$, when does there exist a smallest subset $T \subseteq S$ with $[T] = V$? The answers are closely connected to the important notions of linear independence and linear dependence. As will soon be evident, these are the principal tools that are needed to formalize the more or less intuitive idea of the dimension of a space.

To make all of this precise, we begin with a definition.

DEFINITION 5.6 Let $V(F)$ be a vector space over the field F. A finite set $\{x_1, x_2, \ldots, x_n\}$ of vectors from V is said to be *linearly dependent* (over F) if the zero vector is a nontrivial linear combination of $x_1, x_2, \ldots, x_n$; otherwise, the set $\{x_1, x_2, \ldots, x_n\}$ is termed *linearly independent*.

According to our definition, $\{x_1, x_2, \ldots, x_n\}$ is a linearly dependent set if and only if there exist scalars $c_1, c_2, \ldots, c_n \in F$, not all of which are zero, such that

$$c_1x_1 + c_2x_2 + \cdots + c_nx_n = 0.$$

In the contrary case the zero vector is expressible as a linear combination of independent vectors only in the trivial way in which all the coefficients are zero; that is to say, $c_1x_1 + c_2x_2 + \cdots + c_nx_n = 0$ implies that

$$c_1 = c_2 = \cdots = c_n = 0.$$

It should be kept in mind that the independence or dependence of $\{x_1, x_2, \ldots, x_n\}$ depends on the scalar field F as well as on $\{x_1, x_2, \ldots, x_n\}$.

Although linear dependence and independence are properties of sets of vectors, these terms are frequently applied to the vectors themselves. Thus we shall speak of $x_1, x_2, \ldots, x_n$ as being linearly dependent or independent according as the set $\{x_1, x_2, \ldots, x_n\}$ is dependent or independent. Whether or not the vectors $x_1, x_2, \ldots, x_n$ are linearly independent is unaffected by writing these vectors in a different order.

Two examples should help to clarify matters.

Example 5.11

In $R^{\#}[x]$, the vector space of polynomials in x over $R^{\#}$, consider the three polynomials

$$1 + x + 2x^2, \qquad 2 - x + x^2, \qquad -4 + 5x + x^2$$

of degree two. Since

$$2(1 + x + 2x^2) + (-3)(2 - x + x^2) + (-1)(-4 + 5x + x^2) = 0,$$

the given polynomials are linearly dependent over $R^{\#}$.

Example 5.12
The so-called *unit vectors* $e_1, e_2, \ldots, e_n$, where

$$e_k = (\delta_{1k}, \delta_{2k}, \ldots, \delta_{nk}), \qquad k = 1, 2, \ldots, n,$$

form a linearly independent subset of $V_n(F)$. In this case,

$$\begin{aligned} c_1e_1 &+ c_2e_2 + \cdots + c_ne_n \\ &= (c_1, 0, \ldots, 0) + (0, c_2, \ldots, 0) + \cdots + (0, 0, \ldots, c_n) \\ &= (c_1, c_2, \ldots, c_n); \end{aligned}$$

hence the condition $\sum_{k=1}^{n} c_k e_k = (0, 0, \ldots, 0)$ implies that $c_1 = c_2 = \cdots = c_n = 0$.

The essence of this concept lies in the fact that if $\{x_1, x_2, \ldots, x_n\}$ is a linearly independent set, then any vector $x \in [x_1, x_2, \ldots, x_n]$ is uniquely expressible as a linear combination of the x_k's. Indeed, suppose there are two such representations,

$$x = b_1x_1 + b_2x_2 + \cdots + b_nx_n = c_1x_1 + c_2x_2 + \cdots + c_nx_n;$$

this leads to the relation

$$(b_1 - c_1)x_1 + (b_2 - c_2)x_2 + \cdots + (b_n - c_n)x_n = 0.$$

The hypothesized linear independence of the vectors $x_1, x_2, \ldots, x_n$ then forces

$$b_k - c_k = 0 \qquad \text{or} \qquad b_k = c_k \qquad \text{for all } k.$$

Definition 5.6, as stated, applies only to finite sets of vectors. It is no great step to extend these concepts: Given a vector space $V(F)$, a nonempty subset $S \subseteq V$ is linearly dependent (over F) if it contains some finite subset that is linearly dependent. In the contrary case, S is a linearly independent set; that is to say, S is linearly independent if every finite subset is linearly independent. While S may be infinite, linear dependence is a property of finite character.

The following lemma is inherent in our definitions.

LEMMA *Let V be a vector space and $\varnothing \neq T \subseteq S \subseteq V$. If S is a linearly independent set, so also is the set T; conversely, if T is a linearly dependent set, so also is the set S.*

There is a minor point to be made here, namely, that any set that contains the zero vector is always linearly dependent. This follows from the fact that

$$1(0) + 0x_1 + \cdots + 0x_n = 0$$

is a nontrivial linear combination for arbitrary vectors $x_1, x_2, \ldots, x_n$. Accordingly, all vectors in a linearly independent set must be different from zero. The set consisting of a single vector x is, in particular, linearly independent if and only if $x \neq 0$.

The next theorem in some sense justifies the use of the word *dependent*, since it asserts that in a linearly dependent set one of the vectors belongs to the subspace spanned by the remaining ones; putting the matter informally, one of the vectors *depends* on the others.

THEOREM 5.8 *A set of nonzero vectors $\{x_1, x_2, \ldots, x_n\}$ is linearly dependent if and only if some vector x_k $(k > 1)$ is a linear combination of the preceding ones, $x_1, x_2, \ldots, x_{k-1}$.*

Proof. Suppose the vectors $x_1, x_2, \ldots, x_n$ are linearly dependent, so there is a nontrivial linear relation

$$c_1x_1 + c_2x_2 + \cdots + c_nx_n = 0,$$

where not all the c's are zero. Let k be the largest integer for which $c_k \neq 0$. If $k = 1$, we would have $c_1x_1 = 0$ and hence $x_1 = 0$, contrary to the dependence assumption. Thus $k > 1$ and

$$c_1x_1 + c_2x_2 + \cdots + c_kx_k = 0.$$

Since c_k^{-1} exists in F, it follows that

$$\begin{aligned} x_k &= c_k^{-1}(-1)(c_1x_1 + c_2x_2 + \cdots + c_{k-1}x_{k-1}) \\ &= (-c_k^{-1}c_1)x_1 + (-c_k^{-1}c_2)x_2 + \cdots + (-c_k^{-1}c_{k-1})x_{k-1}. \end{aligned}$$

The vector x_k is therefore expressible as a linear combination of its predecessors, as claimed.

The converse is almost obvious: If the vector x_k depends linearly on $x_1, x_2, \ldots, x_{k-1}$, so that

$$x_k = b_1x_1 + b_2x_2 + \cdots + b_{k-1}x_{k-1}$$

for suitable scalars $b_k \in F$, then

$$b_1x_1 + \cdots + b_{k-1}x_{k-1} + (-1)x_k + 0x_{k-1} + \cdots + 0x_n = 0.$$

Because the coefficient of x_k is nonzero, $\{x_1, x_2, \ldots, x_n\}$ constitutes a dependent set of vectors.

THEOREM 5.9 *If V is a finitely generated vector space, say*

$$V = [x_1, x_2, \dots, x_n],$$

then V is spanned by a linearly independent subset of these vectors.

Proof. If the set $\{x_1, x_2, \dots, x_n\}$ is already independent, nothing needs to be proved. Otherwise, Theorem 5.8 implies that some vector x_k is a linear combination of $x_1, x_2, \dots, x_{k-1}$. By hypothesis, any vector x in V can be written as a linear combination of the n vectors $x_1, x_2, \dots, x_n$; in this combination, x_k may be replaced by a linear combination of $x_1, x_2, \dots, x_{k-1}$, thereby showing that $x \in [x_1, \dots, x_{k-1}, x_{k+1}, \dots, x_n]$. The net result is that the $n-1$ vectors $x_1, \dots, x_{k-1}, x_{k+1}, \dots, x_n$ generate the space V.

Next examine the set $\{x_1, \dots, x_{k-1}, x_{k+1}, \dots, x_n\}$ and repeat the process of removing a vector if it can be written as a linear combination of its predecessors. Continuing in this way, we eventually reach a subset

$$\{x_{i_1}, x_{i_2}, \dots, x_{i_m}\},$$

where $1 \leqslant i_1 < i_2 < \cdots < i_m \leqslant n$, of the original set of n vectors, still having linear span V and such that no x_{i_k} is a linear combination of the preceding vectors. That the set $\{x_{i_1}, x_{i_2}, \dots, x_{i_m}\}$ is linearly independent follows immediately from Theorem 5.8.

Example 5.13

As a simple illustration of some of these ideas, let us observe that

$$V_3(F) = [(1, 1, 0), (1, 0, 1), (0, 1, 0), (1, 1, 1)].$$

The reasoning here is justified by the fact that each of the unit vectors e_1, e_2, e_3 is a linear combination of these vectors:

$$e_1 = (1, 1, 0) - (0, 1, 0), \qquad e_2 = (0, 1, 0), \qquad e_3 = (1, 1, 1) - (1, 1, 0).$$

Since $V_3(F)$ is spanned by the unit vectors, we infer that every element of $V_3(F)$ must be a linear combination of $(1, 1, 0)$, $(1, 0, 1)$, $(0, 1, 0)$, $(1, 1, 1)$.

The linear dependence or independence of this set of vectors is equivalent to the existence or nonexistence of scalars c_1, c_2, c_3, c_4 (not all zero) such that

$$c_1(1, 1, 0) + c_2(1, 0, 1) + c_3(0, 1, 0) + c_4(1, 1, 1) = 0.$$

Checking components, we note that the c_k's must satisfy the three equations

$$c_1 + c_2 + c_4 = 0, \qquad c_1 + c_3 + c_4 = 0, \qquad c_2 + c_4 = 0.$$

In terms of an arbitrary choice of c_4 a solution is $c_1 = 0$, $c_2 = -c_4$,

$c_3 = -c_4$; for instance, $c_4 = 1$ leads to

$$0(1, 1, 0) + (-1)(1, 0, 1) + (-1)(0, 1, 0) + 1(1, 1, 1) = (0, 0, 0),$$

which, of course, implies that the given set of generators is linearly dependent.

To obtain a linearly independent subset of the aforementioned vectors, we need only remove the element (1, 1, 1). In fact, (1, 1, 1) is the only vector that can be written as a linear combination of those preceding it:

$$(1, 1, 1) = (1, 0, 1) + (0, 1, 0).$$

Any member (a_1, a_2, a_3) of $V_3(F)$ can obviously be expressed as

$$(a_1, a_2, a_3) = (a_1 - a_3)(1, 1, 0) + a_3(1, 0, 1) + (a_2 - a_1 + a_3)(0, 1, 0),$$

whence

$$V_3(F) = [(1, 1, 0), (1, 0, 1), (0, 1, 0)].$$

The next, somewhat technical, result is the key to all that follows.

THEOREM 5.10 (Steinitz Replacement Theorem) *Let W be a finitely generated subspace of the vector space V, $W = [x_1, x_2, \ldots, x_n]$, and let $\{y_1, y_2, \ldots, y_m\}$ be any linearly independent subset of W. Then m of the x_k's, say (after suitable renumbering) $x_1, x_2, \ldots, x_m$, may be replaced by $y_1, y_2, \ldots, y_m$, so that $W = [y_1, \ldots, y_m, x_{m+1}, \ldots, x_n]$; in particular, $m \leqslant n$.*

Proof. Since $\{x_1, x_2, \ldots, x_n\}$ spans W, the vector $y_1 \in W$ can be expressed as a linear combination of the x_k's:

$$y_1 = a_1x_1 + a_2x_2 + \cdots + a_nx_n.$$

Not all the coefficients $a_k = 0$; otherwise, $y_1 = 0$, contradicting the linear independence of the set $\{y_1, y_2, \ldots, y_m\}$. Re-indexing, if necessary, we may assume that $a_1 \neq 0$. Now solve for the vector x_1 in terms of $y_1, x_2, \ldots, x_n$:

$$x_1 = a_1^{-1}y_1 + (-a_1^{-1}a_2)x_2 + \cdots + (-a_1^{-1}a_n)x_n.$$

This relation permits us to replace a linear combination of x_1, $x_2, \ldots, x_n$ by a linear combination of the vectors $y_1, x_2, \ldots, x_n$ and leads to the conclusion that $W = [y_1, x_2, \ldots, x_n]$.

Repeat the replacement process with the vector y_2 and the set $\{y_1, x_2, \ldots, x_n\}$. Because y_2 belongs to the subspace spanned by y_1, $x_2, \ldots, x_n$, we must have

$$y_2 = b_1y_1 + b_2x_2 + \cdots + b_nx_n$$

for suitable scalars $b_1, b_2, \ldots, b_n \in F$. The coefficients $b_2, \ldots, b_n$ cannot all be zero, since this would imply that $b_1 \neq 0$ and, in turn, that

$$b_1 y_1 + (-1)y_2 + 0y_3 + \cdots + 0y_m = 0,$$

contrary to the independence hypothesis. Hence one of the coefficients $b_2, \ldots, b_n$ is nonzero; let us, for simplicity, take this to be b_2. As before, we can solve for x_2 in terms of the vectors $y_1, y_2, x_3, \ldots, x_n$ to obtain

$$W = [y_1, y_2, x_3, \ldots, x_n].$$

Continue in this manner: At each stage a y-vector can be introduced and an x-vector deleted so that the new set still spans W. If m were larger than n, after n steps all the x_k's could be removed and the set $\{y_1, y_2, \ldots, y_n\}$ would span W:

$$y_{n+1} = c_1 y_1 + c_2 y_2 + \cdots + c_n y_n$$

with not all the c_k being zero, since $y_{n+1} \neq 0$. Once again a contradiction to the independence of $y_1, y_2, \ldots, y_m$ would arise. It follows that the y's must be exhausted before the x's (that is, $n \geqslant m$) and $W = [y_1, \ldots, y_m, x_{m+1}, \ldots, x_n]$. Of course, the possibility that $n = m$ is not excluded; in this situation the set $\{x_{m+1}, \ldots, x_n\}$ is empty, and the vectors $y_1, y_2, \ldots, y_n$ themselves span W.

COROLLARY *If a vector space V is spanned by n vectors, then any set of $n + 1$ vectors from V is linearly dependent; in particular, any $n + 1$ vectors of the n-tuple space $V_n(F)$ are dependent.*

Proof. A set of $n + 1$ linearly independent vectors from V would be impossible, since the theorem would imply that $n \geqslant n + 1$, an obvious contradiction.

We now come to one of the concepts from which this section gets its title.

DEFINITION 5.7 A *basis* for a vector space V is a linearly independent subset of V that spans the entire space V.

Example 5.14
The familiar unit vectors $e_1, e_2, \ldots, e_n$ of $V_n(F)$, the space of n-tuples of scalars, form a basis (see Examples 5.10 and 5.12). Hereafter, we shall refer to this particular basis as the *natural* or *standard basis* for $V_n(F)$.

Example 5.15
For a more general example, consider the vector space $V_\infty(F)$ of infinite sequences of elements from F. If

$$e_k = (\delta_{1k}, \delta_{2k}, \delta_{3k}, \ldots), \qquad k = 1, 2, \ldots,$$

then the set $S = \{e_1, e_2, e_3, \ldots\}$ is linearly independent. These vectors do not constitute a basis for $V_\infty(F)$, however, since it was shown earlier that the linear span $[S]$ is a proper subset of $V_\infty(F)$.

Example 5.16
Let $M_n(F)$ be the vector space of $n \times n$ matrices over a field F. This space has a basis consisting of the n^2 matrices E_{ij}, where E_{ij} is the square matrix of order n having 1 as its ijth entry and zeros elsewhere. Any matrix $(a_{ij}) \in M_n(F)$ can obviously be written as

$$(a_{ij}) = a_{11}E_{11} + a_{12}E_{12} + \cdots + a_{nn}E_{nn}.$$

Moreover, $(a_{ij}) = 0$ if and only if $a_{11} = a_{12} = \cdots = a_{nn} = 0$; hence the matrices $E_{11}, E_{12}, \ldots, E_{nn}$ are linearly independent over F.

Example 5.17
One final example: Consider $F[x]$, the vector space of polynomials in x with coefficients from F. A basis for $F[x]$ is formed by the set

$$S = \{1, x, x^2, \ldots, x^n, \ldots\}.$$

By definition, each polynomial $p(x) = a_0 1 + a_1 x + \cdots + a_n x^n$ of $F[x]$ is a linear combination of elements from S. The independence of S follows from the fact that, for any finite subset

$$\{x^{n_1}, x^{n_2}, \ldots, x^{n_k}\}, \qquad 0 \leqslant n_1 < n_2 < \cdots < n_k,$$

the relation

$$c_1 x^{n_1} + c_2 x^{n_2} + \cdots + c_k x^{n_k} = 0$$

holds if and only if $c_1 = c_2 = \cdots = c_k = 0$.

These are but a few examples of the more frequently encountered bases and should amply illustrate the concept; as we continue our discussion, additional examples will appear.

We ought to point out several things. First, Theorem 5.9 may be rephrased so as to assert that any vector space that is spanned by a finite subset, linearly independent or not, possesses a finite basis. Since a basis S for a vector space V is of necessity a linearly independent subset of V, it is possible to express each vector $x \in V$ as a linear combination of elements from S in exactly one way. The unique scalar coefficients that occur in this representation are called the *coordinates* of x with respect to the given basis. Thus the notion of a basis enables us to *coordinatize* the space.

Finally, let us observe that a given vector space may have more than one basis. Example 5.13, for instance, shows that the vectors

$$x_1 = (1, 1, 0), \qquad x_2 = (1, 0, 1), \qquad x_3 = (0, 1, 0)$$

constitute a second basis for $V_3(F)$. In this case an arbitrary vector (a_1, a_2, a_3) of $V_3(F)$ can be written as

$$(a_1, a_2, a_3) = (a_1 - a_3)x_1 + a_3x_2 + (a_2 - a_1 + a_3)x_3.$$

While the coefficients in the above linear combination are uniquely determined, they obviously differ from those which represent the same vector relative to the standard basis e_1, e_2, e_2; roughly speaking, a vector has different coordinates with respect to different bases.

We next dispose of a natural question that arises here: Is it possible to obtain a basis for a given vector space? A closely related question is this: Presuming that one can select two different bases for a space, must each contain the same number of elements? When our vector space is the zero space $\{0\}$, no subset is linearly independent and certainly no basis exists. On the other hand, the next theorem guarantees that a nonzero space will always have a basis. The proof is a straightforward application of Zorn's Lemma.

THEOREM 5.11 (Basis Theorem) *Every nonzero vector space V possesses a basis.*

Proof. Let $\mathscr{A}$ be the family of all linearly independent subsets of V. Our immediate aim is to show that for any chain of sets $\{A_i\}$ in $\mathscr{A}$, their union $\cup A_i$ also belongs to $\mathscr{A}$. To do so, we assume the vectors $x_1, x_2, \ldots, x_n \in \cup A_i$ and that the linear combination $c_1x_1 + c_2x_2 + \cdots + c_nx_n = 0$. Now each vector x_k lies in some member A_{i_k} of $\{A_i\}$. As $\{A_i\}$ forms a chain, one of the sets $A_{i_1}, A_{i_2}, \ldots, A_{i_n}$ contains all the others, call it $A_{i'}$. This means that the given vectors $x_1, x_2, \ldots, x_n$ are all in $A_{i'}$. But the linear independence of $A_{i'}$ then implies that the scalar coefficients $c_k = 0$ for all $k = 1, 2, \ldots, n$. Thus the union $\cup A_i$ is itself a linearly independent subset of V, whence a member of the family $\mathscr{A}$.

The hypothesis of Zorn's Lemma being satisfied, there exists a maximal element S in $\mathscr{A}$. As a member of $\mathscr{A}$, S is a linearly independent subset of V; so to complete the proof, it remains simply to establish $[S] = V$. To see this, let x be any vector of V not in S. Because the set $S' = S \cup \{x\}$ properly contains S, it must be linearly dependent (the maximality of S enters here). Therefore for some finite subset $\{y_1, y_2, \ldots, y_m\}$ of S, a dependence relation

$$ax + a_1y_1 + a_2y_2 + \cdots + a_my_m = 0$$

exists in which not all the coefficients are zero. Were $a = 0$, we would contradict the linear independence of S; hence $a \neq 0$, which leads to

$$x = (-a^{-1}a_1)y_1 + (-a^{-1}a_2)y_2 + \cdots + (-a^{-1}a_m)y_m$$

or $x \in [S]$. We conclude from this that the linear span of S is the whole space V, finishing the proof.

The argument used in Theorem 5.11 suggests the following characterization of a basis: S is a basis for the vector space V if and only if S is a maximal linearly independent subset of V (maximal, in the sense that any subset of V that properly contains S is itself linearly dependent). One direction having already been established, let us observe that if S is a basis for V, then any set $S \cup \{x\}$, where $x \in V - S$, is linearly dependent. In fact, since $V = [S]$,

$$x = c_1x_1 + c_2x_2 + \cdots + c_nx_n$$

for some finite set $\{x_1, x_2, \ldots, x_n\}$ of vectors from S and suitable scalars $c_k \in F$. But then

$$(-1)x + c_1x_1 + c_2x_2 + \cdots + c_nx_n = 0$$

is a nontrivial dependence relation among vectors of $S \cup \{x\}$ and leads to the conclusion that $S \cup \{x\}$ is linearly dependent.

A set S of generators for a vector space V is termed *minimal* if every subset formed from S by removing one or more vectors fails to span V. This idea gives rise to another characterization of a basis:

THEOREM 5.12 *A set S is a basis for a vector space V if and only if S is a minimal generating set of V.*

The proof of the theorem is routine, and we leave the details to the reader.

We are now prepared to establish the main theorem of this section: All bases for a given vector space have the same number of elements (for the astute reader, the same cardinality). Because the proof in full generality is somewhat sophisticated, we shall deal with the finite case only.

THEOREM 5.13 *If V is a finitely generated nonzero vector space, then any two bases for V have precisely the same number of elements.*

Proof. Let $\{x_1, x_2, \ldots, x_n\}$ be one basis for the space V and $\{y_1, y_2, \ldots, y_m\}$ be another. Because the vectors $y_1, y_2, \ldots, y_m$ are linearly independent and V is spanned by $x_1, x_2, \ldots, x_n$, it follows from the Replacement Theorem that the integer m is no larger than n, that is, $m \leqslant n$. Simply reversing the roles of $\{x_1, x_2, \ldots, x_n\}$ and $\{y_1, y_2, \ldots, y_m\}$ in the argument, we also conclude that $n \leqslant m$, whence $n = m$.

Theorem 5.13 can be used to define dimension. If a vector space V has a finite basis, then every basis for the space will have the same finite number of elements. This unique integer is called the *dimension* of V and designated by dim V. Although the notion of basis is not applicable to the zero space, it is customary to treat $\{0\}$ as finite-dimensional, with zero dimension. A (nonzero) vector space is said to be *infinite-dimensional* if it is not spanned by any finite subset, that is, if it is not finite-dimensional. Not every space is of finite dimension, as evidenced by $F[x]$, the vector space of polynomials in x with coefficients from F.

In most of what follows, we shall for the sake of conceptual simplicity confine our attention to finite-dimensional vector spaces.

The next theorem indicates that at least for finite-dimensional spaces, one can form bases without too much difficulty.

THEOREM 5.14 *Let V be a finite-dimensional vector space, say* dim $V = n$. *Then*

a) *every set of n vectors that span V is a basis,*
b) *every set of n linearly independent vectors from V is a basis.*

Proof. Suppose $V = [x_1, x_2, \ldots, x_n]$. According to Theorem 5.9, some subset of $\{x_1, x_2, \ldots, x_n\}$ is a basis for V. But the previous result implies that this subset must contain n elements and accordingly is the entire set $\{x_1, x_2, \ldots, x_n\}$.

For a proof of the second assertion, let $x_1, x_2, \ldots, x_n$ be n linearly independent vectors of V. If $x \in V$ is arbitrary, the set $\{x, x_1, \ldots, x_n\}$ is dependent, since the maximal number of linearly independent vectors in V is n. Thus some nontrivial dependence relation exists among these vectors:

$$cx + c_1x_1 + \cdots + c_nx_n = 0.$$

Were the coefficient $c = 0$, a contradiction to the linear independence of $x_1, x_2, \ldots, x_n$ would arise. Hence $c \neq 0$, and we may solve for the vector x in terms of $x_1, x_2, \ldots, x_n$ to obtain

$$x = (-c^{-1}c_1)x_1 + (-c^{-1}c_2)x_2 + \cdots + (-c^{-1}c_n)x_n.$$

This argument shows that $V = [x_1, x_2, \ldots, x_n]$, making the set $\{x_1, x_2, \ldots, x_n\}$ a basis for V.

Theorem 5.9 told us that it is possible to choose a basis for a vector space V from any set of generators of V. In the opposite direction the next theorem asserts that any linearly independent subset of V is either a basis or else can be extended to a basis for V, in the sense that vectors may be added to it to form a basis.

THEOREM 5.15 *If V is a finite-dimensional vector space and $\{x_1, x_2, \ldots, x_n\}$ is a linearly independent subset of V, then there exist vectors $y_{n+1}, \ldots, y_m$ such that $\{x_1, \ldots, x_n, y_{n+1}, \ldots, y_m\}$ forms a basis for V.*

Proof. The proof is short: Since V is finite-dimensional, it has a finite basis $y_1, y_2, \ldots, y_m$. As $V = [y_1, y_2, \ldots, y_m]$, we may apply the Steinitz Theorem to replace n of the y's by x's and obtain a set $\{x_1, \ldots, x_n, y_{n+1}, \ldots, y_m\}$ whose linear span is still V. But any generating set with m elements in an m-dimensional space is a basis.

COROLLARY *Every basis for a subspace of a finite-dimensional vector space can be extended to a basis for the entire space.*

Example 5.18

As a particular case of this last point, consider the vector space $M_2(F)$ and the three matrices

$$\begin{bmatrix} 1 & 0 \\ 0 & 0 \end{bmatrix}, \quad \begin{bmatrix} 1 & 1 \\ 0 & 0 \end{bmatrix}, \quad \begin{bmatrix} 1 & 1 \\ 1 & 0 \end{bmatrix}.$$

To check the linear independence of these matrices, let

$$c_1\begin{bmatrix} 1 & 0 \\ 0 & 0 \end{bmatrix} + c_2\begin{bmatrix} 1 & 1 \\ 0 & 0 \end{bmatrix} + c_3\begin{bmatrix} 1 & 1 \\ 1 & 0 \end{bmatrix} = \begin{bmatrix} 0 & 0 \\ 0 & 0 \end{bmatrix}.$$

On equating corresponding entries we see that

$$c_1 + c_2 + c_3 = 0, \qquad c_2 + c_3 = 0, \qquad c_3 = 0,$$

which imply that $c_1 = c_2 = c_3 = 0$.

Theorem 5.15 now tells us that the given matrices form at least part of a basis for $M_2(F)$. We shall leave the details to the reader and content ourselves with this one comment: The addition of any fourth matrix having a nonzero element in the $(2, 2)$-position will yield a linearly independent set. For instance, among the many possibilities, the matrices

$$\begin{bmatrix} 1 & 0 \\ 0 & 0 \end{bmatrix}, \quad \begin{bmatrix} 1 & 1 \\ 0 & 0 \end{bmatrix}, \quad \begin{bmatrix} 1 & 1 \\ 1 & 0 \end{bmatrix}, \quad \begin{bmatrix} 0 & 0 \\ 0 & 1 \end{bmatrix}$$

are linearly independent and, being four in number, must therefore be a basis for the four-dimensional space $M_2(F)$. This illustrates the wide latitude of choice for bases of $M_2(F)$.

We terminate the present section by giving some useful results concerning the dimensions of subspaces of a given vector space. The reader should first prove the following theorem.

THEOREM 5.16 *If W is a subspace of a finite-dimensional vector space V, then*

a) W is also finite-dimensional with $\dim W \leqslant \dim V$;
b) $\dim W = \dim V$ *if and only if* $W = V$.

The next theorem relates the dimensions of the sum and intersection of two subspaces.

THEOREM 5.17 *If U and W are subspaces of a finite-dimensional vector space V, then*

$$\dim(U + W) = \dim U + \dim W - \dim(U \cap W).$$

Proof. Before entering into the details of the proof we observe that by Theorem 5.16 the four subspaces involved in the statement of our theorem all have finite dimension. Now let $\{x_1, x_2, \ldots, x_n\}$ be a basis for $U \cap W$. According to Theorem 5.15, vectors $u_1, u_2, \ldots, u_m$ will exist such that $\{x_1, \ldots, x_n, u_1, \ldots, u_m\}$ is a basis for the subspace U and vectors $w_1, w_2, \ldots, w_r$ such that $\{x_1, \ldots, x_n, w_1, \ldots, w_r\}$ is a basis for the subspace W. Combine these two bases into a single set

$$\{u_1, \ldots, u_m, x_1, \ldots, x_n, w_1, \ldots, w_r\}.$$

Because the first $m + n$ vectors of the foregoing set are a basis for U and the last $n + r$ vectors are a basis for W, any element of the sum $U + W$ may be expressed as a linear combination of the vectors of this set, that is,

$$U + W = [u_1, \ldots, u_m, x_1, \ldots, x_n, w_1, \ldots, w_r].$$

We wish to show that the vectors on the right are also linearly independent and consequently a basis for the subspace $U + W$; once this is established, it would follow that

$$\begin{aligned}\dim(U + W) &= m + n + r = (m + n) + (n + r) - n \\ &= \dim U + \dim W - \dim(U \cap W).\end{aligned}$$

Therefore let us suppose that

$$a_1u_1 + \cdots + a_mu_m + b_1x_1 + \cdots + b_nx_n + c_1w_1 + \cdots + c_rw_r = 0.$$

Setting $z = c_1w_1 + \cdots + c_rw_r$, we would then have

$$z = -(a_1u_1 + \cdots + a_mu_m + b_1x_1 + \cdots + b_nx_n)$$

or $z \in [x_1, \ldots, x_n, u_1, \ldots, u_m] = U$. Since the vector z also belongs to W, it must be a linear combination of the basis elements $x_1, x_2, \ldots, x_n$ of the subspace $U \cap W$, say,

$$z = c_1w_1 + \cdots + c_rw_r = d_1x_1 + \cdots + d_nx_n,$$

so

$$d_1x_1 + \cdots + d_nx_n - c_1w_1 - \cdots - c_rw_r = 0.$$

But the set $\{x_1, \ldots, x_n, w_1, \ldots, w_r\}$ is linearly independent, being a basis for the subspace W, and hence $d_1 = \cdots = d_n = c_1 = \cdots = c_r = 0$. In conjunction with the linear independence of $\{x_1, \ldots, x_n, u_1, \ldots, u_m\}$ this forces the equality $a_1 = \cdots = a_m = b_1 = \cdots = b_n = 0$. We have thus succeeded in proving that the vectors $u_1, \ldots, u_m, x_1, \ldots, x_n, w_1, \ldots, w_r$ are linearly independent, as required.

PROBLEMS 5.2

In the problems below, the symbol V denotes a vector space over an arbitrary field F; that is, $V = V(F)$.

1. Prove that in any vector space V, if $x \in V$ and $c \in F$, then

a) $(-c)(-x) = cx$,

b) $n(cx) = (nc)x = c(nx)$ for all $n \in Z$,

c) $cx = 0$ implies that either $c = 0$ or $x = 0$.

2. Establish the following cancellation laws for any vector space V:

a) If $0 \neq x \in V$ and $c_1, c_2 \in F$, then $c_1x = c_2x$ implies that $c_1 = c_2$.

b) If $x, y \in V$ and $0 \neq c \in F$, then $cx = cy$ implies that $x = y$.

3. Prove that W is a subspace of the vector space V if and only if $\varnothing \neq W \subseteq V$ and $cx + y \in W$ for all $x, y \in W$ and $c \in F$.

4. For each of the following sets W, determine whether W is a subspace of the vector space $V_n(R^\#)$.

a) $W = \{(a_1, a_2, \ldots, a_n) | a_1 + a_2 + \cdots + a_n \neq 0\}$.

b) $W = \{(a_1, a_2, \ldots, a_n) | a_1 = a_2 = \cdots = a_n\}$.

c) $W = \{(a_1, a_2, \ldots, a_n) | a_1a_2 = 0\}$.

d) $W = \{(a_1, a_2, \ldots, a_n) | a_k \in Z \text{ for all } k\}$.

e) $W = \{(a_1, a_2, \ldots, a_n) | a_k \geqslant 0 \text{ for all } k\}$.

f) $W = \{(a_1, a_2, \ldots, a_n) | a_n = 0\}$.

g) $W = \{(a_1, a_2, \ldots, a_n) | a_1 + a_2 + \cdots + a_n = 1\}$.

5. Let U and W be subspaces of the vector space V. Prove that $U \cup W$ forms a subspace of V if and only if $U \subseteq W$ or $W \subseteq U$.

6. Determine whether W is a subspace of the indicated vector space:

a) $V_3(F)$; for fixed scalars $a_1, a_2, a_3 \in F$,

$$W = \{(x_1, x_2, x_3) | a_1x_1 + a_2x_2 + a_3x_3 = 0\}.$$

b) $V_\infty(F)$; W consists of sequences where all but a finite number of terms are equal to zero; that is,

$$W = \{(a_1, \ldots, a_n, 0, 0, \ldots) | a_k \in F,\ n \in Z_+\}.$$

c) $F[x]$; W consists of all polynomials of degree greater than 4.

d) $V_n(F)$; for a fixed $n \times n$ matrix (a_{ij}),

$$W = \{x \in V_n(F) | x(a_{ij}) = 0\}.$$

e) $M_3(F)$; W consists of all matrices of the form

$$\begin{bmatrix} a & b & 0 \\ 0 & a+b & 0 \\ 0 & a & b \end{bmatrix} \qquad a, b \in F.$$

f) $F[x]$; W consists of all polynomials whose constant term is zero.

g) $M_2(F)$; W consists of all matrices of the form

$$\begin{bmatrix} a & b \\ -b & a \end{bmatrix} \qquad a, b \in F.$$

7. Which of the following sets are subspaces of the vector space $M_n(F)$?

a) All upper triangular matrices of order n. (A square matrix (a_{ij}) is said to be *upper triangular* if $a_{ij} = 0$ for $i > j$ and *strictly upper triangular* if $a_{ij} = 0$ for $i \geqslant j$; analogous definitions hold for lower triangular matrices.)

b) All invertible matrices of order n.

c) All matrices of order n the sum of whose diagonal entries is zero.

8. Find all of the subspaces of $V_2(Z_3)$ and $V_3(Z_2)$.

9. Suppose U, W, and Y are subspaces of the vector space V. Prove that:

a) If $U \subseteq Y$, $W \subseteq Y$, then $U + W \subseteq Y$;

b) $(U \cap Y) + (W \cap Y) \subseteq (U + W) \cap Y$;

c) If $U \subseteq Y$, then $U + (W \cap Y) = (U + W) \cap Y$.

10. Let $V_0(R^\#)$ be the vector space of real-valued functions on the interval $[a, b]$ under pointwise addition and multiplication by a real number.

a) If W is the set of all functions f in V_0 such that $f(a) = 0$, is W a subspace of V_0?

b) If W is the set of all functions f in V_0 for which $f(a) = f(b)$, is W a subspace of V_0?

11. In $V(Z_3)$, determine all of the vectors in the subspace spanned by two vectors $x_1, x_2 \in V$.

12. If S and T are nonempty subsets of the vector space V, establish that:

a) $S \subseteq T$ implies $[S] \subseteq [T]$;

b) $[S \cup T] = [S] + [T]$;

c) $S = [S]$ if and only if S is a subspace of V;

d) $[[S]] = [S]$.

13. If the set S spans V and S is contained in the subspace W, prove that $W = V$.

14. Show that the polynomial space $F_n[x]$ is finitely generated with $F_n[x] = [1, x, x^2, \ldots, x^n]$.

15. For each of the following vector spaces, determine whether the sets listed are linearly dependent or independent.

a) $V_4(F)$: $\{(1, 0, 0, 0), (1, 1, 0, 0), (1, 1, 1, 0), (1, 1, 1, 1)\}$

b) $F[x]$: $\{x^2 + x - 1, x^2 - x - 2, x^2 + x + 1\}$

c) $V_3(Z_5)$: $\{(4, 1, 3), (2, 3, 1), (4, 1, 0)\}$

d) $V_3(C)$: $\{(1, 2 + i, 3), (2 - i, i, 1), (i, 2 + 3i, 2)\}$

16. Prove that $\{x_1, x_2\}$ is a linearly dependent set if and only if $x_1 = cx_2$ or $x_2 = cx_1$ for some $c \in F$.

17. Prove that if each vector $x \in [x_1, x_2, \ldots, x_n]$ is uniquely representable in the form $x = a_1x_1 + a_2x_2 + \cdots + a_nx_n$ $(a_k \in F)$, then the vectors $x_1, x_2, \ldots, x_n$ are linearly independent.

18. Let $(a_{ij}) \in M_n(F)$ be an upper triangular matrix with the property that $a_{ii} \neq 0$ for each i. Show that the columns of (a_{ij}), regarded as vectors in $V_n(F)$, are linearly independent.

19. Given vectors $x_1, x_2, \ldots, x_n \in V$, establish the assertions below:

a) If $x_i = x_j$ for some $i \neq j$, the set $\{x_1, x_2, \ldots, x_n\}$ is linearly dependent.

b) If $\{x_1, x_2, \ldots, x_n\}$ forms a linearly independent set and $a_1, a_2, \ldots, a_{n-1} \in F$, then

$$\{x_1 - a_1x_n, x_2 - a_2x_n, \ldots, x_{n-1} - a_{n-1}x_n\}$$

is also an independent set of vectors.

c) If $\{x_1, x_2, \ldots, x_n\}$ is linearly independent while $\{x_1, \ldots, x_n, x\}$ is linearly dependent, then the vector $x \in [x_1, x_2, \ldots, x_n]$.

d) The set $\{x_1, x_2, \ldots, x_n\}$ is a basis for V if and only if each vector in V is a unique linear combination of $x_1, x_2, \ldots, x_n$.

20. If $[x_1, x_2, \ldots, x_n] = [y_1, y_2, \ldots, y_m]$, where $n \neq m$, show that at least one of the sets $\{x_1, x_2, \ldots, x_n\}$ or $\{y_1, y_2, \ldots, y_m\}$ is linearly dependent.

21. Show that the vectors $(3 - i, 2 + 2i, 4)$, $(2, 2 + 2i, 3)$, and $(1 - i, -2i, 1)$ form a basis for the space $V_3(C)$ and determine the coordinates of each of the standard basis vectors $(1, 0, 0)$, $(0, 1, 0)$, and $(0, 0, 1)$ with respect to this basis.

22 a) Find a basis for the vector space $C(R^{\#})$ and all bases for the spaces $V_3(Z_2)$ and $V_2(Z_3)$.

b) For what values of a do the sets $\{(a, 1, 0), (1, a, 1), (0, 1, a)\}$ and $\{(1 + a, 1, 1), (1, 1 + a, 1), (1, 1, 1 + a)\}$ form bases for $V_3(R^{\#})$?

c) If $\{x_1, x_2, x_3\}$ is a basis for the vector space $V_3(F)$, does it follow that $\{x_1 + x_2, x_2 + x_3, x_3 + x_1\}$ and $\{x_1, x_1 + x_2, x_1 + x_3\}$ are necessarily bases?

23. Assume that $\{x_1, x_2, \ldots, x_n\}$ is a basis for the finite-dimensional vector space V. Establish that:

a) The set $\{x_1, x_1 + x_2, x_1 + x_2 + x_3, \ldots, x_1 + x_2 + \cdots + x_n\}$ is also a basis for V.

b) For any nonzero vector $x \in V$ there exists an index i $(1 \leqslant i \leqslant n)$ such that the set $\{x_1, \ldots, x_{i-1}, x, x_{i+1}, \ldots, x_n\}$ serves as a basis.

24. Determine the dimensions of the following vector spaces:

a) The set of vectors $(a_1, a_2, \ldots, a_n)$ of $V_n(F)$ such that $\sum_{k=1}^{n} a_k = 0$.

b) The set $F_n[x]$ of polynomials of degree n or less.

c) The set of matrices in $M_2(F)$ of the form

$$\begin{bmatrix} a & b \\ -b & a \end{bmatrix}.$$

d) The set $D_n(F)$ of diagonal matrices of order n. (A square matrix $(a_{ij}) \in M_n(R)$ is said to be *diagonal* if $a_{ij} = 0$ for $i \neq j$; in other words, $(a_{ij}) = (a_i \delta_{ij})$.)

25. Let W be the subset of the vector space map(X, F) consisting of functions f such that $f(x) = 0$ for all but finitely many elements $x \in X$. Prove that:

a) W is a subspace of map(X, F).

b) If for each $a \in X$, the function f_a is defined by

$$f_a(x) = \begin{cases} 1 & \text{if } x = a, \\ 0 & \text{if } x \neq a, \end{cases}$$

then the set $S = \{f_a | a \in X\}$ is a basis for W.

26. Show that the vector spaces $F[x]$ and $V_\infty(F)$ are infinite-dimensional.

27. Given a vector space V, prove that:

a) If dim $V = n$, then V contains subspaces of dimension k, for all $k \leqslant n$.

b) If V is infinite-dimensional, then V has a linearly independent subset containing k elements, for all $k \geqslant 1$.

28. If W is a subspace of the finite-dimensional vector space V, verify that:

a) W is also finite-dimensional, with dim $W \leqslant$ dim V.

b) If dim W = dim V, then $W = V$.

29. Let $\{x_1, x_2, \ldots, x_n\}$ be a basis for the complex vector space $V(C)$. Regarding V as a vector space over the field $R^\#$, show that $\{x_1, \ldots, x_n, ix_1, \ldots, ix_n\}$ forms a basis; hence

$$\dim V(R^\#) = 2 \dim V(C).$$

30. Assuming that U and V are subspaces of $V_n(F)$ with dim $U > n/2$, dim $W > n/2$, verify that $U \cap W \neq \{0\}$.

5.3 EXTENSIONS OF FIELDS

The bulk of this section deals with various types of extensions of fields and the extension of isomorphisms between fields. One question on which we wish to focus is the following: Given an arbitrary polynomial of positive degree in $F[x]$, is it possible to find a field F' containing F such that the polynomial in question has a root in F'? If the answer is in the affirmative—as will be revealed in Theorem 5.22—then the obvious goal is to construct these fields F'. A second concern is with the structure of fields having finitely many elements. In this connection it is reasonable to ask whether there exist finite fields other than the fields Z_p (p a prime) and, if so, whether they can be completely described. In actual fact, such fields exist in abundance. We shall prove that there is essentially one field with p^n elements for each prime p and $n > 0$.

By an *extension* F' of a field F we simply mean any field that contains F as a subfield. For instance, the field of real numbers is an extension of Q, the rational number field. In view of Theorem 4.34 we may remark that every field F is an extension of a field isomorphic to Q or to Z_p, according as the characteristic of F is 0 or a prime p.

Assume that F' is an extension field of a field F and let the elements $r_1, r_2, \ldots, r_n$ all lie in F'. The subfield of F' generated by the subset $F \cup \{r_1, r_2, \ldots, r_n\}$ is customarily denoted by $F(r_1, r_2, \ldots, r_n)$:

$$F(r_1, r_2, \ldots, r_n) = \cap\{K | K \text{ is a subfield of } F';\ F \subseteq K;\ r_i \in K\}.$$

Thus $F(r_1, r_2, \ldots, r_n)$ is an extension field of F containing the elements r_i (clearly, it is the smallest such extension), and one speaks of $F(r_1, r_2, \ldots, r_n)$ as being obtained by *adjoining* the r_i to F, or by adjunction of the elements r_i to F. The purpose of the coming theorem is to determine, up to isomorphism, the structure of all *simple extension fields*, that is, extension fields $F(r)$ arising from a field F by the adjunction of a single element r.

Now for each element $r \in F'$ we have at our disposal the substitution homomorphism $\phi_r: F[x] \to F'$; the reader will recall that this is defined by taking $\phi_r f(x) = f(r)$. As before, the image of $F[x]$ under ϕ_r is represented by the symbol $F[r]$:

$$F[r] = \{f(r) | f(x) \in F[x]\}.$$

The set $F[r]$ forms an integral domain (being a subring of the field F') and therefore has a field of quotients $K = Q_{cl}(F[r])$ in F'. It is apparent that $F \cup \{r\} \subseteq F[r] \subseteq K$. But $F(r)$ is the smallest subfield of F' to contain both F and r, whence $F(r) \subseteq K$. On the other hand, any subring of F' that contains F and r necessarily contains the elements of $F[r]$; in particular, $F[r] \subseteq F(r)$. Since $F(r)$ is a subfield of F', it must also contain all the quotients of elements in $F[r]$. Thus $K \subseteq F(r)$, and equality follows. This leads to the more constructive description of $F(r)$ as $F(r) = Q_{cl}(F[r])$. The key to classifying simple extensions is the nature of the kernel of the substitution homomorphism. Bear in mind that $\ker \phi_r$ consists of all polynomials in $F[x]$ having r as a root.

THEOREM 5.18 *Let F' be an extension field of the field F and let $r \in F'$. Then either*

a) $F(r) \simeq Q_{cl}(F[x])$ *or else*
b) $F(r) \simeq F[x]/(f(x))$ *for some monic irreducible polynomial* $f(x) \in F[x]$ *such that* $f(r) = 0$*; this polynomial is uniquely determined.*

Proof. By the Fundamental Homomorphism Theorem we know that

$$F[r] \simeq F[x]/\ker \phi_r.$$

Since $F[r]$ is a subring of a field, it must be an integral domain. Hence $\ker \phi_r$ constitutes a prime ideal of the principal ideal domain $F[x]$. One observation is quite pertinent: $\ker \phi_r$ cannot be all of $F[x]$, since the identity element of $F[x]$ is not mapped onto zero. Now two possibilities arise: Either $\ker \phi_r = \{0\}$ or $\ker \phi_r \neq \{0\}$.

Suppose first that $\ker \phi_r = \{0\}$. Then the homomorphism ϕ_r is actually an isomorphism, and $F[r] \simeq F[x]$. In the situation at hand, $F[r]$ is not a field, since $F[x]$ fails to be one. However, from Theorem 4.32 we know that the quotient fields of $F[r]$ and $F[x]$ are isomorphic under a mapping induced by the substitution homomorphism. Since $F(r)$ is the quotient field of $F[r]$, one thus obtains

$$F(r) \simeq Q_{cl}(F[x]).$$

If the kernel of ϕ_r is nonzero, then $\ker \phi_r = (f(x))$ for some irreducible polynomial (prime element) of $F[x]$, where $f(x)$ can be taken to be monic. Because every nonzero prime ideal of $F[x]$ is maximal, $F[x]/(f(x))$ forms a field, and the same will be true of its isomorphic image $F[r]$. But $F(r)$ is the smallest field to contain both F and r, from which it follows that $F[r] = F(r)$; this leads to the isomorphism

$$F(r) \simeq F[x]/(f(x)).$$

Since $\ker \phi_r = (f(x))$, a polynomial $g(x) \in F[x]$ has r as a root if and only if $g(x)$ is divisible by $f(x)$. Accordingly, if $g(x)$ is any monic irreducible polynomial in $F[x]$ having r as a root, each of $f(x)$ and $g(x)$ divides the other; since both of these polynomials are monic, this is possible only if $f(x) = g(x)$. Thus $f(x)$ is unique, as is asserted in the theorem.

Another virtue of the substitution homomorphism is that it permits us to put the elements of an extension field into one of two essentially different categories:

DEFINITION 5.8 Let F' be an extension field of a field F. An element $r \in F'$ is said to be *algebraic* over F if $\ker \phi_r \neq \{0\}$; otherwise, r is termed *transcendental* over F.

Definition 5.8 in effect asserts that r is algebraic over F if there exists a nonzero polynomial $f(x) \in F[x]$ such that $f(r) = 0$; on the other hand, r is a transcendental element over F in case $f(r) \neq 0$ for every nonzero polynomial $f(x)$ in $F[x]$. As illustrations of these notions, take $F' = R^{\#}$ and $F = Q$; then $\sqrt{2} + \sqrt{3}$ is algebraic over Q, being a root of the polynomial $x^4 - 10x^2 + 1 \in Q[x]$, while π and $2^{\sqrt{2}}$ are both transcendental over Q. Every element of the field F is trivially algebraic over F, since the element $r \in F$ is a root of the linear polynomial $x - r \in F[x]$.

The proof of Theorem 5.18 furnishes us with more detailed information concerning simple extension fields, which we now state as two separate theorems. First, however, let us remark that the field of quotients of the polynomial domain $F[x]$ is traditionally called the *field of rational functions over F*. Thus a rational function (in an indeterminant x) over F can be written as a quotient $f(x)/g(x)$, where $f(x)$ and $g(x) \neq 0$ are both polynomials in $F[x]$.

THEOREM 5.19 (Simple Transcendental Field Extensions) *If $r \in F' \supseteq F$ is transcendental over F, then $F(r)$ is isomorphic to the field of rational functions over F. In fact, there is an isomorphism α_r of $F(r)$ into $Q_{cl}(F[x])$ such that $\alpha_r(r) = x$ and $\alpha_r(a) = a$ for every $a \in F$.*

This theorem completely determines the structure of simple transcendental extensions over F; they are all isomorphic to the field of rational functions over F and hence to each other. Thus, for instance, $Q(\pi) \simeq Q(2^{\sqrt{2}})$.

As regards simple algebraic extensions, we have

THEOREM 5.20 (Simple Algebraic Field Extensions) *If $r \in F' \supseteq F$ is algebraic over F, then a unique monic irreducible polynomial $f(x) \in F[x]$ exists such that $f(r) = 0$. Furthermore, if $g(x)$ is a polynomial in $F[x]$ for which $g(r) = 0$, necessarily $f(x) | g(x)$.*

The unique polynomial $f(x)$ of Theorem 5.20 is referred to as the *minimum polynomial* of (or belonging to) r over F; as the name suggests, the minimum polynomial of r is the monic polynomial in $F[x]$ of minimal degree having r as a root. The *degree* of an algebraic element $r \in F' \supseteq F$ is just the degree of its minimum polynomial. This degree is 1 if and only if $r \in F$.

In the course of proving Theorem 5.18 we established the surprising fact that, whenever r is algebraic over F, the integral domain $F[r]$ becomes a field, so that $F(r) = F[r]$. This amounts to saying that every element of $F(r)$ is of the form $f(r)$, where $f(x)$ is a polynomial in $F[x]$.

Example 5.19

If $n \neq 1$ is any square-free integer, the element $\sqrt{n} \in R^{\#} \supseteq Q$ is a root of the quadratic polynomial $x^2 - n \in Q[x]$ and hence is algebraic over Q. From the preceding paragraph we know that $Q[\sqrt{n}]$ is a field, and so $Q[\sqrt{n}] = Q(\sqrt{n})$; in other words, every member of $Q(\sqrt{n})$ is of the form $f(\sqrt{n})$, where $f(x)$ is a polynomial in $Q[x]$:

$$Q(\sqrt{n}) = \{a_0 + a_1\sqrt{n} + a_2(\sqrt{n})^2 + \cdots + a_k(\sqrt{n})^k | a_i \in Q;\ k \geqslant 0\}.$$

But $(\sqrt{n})^2 = n$, $(\sqrt{n})^3 = n\sqrt{n}$, ..., so $Q(\sqrt{n})$ can be described more simply as

$$Q(\sqrt{n}) = \{a + b\sqrt{n} | a, b \in Q\}.$$

That is to say, the simple field extension $Q(\sqrt{n})$ is what we referred to as a quadratic field in Section 4.5.

Note also that an arbitrary element $a + b\sqrt{n} \in Q(\sqrt{n})$ satisfies the polynomial

$$x^2 - 2ax + (a^2 - b^2n) \in Q[x].$$

The implication is that not only $\sqrt{n}$, but every member of $Q(\sqrt{n})$ is algebraic over Q; the corollary to our next-proved theorem demonstrates that this is no accident.

We still have a few loose ends to tie together, including a more precise description of $F(r)$, when r is algebraic over F.

THEOREM 5.21 *Let F' be an extension of F and $r \in F'$ be algebraic over F of degree n. Then every element of $F(r)$ can be uniquely represented in the form*

$$c_0 + c_1r + c_2r^2 + \cdots + c_{n-1}r^{n-1}$$

where the coefficients $c_i \in F$.

Proof. Let a be any element of $F(r) = F[r]$. Then there exists a polynomial $g(x) \in F[x]$ such that $a = g(r)$. Applying the Division Algorithm to $g(x)$ and the minimum polynomial $f(x)$ of r, we can find $q(x)$ and $s(x)$ in $F[x]$ satisfying

$$g(x) = q(x)f(x) + s(x)$$

where either $s(x) = 0$ or $\deg s(x) < n$. Since $f(r) = 0$, it follows that $s(r) = g(r) = a$. If $s(x) = 0$, then necessarily $a = 0$, while if $s(x) = c_0 + c_1x + \cdots + c_mx^m$ is a nonzero polynomial of degree $m < n$, then $a = c_0 + c_1r + \cdots + c_mr^m$.

It remains to show that the representation of an element of $F(r)$ in the indicated form is unique. Pursuing this aim, let us assume that

$$c_0 + c_1r + \cdots + c_{n-1}r^{n-1} = d_0 + d_1r + \cdots + d_{n-1}r^{n-1}$$

where the c_i and d_i lie in F. Then the polynomial

$$h(x) = (c_0 - d_0) + (c_1 - d_1)x + \cdots + (c_{n-1} - d_{n-1})x^{n-1} \in F[x],$$

and clearly $h(r) = 0$, so $h(x) \in \ker \phi_r = (f(x))$. This being the case, $h(x) = f(x)k(x)$ for some polynomial $k(x)$ in $F[x]$. But if $h(x) \neq 0$, we obtain

$$n > \deg h(x) = \deg f(x) + \deg k(x) \geqslant \deg f(x) = n,$$

a manifestly false conclusion. Thus the polynomial $h(x) = 0$, which forces $c_i - d_i = 0$ or $c_i = d_i$ for all i. The coefficients c_i are therefore uniquely determined.

COROLLARY *Let $r \in F' \supseteq F$ be algebraic over F of degree n. Then each element of the simple extension $F(r)$ is also algebraic over F.*

Proof. Suppose that $a \in F(r)$ and consider the elements 1, a, $a^2, \ldots, a^n$; these all belong to the field $F(r)$. By the theorem, any given member $s \in F(r)$ is uniquely expressible in the form $x = c_0 + c_1r + \cdots + c_{n-1}r^{n-1}$, where $c_i \in F$. In particular, each of the powers a^k $(0 \leqslant k \leqslant n)$ may be so written

$$a^k = c_{0k} + c_{1k}r + \cdots + c_{n-1k}r^{n-1} \qquad (k = 0, 1, \ldots, n).$$

Thus we have $n + 1$ equations in the n elements $1, r, \ldots, r^{n-1}$, and by eliminating these latter elements it follows that there exist b_0, $b_1, \ldots, b_n \in F$, not all zero, for which

$$b_0 1 + b_1 a + \cdots + b_n a^n = 0.$$

But then $f(x) = b_0 + b_1x + \cdots + b_nx^n$ is a nonzero polynomial in $F[x]$ with $f(a) = 0$, implying that a is algebraic over F.

To recapitulate: If $r \in F' \supseteq F$ is algebraic over F of degree n, then every element of $F(r)$ is of the form $f(r)$, where $f(x) \in F[x]$ and either $f(x) = 0$ or $\deg f(x) < n$; in symbols,

$$F(r) = \{c_0 + c_1r + \cdots + c_{n-1}r^{n-1} | c_i \in F\}.$$

Example 5.20

Consider the element $r = \sqrt{2} + i \in C \supseteq Q$, C as usual being the complex number field. Then $r^2 = 1 + 2\sqrt{2}i$, so $(r^2 - 1)^2 = -8$ or, equivalently, $r^4 - 2r^2 + 9 = 0$. Thus r is a root of the polynomial

$$f(x) = x^4 - 2x^2 + 9 \in Q[x]$$

and hence is an algebraic element over Q. Now $f(x)$ has the irreducible factorization over C,

$$f(x) = (x - \sqrt{2} + i)(x - \sqrt{2} - i)(x + \sqrt{2} + i)(x + \sqrt{2} - i),$$

which indicates that $f(x)$ has no linear or quadratic factors in $Q[x]$. Therefore $f(x)$ is irreducible as a member of $Q[x]$ and serves as the minimum polynomial of r over Q; in particular, the element r has degree 4. By Theorem 5.21 the simple extension $Q(r)$ can be described as

$$Q(r) = \{c_0 + c_1 r + c_2 r^2 + c_3 r^3 | c_i \in Q\}.$$

Since $r = \sqrt{2} + i$, $r^2 = 1 + 2\sqrt{2}i$, and $r^3 = -\sqrt{2} + 5i$, this simplifies to

$$Q(\sqrt{2} + i) = \{(a + b\sqrt{2}) + (c + d\sqrt{2})i | a, b, c, d \in Q\}.$$

At the same time, r is a root of the polynomial $x^2 - 2\sqrt{2}x + 3 \in R^{\#}[x]$, with $x^2 - 2\sqrt{2}x + 3$ irreducible over $R^{\#}$; thus r is of degree 2 over $R^{\#}$.

Example 5.21

For a second illustration we turn to the extension field $Q(\sqrt{2}, \sqrt{3})$. The elements $\sqrt{2}$ and $\sqrt{3}$ are clearly algebraic over Q, being roots of the polynomials $x^2 - 2$, $x^2 - 3 \in Q[x]$, respectively. Our contention is that the field $Q(\sqrt{2}, \sqrt{3})$ is actually a simple algebraic extension of Q; in fact, $Q(\sqrt{2}, \sqrt{3}) = Q(\sqrt{2} + \sqrt{3})$, with $\sqrt{2} + \sqrt{3}$ algebraic over Q.

Since the element $\sqrt{2} + \sqrt{3}$ belongs to the field $Q(\sqrt{2}, \sqrt{3})$, we certainly must have $Q(\sqrt{2} + \sqrt{3} \subseteq Q(\sqrt{2}, \sqrt{3})$. As regards the reverse inclusion, a simple computation shows that

$$2\sqrt{2} = (\sqrt{2} + \sqrt{3})^3 - 9(\sqrt{2} + \sqrt{3})$$

is a member of $Q(\sqrt{2} + \sqrt{3})$, and therefore so is $\sqrt{2}$. But then

$$\sqrt{3} = (\sqrt{2} + \sqrt{3}) - \sqrt{2}$$

also lies in $Q(\sqrt{2} + \sqrt{3})$. This leads to the inclusion $Q(\sqrt{2}, \sqrt{3}) \subseteq Q(\sqrt{2} + \sqrt{3})$ and the asserted equality.

To see that $r = \sqrt{2} + \sqrt{3}$ is an algebraic element over Q, note that $r^2 = 5 + 2\sqrt{6}$, $(r^2 - 5)^2 = 24$, and hence the polynomial $f(x) = x^4 - 10x^2 + 1$ has r as a root. One may verify that $f(x)$ is irreducible in $Q[x]$, making it the minimum polynomial of r. Perhaps the quickest way to see this is to view $f(x)$ as a polynomial in $Z[x]$ and apply the Eisenstein Criterion with $p = 5$.

Kronecker's theorem assures us that, given any nonconstant polynomial over a field, there will always be an extension field in which the polynomial has a root. The proof of the theorem, in his 1887 paper "A Fundamental Theorem in General Arithmetic," does more than simply establish existence; it also provides a method for constructing the required field.

Since his name runs like a thread through the text, it is time to devote a few words to Leopold Kronecker (1823–1891). The son of wealthy parents, Kronecker was born near Breslau, Germany, where Ernst Kummer happened to be one of his secondary school teachers. Entering the University of Berlin in 1841, Kronecker completed his doctorate under Dirichlet with a thesis on algebraic number theory (1845). For the next ten years, Kronecker renounced mathematics to devote himself to managing the family business and agricultural affairs. In 1855, when Kummer and Karl Weierstrass were called to the University of Berlin, Kronecker settled there too. As a financially independent private scholar, he was able to devote the rest of his life to his hobby, mathematics. He began lecturing regularly at the University when his election (1861) to the Berlin Academy afforded him this privilege. These lectures, which dealt mainly with algebra and number theory, soon attracted gifted young students from other German universities as well as foreign countries. In the last third of the nineteenth century the triumvirate of Kummer, Kronecker, and Weierstrass, with their emphasis upon severe rigor of demonstration, made Berlin the center of advanced work in mathematics. In 1883, Kummer, feeling that his mathematical productivity was declining, suddenly decided to retire. Kronecker was named to replace his lifelong friend and mentor, thus becoming the first person to hold a Berlin professorship who had also earned his doctoral degree there. He held this position until his death in 1891.

Kronecker disturbed his mathematical followers by advocating the universal "arithmetization" of mathematics—that is, resting the foundations of the subject upon the integers alone. He was willing to accept the "reality" of the integers because these are clear to the intuition; as he said, "God created the integers, all the rest is the work of man." He held that the rational and irrational numbers must be built up, in a finite number of steps, from the integers. It is hardly surprising that he found the contemporary theory of irrational numbers highly suspect. He is reputed to have demanded of Ferdinand Lindemann, who had proved (1882) that π is a transcendental irrational, "Of what use is your beautiful research on π? Why study such problems, since there are no irrational numbers at all?"

Kronecker further insisted that nothing could be said to have mathematical existence unless it could actually be constructed. His wrath was particularly directed at existence proofs that did not permit the construction, in a finite way, of the object in question. He insisted that to countenance such "theological" existence proofs would undermine the entire edifice of mathematical truth. Yet to purge the discipline of all nonconstructive existence proofs would mean discarding vast tracts of mathematics, a sacrifice that few practicing mathematicians would approve, either then or now.

Until now, we have always begun by assuming the existence of an extension field F' of F and then studied the structure of simple extensions $F(r)$ within F'. The subject can be approached from a somewhat different standpoint. Given a field F and an irreducible polynomial $f(x) \in F[x]$, one may ask whether it is possible to construct a simple extension F' of F in which $f(x)$, thought of as a member of $F'[x]$, has a root. If $\deg f(x) = 1$, then in a trivial sense, F is itself the required extension.

To answer this question, we take our cue from Theorem 5.18. If such an extension of F can be found at all, it must be of the form $F(r)$, with r algebraic over F. As was pointed out in our earlier discussion, r will possess a minimum polynomial $g(x)$ that is irreducible in $F[x]$ and such that $F(r) \simeq F[x]/(g(x))$. This suggests that, when starting with a prescribed irreducible polynomial $f(x) \in F[x]$, the natural object of interest should be the associated quotient ring $F[x]/(f(x))$.

After this preamble, let us proceed to some pertinent details.

THEOREM 5.22 (Kronecker) *If $f(x)$ is an irreducible polynomial in $F[x]$, then there is an extension field of F in which $f(x)$ has a root.*

Proof. For brevity we shall write I in place of the principal ideal of $F[x]$ generated by polynomial $f(x)$; that is to say, $I = (f(x))$. Since $f(x)$ is assumed to be irreducible, the associated quotient ring $F' = F[x]/I$ is a field. To see that F' constitutes an extension of F, consider the natural mapping $\mathrm{nat}_I : F[x] \to F'$. According to Theorem 4.29, either the restriction $\mathrm{nat}_I \mid F$ is the trivial homomorphism or else $\mathrm{nat}_I(F)$ forms a field isomorphic to F, where as usual

$$\mathrm{nat}_I(F) = \{a + I \mid a \in F\}.$$

The first possibility is immediately excluded by the fact that

$$\mathrm{nat}_I(1) = 1 + I \neq I,$$

where I is the zero element of F'. Therefore F is imbeddable in the (quotient) field F', and, in this sense, F' becomes an extension of F.

We have yet to establish that the polynomial $f(x)$ actually has a root in F'. Let us assume that $f(x) = a_0 + a_1x + \cdots + a_nx^n$; then, from the definitions of coset addition and multiplication,

$$(a_0 + I) + (a_1 + I)(x + I) + \cdots + (a_n + I)(x + I)^n$$
$$= a_0 + a_1x + \cdots + a_nx^n + I = f(x) + I = 0 + I.$$

If we now identify an element $a_k \in F$ with the coset $a_k + I$ that it determines in F' (the fact that F is isomorphic to $\text{nat}_I(F)$ permits this), we obtain

$$a_0 + a_1(x + I) + \cdots + a_n(x + I)^n = 0,$$

which is equivalent to asserting that $f(x + I) = 0$. In other words, the coset $x + I = 1x + I$ is the root of $f(x)$ sought in F'.

Since each polynomial of positive degree has an irreducible factor (Theorem 4.66), we may drop the restriction that $f(x)$ be irreducible.

COROLLARY *If the polynomial $f(x) \in F[x]$ is of positive degree, then there exists an extension field of F containing a root of $f(x)$.*

Let us go back to Theorem 5.22 for a moment and take a closer look at the nature of the cosets of $I = (f(x))$ in $F[x]$ with the aim of expressing the extension field $F' = F[x]/I$ in a more convenient way. As usual, these cosets are of the form $g(x) + I$, with $g(x) \in F[x]$. By invoking the division algorithm, for each such $g(x)$ there is a unique polynomial $r(x)$ in $F[x]$ satisfying $g(x) = q(x)f(x) + r(x)$, where $r(x) = 0$ or $\deg r(x) < \deg f(x)$. Now $g(x) - r(x) = q(x)f(x) \in I$, so $g(x)$ and $r(x)$ determine the same coset; $g(x) + I = r(x) + I$. From this it is possible to draw the following conclusion: Each coset of I in $F[x]$ contains exactly one polynomial that has degree less than that of $f(x)$ or else is the zero polynomial. In fact, the cosets of I are uniquely determined by remainders on division by $f(x)$ in the sense that $g(x) + I = h(x) + I$ if and only if $g(x)$ and $h(x)$ leave the same remainder when divided by $f(x)$.

This means that if $\deg f(x) = n > 1$ (for instance, $f(x) = a_0 + a_1x + \cdots + a_nx^n$), then the extension field F' may be described by

$$F' = \{b_0 + b_1x + \cdots + b_{n-1}x^{n-1} + I \mid b_k \in F\}.$$

Identifying $b_k + I$ with the element b_k, we see, as before, that a typical coset can be uniquely represented in the form

$$b_0 + b_1(x + I) + \cdots + b_{n-1}(x + I)^{n-1}.$$

As a final simplification, let us replace $x + I$ by some new symbol λ, so that the elements of F' become polynomials in λ:

$$F' = \{b_0 + b_1\lambda + \cdots + b_{n-1}\lambda^{n-1} \mid b_k \in F\}.$$

Observe that since $\lambda = x + I$ is a root of $f(x)$ in F', calculations are carried out with the aid of the relation $a_0 + a_1\lambda + \cdots + a_n\lambda^n = 0$.

We pause now to examine two concrete examples of the ideas just presented.

Example 5.22

Consider Z_2, the field of integers modulo 2, and the polynomial $f(x) = x^3 + x + 1 \in Z_2[x]$. Since neither of the elements 0 and 1 is a root of $x^3 + x + 1$, $f(x)$ must be irreducible in $Z_2[x]$.

Theorem 5.22 thus guarantees the existence of an extension of Z_2, specifically, the field $Z_2[x]/(f(x))$, in which the given polynomial has a root. If we denote this root by λ, the discussion above tells us that

$$\begin{aligned} Z_2[x]/(f(x)) &= \{a + b\lambda + c\lambda^2 \mid a, b, c \in Z_2\} \\ &= \{0, 1, \lambda, 1 + \lambda, \lambda^2, 1 + \lambda^2, \lambda + \lambda^2, 1 + \lambda + \lambda^2\}, \end{aligned}$$

where, of course, $\lambda^3 + \lambda + 1 = 0$.

As an example of operating in this field, let us calculate the inverse of $1 + \lambda + \lambda^2$. Before starting, observe that by using the relations

$$\lambda^3 = -(\lambda + 1) = \lambda + 1, \qquad \lambda^4 = \lambda^2 + \lambda$$

(our coefficients come from Z_2, where $-1 = 1$), the degree of any product can be kept less than 3. Now the problem is to determine elements $a, b, c \in Z_2$ for which

$$(1 + \lambda + \lambda^2)(a + b\lambda + c\lambda^2) = 1.$$

Carrying out the multiplication and substituting for λ^3, λ^4 in terms of 1, λ, and λ^2, we obtain

$$(a + b + c) + a\lambda + (a + b)\lambda^2 = 1.$$

This yields the system of linear equations

$$a + b + c = 1, \qquad a = 0, \qquad a + b = 0,$$

with solution $a = b = 0$, $c = 1$; therefore $(1 + \lambda + \lambda^2)^{-1} = \lambda^2$.

The full multiplication table is given below:

$\cdot$	0	1	λ	$1+\lambda$	λ^2	$1+\lambda^2$	$\lambda+\lambda^2$	$1+\lambda+\lambda^2$
0	0	0	0	0	0	0	0	0
1	0	1	λ	$1+\lambda$	λ^2	$1+\lambda^2$	$\lambda+\lambda^2$	$1+\lambda+\lambda^2$
λ	0	λ	λ^2	$\lambda+\lambda^2$	$1+\lambda$	1	$1+\lambda+\lambda^2$	$1+\lambda^2$
$1+\lambda$	0	$1+\lambda$	$\lambda+\lambda^2$	$1+\lambda^2$	$1+\lambda+\lambda^2$	λ^2	1	λ
λ^2	0	λ^2	$1+\lambda$	$1+\lambda+\lambda^2$	$\lambda+\lambda^2$	λ	$1+\lambda^2$	1
$1+\lambda^2$	0	$1+\lambda^2$	1	λ^2	λ	$1+\lambda+\lambda^2$	$1+\lambda$	$\lambda+\lambda^2$
$\lambda+\lambda^2$	0	$\lambda+\lambda^2$	$1+\lambda+\lambda^2$	1	$1+\lambda^2$	$1+\lambda$	λ	λ^2
$1+\lambda+\lambda^2$	0	$1+\lambda+\lambda^2$	$1+\lambda^2$	λ	1	$\lambda+\lambda^2$	λ^2	$1+\lambda$

It is worth noting that $x^3 + x + 1$ factors completely into linear factors in $Z_2[x]/(f(x))$ and has the three roots λ, λ^2, and $\lambda + \lambda^2$:

$$x^3 + x + 1 = (x - \lambda)(x - \lambda^2)(x - (\lambda + \lambda^2)).$$

Example 5.23

The quadratic polynomial $x^2 + 1$ is irreducible in $R^{\#}[x]$. If $x^2 + 1$ were reducible, it would be of the form

$$\begin{aligned} x^2 + 1 &= (ax + b)(cx + d) \\ &= acx^2 + (ad + bc)x + bd, \end{aligned}$$

where $a, b, c, d \in R^{\#}$. It follows at once that $ac = bd = 1$ and $ad + bc = 0$. Therefore $bc = -(ad)$, and

$$1 = (ac)(bd) = (ad)(bc) = -(ad)^2,$$

or rather $(ad)^2 = -1$, which is impossible.

In this instance the extension field $R^{\#}[x]/(x^2 + 1)$ is described by

$$R^{\#}[x]/(x^2 + 1) = \{a + b\lambda | a, b \in R^{\#}; \lambda^2 + 1 = 0\}.$$

Performing the usual operations for polynomials, we see that

$$(a + b\lambda) + (c + d\lambda) = (a + c) + (b + d)\lambda$$

and

$$\begin{aligned} (a + b\lambda)(c + d\lambda) &= ac + (ad + bc)\lambda + bd\lambda^2 \\ &= (ac - bd) + (ad + bc)\lambda. \end{aligned}$$

The similarity of these formulas to the usual rules for addition and multiplication of complex numbers should be apparent to the reader. As a matter of fact, $R^{\#}[x]/(x^2 + 1)$ is isomorphic to the field C of complex numbers under the mapping $\Phi: R^{\#}[x]/(x^2 + 1) \to C$ given by $\Phi(a + b\lambda) = a + bi$. This provides an elegant way of constructing C from $R^{\#}$.

Before we proceed further, two comments are in order. First, Example 5.22 shows that there exist finite fields other than the fields Z_p of integers modulo a prime p. The fact that the field of this example has $2^3 = 8$ elements is typical of the general situation: If F is a finite field, then F contains p^n elements, where the prime p is the characteristic of F (Theorem 5.32).

In the second place the construction of Theorem 5.22 yields an extension of the field F in which a given (nonconstant) polynomial $f(x) \in F[x]$ splits off one linear factor. By repeated application of this procedure we can build up an extension F' of F in which $f(x)$, thought of as a member of $F'[x]$, factors into a product of linear factors; that is, the field F' is large enough to contain all the roots of $f(x)$. Technically speaking, the polynomial *splits completely* in $F'[x]$. We present this result in the form of an existence theorem.

THEOREM 5.23 *If $f(x) \in F[x]$ is a polynomial of positive degree, then an extension field F' of F exists in which $f(x)$ factors completely into linear polynomials.*

Proof. The proof is by induction on $n = \deg f(x)$. If $n = 1$, then $f(x)$ is already linear, and F itself is the required extension. Therefore assume that $n > 1$ and that the theorem is true for all fields and for all polynomials of degree less than n. Now the polynomial $f(x)$ must have some irreducible factor $g(x)$. By Theorem 5.22 there is an extension field K of F in which $g(x)$ and hence $f(x)$ has a root r_1, specifically, the field $K = F[x]/(g(x))$. Thus $f(x)$ can be written in $K[x]$ as $f(x) = (x - r_1)h(x)$, where $\deg h(x) = n - 1$. By our induction assumption there is an extension F' of K in which $h(x)$ splits completely, say $h(x) = a(x - r_2)(x - r_3)\cdots(x - r_n)$, with $r_i \in F'$, $a \neq 0$. From this we see that $f(x)$ can be factored into linear factors in $F'[x]$.

COROLLARY *Let $f(x) \in F[x]$, $\deg f(x) = n > 0$. Then an extension of F exists in which $f(x)$ has n (not necessarily distinct) roots.*

Example 5.24

To illustrate the foregoing situation, let us look at the polynomial $f(x) = x^4 - 2$ in $Z_5[x]$. Since $f(x)$ has no roots in Z_5, it can have no linear factors in $Z_5[x]$. As regards a possible quadratic factorization, suppose that

$$x^4 - 2 = (x^2 + ax + b)(x^2 + cx + d),$$

where $a, b, c, d \in Z_5$. Then, on equating corresponding coefficients,

$$a + c = 0, \qquad ac + b + d = 0, \qquad ad + bc = 0, \qquad bd = -2.$$

Substitution of $c = -a$ into $ad + bc = 0$ gives $a(d - b) = 0$. If $a \neq 0$, then $d = b$, which leads to $b^2 = -2$, an impossibility in Z_5. On the other hand, if $a = 0$, then the system of equations reduces to

$$b + d = 0, \qquad bd = -2.$$

These in their turn yield $b^2 = 2$, which again is impossible. The implication of our reasoning is that $f(x)$ cannot be factored into either linear or quadratic polynomials and is therefore irreducible over Z_5.

Appealing to Theorem 5.22, we see that the quotient ring

$$Z_5[x]/(x^4 - 2) = \{a + b\lambda + c\lambda^2 + d\lambda^3 | a, b, c, d \in Z_5;\ \lambda^4 = 2\}$$

is a field containing $5^4 = 625$ elements. Notice that because $2^4 \equiv 3^4 \equiv 4^4 \equiv 1 \pmod 5$,

$$\lambda^4 = (2\lambda)^4 = (3\lambda)^4 = (4\lambda)^4 = 2;$$

and so $\lambda, 2\lambda, 3\lambda, 4\lambda$ are all roots of $f(x)$ in this extension field of Z_5. Here we obtain a complete factorization into linear factors

$$f(x) = (x - \lambda)(x - 2\lambda)(x - 3\lambda)(x - 4\lambda)$$

or, by conveniently writing λ as $\sqrt[4]{2}$,

$$f(x) = (x - \sqrt[4]{2})(x - 2\sqrt[4]{2})(x - 3\sqrt[4]{2})(x - 4\sqrt[4]{2}).$$

Example 5.25

A second example is afforded by considering the polynomial $f(x) = (x^2 - 2)(x^2 - 3)$ over the field Q of rational numbers. From Example 4.38, $x^2 - 2$ (and by similar reasoning, $x^2 - 3$) is already known to be irreducible in $Q[x]$. So we begin by extending Q to the field F_1, where

$$F_1 = Q[x]/(x^2 - 2) = \{a + b\lambda | a, b \in Q;\ \lambda^2 - 2 = 0\};$$

and obtain the factorization

$$\begin{aligned} f(x) &= (x - \lambda)(x + \lambda)(x^2 - 3) \\ &= (x - \sqrt{2})(x + \sqrt{2})(x^2 - 3). \end{aligned}$$

(As $\lambda^2 = 2$, one customarily identifies λ with $\sqrt{2}$.)

However, $f(x)$ does not split completely, since the polynomial $x^2 - 3$ remains irreducible in $F_1[x]$. Suppose to the contrary that $x^2 - 3$ has a root in F_1, say $c + d\sqrt{2}$, with $c, d \in Q$. Substituting, we find that

$$(c^2 + 2d^2 - 3) + 2cd\sqrt{2} = 0,$$

or, what amounts to the same thing,

$$c^2 + 2d^2 - 3 = 0, \qquad cd = 0.$$

The latter equation implies that either $c = 0$ or $d = 0$. But neither c nor d can be zero, since this would mean that $d^2 = 3/2$ or $c^2 = 3$, which is clearly impossible. Accordingly, $x^2 - 3$ does not split in $F_1[x]$.

To factor $f(x)$ into linear factors, it becomes necessary to extend the coefficient field further. We therefore construct a second extension F_2, where

$$F_2 = F_1[x]/(x^2 - 3) = \{\alpha + \beta\mu | \alpha, \beta \in F_1;\ \mu^2 - 3 = 0\}.$$

The elements of F_2 can be expressed alternatively in the form

$$(a + b\sqrt{2}) + (c + d\sqrt{2})\sqrt{3} = a + b\sqrt{2} + c\sqrt{3} + d\sqrt{6},$$

where, of course, the coefficients a, b, c, d all lie in Q. It follows without difficulty that the original polynomial now factors in $F_2[x]$ as

$$\begin{aligned} f(x) &= (x - \lambda)(x + \lambda)(x - \mu)(x + \mu) \\ &= (x - \sqrt{2})(x + \sqrt{2})(x - \sqrt{3})(x + \sqrt{3}). \end{aligned}$$

Let a field F be given and consider a nonconstant polynomial $f(x) \in F[x]$. An extension field F' of F is said to be a *splitting field for $f(x)$ over F* provided that $f(x)$ can be factored completely into linear factors in $F'[x]$ but not so factored over any proper subfield of F' containing F (this minimum nature of the splitting field is not required by all authors). Loosely speaking, a splitting field is the smallest extension field F' in which the prescribed polynomial factors linearly:

$$f(x) = a(x - r_1)(x - r_2)\cdots(x - r_n), \qquad r_i \in F'.$$

To obtain a splitting field for $f(x)$, we need only consider the family $\{F_i\}$ of all extension fields F_i in which $f(x)$ can be decomposed as a product of linear factors—Theorem 5.22 guarantees the existence of such extensions. Then $\cap F_i$ serves as a splitting field for $f(x)$ over F.

The existence of a splitting field for an arbitrary polynomial in $F[x]$ thus being indicated, it is natural to follow this up with a query as to uniqueness. We shall prove that any two splitting fields of the same nonconstant polynomial are isomorphic; this being so, one is justified in using the definite article and speaking of *the* splitting field of a given polynomial.

Before we present the main theorem, two preparatory results of a somewhat technical nature are needed.

LEMMA *Let $f(x)$ be an irreducible polynomial in $F[x]$ and r be a root of $f(x)$ in some extension field K of F. Then $F(r) \simeq F[x]/(f(x))$ under an isomorphism whereby the element r corresponds to the coset $x + (f(x))$.*

Proof. Since the element r is algebraic over F, it follows directly from Theorem 5.18 that $F(r) \simeq F[x]/(f(x))$ via an isomorphism θ with the property that $\phi_r = \theta \circ \mathrm{nat}_{(f(x))}$. (As usual, $\phi_r : F[x] \to K$ is the substitution homomorphism induced by r.) Regarding the last statement of the lemma, we necessarily have

$$r = \phi_r(x) = (\theta \circ \mathrm{nat}_{(f(x))})(x) = \theta(x + (f(x))).$$

The chief value of this lemma is that it leads almost immediately to the following theorem.

THEOREM 5.24 (Isomorphism Extension Theorem) *Let σ be an isomorphism from the field F onto the field F'. Also let $f(x) = a_0 + a_1x + \cdots + a_nx^n$ be an irreducible polynomial in $F[x]$ and $f'(y) = \sigma(a_0) + \sigma(a_1)y + \cdots + \sigma(a_n)y^n$ be the corresponding polynomial in $F'[y]$. Then $f'(y)$ is likewise irreducible. Furthermore, if r is a root of $f(x)$ in some extension field of F and r' is a root of $f'(y)$ in some extension field of F', then σ can be extended to an isomorphism Φ of $F(r)$ onto $F'(r')$ with $\Phi(r) = r'$.*

Proof. Let us first extend σ to a mapping $\bar{\sigma}$ between the polynomial rings $F[x]$ and $F'[y]$ by setting

$$\bar{\sigma}g(x) = \bar{\sigma}(b_0 + b_1x + \cdots + b_nx^n) = \sigma(b_0) + \sigma(b_1)y + \cdots + \sigma(b_n)y^n$$

for any polynomial $g(x) = b_0 + b_1x + \cdots + b_nx^n \in F[x]$. We bequeath to the reader the task of supplying the necessary details that $\bar{\sigma}$ is an isomorphism of $F[x]$ onto $F'[y]$. It is important to notice that for any polynomial $g(x)$ in $F[x]$, an element $a \in F$ is a root of $g(x)$ if and only if $\sigma(a)$ is a root of $\bar{\sigma}g(x)$. Indeed, if, as before, $g(x) = b_0 + b_1x + \cdots + b_nx^n$, then, when we evaluate $\bar{\sigma}g(x)$ at $\sigma(a)$,

$$\begin{aligned}\bar{\sigma}g(\sigma(a)) &= \sigma(b_0) + \sigma(b_1)\sigma(a) + \cdots + \sigma(b_n)\sigma(a)^n \\ &= \sigma(b_0 + b_1a + \cdots + b_na^n) \\ &= \sigma(g(a)),\end{aligned}$$

from which our assertion follows. In particular, we infer that the polynomials $g(x)$ and $\bar{\sigma}g(x)$ are simultaneously reducible or irreducible in $F[x]$ and $F'[y]$, respectively. This being so, $f'(y) = \bar{\sigma}f(x)$ is irreducible in $F'[y]$.

By the foregoing lemma we know that there exist isomorphisms $\alpha: F(r) \to F[x]/(f(x))$ and $\beta: F'(r') \to F'[y]/(f'(y))$, with

$$\alpha(r) = x + (f(x)), \qquad \beta(r') = y + (f'(y)).$$

Moreover, it is an easy matter to show that there is also an isomorphism τ of $F[x]/(f'(x))$ onto $F'[y]/(f'(y))$ defined by

$$\tau(g(x) + (f(x))) = \bar{\sigma}g(x) + (f'(y)) \qquad (g(x) \in F[x]).$$

Observe particularly that τ carries the coset $x + (f(x))$ onto $y + (f'(y))$. We contend that $F(r) \simeq F'(r')$ under the composition of maps

$$\Phi = \beta^{-1} \circ \tau \circ \alpha,$$

where $\Phi: F(r) \to F'(r')$; this situation is portrayed in the diagram below:

$$\begin{array}{ccc} F(r) & \xrightarrow{\Phi} & F'(r') \\ \alpha \downarrow & & \uparrow \beta^{-1} \\ F[x]/(f(x)) & \xrightarrow{\tau} & F'[y]/(f'(y)) \end{array}$$

Certainly, Φ is an isomorphism of $F(r)$ onto $F'(r')$, for the individual mappings α, τ, β^{-1} are themselves isomorphisms. If a is an arbitrary element of F, then

$$\begin{aligned} \Phi(a) &= (\beta^{-1} \circ \tau)(\alpha(a)) = (\beta^{-1} \circ \tau)(a + (f(x))) \\ &= \beta^{-1}(\sigma(a) + (f'(y))) = \sigma(a), \end{aligned}$$

whence Φ is actually an extension of σ to all of $F(r)$. Finally, we point out that

$$\begin{aligned} \Phi(r) &= (\beta^{-1} \circ \tau)(\alpha(r)) = (\beta^{-1} \circ \tau)(x + (f(x)) \\ &= \beta^{-1}(y + (f'(y))) = r', \end{aligned}$$

as required, and the theorem is proved in its entirety.

For a simple, but nonetheless satisfying, illustration of this last result, take both F and F' to be the real number field $R^\#$; let $f(x) \in R^\#[x]$ be the irreducible polynomial $f(x) = x^2 + 1$, so that $f'(y) = y^2 + 1$ (recall that the identity map is the only isomorphism of $R^\#$ onto itself). Finally, choose $r = i$ and $r' = -i$. Theorem 5.24 then asserts that $R^\#(i) \simeq R^\#(-i)$ under an isomorphism that carries i onto $-i$. Inasmuch as $R^\#(i) = R^\#(-i) = C$, the isomorphism in question is just the correspondence between a complex number and its conjugate.

We now have the mathematical machinery to show the uniqueness to within isomorphism of splitting fields. Actually, we shall prove a somewhat more general result.

THEOREM 5.25 *Let σ be an isomorphism of the field F onto the field F'. Let $f(x) = a_0 + a_1 x + \cdots + a_n x^n \in F[x]$ and $f'(y) = \sigma(a_0) + \sigma(a_1)y + \cdots + \sigma(a_n)y^n$ be the corresponding polynomial in $F'[y]$. If K is a splitting field of $f(x)$ and K' a splitting field of $f'(y)$, then σ can be extended to an isomorphism Φ of K onto K'.*

Proof. Our argument will be by induction on the number n of roots of $f(x)$ that lie outside F, but (needless to say) in K. When $n = 0$, all the roots of $f(x)$ belong to F, and F is itself the splitting field of $f(x)$;

that is, $K = F$. This in turn induces a splitting of the polynomial $f'(y)$ into a product of linear factors in $F'[y]$, so that $K' = F'$. Thus when it happens that $n = 0$, the isomorphism σ is, in a trivial sense, the desired extension to the splitting fields.

Let us next assume, inductively, that the theorem holds true for any pair of corresponding polynomials $f(x)$ and $f'(y)$ over isomorphic fields E and E', provided that the number of roots of $f(x)$ outside of E is less than n ($n \geqslant 1$).

If $f(x) \in F[x]$ is a polynomial having n roots outside of F, then not all the irreducible factors of $f(x)$ can be linear in $F[x]$; otherwise, $f(x)$ would split completely in F, contrary to assumption. Accordingly, $f(x)$ must have some factor $g(x)$ of degree $m > 1$ that is irreducible in $F[x]$. Let $g'(y)$ denote the corresponding irreducible factor of $f'(y)$. Since K is a splitting field of $f(x)$ over F, $g(x)$ in particular must have a root in K; call it r. Similarly, one of the roots of the polynomial $f'(y)$, say r', is a root of $g'(y)$ in K'. By Theorem 5.24, σ can be extended to an isomorphism σ' between the fields $F(r)$ and $F'(r')$. Now K is a splitting field of $f(x)$, viewed as a polynomial with coefficients from $F(r)$; in a like manner, K' can be regarded as a splitting field of $f'(y)$ over the field $F'(r')$. Because the number of roots of $f(x)$ lying outside $F(r)$ is less than n, the induction hypothesis permits us to extend σ'—itself an extension of σ—to an isomorphism Φ of K' onto K. This completes the induction step and the proof of the theorem as well, since σ has been suitably extended.

With the corollary below, we achieve our objective.

COROLLARY *Any two splitting fields of a nonconstant polynomial $f(x) \in F[x]$ are isomorphic via an isomorphism Φ such that the restriction $\Phi|F$ is the identity mapping.*

Proof. This is an immediate consequence of the theorem on taking $F = F'$ and σ to be the identity isomorphism i_F.

Another classification of field extensions $K \supseteq F$ comes from regarding K as a vector space over F.

DEFINITION 5.9 An extension K of the field F is said to be a *finite extension* of degree n if K is a finite dimensional vector space of dimension n over F. We shall use the symbol $[K:F]$ to represent the degree of K over F.

As a consequence of our definition, we can recast Theorem 5.21 as follows: If $K = F(r)$ is a simple algebraic extension of F, where the minimum polynomial of r over F has degree n, then $[F(r):F] = n$.

While any simple algebraic extension is finite, the converse is not true without additional hypotheses. It will be shown shortly that all finite extensions of a field of characteristic zero can be realized as simple extensions. What we can say at the moment is

THEOREM 5.26 *If K is a finite extension of a field F, then K is algebraic over F.*

Proof. Let $[K:F] = n$ and $a \in K$. The elements $1, a, a^2, \ldots, a^n$, being $n + 1$ in number, must be a linearly dependent set of vectors in the vector space K over F. Thus there exist scalars (that is, elements of F), say $c_0, c_1, \ldots, c_n$, not all zero, such that

$$c_0 + c_1 a + \cdots + c_n a^n = 0.$$

In view of this, a is a root of the nonzero polynomial

$$c_0 + c_1 x + \cdots + c_n x^n \in F[x].$$

It follows that a is an algebraic element over F of degree at most n; and so K is algebraic over F.

To give an illustration, the field C of complex numbers has degree 2 over the real field $R^{\#}$, with $\{1, i\}$ serving as a basis. Therefore C is an algebraic extension of $R^{\#}$.

The next theorem indicates a transitivity property of finite extensions.

THEOREM 5.27 *If K is a finite extension of E and E is a finite extension of F, then K is a finite extension of F. Furthermore,*

$$[K:F] = [K:E][E:F].$$

Proof. Let $\{a_1, a_2, \ldots, a_n\}$ be a basis for K over E and $\{b_1, b_2, \ldots, b_m\}$ be a basis for E over F. Our strategy is to show that the nm products

$$a_i b_j \qquad (i = 1, 2, \ldots, n; \quad j = 1, 2, \ldots, m)$$

form a basis for K over F. Suppose, then, that some linear combination of them is equal to 0:

$$\sum_{i,j} c_{ij} a_i b_j = 0$$

where each $c_{ij} \in F$. This can be rearranged as

$$\sum_{i=1}^{n} \left(\sum_{j=1}^{m} c_{ij} b_j \right) a_i = 0.$$

Since the coefficients $c_{ij}b_j$ lie in E and $\{a_1, a_2, \ldots, a_n\}$ is a linearly independent set over E, we have

$$\sum_{j=1}^{m} c_{ij}b_j = 0$$

for each i. Now the linear independence of $\{b_1, b_2, \ldots, b_m\}$ over F implies in turn that every $c_{ij} = 0$. In other words, the elements a_ib_j are linearly independent over F. To see that they also span K over F, select an arbitrary member a of K. Then because the a_i span K over E, we can write

$$a = \sum_{i=1}^{n} d_ia_i$$

for suitable choices of $d_i \in E$. But each element d_i, being in E, can be expressed as a linear combination of the b_j over F, say,

$$d_i = \sum_{j=1}^{m} e_{ij}b_j,$$

where $e_{ij} \in F$. Combining these equations, we get

$$a = \sum_{i=1}^{n} d_ia_i = \sum_{i=1}^{n} \left(\sum_{j=1}^{m} e_{ij}b_j \right) a_i = \sum_{i,j} e_{ij}(a_ib_j),$$

whence the a_ib_j satisfy the conditions that make them a basis for K over the field F.

For example, let $E = Q(\sqrt{2})$ and $K = E(i)$. Then $[E:Q] = \deg(x^2 - 2) = 2$, with $\{1, \sqrt{2}\}$ serving as a basis for E over Q. Similarly, since $x^2 + 1$ remains irreducible over E, we have $[K:E] = \deg(x^2 + 1) = 2$ with a basis for K over E given by the set $\{1, i\}$. Thus the four-element set

$$\{1, \sqrt{2}, i, i\sqrt{2}\}$$

acts as a basis for the extension $K = Q(\sqrt{2})(i) = Q(\sqrt{2}, i)$.

A special type of extension that will stand us in fine stead when we study Galois theory is the normal extension. These are defined by a condition that will ensure good behavior with respect to the splitting of polynomials.

DEFINITION 5.10 An extension K of a field F is said to be *normal* over F if every irreducible polynomial $f(x) \in F[x]$ that has one root in K splits completely in K.

We should observe that not all finite extension fields are normal. For instance, $Q(\sqrt[3]{2})$ is not a normal extension of Q, where $\sqrt[3]{2}$ is the real cube root of 2. The irreducible polynomial $x^3 - 2 \in Q[x]$ has one root in $Q(\sqrt[3]{2})$, namely $\sqrt[3]{2}$, but does not split completely in $Q(\sqrt[3]{2})$. If $\omega = (-1 + i\sqrt{3})/2$, then the other two roots of $x^3 - 2$ in a splitting field are $\omega\sqrt[3]{2}$ and $\omega^2\sqrt[3]{2}$ (since $\omega^3 = 1$); $Q(\sqrt[3]{2})$, being a subfield of the real field $R^\#$, contains neither of these complex roots.

It will be seen that, as long as we deal with a finite extension of a field F, the concept of a normal extension turns out to be equivalent to that of a splitting field of a polynomial over F. This is proved in two stages, the first being

THEOREM 5.28 *Let K be a finite extension of a field F. If K is the splitting field of a polynomial $f(x) \in F[x]$, then K is a normal extension of F.*

Proof. Let $r_1, \ldots, r_n$ be the distinct roots of $f(x)$ in K and write $K = F(r_1, \ldots, r_n)$. Given an irreducible polynomial $g(x) \in F[x]$ that has a root s in K, we propose to show that $g(x)$ splits in K. Suppose that s' is another root of $g(x)$ in some extension of K. By Theorem 5.24 the identity mapping on F can be extended to an isomorphism ϕ of $F(s)$ onto $F(s')$. Clearly,

$$K = F(s, r_1, \ldots, r_n) = F(s)(r_1, \ldots, r_n)$$

is the splitting field for $f(x)$ considered as a polynomial over $F(s)$; on the other hand, the splitting field for $f(x)$ over $F(s')$ is

$$F(s')(r_1, \ldots, r_n) = F(s', r_1, \ldots, r_n) = K(s').$$

We appeal to Theorem 5.25 to see that there is an extension of ϕ to an isomorphism Φ of K onto $K(s')$ and therefore that $[K:F(s)] = [K(s'):F(s')]$. To complete the argument, we calculate degrees:

$$\begin{aligned}[K:F] &= [K:F(s)][F(s):F] \\ &= [K(s'):F(s')][F(s'):F] = [K(s'):F].\end{aligned}$$

But K is a subspace of $K(s')$. Since both have the same degree over F, $K = K(s')$ and so $s' \in K$. Hence all the roots of $g(x)$ lie in K, making K normal over F.

The converse to this last theorem appears next.

THEOREM 5.29 *If K is a finite normal extension of a field F, then K is the splitting field of some polynomial $f(x)$ in $F[x]$.*

Proof. If K is normal over F and $[K:F]$ is finite, then by Theorem 5.26, $K = F(r_1, r_2, \ldots, r_n)$, where each r_i is algebraic over F.

Let $f_i(x)$ be the minimum polynomial of r_i over F and put $f(x) = f_1(x)f_2(x)\cdots f_n(x)$. Each polynomial $f_i(x)$ has a root in K (namely, r_i) and hence, by normality, splits completely in K. This means that all the roots of $f(x)$ lie in K or that K contains a splitting field L of $f(x)$. Since K is generated by F and the roots $r_1, r_2, \ldots, r_n$ of $f(x)$, it follows that $L = K$. Simply stated, K is a splitting field of $f(x) \in F[x]$.

Taken together, our last two results characterize normal extensions among the finite extensions of F as being splitting fields of polynomials in $F[x]$. This is summed up in

THEOREM 5.30 *Let K be a finite extension of a field F. Then K is a normal extension of F if and only if K is a splitting field of some polynomial in $F[x]$.*

One notable consequence of Theorem 5.30 is that a normal extension K of F is also a normal extension of any field between K and F.

COROLLARY *Let K be a finite normal extension of a field F. If E is any subfield of K with $K \supseteq E \supseteq F$, then K is a normal extension of E.*

Proof. In view of the theorem, K is the splitting field of some polynomial $f(x) \in F[x]$. Now $f(x)$ may be considered as belonging to $E[x]$, since each coefficient lies in $F \subseteq E$. Consequently, K is also the splitting field of $f(x)$ over E and therefore must be a normal extension of E.

We next introduce a convenient term that is connected with the most elementary type of field extension.

DEFINITION 5.11 If K is a simple extension of F, say $K = F(r)$, then such a generating element r is said to be a *primitive element* of K over F.

For instance, $\sqrt{2} + \sqrt{3}$ is a primitive element of $Q(\sqrt{2}, \sqrt{3})$, since $Q(\sqrt{2}, \sqrt{3}) = Q(\sqrt{2} + \sqrt{3})$. While a finite extension need not be realizable as a simple extension (and hence have a primitive element), the situations in which this does occur are highly desirable ones. For our purposes the most important case is that of an extension of a field of characteristic zero; the lemma below, aside from its importance to the result we seek, has an interest all its own.

LEMMA *If F is a field of characteristic zero, then an irreducible polynomial $f(x) \in F[x]$ has no multiple roots in any extension field $K \supseteq F$.*

Proof. Suppose that $f(x)$ has a root r of multiplicity $n > 1$ in K. Then we can write

$$f(x) = (x - r)^n g(x)$$

for some polynomial $g(x) \in K[x]$ where $g(r) \neq 0$. Calculating the derivative of $f(x)$ (see Problem 21, Section 4.8), we get

$$\delta f(x) = n(x - r)^{n-1} g(x) + (x - r)^n \delta g(x).$$

Hence $\delta r(r) = 0$, and r is a root of $\delta f(x)$. Let $p(x)$ be the minimum polynomial of r over F. Clearly, $f(x)$ and $\delta f(x)$ are both multiples of $p(x)$ because r is a root of each. Since $f(x)$ is assumed to be irreducible, it follows that $f(x)$ must equal $p(x)$. Thus there is no polynomial in $F[x]$ of smaller degree with r as a root. Since $\delta f(x)$ has its coefficients in F and, if nonzero, satisfies $\deg \delta f(x) < \deg f(x)$, the only possibility is that $\delta f(x)$ is the zero polynomial.

If we assume that $f(x)$ has the form

$$f(x) = a_0 + a_1 x + \cdots + a_k x^k,$$

then $\delta f(x) = a_1 + 2a_2 + \cdots + k a_k x^{k-1}$. To say that $\delta f(x) = 0$ means that $i a_i = 0$ for $1 \leqslant i \leqslant k$. For a field of characteristic zero this can happen only if $a_i = 0$. Thus $f(x)$ is a constant polynomial, contradicting the fact that $f(x)$ has roots.

This brings us to the previously mentioned result, sometimes called the Theorem of the Primitive Element, which asserts that any finite extension of a field of characteristic zero is actually simple.

THEOREM 5.31 *If F is a field of characteristic zero containing elements r and s algebraic over F, then an element $t \in F(r, s)$ exists such that $F(r, s) = F(t)$.*

Proof. Let $f(x)$ and $g(x)$ be the minimum polynomials over F of r and s, with roots $r = r_1, r_2, \ldots, r_n$ and $s = s_1, s_2, \ldots, s_m$, respectively, in some extension field K of F. These roots are all distinct, as K is of characteristic zero over F. Consider the following $n(m - 1)$ linear equations in λ:

$$r_i + \lambda s_j = r + \lambda s \qquad (1 \leqslant i \leqslant n;\ 2 \leqslant j \leqslant m).$$

Each equation has a unique solution in K, namely, $\lambda = (r_i - r)(s - s_j)^{-1}$ and hence at most one solution in F. Since F is assumed to have infinitely many elements, it is possible to choose an element $\mu \in F$ that is not one of these solutions. Then $r_i + \mu s_j \neq r + \mu s$ for all i and $j \neq 1$.

Setting $t = r + \mu s$, we claim that t is a primitive element of

$F(r, s)$. Because $t \in F(r, s)$, the inclusion $F(t) \subseteq F(r, s)$ is obvious. Now look at the polynomials $g(x)$ and $h(x) = f(t - \mu x)$ over $F(t)$. The element s satisfies both, since $h(s) = f(t - \mu s) = f(r) = 0$. By our choice of t these polynomials have no other common root in $F(t)$; indeed, the other roots of $g(x)$ are s_j $(2 \leqslant j \leqslant m)$ and

$$t - us_j = r + \mu(s - s_j) \neq r_i.$$

Thus $x - s$ is the greatest common divisor of $g(x)$ and $h(x)$ in K. The fact that the coefficients of these polynomials are in $F(t)$ implies that the same is true for $x - s$; consequently, $s \in F(t)$ and, in its turn, $r = t - \mu s \in F(t)$. This gives $F(r, s) \subseteq F(t)$, yielding the desired equality.

Our theorem has an immediate consequence.

COROLLARY *Any finite extension K of a field F of characteristic zero is a simple extension. In particular, the splitting field of a polynomial $f(x) \in F[x]$ is a simple extension of F.*

Proof. Write $K = F(r_1, r_2, \ldots, r_n)$ and note that $r_1, r_2, \ldots, r_n$ are algebraic over F. The argument proceeds by induction on the number n of elements r_i. Suppose that the result is valid for $F(r_1, \ldots, r_k)$, $k < n$, so that $F(r_1, \ldots, r_k) = F(s)$ for some $s \in F(r_1, \ldots, r_k)$. Now $F(r_1, \ldots, r_k, r_{k+1}) = F(s, r_{k+1})$. The theorem assures us that there is an element $t \in F(s, r_{k+1})$ satisfying $F(s, r_{k+1}) = F(t)$. Knowing this, we reach $F(r_1, \ldots, r_k, r_{k+1}) = F(t)$, which completes the induction step and the proof.

We now change direction and examine the structure of fields with a finite number of elements. Any finite field F is an extension of a prime field isomorphic to the field Z_p, where p is the characteristic of F. Nothing will be lost in assuming, as we shall hereafter, that the prime subfield of F is actually Z_p.

Let us launch our inquiry by proving that every finite field is of prime-power order.

THEOREM 5.32 *If F is a finite field, then F has exactly p^n elements for some prime p and $n \in Z_+$.*

Proof. Since F is finite, it must have characteristic p for some prime p. This means that every nonzero element of the additive group of F has (additive) order p; in other words, $(F, +)$ is a finite commutative p-group. By virtue of Theorem 4.72, F contains p^n elements where $n > 0$.

The significance of Theorem 5.32 is that one cannot construct finite fields with q elements unless q is a power of a prime. This immediately raises a question: Given q, a prescribed power of a prime, do fields with q elements actually exist? In obtaining an answer it is crucial to know the following.

LEMMA *Every element of a finite field F with p^n elements is a root of the polynomial $f(x) = x^{p^n} - x \in F[x]$.*

Proof. The nonzero elements of F form a multiplicative group of order $p^n - 1$, so that for each such $a \in F$ we must have $a^{p^n - 1} = 1$. Since

$$a^{p^n} - a = a(a^{p^n - 1} - 1),$$

it follows that every element $a \in F$ (zero or nonzero) satisfies the equation $a^{p^n} - a = 0$.

COROLLARY 1 *Any finite field F with p^n elements is the splitting field of the polynomial $f(x) = x^{p^n} - x \in Z_p[x]$.*

Proof. We begin by recalling that $f(x)$ can have at most $\deg f(x) = p^n$ distinct roots in F. But p^n roots are already known, namely, the elements of F; whence the polynomial $f(x)$ splits completely into linear factors in $F[x]$. Needless to say, it cannot split in any proper subfield of F, since no proper subfield contains p^n elements. Thus we conclude that F is the splitting field of $f(x)$ over Z_p and

$$x^{p^n} - x = \prod_{r_i \in F} (x - r_i),$$

where the r_i are the distinct elements of F.

Since any two splitting fields of a given nonconstant polynomial over a given field are isomorphic (Theorem 5.25), Corollary 1 leads to

COROLLARY 2 *Any two finite fields having the same number of elements are isomorphic.*

Before going on, let us digress to point out that this lemma has an interesting application in the theory of numbers. We wish to establish what, in the literature, goes by the name of *Wilson's Theorem*: If p is a prime number, then

$$(p - 1)! \equiv -1 \pmod{p}.$$

Since this holds trivially for $p = 2$, one may assume that $p > 2$; thus p is an odd prime. We take for F the field Z_p. Then, from above, every nonzero element of Z_p is a root of the polynomial $x^{p-1} - 1 \in Z_p[x]$:

$$x^{p-1} - 1 \equiv (x - 1)(x - 2) \cdots (x - (p - 1)) \pmod{p}.$$

Putting $x \equiv 0 \pmod{p}$ in the just-written equation, it follows that

$$-1 \equiv (-1)(-2)\cdots(-(p-1)) = (-1)^{p-1}(p-1)! \pmod{p}.$$

Since $p - 1$ is even, this leads directly to Wilson's Theorem.

We now settle the question of the existence of finite fields.

THEOREM 5.33 *For any prime number p and positive integer n there exists a field (unique up to isomorphism) with exactly p^n elements.*

Proof. Consider the splitting field F' of the polynomial $f(x) = x^{p^n} - x$ in $Z_p[x]$. Since F' contains Z_p, it has finite characteristic p. Suppose that r is a root of $f(x)$ in F'. Then

$$(x - r)^{p^n} - (x - r) = (x^{p^n} - r^{p^n}) - (x - r) = x^{p^n} - x,$$

since $r^{p^n} = r$ and the binomial coefficients $\binom{p^n}{k}$ are divisible by p for $k = 1, 2, \ldots, p^n - 1$. This allows us to write $f(x)$ as

$$f(x) = (x - r)[(x - r)^{p^n - 1} - 1].$$

But $(x - r)^{p^n - 1} - 1$ is clearly not divisible by $x - r$, which means that r is a simple root of $f(x)$; that is, the polynomial has no repeated roots in F'. Let the subset $F \subseteq F'$ consist of the p^n distinct roots of $f(x)$ in F':

$$F = \{a \in F' | a^{p^n} = a\}.$$

Clearly, the elements 0 and 1 lie in F. If $a, b \in F$, with $b \neq 0$, then we have

$$(a - b)^{p^n} = a^{p^n} - b^{p^n} = a - b,$$

$$(ab^{-1})^{p^n} = a^{p^n}(b^{p^n})^{-1} = ab^{-1}$$

(again using the fact that char $F' = p$), so that both $a - b$, $ab^{-1} \in F$. Consequently, the set F constitutes a subfield of F', and thus a field with p^n elements. From Corollary 1 of the last theorem we infer that $F = F'$ and F' is the desired field. Uniqueness follows from the result that any two fields having p^n elements are isomorphic (Corollary 2).

We next examine the multiplicative structure of a finite field. The finiteness assumption leads to a particularly simple description: The nonzero elements form a cyclic group under multiplication.

THEOREM 5.34 *The multiplicative group of a finite field is cyclic.*

Proof. Let F be a finite field with p^n elements and F^* be its multiplicative group of nonzero elements; this group has order $p^n - 1$. The argument about to be presented hinges on our being able to find an element in F^* of order $h = p^n - 1$. To this end, we first consider the prime factorization of h:

$$h = q_1^{r_1} q_2^{r_2} \cdots q_m^{r_m},$$

where the q_i are distinct primes and $r_i \in Z_+$. For $i = 1, 2, \ldots, m$, set $h_i = h/q_i$. Now a nonzero element $a_i \in F$ exists that is not a root of the polynomial $x^{h_i} - 1 \in F[x]$; this polynomial has at most h_i distinct roots in F and $h_i < h$, the number of nonzero elements of F.

Next, take

$$b_i = a_i^{h/s_i}, \qquad \text{where} \qquad s_i = q_i^{r_i} \qquad (i = 1, 2, \ldots, m)$$

and define $b = b_1 b_2 \cdots b_m$. We certainly have

$$b_i^{s_i} = a_i^h = 1,$$

so the order of b_i divides $s_i = q_i^{r_i}$ and thus is a power $q_i^{t_i}$, where $t_i \leqslant r_i$. If

$$b_i^{s_i/q_i} = 1,$$

then

$$a_i^{h_i} = a^{h/q_i} = b_i^{s_i/q_i} = 1,$$

contrary to our original choice of the element a_i. The implication is that b_i has order $q_i^{r_i}$.

To settle the whole affair, we will show that the element b is of order h. In the contrary case the order of b must be a proper divisor of h (since $b^h = 1$, the order of b certainly divides h) and therefore divides at least one of the integers h_i ($i = 1, 2, \ldots, m$), say h_1. We then have

$$1 = b^{h_1} = b_1^{h_1} b_2^{h_1} \cdots b_m^{h_1}.$$

If $2 \leqslant i \leqslant m$, then $q_i^{r_i} | h_1$, which implies that $b_i^{h_1} = 1$ and so $b_1^{h_1} = 1$. This means that $q_1^{r_1}$ (the order of b_1) divides h_1, which is impossible. Thus the element b has order h, and in consequence the cyclic subgroup of F^* generated by b will also be of order h; since F^* contains only $p^n - 1 = h$ elements, this cyclic group must be all of F^*.

The following is not surprising and is quite easy to prove.

Corollary *Any finite field F with p^n elements is a simple extension of the field Z_p.*

Proof. We already know that F is an extension of its prime subfield Z_p. The theorem above indicates that the p^n elements of F can be written as $0, 1, b, b^2, \ldots, b^{p^n-2}$ for some $b \in F^*$; in other words, the field $F = Z_p(b)$.

As an application of these ideas, let us prove a statement made earlier to the effect that, for any finite field F, the polynomial domain $F[x]$ contains irreducible polynomials of arbitrary degree.

THEOREM 5.35 *For each positive integer n there exists an irreducible polynomial* $f(x) \in Z_p[x]$ *with* $\deg f(x) = n$.

Proof. The argument of Theorem 5.33 tells us that Z_p has an extension field F' with p^n elements. As was just seen, some element b in F' exists such that $F' = Z_p(b)$. Now consider the minimum polynomial $f(x)$ of b over Z_p, say $\deg f(x) = k$. By Theorem 5.21, each member of $Z_p(b)$, and hence of F', is uniquely representable in the form $c_0 + c_1 b + \cdots + c_{k-1} b^{k-1}$, where the $c_i \in Z_p$. Since each coefficient c_i can assume p possible values, it follows that the number of elements in F' is exactly p^k. Thus $k = n$ and $f(x)$ is the required irreducible polynomial of degree n.

COROLLARY *If F is any finite field and* $n \in Z_+$, *then there is an irreducible polynomial in* $F[x]$ *of degree n.*

Finite fields are called *Galois fields* after the French mathematician Evariste Galois, who first discovered the existence of finite fields aside from those of the form Z_p. The (essentially unique) field with p^n elements is commonly denoted by the symbol $\mathrm{GF}(p^n)$. To construct $\mathrm{GF}(p^n)$, we need only determine an irreducible polynomial $f(x)$ of degree n in $Z_p[x]$; then $Z_p[x]/(f(x))$ is the required Galois field with p^n elements.

Example 5.26

Let us construct a field having four elements. Since $\mathrm{GF}(2^2)$ is in the sense of isomorphism the unique field with four elements, it suffices to find an irreducible polynomial of degree 2 over Z_2 and then to form $Z_2[x]/(f(x))$. The only choice for $f(x)$ is the polynomial $x^2 + x + 1$. Thus

$$\begin{aligned}\mathrm{GF}(2^2) &= Z_2[x]/(x^2 + x + 1)\\ &= \{a + b\lambda \mid a, b \in Z_2;\ \lambda^2 + \lambda + 1 = 0\},\end{aligned}$$

λ playing the role of the coset $x + (x^2 + x + 1)$. We see that the four elements of $\mathrm{GF}(2^2)$ can be written as

$$\mathrm{GF}(2^2) = \{0, 1, \lambda, 1 + \lambda\}.$$

Using the relationship $\lambda^2 + \lambda + 1 = 0$, we compute the following operation tables:

$+$	0	1	λ	$1+\lambda$
0	0	1	λ	$1+\lambda$
1	1	0	$1+\lambda$	λ
λ	λ	$1+\lambda$	0	1
$1+\lambda$	$1+\lambda$	λ	1	0

$\cdot$	0	1	λ	$1+\lambda$
0	0	0	0	0
1	0	1	λ	$1+\lambda$
λ	0	λ	$1+\lambda$	1
$1+\lambda$	0	$1+\lambda$	1	λ

Observe that $\lambda^2 = -1 - \lambda = 1 + \lambda$, so these tables could be expressed as follows:

$+$	0	1	λ	λ^2
0	0	1	λ	λ^2
1	1	0	λ^2	λ
λ	λ	λ^2	0	1
λ^2	λ^2	λ	1	0

$\cdot$	0	1	λ	λ^2
0	0	0	0	0
1	0	1	λ	λ^2
λ	0	λ	λ^2	1
λ^2	0	λ^2	1	λ

This confirms that $GF(2^2)^*$ is a cyclic group of order 3; that is, there is some $\lambda \in GF(2^2)^*$ with $\lambda^3 = 1$ and $\lambda \neq 1$.

The proof of the existence of Galois fields with p^n elements appears in Camille Jordan's classic work *Traité des Substitutions* (1870). In 1893, E. H. Moore established that any finite field is isomorphic to some $GF(p^n)$, a result that has been hailed as the beginning of abstract algebra in America.

During the final decade of the nineteenth century and the first decade of the twentieth, advanced training in mathematics firmly took hold in many U.S. universities. The commitment to mathematics at the research level was largely the result of the efforts of Eliakim Hastings Moore (1862–1932). When the University of Chicago was founded in 1892, Moore was called from Northwestern University to take charge of the mathematics department of this newest seat of learning. The first major undertaking of the young department involved the organization of an International Congress of Mathematicians to be held in connection with the World's Fair in Chicago in 1893 (commemorating the 400th anniversary of the discovery of America). Invited papers were given by illustrious Europeans, among them Felix Klein, then at the height of his powers and fame. Afterwards, he delivered a two-week series of lectures at Northwestern University, the first colloquium of American mathematicians.

At the beginning of the 1890's the organized mathematical interests of the country were centered in New York in a group known as the New York Mathematical Society. It was in their journal, the *Bulletin of the New York Mathematical Society*, that Moore published his theorem on fields containing a finite number of elements. When it became evident that the name failed to indicate its widening membership, the Society adopted (1894) its present title, the American Mathematical Society; simultaneously, its periodical became the *Bulletin of the American Mathematical Society*. To provide mathematical interaction for those located in the Midwest, a Chicago section of the Society was sanctioned in 1887 with Moore as chairman. The increase in the activity of the members of the Society and the quality of the papers presented at its section meetings led to a new publication outlet, the *Transactions of the American Mathematical Society*. It began publication in 1900, and Moore served as editor-in-chief for the first eight years. Later that year his colleagues awarded Moore their highest honor by electing him President of the Society.

Under the chairmanship of Moore the mathematics department of Chicago became the source of the first generation of U.S.-trained mathematicians to attain international reputation. Four of Moore's early students, who took off in different mathematical directions, were to become the brightest stars of the twentieth century: the algebraist L. E. Dickson (1874–1954), the geometer O. Veblen (1880–1960), the mathematical physicist G. D. Birkhoff (1884–1944), and the topologist R. L. Moore (1882–1974). Together these four mathematicians published 30 books and over 600 papers in addition to directing the research of almost 200 doctoral candidates. It is not surprising that many consider E. H. Moore to be the driving force who changed the United States from a mathematical backwater into a world leader in the field.

PROBLEMS 5.3

1. a) Prove that if $n > 1$ is a square-free integer, then

$$Q(\sqrt{n}) = \{a + b\sqrt{n} \mid a, b \in Q\}$$

is a field extension of Q.

b) If F is a subfield of the real field $R^{\#}$, show that

$$F(i) = \{a + bi \mid a, b \in F\}$$

forms a subfield of C containing Q; in particular, $R^{\#}(i) = C$.

2. Suppose that F' is an extension field of the field F and let the elements r_1, $r_2 \in F'$. Verify that $F(r_1, r_2) = (F(r_1))(r_2) = (F(r_2))(r_1)$.

3. Establish that the element $r = \sqrt{1 + \sqrt{3}}$ is algebraic over Q by exhibiting a polynomial $f(x) \in Q[x]$ for which $f(r) = 0$. Do the same for $r = \sqrt{5} + \sqrt[3]{7}$.

4. Show that the minimum polynomial of $\sqrt{3} + i$ is:

a) $x^4 - 4x + 16$ over the rational field Q;

b) $x^2 - 2\sqrt{3}x + 4$ over the real field $R^{\#}$.

5. Given that F' is an extension field of the field F and $r \in F'$, prove that:

a) If r is algebraic over F, then r is algebraic over every field K with $F \subseteq K \subseteq F'$.

b) The element $r \in F$ if and only if r is algebraic over F of degree 1.

c) Each element of $F[r]$ is uniquely representable as a polynomial in r with coefficients from F if and only if r is transcendental over F.

6. Let the element r be algebraic over the field F and let $f(x) \in F[x]$ be a monic polynomial such that $f(r) = 0$. Prove that $f(x)$ is the minimum polynomial of r over F if and only if $f(x)$ is irreducible in $F[x]$.

7. For an extension K of the field F and $r \in K$, establish each of the following:

a) If r^2 is algebraic over F, then r is algebraic over F.

b) If $r \neq 0$ is algebraic over F, then r^{-1} is algebraic over F.

c) If $f(x) \in F[x]$ and $f(r)$ is algebraic over F, then r is algebraic over F.

8. Let K be an extension of the field F and $r, s \in K$. Prove that if rs is algebraic over F, then r is algebraic over $F(s)$.

9. Describe the elements in the following fields:

a) $Q(i)$;

b) $Q(\sqrt{1 + \sqrt{3}})$;

c) $Z_2(r)$, where the minimum polynomial of r over Z_2 is $x^3 + x^2 + x$;

d) $Q(r^2)$, where the minimum polynomial of r over Q is $x^4 - 2$.

10. Let r be a root of the polynomial $f(x) = x^3 - x + 1 \in Q[x]$. Find the inverse of the element $1 - 2r + 3r^2$ in $Q(r)$.

11. a) If $r, s \in Q$ with $r > 0$, establish that r^s is algebraic over Q. [*Hint:* Let $s = m/n$ where $m, n \in Z$ and $n > 0$; consider $x^n - r^m \in Q[x]$.]

b) Show that if r is algebraic over Q, then $\sqrt{r}$ is also algebraic over Q.

12. Prove each of the statements below:

a) Let K be an extension of the field F and $a, b \in F$ with $b \neq 0$. If $r \in K$, then $F(a + br) = F(r)$.

b) If $a, b \in R^{\#}$ with $b \neq 0$, then $R^{\#}(a + bi) = C$.

c) $Q(4 - i) = Q(1 + i)$.

13. Suppose that $a, b \in Q$ are such that $\sqrt{a}, \sqrt{b} \notin Q$. Show that $Q(\sqrt{a}) = Q(\sqrt{b})$ if and only if $a = bc^2$ for some $c \in Q$. [*Hint:* First calculate $(\sqrt{a} + \sqrt{b})^2$.]

14. For each of the following extension fields, find an element r such that $Q(r_1, r_2) = Q(r)$:

$$Q(\sqrt{3}, \sqrt{7}), \quad Q(\sqrt{3}, i), \quad Q(\sqrt{2}, \sqrt[3]{5}).$$

15. Let K be an extension of the field F with the property that every subring of K containing F is a field. Prove that any element of K is algebraic over F. [*Hint:* Given an element $0 \neq r \in K$, $r^{-1} \in F[r]$.]

16. If the element r is algebraic over Q, show that there is a positive integer n for which nr is a root of some monic polynomial in $Z[x]$.

17. Find the integers n with $2 \leqslant n \leqslant 7$ such that $Z_n[x]/(x^2 + x + 1)$ is a field.

18. Show that $f(x) = x^3 + x + 1 \in Z_5[x]$ is irreducible in $Z_5[x]$ and that $Z_5[x]/(f(x))$ is a field with 125 elements.

19. a) Given that $f(x) = x^2 + x + 2$ is an irreducible polynomial in $Z_3[x]$, construct the addition and multiplication tables for the field $Z_3[x]/(f(x))$.

b) Write out the addition and multiplication tables for $Z_3[x]/(x^2 + 2)$. Determine whether or not this ring is a field.

20. Establish that $f(x) = x^4 + x + 1 \in Z_2[x]$ is irreducible in $Z_2[x]$, and write out the multiplication table for the field $Z_2[x]/(f(x))$.

21. a) Consider the irreducible polynomial $f(x) = x^3 + 4x + 2$ in $Z_5[x]$. If λ is a root of $f(x)$ in the field $Z_5[x]/(f(x))$, find the multiplicative inverse of $\lambda + 1$.

b) If λ is a root of the polynomial $f(x) = x^3 + x^2 + 1$ in the field $Q[x]/(f(x))$, confirm that $f(x) = (x - \lambda)(x^2 + (1 + \lambda)x + (\lambda + \lambda^2))$.

22. If $f(x) = x - a \in F[x]$, where F is a field, prove that $F[x]/(f(x)) \simeq F$; in particular, conclude that $R^{\#}[x]/(x - \pi) \simeq R^{\#}$.

23. a) Show that the polynomial $f(x) = x^3 + x^2 + 1 \in Z_2[x]$ factors into linear factors in $Z_2[x]/(f(x))$ by actually finding the factors.

b) Factor the polynomial $f(x) = x^3 + 2x + 1$ in the field $Z_3[x]/(f(x))$.

24. Prove that the set of algebraic numbers

$$A = \{r \in C \mid r \text{ is algebraic over } Q\}$$

forms a subfield of C containing Q. [*Hint:* If $r, s \in A$, then the finite extension $Q(r, s)$ is algebraic over Q.]

25. If $n \neq 1$ is a (nonzero) square-free integer, verify that $Q[x]/(x^2 - n)$ forms a field that is isomorphic to the quadratic field $Q(\sqrt{n})$.

26. Let $f(x) \in Z_p[x]$ be an irreducible polynomial of degree n, p a prime. Establish that the field $F = Z_p[x]/(f(x))$ contains p^n elements.

27. a) Let F' be an extension of the field F and $r \in F'$ be a root of the irreducible polynomial $f(x) \in F[x]$. If σ is an automorphism of F' that leaves each element of the field F fixed, show that $\sigma(r)$ is also a root of $f(x)$.

b) Make use of part (a) to establish that the complex roots of polynomials with real coefficients occur in conjugate pairs.

28. Verify that the polynomial $x^2 - 3 \in Q[x]$ splits completely over $R^\#$ but that its splitting field is $Q(\sqrt{3})$.

29. a) Show that $Q(\sqrt{3}, i)$ is the splitting field for $x^4 - 4x + 16 \in Q[x]$ and also for $x^2 - 2\sqrt{3}x + 4 \in Q(\sqrt{3})[x]$.

b) Prove that the polynomials $f(x) = (x^2 - 3)(x^3 + 1)$ and $g(x) = x^4 - 9$ have the same splitting field over Q.

30. Confirm that $Q(\sqrt[4]{2}, i)$ is the splitting field for the polynomial $x^4 - 2 \in Q[x]$. [*Hint:* The complex roots of $x^4 - 2$ are $\pm\sqrt[4]{2}$ and $\pm i\sqrt[4]{2}$.]

31. Prove that the splitting field of the polynomial $ax^2 + bx + c \in Q[x]$ is the field $Q(\sqrt{b^2 - 4ac})$.

32. Suppose that F' is the splitting field for the polynomial $f(x) \in F[x]$, say

$$f(x) = a(x - r_1)(x - r_2)\cdots(x - r_n) \qquad (r_i \in F',\ a \neq 0).$$

Prove that $F' = F(r_1, r_2, \ldots, r_n)$. As a particular illustration, establish that $Q(\sqrt{2}, \sqrt{3})$ is the splitting field of $(x^2 - 2)(x^2 - 3) \in Q[x]$.

33. A field F is said to be *algebraically closed* if every nonconstant polynomial in $F[x]$ has a root in F. Prove the equivalence of the following statements:

a) F is algebraically closed.

b) Every nonconstant polynomial in $F[x]$ splits in F.

c) Every irreducible polynomial in $F[x]$ is linear.

34. Show that $[Q(2i, \sqrt{3}) : Q] = 4$ by first establishing that the extension $Q(2i, \sqrt{3}) = Q(2i + 4\sqrt{3})$.

35. If $r = \sqrt{1 + \sqrt[3]{2}}$, show that $[Q(r) : Q] = 6$ and that

$$\{1, \sqrt[3]{2}, (\sqrt[3]{2})^2, r, r\sqrt[3]{2}, r(\sqrt[3]{2})^2\}$$

forms a basis for $Q(r)$ over Q.

36. For a finite extension K of a field F, prove the assertions below:

a) $[K : F] = 1$ if and only if $K = F$.

b) If $[K : F] = p$, where p is a prime, then no field is properly contained between K and F.

c) If $[K : F] = p$, p a prime, then $K = F(r)$ for every element r in K that is not in F.

d) If $[K : F] = 2^n$ and the field E satisfies $K \supseteq E \supseteq F$, then $[K : E] = 2^m$ for some $m \leqslant n$.

37. a) If $F \subseteq C$ is a finite extension of the real field $R^\#$, prove that either $F = R^\#$ or $F = C$. [*Hint:* See part (b) of the previous problem.]

b) For distinct integers $n_1, n_2, \ldots, n_r$, show that

$$[Q(\sqrt{n_1}, \sqrt{n_2}, \ldots, \sqrt{n_r}):Q] \leqslant 2^r.$$

38. Prove that if K is the splitting field of $f(x) \in F[x]$, where $\deg f(x) = n$, then $[K:F]$ divides $n!$.

39. Show that each of the following extensions is normal over the given field:

a) $Q(7i)$ over Q;

b) $R^{\#}(\sqrt{5}i)$ over $R^{\#}$;

c) $Q(\sqrt{2}, i)$ over Q.

40. If K is an extension of the field F with $[K:F] = 2$, verify that K is normal over F; in particular, C is normal over $R^{\#}$.

41. a) Suppose that K is an extension of the field F and $r, s \in K$ are algebraic over F with degrees m and n, respectively. If $\gcd(m, n) = 1$, show that $[F(r, s):F] = mn$.

b) Give an example of two algebraic elements r and s of degrees 2 and 4, respectively, such that $[Q(r, s):Q] < 8$.

42. Confirm that the field $Q(\sqrt[4]{2})$ is a normal extension of $Q(\sqrt{2})$ and that $Q(\sqrt{2})$ is a normal extension of Q, but that $Q(\sqrt[4]{2})$ is not a normal extension of Q.

43. Let $f(x)$ be an irreducible polynomial over the field F. If K is a finite extension of F and $\deg f(x) \nmid [K:E]$, prove that $f(x)$ has no root in K.

44. Find a primitive element for $Q(\sqrt{2}, \sqrt{3}, \sqrt{5})$; that is, find r such that $Q(\sqrt{2}, \sqrt{3}, \sqrt{5}) = Q(r)$. Do the same for $Q(\sqrt{2}, \sqrt{3}, \sqrt{-5})$.

45. Let $f(x) = x^3 + x + 1$ and $g(x) = x^3 + x^2 + 1$ belong to $Z_5[x]$. Show that $f(x)$ and $g(x)$ are irreducible in $Z_5[x]$ and that the fields $Z_5[x]/(f(x))$ and $Z_5[x]/(g(x))$ are isomorphic.

46. Establish that

a) $[\mathrm{GF}(p^n):\mathrm{GF}(p)] = n$;

b) If $m|n$, then $[\mathrm{GF}(p^n):\mathrm{GF}(p^m)] = n/m$.

47. Prove that for any field F, any finite subgroup of the multiplicative group F^* is cyclic.

48. a) If p is a prime, verify that the multiplicative group Z_p^* is a cyclic group.

b) Find a generator for the cyclic group Z_7^*, as well as for Z_{11}^*.

49. a) If F' is a subfield of $\mathrm{GF}(p^n)$, prove that $F' \simeq \mathrm{GF}(p^m)$ where $m|n$. [*Hint:* By Theorem 5.27, $[\mathrm{GF}(p^n):Z_p] = [\mathrm{GF}(p^n):\mathrm{GF}(p^m)][\mathrm{GF}(p^m):Z_p]$.]

b) Show that $\mathrm{GF}(p^n)$ has a subfield $F' \simeq \mathrm{GF}(p^m)$ if and only if $m|n$. [*Hint:* If $m|n$, then $a - 1 | a^k - 1$ for $k > 1$ implies that $p^m - 1 | p^n - 1$; hence $x^{p^m} - x | x^{p^n} - x$ in $Z_p[x]$.]

50. Find all the subfields of the Galois field $\mathrm{GF}(2^6)$.

51. a) Write out the addition and multiplication tables for the field $GF(3^2)$.

b) Find an irreducible polynomial of degree 4 in $Z_2[x]$ and use this to construct the field $GF(2^4)$.

52. Let F be a finite field with p^n elements. Show that the mapping $\sigma_p: F \to F$ defined by taking $\sigma_p(a) = a^p$ is an automorphism, the so-called *Frobenius automorphism* of F. Furthermore, $\sigma_p^n = i_F$. [*Hint:* Recall that $a^{p^n} = a$ for all $a \in F$.]

53. Given that F is a field of characteristic $p \neq 0$, show that every element of F is the pth power of some element in F; in other words, if $a \in F$, some $b \in F$ exists for which $a = b^p$. [*Hint:* Use Problem 52.]

54. Establish that no finite field F is algebraically closed. [*Hint:* If $a_0 = 0$, $a_1, \ldots, a_k$ are the elements of F, consider the polynomial $f(x) = a_1 + (x - a_0)(x - a_1)\cdots(x - a_k)$ in $F[x]$.]

55. Prove that no element of $GF(2^n)$ has multiplicative order 2 and that in $GF(p^n)$, where $p > 2$, the only element of multiplicative order 2 is -1.

56. Let P be a prime ideal of R, a commutative ring with identity. If the quotient ring R/P has only a finite number of elements, verify that R/P is a Galois field.

5.4 THE GALOIS THEORY OF FIELDS

By 1843, Galois's manuscripts had found their way to Joseph Liouville (1809–1882), who, after spending several months in the attempt to understand them, became convinced of their importance. He addressed the Academy of Sciences on July 4, 1843, opening with the words:

> I hope to interest the Academy in announcing that among the papers of Evariste Galois I have found a solution, as precise as it is profound, of this beautiful problem: whether or not the general equation of fifth degree is solvable by radicals.

Liouville announced that he would publish Galois's papers in the December 1843 issue in his recently founded periodical *Journal de Mathématiques Pures et Appliquées*. But for some reason, publication of the heavily edited version of the celebrated 1831 memoir did not occur until the October–November 1846 issue.

Although no trace of Galois's grave remains, his enduring monument lies in his ideas. During the late 1800's, Galois's theory—as well as a new topic it brought to life, group theory—became an integral and accepted part of mathematics. Galois's theory appears to have been taught in the

German universities for the first time by Richard Dedekind, who lectured on the topic at Göttingen in the winter of 1856–1857; it is said that only two students came to hear him. The first full and clear presentation of the Galois theory was given by Camille Jordan in his book *Traité des Substitutions et des Equations Algébriques* (1870).

Galois theory is concerned with normal extensions $K \supseteq F$ of a field F and, more particularly, with the relationship between subfields of K and subgroups of the group of automorphisms of K. The vitality of the subject derives not only from the interesting questions generated by the theory itself, but from its application to a variety of classical problems in algebra, number theory, and geometry. The Fundamental Theorem of Galois Theory sets up a one-to-one correspondence between subfields of K that contain F and subgroups of the group $G(K/F)$ of automorphisms of K that leave elements of F fixed. This duality allows us to rephrase certain questions about fields (for instance, the problem of deciding whether a given polynomial can be solved in radicals) as more tractable group-theoretic questions. It is the interplay of group and field theory that makes the theory evolving from the work of Galois such a fascinating part of modern algebra.

To get down to particulars, let K be an extension field of the field F and consider the set $\operatorname{Aut} K$ of all automorphisms of K. Recall that $\operatorname{Aut} K$ forms a group, the group operation being composition of functions. An element r of K is left *fixed* by $\sigma \in \operatorname{Aut} K$ provided that $\sigma(r) = r$. We say that σ leaves F fixed if it leaves every member of F fixed; that is, if $\sigma(r) = r$ for all $r \in F$. From now on, the notation $G(K/F)$ will be used to represent the set of all automorphisms of K that leave the field F fixed:

$$G(K/F) = \{\sigma \in \operatorname{Aut} K \mid \sigma(r) = r \text{ for all } r \in F\}.$$

The algebraic nature of $G(K/F)$ is described by

THEOREM 5.36 *If K is an extension field of F, then $G(K/F)$ forms a subgroup of the group* $\operatorname{Aut} K$, *the so-called Galois group of K over F.*

Proof. Clearly, $G(K/F)$ is nonempty, since it must contain the identity automorphism ι. Choose $\sigma_1, \sigma_2 \in G(K/F)$. By the definition of $G(K/F)$ we have $\sigma_1(r) = r$ and $\sigma_2(r) = r$ for all $r \in F$ and hence $\sigma_2^{-1}(r) = r$. Then

$$(\sigma_1 \circ \sigma_2^{-1})(r) = \sigma_1(\sigma_2^{-1}(r)) = \sigma_1(r) = r$$

for an arbitrary element $r \in F$, thereby putting $\sigma_1 \circ \sigma_2^{-1}$ in $G(K/F)$. This makes $G(K/F)$ a subgroup of $\operatorname{Aut} K$.

COROLLARY *If $K \supseteq E \supseteq F$ are all fields, then $G(K/E)$ is a subgroup of the Galois group $G(K/F)$.*

Having just seen that each subfield of K has associated with it a certain subgroup of Aut K, let us turn the focus around. We now start with any subgroup H of Aut K and consider the set $H^{\blacktriangle}$ consisting of those elements of K that are left fixed by every automorphism belonging to H; that is,

$$H^{\blacktriangle} = \{r \in K \mid \sigma(r) = r \text{ for all } \sigma \in H\}.$$

Despite its simplicity, the following theorem is crucial.

THEOREM 5.37 *If H is a subgroup of* Aut K, *then $H^{\blacktriangle}$ forms a subfield of the field K, known as the fixed field of H.*

Proof. Notice that $H^{\blacktriangle} \neq \emptyset$, since it contains both the elements 0 and 1. Pick $r, s \in H^{\blacktriangle}$, so that $\sigma(r) = r$ and $\sigma(s) = s$ for all σ in H. Then

$$\sigma(r - s) = \sigma(r) - \sigma(s) = r - s,$$

which implies that $r - s \in H^{\blacktriangle}$. In short, $H^{\blacktriangle}$ is a subgroup of the additive group of K. If the element $s \neq 0$, we get

$$\sigma(rs^{-1}) = \sigma(r)\sigma(s)^{-1} = rs^{-1}$$

telling us that rs^{-1} lies in $H^{\blacktriangle}$. As a consequence, the nonzero elements of $H^{\blacktriangle}$ are a subgroup of the multiplicative group of K. Our argument, taken as a whole, shows $H^{\blacktriangle}$ to be a subfield of the field K.

COROLLARY *For any subgroup H of $G(K/F)$ the fixed field $H^{\blacktriangle}$ satisfies $K \supseteq H^{\blacktriangle} \supseteq F$.*

Let us collect some immediate consequences of the preceding definitions: Given subfields E and E' of a field K and subgroups H' and H of the group Aut K:

1. if $E \subseteq E'$, then $G(K/E') \subseteq G(K/E)$;
2. if $H \subseteq H'$, then $H'^{\blacktriangle} \subseteq H^{\blacktriangle}$;
3. $E \subseteq G(K/E)^{\blacktriangle}$;
4. $H \subseteq G(K/H^{\blacktriangle})$.

A major goal of the next few pages is to obtain conditions that will ensure that the inclusions in (3) and (4) are actually equalities. Another essential concern is to determine when the two mappings $\Gamma(H) = H^{\blacktriangle}$ and $\Psi(E) = G(K/E)$ are one-to-one functions.

This seems an opportune moment to pause and present several examples.

Example 5.27
Let us begin by determining the Galois group of the complex field C over the field $R^{\#}$ of real numbers. If $\sigma \in G(C/R^{\#})$, then by exploiting the usual properties of an automorphism

$$\sigma(i)^2 = \sigma(i^2) = \sigma(-1) = -\sigma(1) = -1,$$

so that $\sigma(i) = \pm i$. Since σ leaves each element of $R^{\#}$ fixed,

$$\sigma(a + bi) = \sigma(a) + \sigma(b)\sigma(i) = a \pm bi,$$

for all $a, b \in R^{\#}$. What we have shown is that the identity mapping ι and the conjugate mapping $\bar{\sigma}$ defined by

$$\bar{\sigma}(a + bi) = a - bi$$

are the only two automorphisms of C leaving $R^{\#}$ fixed; in other words,

$$G(C/R^{\#}) = \{\iota, \bar{\sigma}\}.$$

Now suppose the complex number $a + bi$ lies in the fixed field of $G(C/R^{\#})$. Then

$$a + bi = \iota(a + bi) = \bar{\sigma}(a + bi) = a - bi,$$

yielding as a consequence $b = 0$. Hence $G(C/R^{\#})^{\blacktriangle} \subseteq R^{\#}$. Because the opposing inclusion always holds, we obtain

$$G(C/R^{\#})^{\blacktriangle} = (\iota, \bar{\sigma}\}^{\blacktriangle} = R^{\#}.$$

Example 5.28
To see that the Galois group of a field over a proper subfield may be trivial, consider $Q(\sqrt[3]{2})$, where $\sqrt[3]{2}$ denotes the real cube root of 2. A typical element of $Q(\sqrt[3]{2})$ may be written in the form $a + b\sqrt[3]{2} + c(\sqrt[3]{2})^2$, where $a, b, c \in Q$. If σ is an automorphism in $G(Q(\sqrt[3]{2})/Q)$, then

$$\sigma(\sqrt[3]{2})^3 = \sigma((\sqrt[3]{2})^3) = \sigma(2) = 2,$$

from which we deduce that $\sigma(\sqrt[3]{2})$ is also a cube root of 2. Since $Q(\sqrt[3]{2}) \subseteq R^{\#}$, no cube roots of 2 are contained in $Q(\sqrt[3]{2})$ other than $\sqrt[3]{2}$ itself; hence $\sigma(\sqrt[3]{2}) = \sqrt[3]{2}$. It follows that the only automorphism of $Q(\sqrt[3]{2})$ leaving Q fixed is the identity automorphism ι:

$$\begin{aligned}\sigma(a + b\sqrt[3]{2} + c(\sqrt[3]{2})^2) &= a + b\sigma(\sqrt[3]{2}) + c\sigma(\sqrt[3]{2})^2 \\ &= a + b\sqrt[3]{2} + c(\sqrt[3]{2})^2 \\ &= \iota(a + b\sqrt[3]{2} + c(\sqrt[3]{2})^2).\end{aligned}$$

Thus $G(Q(\sqrt[3]{2})/Q) = \{\imath\}$, and so

$$G(Q(\sqrt[3]{2})/Q)^{\blacktriangle} = \{\imath\}^{\blacktriangle} = Q(\sqrt[3]{2}).$$

Example 5.29

For our last illustration we turn to the field $Q(\sqrt{2}, \sqrt{3})$. Because $\{1, \sqrt{2}, \sqrt{3}, \sqrt{6}\}$ serves as a basis for $Q(\sqrt{2}, \sqrt{3})$, an arbitrary member can be expressed uniquely as $a + b\sqrt{2} + c\sqrt{3} + d\sqrt{6}$ for suitable rational numbers a, b, c, d. Thus an automorphism $\sigma \in G(Q(\sqrt{2}, \sqrt{3})/Q)$ is completely determined by how it acts on the basis elements $\sqrt{2}$ and $\sqrt{3}$. Now

$$\sigma(\sqrt{2})^2 = \sigma((\sqrt{2})^2) = \sigma(2) = 2,$$
$$\sigma(\sqrt{3})^2 = \sigma((\sqrt{3})^2) = \sigma(3) = 3.$$

This implies that the only possible values of $\sigma(\sqrt{2})$ are ± 2, while those of $\sigma(\sqrt{3})$ are ± 3. Consequently, the Galois group $G(Q(\sqrt{2}, \sqrt{3})/Q)$ consists of four automorphisms $\sigma_1 = \imath, \sigma_2, \sigma_3, \sigma_4$, where

$$\begin{aligned} \sigma_1(\sqrt{2}) &= \sqrt{2}, & \sigma_1(\sqrt{3}) &= \sqrt{3} \\ \sigma_2(\sqrt{2}) &= -\sqrt{2}, & \sigma_2(\sqrt{3}) &= \sqrt{3}, \\ \sigma_3(\sqrt{2}) &= \sqrt{2}, & \sigma_3(\sqrt{3}) &= -\sqrt{3}, \\ \sigma_4(\sqrt{2}) &= -\sqrt{2}, & \sigma_4(\sqrt{3}) &= -\sqrt{3}. \end{aligned}$$

Certainly the fixed field of $G(Q(\sqrt{2}, \sqrt{3})/Q)$ always contains Q, but does it include anything else? Since $\sigma_2(\sqrt{2}) = -\sqrt{2} = \sigma_4(\sqrt{2})$ and $\sigma_3(\sqrt{3}) = -\sqrt{3}$, it is apparent that only the elements of Q remain fixed by all the automorphisms in $G(Q(\sqrt{2}, \sqrt{3})/Q)$; simply put,

$$G(Q(\sqrt{2}, \sqrt{3})/Q)^{\blacktriangle} = \{\sigma_1 = \imath, \sigma_2, \sigma_3, \sigma_4\}^{\blacktriangle} = Q.$$

Let us go on to obtain some information about the size of Galois groups. We begin with

THEOREM 5.38 *If the element r is algebraic over F, with minimum polynomial $f(x) \in F[x]$, then*

$$o(G(F(r)/F)) \leqslant \deg f(x).$$

Proof. For the proof, let $\deg f(x) = n$. Then, by appealing to Theorem 5.21 a typical element a of $F(r)$ can be written uniquely in the form

$$a = a_0 + a_1 r + \cdots + a_{n-1} r^{n-1} \qquad a_i \in F.$$

Given any automorphism $\sigma \in G(F(r)/F)$, we have

$$\begin{aligned}\sigma(a) &= \sigma(a_0) + \sigma(a_1)\sigma(r) + \cdots + \sigma(a_{n-1})\sigma(r)^{n-1}\\ &= a_0 + a_1\sigma(r) + \cdots + a_{n-1}\sigma(r)^{n-1},\end{aligned}$$

which shows that σ will be determined once its effect upon r is known.

Now if the polynomial $f(x) = c_0 + c_1x + \cdots + c_nx^n$, we would know that $0 = c_0 + c_1r + \cdots + c_nr^n$. Then

$$\begin{aligned}0 = \sigma(0) &= \sigma(c_0 + c_1r + \cdots + c_nr^n)\\ &= c_0 + c_1\sigma(r) + \cdots + c_n\sigma(r)^n = f(\sigma(r)).\end{aligned}$$

This relation states that $\sigma(r)$ is also a root of $f(x)$ in $F(r)$. But $f(x)$ cannot have more than n roots in $F(r)$—or, for that matter, in any splitting field of $f(x)$. Hence there are at most n possible choices for the image $\sigma(r)$ and, in its turn, at most n automorphisms in $G(F(r)/F)$.

COROLLARY 1 *If K is a finite extension of F, a field of characteristic zero, then $G(K/F)$ is a finite group with $o(G(K/F)) \leqslant [K:F]$.*

Proof. By Theorem 5.31, any finite extension is a simple algebraic extension $K = F(r)$. If $f(x)$ is the minimum polynomial of r over F, then $\deg f(x) = [F(r):F]$.

To illustrate this corollary, consider $G(C/R^{\#})$. By Example 5.27, $o(G(C/R^{\#})) = 2$ and $[C:R^{\#}] = 2$ also, since $\{1, i\}$ is a basis for C over R.

COROLLARY 2 *Let $f(x) \in F[x]$ have n distinct roots in a splitting field K over F. Then the Galois group $G(K/F)$ is isomorphic to a subgroup of the symmetric group S_n.*

Proof. By virtue of the argument of Theorem 5.38, each automorphism $\sigma \in G(K/F)$ induces a permutation of the distinct roots r_1, $r_2, \ldots, r_n$ of $f(x)$. We may therefore define a function Λ from $G(K/F)$ into the group of permutations of $\{r_1, r_2, \ldots, r_n\}$ by means of

$$\Lambda(\sigma) = \begin{pmatrix} r_1 & r_2 & \cdots & r_n \\ \sigma(r_1) & \sigma(r_2) & \cdots & \sigma(r_n)\end{pmatrix}$$

The calculation

$$\begin{aligned}\Lambda(\sigma) \circ \Lambda(\tau) &= \begin{pmatrix} r_1 & r_2 & \cdots & r_n \\ \sigma(r_1) & \sigma(r_2) & \cdots & \sigma(r_n)\end{pmatrix} \circ \begin{pmatrix} r_1 & r_2 & \cdots & r_n \\ \tau(r_1) & \tau(r_2) & \cdots & \tau(r_n)\end{pmatrix}\\ &= \begin{pmatrix} r_1 & r_2 & \cdots & r_n \\ (\sigma \circ \tau)(r_1) & (\sigma \circ \tau)(r_2) & \cdots & (\sigma \circ \tau)(r_n)\end{pmatrix} = \Lambda(\sigma \circ \tau)\end{aligned}$$

demonstrates that this mapping is a group homomorphism. Observe further that if $\Lambda(\sigma) = \Lambda(\tau)$, then $\sigma(r_i) = \tau(r_i)$ for each i; since the field $K = F(r_1, r_2, \ldots, r_n)$ is generated by $r_1, r_2, \ldots, r_n$, we get $\sigma = \tau$. Hence Λ is a one-to-one mapping, making it an isomorphism of $G(K/F)$ onto a subgroup of the symmetric group of $\{r_1, r_2, \ldots, r_n\}$. The latter group is, of course, isomorphic to S_n.

Corollary 1 furnishes a crude bound to the number of automorphisms of an extension K that fix F. The next theorem imposes a condition on K that will sharpen the inequality to an equality.

THEOREM 5.39 *If K is a normal extension of F, where F is a field of characteristic zero, then*

$$o(G(K/F)) = [K:F].$$

Proof. Theorem 5.31 guarantees that K is a simple algebraic extension of F; that is, we can write $K = F(r)$ for some $r \in K$. Let $f(x)$ be the minimum polynomial of r over F. Since K is normal over F and $f(x)$ has one root in K, $f(x)$ must split completely in K. It cannot split in any proper subfield of K; any subfield containing the root r will include all of $F(r) = K$. All this assures us that K is the splitting field for the polynomial $f(x)$.

Since K is of characteristic zero, $f(x)$ has $n = \deg f(x)$ distinct roots in K. Call them $r_1 = r, r_2, \ldots, r_n$. By using Theorem 5.24 the identity automorphism on F can be extended uniquely to an isomorphism $\sigma_i : F(r_1) \to F(r_i)$ taking r_1 onto r_i and leaving every element of F fixed. By normality, $F(r_1) = F(r_2) = \cdots = F(r_n)$. Thus each $F(r_i)$ coincides with K, and σ_i is actually an automorphism of K; that is, $\sigma_i \in G(K/F)$ for $i = 1, 2, \ldots, n$. These n mappings are the only members of $G(K/F)$, since any automorphism in $G(K/F)$ must send r onto a root of $f(x)$. As a result,

$$o(G(K/F)) = n = [K:F],$$

thereby proving the theorem.

COROLLARY *If r is a root of an irreducible polynomial $f(x) \in F[x]$ and $F(r)$ is a splitting field of $f(x)$, then*

$$o(G(F(r)/F)) = \deg f(x).$$

Although F is left unaltered by the Galois group $G(K/F)$, it is possible that the automorphisms that fix F also fix other elements of K that are not in F. Another way of saying the same thing is: F may be smaller than the fixed field of $G(K/F)$. This cannot happen when K is taken to be a normal extension of a field of characteristic zero, as the following theorem shows.

THEOREM 5.40 *Let K be a finite extension of a field F whose characteristic is zero. Then K is a normal extension of F if and only if F is the fixed field of $G(K/F)$.*

Proof. Since K has characteristic zero, $K = F(r)$ for some $r \in K$. Let $f(x) \in F[x]$ be the minimum polynomial of r. We assume first that K is a normal extension of F. Then $f(x)$ splits completely in K, because it admits one root $r \in K$. Thus $f(x)$ has $r = r_1, r_2, \ldots, r_n$ roots in K. The last part of the proof of Theorem 5.39 carries over verbatim and shows that corresponding to r_i there is an automorphism σ_i of K that sends r to r_i and leaves F fixed. Now pick $a \in K = F(r)$ to be an arbitrary element in the fixed field of $G(K/F)$. Then a has the form

$$a = a_0 + a_1 r + \cdots + a_{n-1} r^{n-1},$$

where each $a_i \in F$. Applying the automorphisms $\sigma_1, \sigma_2, \ldots, \sigma_n$ to this relation, we obtain

$$a = \sigma_i(a) = a_0 + a_1 r_i + \cdots + a_{n-1} r_i^{n-1} \qquad (i = 1, 2, \ldots, n).$$

The implication is that the polynomial

$$g(x) = (a_0 - a) + a_1 x + \cdots + a_{n-1} x^{n-1}$$

has n distinct roots $r_1, r_2, \ldots, r_n$ in K. But this contradicts Theorem 4.62 unless $g(x)$ is the zero polynomial, that is, unless all the coefficients of $g(x)$ are zero. In particular, $a_0 - a = 0$, and so $a = a_0 \in F$. It turns out that the fixed field of $G(K/F)$ is contained in F, leading to their equality.

To establish the converse, suppose that F is the fixed field of $G(K/F)$. As was explained above, $K = F(r)$ where $r \in K$. We let

$$G(K/F) = \{\sigma_1 = \iota, \sigma_2, \ldots, \sigma_n\}$$

and consider the polynomial

$$f(x) = (x - \sigma_1(r))(x - \sigma_2(r)) \cdots (x - \sigma_n(r)) \in K[x].$$

Clearly, this polynomial splits completely in K. Since the element r is a root of $f(x)$ and r generates K over F, $f(x)$ cannot split in any proper subfield of K that contains F. Hence K is the splitting field of $f(x)$.

What is required is to show that $f(x)$ has all its coefficients in F. Toward this end, recall that each automorphism σ_i in $G(K/F)$ induces an isomorphism $\bar{\sigma}_i : K[x] \to K[x]$ upon setting

$$\bar{\sigma}_i(a_0 + a_1 x + \cdots + a_n x^n) = \sigma_i(a_0) + \sigma_i(a_1)x + \cdots + \sigma_i(a_n)x^n.$$

In the situation at hand,

$$\bar{\sigma}_i(f(x)) = (x - \sigma_i(\sigma_1(r)))\cdots(x - \sigma_i(\sigma_n(r))).$$

Now we invoke the fact that $G(K/F)$ is a group to arrive at

$$\sigma_i(\sigma_j(r)) = (\sigma_i \circ \sigma_j)(r) = \sigma_k(r)$$

for some k. That is, any member of $G(K/F)$ simply permutes the elements $\sigma_1(r), \sigma_2(r), \ldots, \sigma_n(r)$ among themselves. This implies that the factors of $\bar{\sigma}_i(f(x))$ are the same as those of $f(x)$, except possibly for order. Put more succinctly, $\bar{\sigma}_i(f(x)) = f(x)$. Because equal polynomials have equal coefficients, we conclude immediately that each automorphism $\sigma_i \in G(K/F)$ leaves the coefficients of $f(x)$ unaltered. This would place the coefficients of $f(x)$ in the fixed field of $G(K/F)$, which, by assumption, is F. To settle the whole affair, we need only observe that, as the splitting field of a polynomial $f(x) \in F[x]$, K must be a normal extension of F. This is the theorem.

Theorem 5.40 may be viewed in another light. It asserts that if K is a normal extension of F, then every element of K that lies outside of F is moved by at least one automorphism in $G(K/F)$. As a very simple illustration, let us take the complex field C, which we know to be normal over the field $R^{\#}$ of real numbers. Given $x = a + bi$, with $b \neq 0$, there is an element in $G(C/R^{\#})$ that moves x; namely, the automorphism that send every complex number to its complex conjugate.

We now have the required information to prove that a completely natural one-to-one correspondence exists between subgroups of $G(K/F)$ and fields intermediate between K and F. This is the content of a result known as the *Fundamental Theorem of Galois Theory*.

THEOREM 5.41 (Fundamental Theorem of Galois Theory) *Let K be a normal extension of a field F of characteristic zero. Then a one-to-one correspondence exists between the fields E such that $K \supseteq E \supseteq F$ and the subgroups of the Galois group $G(K/F)$; this correspondence is given by*

$$E \leftrightarrow G(K/E).$$

Furthermore, $[K:E] = o(G(K/E))$ and $[E:F] = [G(K/F):G(K/E)]$.

Proof. We define a mapping Γ from the set of all subfields between K and F and the set of all subgroups of $G(K/F)$ by the following rule: If $K \supseteq E \supseteq F$, then $\Gamma(E) = G(K/E)$. Since the extension K is normal over F, K is also normal over any intermediate field E satisfying $K \supseteq E \supseteq F$ (corollary to Theorem 5.30). Then, as indicated by Theorem 5.40, $E = G(K/E)^{\blacktriangle}$, the fixed field of $G(K/E)$. The main point is that if $\Gamma(E_1) = \Gamma(E_2)$, or equivalently $G(K/E_1) = G(K/E_2)$, then

$$E_1 = G(K/E_1)^\blacktriangle = G(K/E_2)^\blacktriangle = E_2.$$

This observation establishes the one-to-one character of the mapping Γ.

The next step is to confirm that Γ maps onto the set of subgroups of the Galois group $G(K/F)$. For this, let $H = \{\sigma_1 = \iota, \sigma_2, \ldots, \sigma_n\}$ be an arbitrary subgroup of $G(K/F)$. Our plan is to argue that $H = G(K/H^\blacktriangle)$ and hence $H = \Gamma(H^\blacktriangle)$. According to definition, the automorphisms in H leave fixed the elements of $H^\blacktriangle$, so the inclusion $H \subseteq G(K/H^\blacktriangle)$ certainly holds. Now K can be expressed as a simple algebraic extension $K = H^\blacktriangle(r)$ with $r \in K$. If $f(x)$ is the minimum polynomial of r over $H^\blacktriangle$, then by the corollary to Theorem 5.39, $o(G(K/H^\blacktriangle)) = \deg f(x)$. Using the same proof as we gave in Theorem 5.40, it can be shown that the coefficients of the polynomial

$$g(x) = (x - \sigma_1(r))(x - \sigma_2(r))\cdots(x - \sigma_n(r))$$

must belong to the fixed field of H, which is simply to say that $g(x) \in H^\blacktriangle[x]$. Because $g(x)$ has r as a root, the minimum nature of $f(x)$ over $H^\blacktriangle$ implies that $n = \deg g(x) \geqslant \deg f(x)$. All of this yields

$$o(H) = n \geqslant \deg f(x) = o(G(K/H^\blacktriangle)) \geqslant o(H),$$

leading to the equality $o(H) = o(G(K/H^\blacktriangle))$. But this last relation forces $H = G(K/H^\blacktriangle)$, exactly as asserted.

This brings us to the last contention of the theorem. Since K is also a normal extension of any intermediate field E, $K \supseteq E \supseteq F$, Theorem 5.39 applied to E gives $o(G(K/E)) = [K:E]$. Consequently,

$$o(G(K/F)) = [K:F] = [K:E][E:F] = o(G(K/E))[E:F],$$

and we are left with

$$[E:F] = o(G(K/F))/o(G(K/E)) = [G(K/F):G(K/E)]$$

by Lagrange's Theorem. Thus all claims are established.

The Fundamental Theorem describes a one-to-one correspondence between subfields of K that contain F and subgroups of $G(K/F)$. One refers to this pairing as being the *Galois correspondence*. Observe that inclusion relations are reversed under the Galois correspondence, in the sense that larger subfields correspond to smaller subgroups. More precisely, given $K \supseteq E_1 \supseteq F$ and $K \supseteq E_2 \supseteq F$,

$$E_1 \subseteq E_2 \quad \text{if and only if} \quad G(K/E_2) \subseteq G(K/E_1).$$

Indeed, if $E_1 \subseteq E_2$, then any automorphism $\sigma \in G(K/E_2)$ that leaves E_2 fixed will necessarily fix E_1; hence $G(K/E_2) \subseteq G(K/E_1)$. On the other hand, if

$G(K/E_2) \subseteq G(K/E_1)$, then, because the operation ▲ reverses inclusions,

$$E_1 = G(K/E_1)^{\blacktriangle} \subseteq G(K/E_2)^{\blacktriangle} = E_2.$$

It is not by chance that the adjective "normal," which has a prominent place in group theory, has also been applied to a certain type of extension field. As a corollary to Theorem 5.41, we shall now show that if the intermediate field E is normal over F, then the subgroup $G(K/E)$ corresponding to it is normal, and conversely; in short, the Galois correspondence preserves the relation of normality.

COROLLARY *Under the hypothesis of the theorem, E is a normal extension of F if and only if $G(K/E)$ is a normal subgroup of the Galois group $G(K/F)$. When this is the case,*

$$G(E/F) \simeq G(K/F)/G(K/E).$$

Proof. We start by assuming that E is normal over the field F. Now $E = F(r)$ for some element r in E. If $f(x) \in F[x]$ is the minimum polynomial of r over F, then since $f(x)$ has the root r in E, normality implies that all its roots lie in E. Given an automorphism $\sigma \in G(K/F)$, $\sigma(r)$ is a root of $f(x)$ and so must belong to E. Because r generates E over F, it follows that $\sigma(E) \subseteq E$. We infer from this that the restriction $\sigma|E$ of σ to E is an automorphism of E. Thus a mapping $\Psi: G(K/F) \to G(K/E)$ can be defined by taking $\Psi(\sigma) = \sigma|E$. It is easily verified that Ψ is a group homomorphism. As for its kernel, notice that

$$\begin{aligned}\ker \Psi &= \{\sigma \in G(K/F) | \sigma|E = i_E\}\\ &= \{\sigma \in G(K/F) | \sigma(a) = a \text{ for all } a \in E\}\\ &= G(K/E).\end{aligned}$$

Consequently, $G(K/E)$ forms a normal subgroup of $G(K/F)$, being the kernel of a homomorphism.

The next step is to show that Ψ maps onto the group $G(E/F)$. By a standard homomorphism theorem,

$$\Psi G(K/F) \simeq G(K/F)/G(K/E),$$

which leads us to conclude that the order of $\Psi G(K/F)$ is

$$o(G(K/F))/o(G(K/E)) = [G(K/F): G(K/E)].$$

Appealing to the theorem just proved, we see that the right-hand index has value $[E:F] = o(G(E/F))$. Now $\Psi G(K/F)$ is a subgroup of $G(E/F)$, with order equal to that of $G(E/F)$; therefore $\Psi G(K/F) = G(E/F)$. This yields the desired isomorphism

$$G(E/F) \simeq G(K/F)/G(K/E).$$

For the opposite direction of the corollary, suppose that $G(K/E)$ is a normal subgroup of $G(K/F)$. We contend that the fixed field of $G(E/F)$ is precisely F; hence by Theorem 5.40, E is a normal extension of F. The assumed normality of $G(K/E)$ gives $\sigma^{-1} \circ \tau \circ \sigma \in G(K/E)$ for any σ in $G(K/F)$ and τ in $G(K/E)$. By definition, $G(K/E)$ fixes the field E, so $(\sigma^{-1} \circ \tau \circ \sigma)(a) = a$ for all $a \in E$ or, what is the same thing, $\tau(\sigma(a)) = \sigma(a)$. The implication is that $\sigma(a) \in G(K/E)^{\blacktriangle} = E$ for each $a \in E$. Expressed otherwise, $\sigma(E) \subseteq E$. This allows us to conclude that the restriction $\sigma|E$ is in $G(E/F)$. If the element a lies in the fixed field of $G(E/F)$, it follows that $(\sigma|E)(a) = a$, or rather, $\sigma(a) = a$. Thus any automorphism in $G(K/F)$ fixes a, and so $a \in G(K/F)^{\blacktriangle} = F$. The net result of this is the inclusion $G(E/F)^{\blacktriangle} \subseteq F$, leading to equality; we have proved what we set out to prove.

Remark. Let $f(x)$ be a polynomial in $F[x]$, with a splitting field K over F. The usual practice is to refer to the group $G(K/F)$ as the *Galois group of the polynomial* $f(x)$ or the equation $f(x) = 0$. Since any two splitting fields K and K' of $f(x)$ are isomorphic, the Galois group of $f(x)$ is unique for all purposes. Indeed, if $K \simeq K'$ via the mapping $\alpha : K \to K'$, then there is an isomorphism between $G(K/F)$ and $G(K'/F)$; simply define $\theta : G(K/F) \to G(K'/F)$ by $\theta(\sigma) = \alpha \circ \sigma \circ \alpha^{-1}$.

We pause to look at two examples illustrating the results of this section.

Example 5.30

Once again consider $K = Q(\sqrt{2}, \sqrt{3})$, the splitting field over Q of the polynomial $f(x) = (x^2 - 2)(x^2 - 3)$. As we saw earlier, the Galois group $G(K/Q) = \{\sigma_1 = \iota, \sigma_2, \sigma_3, \sigma_4\}$, where each automorphism σ_i is determined by its effect on the generating elements $\sqrt{2}$ and $\sqrt{3}$ of K:

$$\begin{aligned} \sigma_1(\sqrt{2}) &= \sqrt{2}, & \sigma_1(\sqrt{3}) &= \sqrt{3}, \\ \sigma_2(\sqrt{2}) &= -\sqrt{2}, & \sigma_2(\sqrt{3}) &= \sqrt{3}, \\ \sigma_3(\sqrt{2}) &= \sqrt{2}, & \sigma_3(\sqrt{3}) &= -\sqrt{3}, \\ \sigma_4(\sqrt{2}) &= -\sqrt{2}, & \sigma_4(\sqrt{3}) &= -\sqrt{3}. \end{aligned}$$

The group $G(K/Q)$ has three nontrivial subgroups:

$$H_1 = \{\sigma_1, \sigma_2\}, \qquad H_2 = \{\sigma_1, \sigma_3\}, \qquad H_3 = \{\sigma_1, \sigma_4\}.$$

Since K is a normal extension of Q, the Fundamental Theorem of Galois Theory enables us to conclude that K has precisely three proper subfields larger than Q, namely, the fixed fields of the foregoing subgroups.

An arbitrary element $x \in K$ can be expressed uniquely in the form

$$x = a + b\sqrt{2} + c\sqrt{3} + d\sqrt{6}$$

where a, b, c, $d \in Q$. Now x lies in the fixed field $H_1^\blacktriangle$ if and only if

$$\begin{aligned} x = \sigma_2(x) &= a + b\sigma_2(\sqrt{2}) + c\sigma_2(\sqrt{3}) + d\sigma_2(\sqrt{6}) \\ &= a - b\sqrt{2} + c\sqrt{3} - d\sqrt{6}. \end{aligned}$$

By the uniqueness of the representation of x this will occur if and only if $b = -b$ and $d = -d$; in other words, $b = d = 0$, giving us

$$x = a + c\sqrt{3}.$$

We end up with $Q(\sqrt{3})$ as the fixed field of H_1.

A similar argument can be given for the fixed field of H_2. That is, x is left fixed by H_2 if and only if

$$x = \sigma_3(x) = a + b\sqrt{2} - c\sqrt{3} - d\sqrt{6}.$$

For this equation to be satisfied, we must have $c = -c$ and $d = -d$ or, equivalently, $c = d = 0$. In short, $x = a + b\sqrt{2}$. The subfield of K corresponding to H_2 is therefore $Q(\sqrt{2})$.

Going on to $H_3^\blacktriangle$, observe that $x \in H_3^\blacktriangle$ provided that

$$x = \sigma_4(x) = a - b\sqrt{2} - c\sqrt{3} + d\sqrt{6},$$

which yields $b = c = 0$. Thus $x = a + d\sqrt{6}$, and we obtain $H_3^\blacktriangle = Q(\sqrt{6})$.

All in all, it may be inferred that there are precisely three intermediate fields $Q(\sqrt{2})$, $Q(\sqrt{3})$, and $Q(\sqrt{6})$ between K and Q. The following lattice diagram illustrates the inclusion relationships among the various subfields.

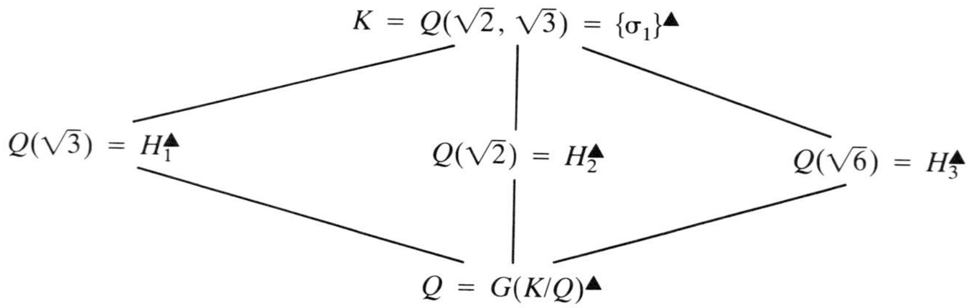

Example 5.31

Let K be the splitting field of the polynomial $x^3 - 2 \in Q[x]$. Then K is a normal extension of Q. The distinct roots of $x^3 - 2$ in the complex number field are $\sqrt[3]{2}$, $\omega\sqrt[3]{2}$, and $\omega^2\sqrt[3]{2}$, where $\omega = (-1 + i\sqrt{3})/2$ and $\sqrt[3]{2}$ is the real cube root of 2. Therefore $K = Q(\omega, \sqrt[3]{2})$.

We first find the degree of K over Q and thus the order of $G(K/Q)$. Clearly, the inclusions

$$Q \subset Q(\sqrt[3]{2}) \subset Q(\sqrt[3]{2})(\omega) = K$$

hold. Since the minimum polynomial of $\sqrt[3]{2}$ over Q is $x^3 - 2$, Theorem 5.39 asserts that $[Q(\sqrt[3]{2}):Q] = 3$. Likewise, $x^2 + x + 1$ being the minimum polynomial of ω over $Q(\sqrt[3]{2})$ implies that $[K:Q(\sqrt[3]{2})] = 2$. In light of Theorem 5.27,

$$[K:Q] = [K:Q(\sqrt[3]{2})][Q(\sqrt[3]{2}):Q] = 2 \cdot 3 = 6,$$

and, because K is normal over Q,

$$o(G(K/Q)) = [K:Q] = 6.$$

Since a basis for $Q(\sqrt[3]{2})$ over Q is given by the set $\{1, \sqrt[3]{2}, (\sqrt[3]{2})^2\}$ and $\{1, \omega\}$ is a basis for K over $Q(\sqrt[3]{2})$, products of these elements yield a basis

$$\{1, \sqrt[3]{2}, (\sqrt[3]{2})^2, \omega, \omega\sqrt[3]{2}, \omega(\sqrt[3]{2})^2\}$$

for K over Q. As a result, any automorphism $\sigma \in G(K/Q)$ will be completely determined by its effect on $\sqrt[3]{2}$ and ω. Now $\sigma(\sqrt[3]{2})$ can be only one of the three roots $\sqrt[3]{2}$, $\omega\sqrt[3]{2}$, $\omega^2\sqrt[3]{2}$ of $x^3 - 2$; similarly, ω is mapped by σ to one of the two roots ω, ω^2 of $x^2 + x + 1$. This means that

$$G(K/Q) = (\sigma_1 = \iota, \sigma_2, \sigma_3, \sigma_4, \sigma_5, \sigma_6\},$$

where the behavior of σ_i is described in the table below:

Automorphism	*Value at* $\sqrt[3]{2}$	*Value at* ω
σ_1	$\sqrt[3]{2}$	ω
σ_2	$\sqrt[3]{2}$	ω^2
σ_3	$\omega\sqrt[3]{2}$	ω
σ_4	$\omega\sqrt[3]{2}$	ω^2
σ_5	$\omega^2\sqrt[3]{2}$	ω
σ_6	$\omega^2\sqrt[3]{2}$	ω^2

The Galois group $G(K/Q)$ possesses four nontrivial subgroups

$$H_1 = \{\sigma_1, \sigma_2\}, \quad H_2 = \{\sigma_1, \sigma_4\}, \quad H_3 = \{\sigma_1, \sigma_6\}, \quad H_4 = \{\sigma_1, \sigma_3, \sigma_5\}$$

with corresponding fixed fields

$$H_1^{\blacktriangle} = Q(\sqrt[3]{2}), \qquad H_2^{\blacktriangle} = Q(\omega^2\sqrt[3]{2}), \qquad H_3^{\blacktriangle} = Q(\omega\sqrt[3]{2}), \qquad H_4^{\blacktriangle} = Q(\omega).$$

This may be visualized by the diagram

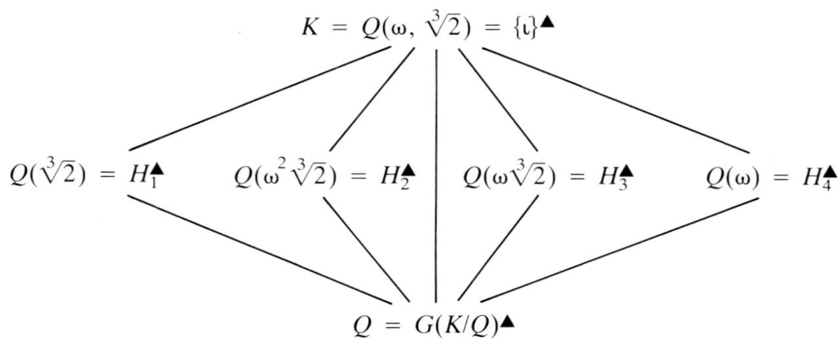

To carry through the details in one instance, let us calculate $H_2^{\blacktriangle}$. Any $x \in K$ can be represented uniquely as a linear combination of the six basic elements:

$$x = a_1 + a_2\sqrt[3]{2} + a_3(\sqrt[3]{2})^2 + a_4\omega + a_5\omega\sqrt[3]{2} + a_6\omega(\sqrt[3]{2})^2,$$

where all of the a_i belong to Q. If x lies in the fixed field of H_2, then

$$\begin{aligned} x &= \sigma_4(x) \\ &= a_1 + a_2\omega\sqrt[3]{2} + a_3(\omega\sqrt[3]{2})^2 + a_4\omega^2 + a_5\omega^2(\omega\sqrt[3]{2}) + a_6\omega^2(\omega\sqrt[3]{2})^2. \end{aligned}$$

This relation, along with the fact that $\omega^2 + \omega + 1 = 0$, yields

$$\begin{aligned} x = (a_1 - a_4) &+ a_5\sqrt[3]{2} - a_3(\sqrt[3]{2})^2 - a_4\omega \\ &+ a_2\omega\sqrt[3]{2} + (a_6 - a_3)\omega(\sqrt[3]{2})^2. \end{aligned}$$

Comparing coefficients of like terms, we obtain

$$a_1 = a_1 - a_4,\ a_2 = a_5,\ a_3 = -a_3,\ a_4 = -a_4,\ a_6 = a_6 - a_3,$$

which means that $a_2 = a_5$, $a_3 = a_4 = 0$. Thus the elements of K that are left fixed by H_2 can be written in the form

$$\begin{aligned} x &= a_1 + a_2(1 + \omega)\sqrt[3]{2} + a_6\omega^4(\sqrt[3]{2})^2 \\ &= a_1 - a_2\omega^2\sqrt[3]{2} + a_6(\omega^2\sqrt[3]{2})^2. \end{aligned}$$

This amounts to asserting that $H_2^{\blacktriangle} = Q(\omega^2\sqrt[3]{2})$.

Before leaving the example we might point out that H_4 is a normal subgroup of $G(K/Q)$, being of index 2 in $G(K/Q)$. Hence

$H_4^\blacktriangle = Q(\omega)$ is a normal extension of Q and so must be the splitting field of some polynomial $f(x)$ in $Q[x]$; the required polynomial is in fact $f(x) = x^2 + x + 1$. We leave it to the reader to show that none of the remaining subfields $Q(\sqrt[3]{2})$, $Q(\omega\sqrt[3]{2})$, or $Q(\omega^2\sqrt[3]{2})$ are normal over Q.

PROBLEMS 5.4

1. Consider an extension K of the field F. Let F_1 and F_2 be subfields of K containing F and H_1 and H_2 be subgroups of the Galois group $G(K/F)$. If F_1F_2 denotes the smallest subfield of K containing both F_1 and F_2, prove that

a) $G(K/(F_1 \cap F_2)) = G(K/F_1)G(K/F_2)$;
$G(K/F_1F_2) = G(K/F_1) \cap G(K/F_2)$.

b) $(H_1H_2)^\blacktriangle = H_1^\blacktriangle \cap H_2^\blacktriangle$; $(H_1 \cap H_2)^\blacktriangle = H_1^\blacktriangle H_2^\blacktriangle$.

2. If K is an extension of the field F and H is a subgroup of $G(K/F)$, show that

$$G(K/H^\blacktriangle)^\blacktriangle = H^\blacktriangle.$$

3. a) Prove that $G(R^\#/Q) = \{i_{R^\#}\}$. [*Hint:* Any $\sigma \in G(R^\#/Q)$ preserves order in $R^\#$; if $r > s$, $\sigma(r) - \sigma(s) = \sigma(\sqrt{r - s})^2 > 0$. If $\sigma(r) \neq r$, say $r < \sigma(r)$, pick $q \in Q$ with $r < q < \sigma(r)$ and argue until a contradiction is reached.]

b) If $K = Q(\sqrt[3]{2})$, where $\sqrt[3]{2}$ is the real cube root of 2, show that $G(K/Q) = \{i_K\}$.

4. Establish that $G(Q(\sqrt{2})/Q) = \{\iota, \sigma\}$, where σ is defined by $\sigma(a + b\sqrt{2}) = a - b\sqrt{2}$.

5. a) If K is the splitting field of a polynomial $f(x) \in F[x]$ of degree n, prove that $o(G(K/F))|n!$.

b) If $K = Q(\sqrt{2}, \sqrt{3}, \sqrt{5})$, show that $o(G(K/Q)) = 8$.

6. For $K = Q(r)$, where $r = (-1 + \sqrt{3}i)/2$, deduce that $G(K/Q) \simeq Z_3$. [*Hint:* Observe that $r^3 = 1$, and hence $\{1, r, r^2\}$ is a basis for $Q(r)$ over Q.]

7. If K is an extension of a field F with $[K:F] = 2$, prove that $G(K/F)^\blacktriangle = F$.

8. Let $K = Q(r)$, where $r = \sqrt[4]{2}$ is the real fourth root of 2. Verify the assertions below:

a) $G(K/Q) = \{i_K, \sigma\}$, with σ defined by $\sigma(r) = -r$;

b) $r^2 \in G(K/Q)^\blacktriangle$, but $r^2 \notin Q$;

c) K is not normal over Q.

9. Given that K is the splitting field of the irreducible polynomial $x^2 + bx + c \in Q[x]$, show that $G(K/Q) \simeq Z_2$.

10. Assume that K is a normal extension of a field F of characteristic zero and that E is an intermediate field between K and F that is normal over F. Use the Fundamental Theorem of Galois Theory to prove that:

a) If $G(K/F)$ is commutative, then $G(K/E)$ and $G(E/F)$ are both commutative.

b) If $G(K/F)$ is cyclic, then $G(K/E)$ and $G(E/F)$ are both cyclic.

11. Let $K = Q(\sqrt{2}, i)$ be the splitting field of the polynomial $(x^2 - 2)(x^2 + 1) \in Q[x]$.

a) Show that $[K:Q] = 4$.

b) Describe the four automorphisms of $G(K/Q)$ by giving their values at $\pm\sqrt{2}$ and $\pm i$.

c) Find all the subgroups of $G(K/Q)$ and show that their fixed fields may be represented schematically by

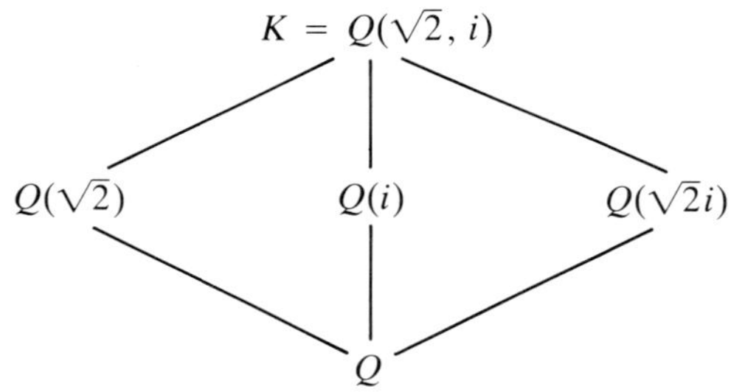

12. Suppose that K is a normal extension of a field F of characteristic zero. Prove the following:

a) If $o(G(K/F)) = pq$, where $p < q$ are both primes, then K has a subfield E such that $[E:F] = p$.

b) If $G(K/F) \simeq A_4$, then there is no intermediate field E, $K \supseteq E \supseteq F$, such that $[E:F] = 2$.

13. Show that the Galois group of the polynomial $x^4 - 2x^2 + 9 \in Q[x]$ is the same as that determined in Problem 11.

14. a) With reference to Example 5.31, show that $G(Q(\sqrt[3]{2}, \omega)/Q) \simeq S_3$, the symmetric group on three symbols.

b) Prove that if K is the splitting field of the polynomial $f(x) = x^3 - 3x - 1 \in Q[x]$, then $G(K/Q) \simeq Z_3$. [*Hint:* If r is one root of $f(x)$, then other roots are $-1/(r + 1)$ and $(1 - r - r^2)/(r + 1)$; hence $K = Q(r)$. Define $\sigma \in G(K/Q)$ by $\sigma(r) = -1/(r + 1)$ and show that $\sigma^3 = i_K$.]

15. If K is the splitting field of the polynomial $(x^3 - 2)(x^2 - 2) \in Q[x]$, determine the subgroups of the Galois group $G(K/Q)$ and the subfields of K.

16. Let $K = Q(r, i)$ be the splitting field of the polynomial $x^4 - 2 \in Q[x]$, where $r = \sqrt[4]{2}$ is the real fourth root of 2.

a) Show that $[K:Q] = 8$. [*Hint:* A basis for $Q(r)$ over Q is $\{1, r, r^2, r^3\}$.]

b) Describe the eight automorphisms in $G(K/Q)$ by giving their values at the elements r and i. [*Hint:* The four roots of $x^4 - 2$ in C are r, $-r$, ir, $-ir$.]

c) Find all the subgroups of $G(K/Q)$ and show that the corresponding fixed fields may be diagrammed as follows:

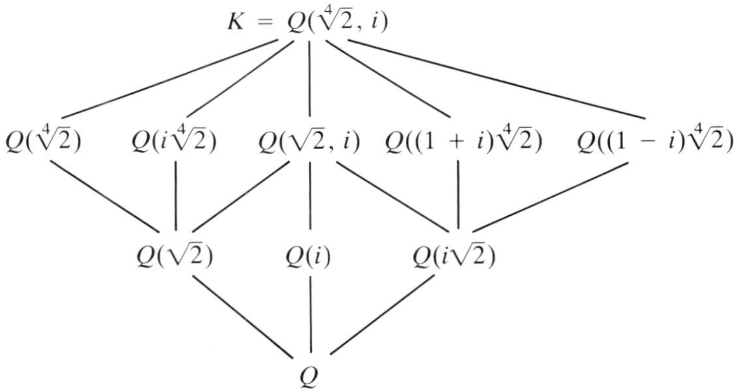

d) Confirm that the isomorphism $G(K/Q) \simeq G_S$ holds, where G_S is the group of symmetries of the square.

e) Which of the above subfields of K are normal extensions of Q?

17. Assume that K is a normal extension of a field F of characteristic zero. If $G(K/F)$ is a commutative group of order 14, diagram the lattice of subfields between K and F. [*Hint:* See Problem 19 of Section 3.2.]

5.5 EQUATIONS SOLVABLE BY RADICALS

This final section has a twofold objective. The first is to investigate several types of commutative Galois groups, the simplest of these being the cyclic ones. The information obtained will be helpful in attaining our second goal, which is to study the question of when an equation is solvable by radicals. Galois showed that a given polynomial equation $f(x) = 0$ can be solved by an explicit radical formula if and only if the associated Galois group is solvable. Thus the problem of solvability can be formulated in group-theoretic terms. Using this approach, we can obtain a large class of quintic equations over Q that cannot be solved by radicals.

Before we embark on our program, some new terminology is required. We start off with:

DEFINITION 5.12 Let F be a field and n be a positive integer. An element $\omega \in F$ is said to be an *nth root of unity* provided that ω is a root of the polynomial $x^n - 1 \in F[x]$.

Suppose that the field F has characteristic zero. Then the derivative of $x^n - 1$, that is, nx^{n-1}, has only the root 0; hence $\delta(x^n - 1)$ has no roots in common with $x^n - 1$. This means that the polynomial $x^n - 1$ possesses no multiple roots in any extension field $K \supseteq F$ (see Problem 22, Section 4.8). Therefore the number of nth roots of unity in a splitting field of $x^n - 1$ over F is precisely n. We are able to assert even more:

LEMMA *Let K be the splitting field of $x^n - 1 \in F[x]$, where F has characteristic zero. Then the nth roots of unity in K form a cyclic group with respect to multiplication.*

Proof. If r and s are nth roots of unity in K, then $r^n = s^n = 1$. Clearly, $(rs^{-1})^n = r^n(s^n)^{-1} = 1$, which implies that rs^{-1} is also an nth root of unity. Thus the nth roots of unity are a subgroup of K^*, the multiplicative group of nonzero elements of K. By Theorem 5.34, any finite multiplicative group in a field is cyclic.

DEFINITION 5.13 A generator of the cyclic group of nth roots of unity is called a *primitive nth root of unity*. Alternatively, an element $\omega \in F$ is a primitive nth root of unity if $\omega^n = 1$, but $\omega^k \neq 1$ for $1 \leqslant k < n$.

If an extension $K \supseteq F$ possesses a primitive nth root of unity ω, then $1, \omega, \omega^2, \ldots, \omega^{n-1}$ make up all the nth roots of unity in K; and $x^n - 1$ splits completely in K:

$$x^n - 1 = \prod_{k=0}^{n-1} (x - \omega^k).$$

Since ω generates a cyclic group of order n, the other primitive nth roots of unity are of the form ω^k, with $1 \leqslant k < n$ and $\gcd(k, n) = 1$ (Theorem 2.35). It follows that there are exactly $\phi(n)$ primitive nth roots of unity in K, where ϕ denotes the Euler phi-function. Thus we have

THEOREM 5.42 *If ω is a primitive nth root of unity, ω^k is also a primitive nth root of unity if and only if* $\gcd(k, n) = 1$. *The number of primitive nth roots of unity is* $\phi(n)$.

The four roots in C of $x^4 - 1 \in Q[x]$ are $1, -1, i, -i$. Since $\phi(4) = 2$, there are two primitive fourth roots of unity; they are $\omega = i$ and $\omega^3 = -i$.

The structure of the Galois group of $x^n - 1 \in F[x]$ will be particularly important in our work. Let us first look at

THEOREM 5.43 *Let K be the splitting field of the polynomial $x^n - 1 \in F[x]$, with F a field of characteristic zero. Then*

a) $K = F(\omega)$, *where ω is any primitive nth root of unity;*

b) the Galois group $G(K/F)$ is isomorphic to a subgroup of the multiplicative group of invertible elements of the ring Z_n.

Proof. The splitting field K can be expressed as $K = F(r_1, r_2, \ldots, r_n)$, where $r_1 = 1, r_2, \ldots, r_n$ are the n distinct roots of $x^n - 1$ in K. Let ω be a primitive nth root of unity. Since every nth root of unity is a power of ω, we have $F(r_1, r_2, \ldots, r_n) = F(\omega)$. Thus each automorphism $\sigma \in G(K/F)$ will be determined by its effect on the generator ω. But $\sigma(\omega)$ is also an nth root of unity, whence $\sigma(\omega) = \omega^k$ for $1 \leqslant k \leqslant n$. We further assert that $\sigma(\omega)$ is primitive; otherwise, $\gcd(k, n) = d > 1$, and

$$\sigma(\omega^{n/d}) = \sigma(\omega)^{n/d} = (\omega^k)^{n/d} = (\omega^n)^{k/d} = 1 = \sigma(1),$$

contrary to the one-to-one nature of σ.

These observations allow us to define a mapping $\chi: G(K/F) \to Z_n^*$ by $\chi(\sigma) = [k]$, where σ is described by $\sigma(\omega) = \omega^k$. Suppose that $\sigma, \tau \in G(K/F)$. Then there are integers k and j, relatively prime to n, satisfying $\sigma(\omega) = \omega^k$ and $\tau(\omega) = \omega^j$. Evidently,

$$(\sigma \circ \tau)(\omega) = \sigma(\tau(\omega)) = \sigma(\omega^j) = \sigma(\omega)^j = \omega^{kj},$$

with $\gcd(kj, n) = 1$. This makes χ a group homomorphism, as is quickly verified by

$$\chi(\sigma \circ \tau) = [kj] = [k] \cdot_n [j] = \chi(\sigma) \cdot_n \chi(\tau).$$

To confirm that χ is also a one-to-one function, let $\chi(\sigma) = \chi(\tau)$. Then $k \equiv j \pmod{n}$, and so

$$\sigma(\omega) = \omega^k = \omega^j = \tau(\omega).$$

Because an element of $G(K/F)$ is uniquely determined by its value at ω, this forces $\sigma = \tau$. The end result is that χ is an isomorphism from $G(K/F)$ onto a subgroup of Z_n^*.

COROLLARY *Under the hypothesis of the theorem, $G(K/F)$ forms a commutative group whose order divides $\phi(n)$.*

As a cautionary note, let us observe that if $K \supseteq F$ is a field containing a primitive nth root of unity ω, then the order of the Galois group $G(F(\omega)/F)$ need not equal $\phi(n)$. This can be seen by considering the eighth roots of unity in C:

$$\pm 1,\ \pm i,\ \pm(1+i)/\sqrt{2},\ \pm(-1+i)/\sqrt{2}.$$

There are $\phi(8) = 4$ primitive eighth roots of unity, specifically the last four of the roots listed above. Putting $\omega = (1+i)/\sqrt{2}$, the minimum polynomial of ω over the field $R^{\#}$ of real numbers is seen to be $f(x) = x^2 - \sqrt{2}x + 1$. According to Theorem 5.39,

$$o(G(R^{\#}(\omega)/R^{\#})) = \deg f(x) = 2 < \phi(8).$$

An interesting point is that in the special case in which the base field F equals Q, the field of rational numbers, $G(F(\omega)/F)$ is actually isomorphic to Z_n^* and hence has order $\phi(n)$. Before turning to this we need a collateral result.

LEMMA *If ω is a primitive nth root of unity in a field $F \supseteq Q$, then $[Q(\omega):Q] = \phi(n)$.*

Proof. Let $f(x)$ be the minimum polynomial of ω over Q. Then $f(x)$ divides $x^n - 1$; say, $x^n - 1 = f(x)g(x)$ for some monic polynomial $g(x)$ in $Q[x]$. By Gauss's lemma it may be assumed that both $f(x)$ and $g(x)$ have integral coefficients. We seek to show that for every prime p, with $\gcd(p, n) = 1$, ω^p is a root of $f(x)$. Suppose, for a contradiction, that $f(\omega^p) \neq 0$. From

$$f(\omega^p)g(\omega^p) = (\omega^p)^n - 1 = (\omega^n)^p - 1 = 0$$

it is clear that $g(\omega^p) = 0$; to put it another way, ω is a root of $g(x^p)$. By the minimal property of $f(x)$ there is a monic polynomial $h(x)$ in $Q[x]$ for which $g(x^p) = f(x)h(x)$. A second application of Gauss's lemma allows us to take $h(x) \in Z[x]$.

By considering the reduction homomorphism v, which sends a polynomial $r(x) = a_0 + a_1x + \cdots + a_kx^k$ to

$$v(r(x)) = (a_0 + (p)) + (a_1 + (p))x + \cdots + (a_k + (p))x^k,$$

we may pass from $Z[x]$ to $Z_p[x]$. Denote $v(r(x))$ simply by $\overline{r(x)}$. Since $a^p \equiv a (\text{mod } p)$ for all $a \in Z$ by Fermat's Little Theorem, we have $(a + (p))^p = a + (p)$. Thus the images of the preceding polynomial relations satisfy

$$x^n - \bar{1} = \overline{f(x)}\,\overline{g(x)} \qquad \text{an} \qquad \overline{g(x)}^p = \overline{g(x^p)} = \overline{f(x)}\,\overline{h(x)}.$$

As $Z_p[x]$ is a unique factorization domain, $\overline{f(x)}$ possesses an irreducible factor. Any such factor divides $\overline{g(x)}^p$ and hence divides $\overline{g(x)}$; and so its square will divide $x^n - \bar{1}$. In short, $x^n - 1$ has a multiple root in some extension of Z_p. This is impossible, since by our choice of p the derivative of $x^n - \bar{1}$ is $nx^{n-1} \not\equiv 0 (\text{mod } p)$. Having arrived at a contradiction, we see that $f(\omega^p) = 0$.

Now any other primitive nth root of unity has the form ω^k for some integer k with $1 \leqslant k < n$ and $\gcd(k, n) = 1$. Write $k = p_1 p_2 \cdots p_r$ as a product of (not necessarily distinct) primes. According to what has just been proved, ω, ω^{p_1}, $\omega^{p_1 p_2}, \ldots, \omega^{p_1 p_2 \cdots p_r} = \omega^k$ are all roots of the polynomial $f(x)$. In other words, each of the $\phi(n)$ primitive nth roots of unity in $Q(\omega)$ is a root of $f(x)$. This amounts to saying that

$$[Q(\omega):Q] = \deg f(x) \geqslant \phi(n).$$

But by Theorem 5.39,

$$[Q(\omega):Q] = o(G(Q(\omega)/Q)) \leqslant \phi(n) \text{ and}$$

hence the asserted equality holds. The lemma is settled.

This gives us all the pieces needed to pin down the structure of $G(Q(\omega)/Q)$, where ω is a primitive nth root of unity. We formulate this result as

THEOREM 5.44 *Let K be the splitting field of $x^n - 1 \in Q[x]$. Then the Galois group $G(K/Q)$ is isomorphic to the multiplicative group of invertible elements in Z_n.*

Proof. From Theorem 5.43 the field $K = Q(\omega)$, where $\omega \in K$ is any primitive nth root of unity; and $G(Q(\omega)/Q)$ is isomorphic to a subgroup of Z_n^*, the group of invertible elements of Z_n. In view of the preceding lemma,

$$o(G(Q(\omega)/Q)) = \phi(n) = o(Z_n^*),$$

from which we immediately secure the isomorphism

$$G(Q(\omega)/Q) \simeq Z_n^*.$$

An example is perhaps in order.

Example 5.32

Suppose that K is the splitting field over Q of the polynomial $x^8 - 1$. Then $K = Q(\omega)$ for some primitive eighth root of unity; for definiteness we let $\omega = (1 + i)/\sqrt{2}$. Since $\phi(8) = 4$, the Galois group

$$G(K/Q) = \{\sigma_1 = \iota, \sigma_2, \sigma_3, \sigma_4\},$$

where each of these automorphisms is determined by its action on ω. Any member of $G(K/Q)$ takes ω to another primitive eighth root of unity ω^k, where $\gcd(k, 8) = 1$; hence $\sigma_1, \sigma_2, \sigma_3, \sigma_4$ are defined by

$$\sigma_1(\omega) = \omega, \quad \sigma_2(\omega) = \omega^3, \quad \sigma_3(\omega) = \omega^5 = -\omega, \quad \sigma_4(\omega) = \omega^7 = \omega^{-1}.$$

We make the observation that

$$\begin{aligned}\sigma_2^2(\omega) &= (\omega^3)^3 = \omega,\\ \sigma_3^2(\omega) &= -(-\omega) = \omega,\\ \sigma_4^2(\omega) &= (\omega^{-1})^{-1} = \omega,\end{aligned}$$

giving rise to $\sigma_1^2 = \sigma_2^2 = \sigma_3^2 = \sigma_4^2$. Thus $G(K/Q)$ is isomorphic to the Klein 4-group K_4, which, in its turn, is isomorphic to the multiplicative group $Z_8^* = \{1, 3, 5, 7\}$. The three nontrivial subgroups of $G(K/Q)$ are

$$H_1 = \{\sigma_1, \sigma_2\}, \qquad H_2 = \{\sigma_1, \sigma_3\}, \qquad H_3 = \{\sigma_1, \sigma_4\}.$$

We wish to determine the intermediate fields between K and Q, that is, the fixed fields $H_i^\blacktriangle$. Now $x^4 + 1$ is the minimum polynomial of ω over Q, so $[K:Q] = 4$ and K has the basis $\{1, \omega, \omega^2, \omega^3\}$ over Q. A typical element $x = a + b\omega + c\omega^2 + d\omega^3$ of K belongs to the fixed field of H_1 if and only if $\sigma_2(x) = x$. A direct calculation yields

$$\sigma_2(x) = a + b\omega^3 + c\omega^6 + d\omega^9 = a + d\omega - c\omega^2 + b\omega^3,$$

whence x is fixed by σ_2 (and thus by H_1) if and only if $b = d$ and $c = 0$. All of this combines to tell us that $x \in H_1^\blacktriangle$ provided that

$$x = a + b(\omega + \omega^3) = a + b\sqrt{2}i,$$

with $a, b \in Q$, which is just another way of saying that $H_1^\blacktriangle = Q(\sqrt{2}i)$. A similar analysis establishes the other two fixed fields to be

$$H_2^\blacktriangle = Q(i) \qquad \text{and} \qquad H_3^\blacktriangle = Q(\sqrt{2}).$$

The subfields of K may be represented schematically by

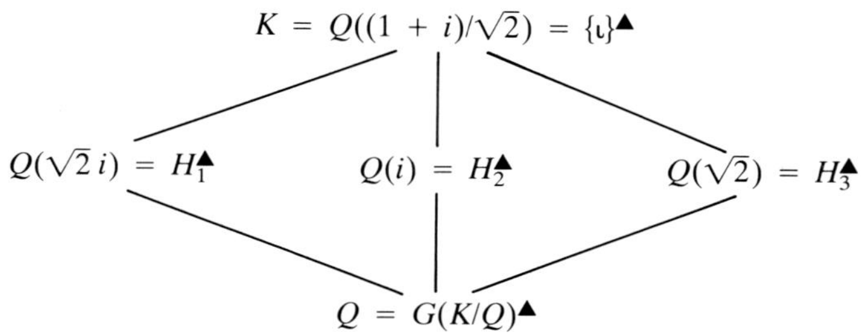

Let us continue in the same spirit by investigating another instance in which the Galois groups are commutative, namely, the Galois groups of polynomials of the form $x^n - a$. Here we impose the additional constraint that the ground field contain a primitive nth root of unity and hence contain all the nth roots of unity. The precise result is contained in the next theorem.

THEOREM 5.45 *Let F be a field of characteristic zero that contains a primitive nth root of unity. If K is the splitting field of $x^n - a \in F[x]$, $a \neq 0$, then*

a) $K = F(r)$, where r is any root of $x^n - a$;

b) the Galois group $G(K/F)$ is isomorphic to a subgroup of the additive group Z_n.

Proof. Let $\omega \in F$ be a primitive nth root of unity and r be a root of $x^n - a$. Then the n distinct roots of $x^n - a$ in K are $r, \omega r, \omega^2 r, \ldots, \omega^{n-1} r$. Obviously,

$$(\omega^k r)^n - a = (\omega^n)^k r^n - a = r^n - a = 0$$

for $0 \leqslant k \leqslant n - 1$. Moreover, if $\omega^k r = \omega^j r$, with $0 \leqslant j < k \leqslant n - 1$, then $\omega^{k-j} = 1$, contradicting the fact that ω is a primitive nth root of unity. Since $\omega \in F$, all these roots lie in $F(r)$, which makes $F(r)$ the splitting field of $x^n - a$ over F; put succinctly, $K = F(r)$.

Given $\sigma \in G(K/F)$, the element $\sigma(r)$ is also a root of $x^n - a$, and so $\sigma(r) = \omega^k r$ for some integer $k\,(0 \leqslant k \leqslant n - 1)$. We may therefore define a mapping $\chi: G(K/F) \to Z_n$ by setting $\chi(\sigma) = [k]$. It is readily verified that χ is a group homomorphism. If σ, $\tau \in G(K/F)$, with $\sigma(r) = \omega^k r$ and $\tau(r) = \omega^j r$, then

$$(\sigma \circ \tau)(r) = \sigma(\omega^j r) = \omega^j \sigma(r) = \omega^{k+j} r;$$

this means that

$$\chi(\sigma \circ \tau) = [k + j] = [k] +_n [j] = \chi(\sigma) +_n \chi(\tau).$$

Next observe that our mapping is a one-to-one function. Indeed, the equality $\chi(\sigma) = \chi(\tau)$ implies that $k \equiv j \pmod n$, leading to

$$\sigma(r) = \omega^k r = \omega^j r = \tau(r).$$

Thus $\sigma = \tau$, since each automorphism in $G(K/F)$ is completely determined once its value at r is known. Combined with what we did above, this yields the full result that χ is an isomorphism from $G(K/F)$ onto a subgroup of the additive group $(Z_n, +_n)$.

COROLLARY *Under the hypothesis of the theorem the Galois group $G(K/F)$ is a cyclic group whose order divides n.*

It might be helpful to see how this works in a specific instance. Take $F = Q(\omega)$, where $\omega = (-1 + i\sqrt{3})/2$ is a primitive cube root of unity. If K is the splitting field of $x^3 - 2 \in F[x]$, then from Example 5.31,

$$K = Q(\omega, \sqrt[3]{2}) = F(\sqrt[3]{2}).$$

In light of the theorem just proved, the Galois group $G(K/F)$ is isomorphic to a subgroup of the additive group Z_3. But the minimum polynomial of $\sqrt[3]{2}$ over F is $x^3 - 2$; hence

$$o(G(K/F)) = \deg(x^3 - 2) = 3,$$

leaving us with $G(K/F) \simeq Z_3$.

Any element of $G(K/F)$ is determined by what it does to $\sqrt[3]{2}$. Thus the automorphisms of K that fix F are ι, the identity mapping, and σ and τ, where

$$\sigma(\sqrt[3]{2}) = \omega\sqrt[3]{2}, \qquad \tau(\sqrt[3]{2}) = \omega^2\sqrt[3]{2}.$$

A direct calculation gives

$$\sigma^3(\sqrt[3]{2}) = \omega\sigma^2(\sqrt[3]{2}) = \omega^2\sigma(\sqrt[3]{2}) = \omega^3\sqrt[3]{2} = \sqrt[3]{2}.$$

In a similar way we get $\tau^3(\sqrt[3]{2}) = \sqrt[3]{2}$. This allows us to conclude that $\sigma^3 = \tau^3 = \iota$; and we perceive, a little more explicitly, that $G(K/F)$ is a cyclic group of order 3.

Every beginning student in algebra is familiar with the quadratic formula, which asserts that the roots of the quadratic equation $x^2 + bx + c = 0$ are given by

$$x = \frac{-b \pm \sqrt{b^2 - 4c}}{2}.$$

This formula was essentially known to the Greeks of 300 B.C. in terms of a geometric construction; in fact, the Babylonians had methods for solving the equation as early as 1800 B.C. Cubic equations were considered by the late Greek mathematician Diophantus (ca. 250), but no significant progress was made on their general solution until the sixteenth century, the great age of elementary algebra. The Italian mathematicians Scipione del Ferro (ca. 1515) and Niccolò Tartaglia (ca. 1535) obtained a complete solution of the general cubic $x^3 + px + q = 0$. (An arbitrary cubic $x^3 + ax^2 + bx + c = 0$ can be reduced, by introducing a new variable $y = x + a/3$, to the simpler form $y^3 + py + q = 0$.) Their method first appeared in Girolamo Cardano's treatise *Ars Magna*, printed in Nuremberg in 1545. It is for this reason that Cardano's name has traditionally been associated with the formula

$$x = \sqrt[3]{-q/2 + \sqrt{d}} + \sqrt[3]{-q/2 - \sqrt{d}}. \qquad \text{where } d = p^3/27 + q^2/4,$$

for the roots of a cubic equation. The next step was taken by Ludovico

Ferrari (1522–1565), a pupil of Cardano who showed that a solution of the general quartic equation can be obtained by first solving a certain cubic; so here too there is an algebraic formula yielding the roots. Cardano had the pleasure of publishing this discovery also in his *Ars Magna.*

Needless to say, mathematicians now directed their efforts toward the quintic equation, with the intent of finding an analogous solution in radicals. They were looking for a general "formula" so that when the coefficients of any given quintic were inserted, out would come the roots, exactly as in the known cases of the quadratic, cubic, and quartic. The spectacular success of the Italian algebraists of the sixteenth century held out strong hope that the desired solution to the quintic was near at hand; but repeated failures led to the gradual realization that such a formula for expressing the roots might, just possibly, not exist. As yet this conviction could not be proved either. In the years between 1798 and 1813, Paolo Ruffini (1765–1822) offered five deficient proofs on the impossibility of solving the general quintic equation by a radical expression. The question was finally settled in 1824, when the gifted young mathematician Niels Abel (1802–1829) succeeded in filling the gaps that Ruffini was not able to eliminate from his argument. Abel published, at his own expense, a six-page pamphlet dealing with the impossibility proof. Gauss tossed the pamphlet aside unread, grumbling, "Here is another of those monstrosities." The appearance two years later of a more elaborate version of Abel's result in the first issue of Crelle's *Journal* encouraged Galois's perseverance in his own investigations. It was Galois who finally established group-theoretic criteria for determining which equations of a given degree admit a solution in radicals. One of the most important and beautiful theories in algebra, it remained unnoticed in Galois's short lifetime and was not published until 1846, fourteen years after his death.

Before we enter into a discussion of Galois's results, it is essential to explain just what is meant by saying that a polynomial equation is solvable by radicals. For this, consider the polynomial

$$f(x) = c_0 + c_1x + \cdots + c_nx^n \in F[x],$$

where F is a field of characteristic zero. Informally, the equation $f(x) = 0$ should be solvable by radicals provided that all the roots of $f(x)$ can be expressed in terms of the coefficients $c_0, c_1, \ldots, c_n$ by using a finite sequence of rational operations (addition, subtraction, multiplication, division by non-zero elements) and taking roots of various orders. The four arithmetic operations can be carried out without leaving the field F to which the coefficients belong; but if the solution calls for the extraction of an nth root of an element $r \in F$, then most likely it will not be possible to do this within F. To perform the operation, the field F will have to be enlarged by adjoining the required nth root of r. This leads to a new field $F' \supseteq F$ whose members can be expressed rationally in terms of $\sqrt[n]{r}$ and the elements of F. If the solution involves taking another root not in F', then it will be necessary to perform another adjunction. The process is continued until we arrive at an extension field that contains all the elements that are involved in the solution of the equation $f(x) = 0$. In other words, a sequence of field extensions of F is built up by successively adjoining roots of previously constructed elements until $f(x)$ splits in some extension.

For instance, the roots of the sixth-degree equation

$$x^6 - 6x^4 + 12x^2 - 13 = (x^2 - 2)^3 - 5 = 0$$

can be given in terms of the radical expression $\sqrt{2 + \sqrt[3]{5}}$. To construct an extension of Q that contains this expression, we first adjoin $\sqrt[3]{5}$ to get a field containing $2 + \sqrt[3]{5}$; then to this field we adjoin $\sqrt{2 + \sqrt[3]{5}}$ to produce a second extension. The roots of the equation would lie in the top field of the sequence of fields

$$Q \subseteq Q(\sqrt[3]{5}) \subseteq Q(\sqrt[3]{5})(\sqrt{2 + \sqrt[3]{5}}).$$

Our next step is to formalize the intuitive ideas expressed above. We do this in terms of the concept of a radical extension.

Definition 5.14 An extension F' of a field F is called a *radical extension* if there is a finite sequence of fields

$$F = F_0 \subseteq F_1 \subseteq F_2 \subseteq \cdots \subseteq F_r = F'$$

such that $F_i = F_{i-1}(a_i)$ with $a_i^{n_i} = b_i \in F_{i-1}$ for some $n_i \in Z_+$ for $i = 1, 2, \ldots, r$; that is, a_i is a root of $x^{n_i} - b_i \in F_{i-1}[x]$.

It is important to notice that the definition does not require the radical extension F' to be a normal extension of F. However, if $F = F_0 \subseteq F_1 \subseteq \cdots \subseteq F_r$ is a radical extension of F, then it is always possible to construct a radical extension $F = F'_0 \subseteq F'_1 \subseteq \cdots \subseteq F'_s$ of F with $F_r \subseteq F'_s$ and F'_s normal over F. Let us accept this without proof.

Additional terminology is needed. For a nonconstant polynomial $f(x) \in F[x]$, we say that the equation $f(x) = 0$ is *solvable by radicals* over F when the splitting field of $f(x)$ is contained in some radical extension F' of F. Since $f(x)$ splits in $F'[x]$, the implication is that any root can be represented (in the notation of Definition 5.14) as a rational function of $a_1, a_2, \ldots, a_r$ where a_i is an n_ith root of an element of F_{i-1} (that is, $a_i^{n_i} \in F_{i-1}$, whence $a_i = \sqrt[n_i]{a_i^{n_i}}$ is a "radical"); this expresses the informal idea of obtaining an explicit "formula" for the solutions of $f(x) = 0$.

Finally, let us recall that by the Galois group of a polynomial $f(x) \in F[x]$ is meant the group $G(K/F)$ where K is the splitting field of $f(x)$ over F. According to Corollary 2 of Theorem 5.38, the Galois group of $f(x)$ can be identified in a natural way with a permutation group: If $f(x)$ has n

distinct roots in the splitting field K, then $G(K/F)$ is isomorphic to a subgroup of the symmetric group S_n. This is essentially the way that Galois originally defined the "group of an equation," the key concept in his theory.

Equipped with these definitions, we now establish a remarkable theorem of Galois connecting the solvability of equations with group theory. It appears in a manuscript "Memoir on the Conditions for Solvability of Equations by Radicals," dated January 16, 1831; this is the manuscript that was retrieved after Galois's death and first published by Liouville in his *Journal de Mathématiques* in 1846. It is to Liouville's credit that he recognized the ground-breaking character of Galois's work and preserved it to the profit of all ages.

For simplicity we assume in the theorem that F is a field that contains the nth roots of unity for every integer n (or as many such roots as are needed).

THEOREM 5.46 (Galois) *Let the nonconstant polynomial $f(x) \in F[x]$, where F is a field of characteristic zero. If the equation $f(x) = 0$ is solvable by radicals, then the Galois group of $f(x)$ is a solvable group.*

Proof. Let K be the splitting field of $f(x)$ over F; then $G(K/F)$ is the Galois group of $f(x)$. Assume that $f(x) = 0$ is solvable by radicals. This would mean that there is a radical extension of F,

$$F = F_0 \subseteq F_1 \subseteq F_2 \subseteq \cdots \subseteq F_r,$$

such that $K \subseteq F_r$; moreover, $F_i = F_{i-1}(a_i)$, where $a_i^{n_i} \in F_{i-1}$ for $i = 1, 2, \ldots, r$. Without loss of generality, F_r may be taken to be a normal extension of F; as such, it is a normal extension of every intermediate field F_i. Furthermore, each F_i is a normal extension of F_{i-1}, being the splitting field of $x^{n_i} - a_i^{n_i} \in F_{i-1}[x]$.

Next consider the descending sequence of subgroups

$$G(F_r/F) = G(F_r/F_0) \supseteq G(F_r/F_1) \supseteq G(F_r/F_2) \supseteq \cdots \supseteq G(F_r/F_r) = \{\iota\}.$$

This is a normal chain for $G(F_r/F)$, since by the corollary to the Fundamental Theorem of Galois Theory, $G(F_r/F_i)$ is a normal subgroup of its predecessor $G(F/F_{i-1})$ for each i. In addition, the same corollary asserts that there is an isomorphism

$$G(F_i/F_{i-1}) \simeq G(F_r/F_{i-1})/G(F_r/F_i).$$

Theorem 5.45 assures us that $G(F_i/F_{i-1})$ is a commutative group, since it is isomorphic to a subgroup of the additive group Z_n. Hence the associated quotient groups of our chain are all commutative, making it a solvable chain for $G(F_r/F)$. All of this combines to tell us that $G(F_r/F)$ is a solvable group.

The solvability of the Galois group $G(K/F)$ can now be

deduced without difficulty. We have $F \subseteq K \subseteq F_r$, with K a normal extension of F (being the splitting field of $f(x)$ over F). Another application of the previously mentioned corollary gives

$$G(K/F) \simeq G(F_r/F)/G(F_r/K).$$

Because $G(K/F)$ is isomorphic to a quotient group of the solvable group $G(F_r/F)$, Theorem 3.29 implies that $G(K/F)$ is itself solvable. This is exactly what is claimed in the theorem.

It should be pointed out that the solvability of the Galois group is a necessary and sufficient condition for the solvability of the polynomial equation by radicals; we have preferred to give the proof in the easiest direction. Further, the restriction that the base field F contain all the roots of unity can be dropped. If F does not contain enough suitable roots of unity, we build the radical extension not on F alone, but on F with the relevant roots of unity adjoined. The argument is outlined, very roughly, as follows: Let

$$F = F_0 \subseteq F_1 \subseteq \cdots \subseteq F_r$$

be a radical extension of F such that F_r contains the splitting field K of $f(x)$ over F. It is understood that $F_i = F_{i-1}(a_i)$ with $a_i^{n_i} \in F_{i-1}$ for $i = 1, 2, \ldots, r$. To each field in this sequence, adjoin a primitive nth root of unity, which we shall call ω, where $n = \operatorname{lcm}(n_1, \ldots, n_r)$. For $i = 1, 2, \ldots, r$, put $F_i' = F_i(\omega) = F_{i-1}(a_i, \omega)$. This leads to a radical extension

$$F(\omega) = F_0' \subseteq F_1' \subseteq \cdots \subseteq F_r'$$

in which the bottom field $F(\omega)$ contains the necessary roots of unity. F_r' is a normal extension of $F(\omega)$; if $g(x) \in F[x]$ is the polynomial that splits in K, then $(x^n - 1)g(x)$ splits in F_r'. The argument employed in Galois's theorem shows that $G(F'/F(\omega))$ is a solvable group. Now $G(F(\omega)/F)$ is solvable, being a commutative group (Theorem 5.43); and by the Fundamental Homomorphism Theorem,

$$G(F(\omega)/F) \simeq G(F_r'/F)/G(F_r'/F(\omega)).$$

The net result is that $G(F_r'/F)$ has a solvable normal subgroup whose quotient group is also solvable. When this happens, Theorem 3.30 ensures that $G(F_r'/F)$ is a solvable group. Finally, being a subgroup of $G(F_r'/F)$, $G(K/F)$ must also be solvable.

To find a polynomial that is not solvable by radicals, it suffices to produce one whose Galois group is not solvable. In our final theorem we shall indicate how to do this for a special type of polynomial with coefficients taken from the field Q of rational numbers.

THEOREM 5.47 *Let p be a prime number and $f(x) \in Q[x]$ be an irreducible polynomial of degree p. If $f(x)$ has exactly $p - 2$ real roots in the complex field C, then the Galois group of $f(x)$ is S_p.*

Proof. Suppose that K denotes the splitting field of $f(x)$ over Q, so that $G(K/Q)$ is the Galois group of $f(x)$. As was already noted, $G(K/Q)$ may be identified with a subgroup of the group S_p of permutations of the roots of $f(x)$. We shall show that $G(K/Q)$ contains a p-cycle and a transposition; since these two permutations generate S_p (Problem 30, Section 2.6), it will follow that $G(K/Q) = S_p$.

Suppose that $r_1, r_2, \ldots, r_p$ are the roots of $f(x)$ in K, where r_1 and r_2 are the two complex, nonreal roots. Because the coefficients of $f(x)$ are rational, the complex conjugate $\bar{r}_1$ of r_1 is also a root of $f(x)$; hence $\bar{r}_1 = r_2$. Now $[Q(r_1):Q] = \deg f(x) = p$, since r_1 is a root of an irreducible polynomial over Q. From the equality $[K:Q] = [K:Q(r_1)][Q(r_1):Q]$ it follows that p divides $[K:Q]$. Notice too that K is a normal extension of Q, so by Theorem 5.39, $[K:Q] = o(G(K/Q))$. All of this allows us to conclude that the prime p divides the order of the Galois group $G(K/Q)$. Appealing to Cauchy's Theorem, we find that $G(K/Q)$ contains an element of order p. This element must be a p-cycle, since these are the only members of S_p having order p.

Finally, to see that $G(K/Q)$ also contains a transposition, observe that the conjugation mapping $\bar{\sigma}: K \to K$ defined by $\bar{\sigma}(a) = \bar{a}$ is an automorphism of K, leaving Q fixed; $\bar{\sigma}$ interchanges the complex roots r_1 and r_2, while sending each of the $p - 2$ other real roots to themselves. Thus $\bar{\sigma}$ may be regarded as the transposition $(1\ \ 2)$, thereby establishing the theorem.

We are now in position to exhibit a polynomial equation that is not solvable by radicals; all that is required is to find a quintic polynomial in $Q[x]$ satisfying the conditions of Theorem 5.47. In fact,

$$f(x) = x^5 - 10x + 5$$

is one such polynomial. The irreducibility of $f(x)$ over Q follows immediately from Eisenstein's criterion. With a little assist from Descartes's "rule of signs," the number of real roots of $f(x)$ can be determined. This result asserts that there cannot be more positive roots than the number of changes of sign presented by the coefficients of $f(x)$—namely, two—and that there cannot be more negative roots than the number of changes of sign presented by the coefficients $f(-x) = -x^5 + 10x + 5$—namely, one. Thus $f(x)$ has at most three real roots. But $f(-2) < 0$, $f(-1) > 0$, $f(1) < 0$, and $f(2) > 0$, so there exist three real roots of multiplicity 1 lying in the intervals $[-2, -1]$, $(-1, 1)$ and $[1, 2]$. It turns out that the Galois group of $f(x)$ is S_5, which we know is not a solvable group (Theorem 3.32). Hence by Galois's theorem the equation $f(x) = 0$ is not solvable by radicals over Q.

PROBLEMS 5.5

1. For a given integer $n > 1$, let $\omega_1, \omega_2, \ldots, \omega_n$ be the nth roots of unity in a field F. Prove that:

a) $\omega_1 + \omega_2 + \cdots + \omega_n = 0$. [*Hint:* If $\omega \neq 1$ is an nth root of unity, then $1 + \omega + \omega^2 + \cdots + \omega^{n-1} = 0$.]

b) $\omega_1 \omega_2 \ldots \omega_n = (-1)^{n-1}$.

2. Show that if ω is an nth root of unity, then ω is a primitive mth root of unity for some m where $m|n$. [*Hint:* Pick m to be the smallest integer in the set $\{k \in Z_+ | \omega^k = 1\}$.]

3. a) Let p be an odd prime and ω be a primitive pth root of unity in C. Prove that ω is algebraic over Q with minimum polynomial $f(x) = x^{p-1} + x^{p-2} + \cdots + x + 1$.

b) Establish that $Q(i) = Q(\omega)$, where ω is a primitive fourth root of unity in C.

4. If K is the splitting field for $x^n - 1 \in Q[x]$, verify that $[K:Q] = \phi(n)$.

5. Assume that K is the splitting field of $x^{12} - 1 \in Q[x]$.

a) Prove that the Galois group $G(K/Q) \simeq (\{1, 5, 7, 11\}, \cdot_{12})$.

b) If ω is a primitive 12th root of unity, describe each of the four automorphisms in $G(K/Q)$ by its effect on ω.

c) Show that $\sigma^2 = i_K$ for each $\sigma \in G(K/Q)$; hence $G(K/Q) \simeq K_4$.

6. For any prime p, prove that the Galois group of $x^p - 1 \in Q[x]$ is a cyclic group of order $p - 1$.

7. a) Show that $x^{10} - 1$ and $x^8 - 1$ over Q have isomorphic Galois groups; in fact, each group is isomorphic to K_4.

b) Show that $x^9 - 1$ and $x^7 - 1$ over Q have isomorphic Galois groups; in fact, each group is isomorphic to Z_6.

8. Establish each of the assertions below:

a) If K is the splitting field of $x^5 - 1 \in Q[x]$, then there is a unique intermediate field E, $K \supseteq E \supseteq F$, such that $[E:Q] = 2$.

b) If K is the splitting field of $x^6 - 1 \in Q[x]$, then there is no intermediate field E with $K \supset E \supset Q$.

c) If K is the splitting field of $x^{11} - 1 \in Q[x]$, then there is an intermediate field E that is normal over Q and satisfies $[E:Q] = 5$.

9. Let K be the splitting field of $x^7 - 1 \in Q[x]$. Find all subgroups of $G(K/Q)$ and intermediate fields between K and Q.

10. Suppose that ω is a primitive 16th root of unity in C.

a) Verify that the Galois group $G(Q(\omega)/Q) \simeq Z_8$.

b) Find all the subfields of $Q(\omega)$.

c) Obtain a primitive element for each subfield of $Q(\omega)$ and determine its minimum polynomial over Q.

11. The *nth cyclotomic polynomial* $\Phi_n(x)$ is the monic polynomial whose roots are the $\phi(n)$ complex primitive nth roots of n; that is,

$$\Phi_n(x) = \prod_{\substack{\gcd(k,n)=1\\ 1 \leqslant k \leqslant n}} (x - \omega^k)$$

where ω is a primitive nth root of unity in C. Prove the following:

a) For any $n \geqslant 1$,

$$x^n - 1 = \prod_{d|n} \Phi_d(x),$$

where the product is over all the divisors d of n. [*Hint:* ω^k is an n/dth root of unity if and only if $\gcd(k, n) = d$.]

b) If p is a prime, then $\Phi_p(n) = x^{p-1} + \cdots + x^2 + x + 1$. [*Hint:* Use part (a).]

c) If $n > 1$ is an odd integer, then $\Phi_{2n}(x) = \Phi_n(-x)$. [*Hint:* Note that $x^{2n} - 1 = -(x^n - 1)((-x)^n - 1)$.]

d) For any $n \geqslant 1$, $\Phi_n(x) \in Z[x]$; that is, $\Phi_n(x)$ has integer coefficients. [*Hint:* Use induction on n; for $n > 1$, write $x^n - 1 = f(x)\Phi_n(x)$, where $f(x) = \prod_{\substack{d|n\\ d<n}} \Phi_d(x)$.]

e) $\deg \Phi_n(x) = \phi(n)$.

12. Problem 11(a) makes it possible to calculate $\Phi_n(x)$ recursively by division:

$$\Phi_1(x) = x - 1$$

$$\Phi_n(x) = \frac{x^n - 1}{\prod_{\substack{d|n\\ d<n}} \Phi_d(x)} \qquad \text{for } n > 1.$$

Use this method to exhibit $\Phi_n(x)$ for $n = 4, 6, 8, 10$, and 12. (For $n < 105$, each of the coefficients of $\Phi_n(x)$ is either 1, 0, or -1.)

13. Assume that $\Phi_n(x)$ is irreducible over Q. If K is the splitting field of $\Phi_n(x)$, prove that $G(K/Q) \simeq Z_n^*$. [*Hint:* $\Phi_n(x)$ and $x^{n-1} - 1 \in Q[x]$ have the same splitting field.]

14. Show that the Galois group of $x^3 - 1 \in Q[x]$ is cyclic, while the Galois group of $x^3 - 2 \in Q[x]$ is not.

15. Let K be the splitting field of the polynomial $x^n - a \in F[x]$, where $a \neq 0$. Prove that K contains all the nth roots of unity. [*Hint:* If $r_1, r_2, \ldots, r_n$ are the n distinct roots of $x^n - 1$ in K, consider $\omega_k = r_1 r_k^{-1}$ for $k = 1, 2, \ldots, n$.]

16. Describe the Galois group of $x^7 - 2$ over $Q(\omega)$, where ω is a primitive seventh root of unity in C.

17. a) Find a radical extension of Q that contains the element $(\sqrt[3]{7} - 3\sqrt{2})/\sqrt[4]{5}$.

b) Show that $Q(\sqrt{2}, \sqrt[3]{\sqrt{2} - 2})$ is a radical extension of Q.

18. For $a \in Q_+$, prove that the polynomial $x^n - a \in Q[x]$ is solvable by radicals over Q. [*Hint:* The splitting field is $K = Q(\sqrt[n]{a}, \omega)$, where ω is an nth root of unity in C.]

19. Prove that every nonconstant polynomial in $R^{\#}[x]$ is solvable by radicals over $R^{\#}$. [*Hint:* The field C is algebraically closed.]

20. Verify that the following polynomials are not solvable by radicals over Q:

a) $x^5 - 9x + 3$;

b) $x^5 - x - 1$;

c) $x^5 + 5x^4 - 5$;

d) $x^5 - 10x^4 + 2x^3 - 24x^2 + 2$. [*Hint:* The polynomial has derivative $x(5x^2 + 6)(x - 8)$.]

Appendix: Zorn's Lemma

In this appendix we give a brief account of some of the axioms of set theory, with the primary purpose of introducing Zorn's Lemma. Our presentation is descriptive, and most of the facts are merely stated.

We begin by defining another type of relation that occurs in various branches of mathematics, the so-called *partial order relation*. Just as equivalence generalizes equality, this relation (as we define it below) generalizes the idea of "less than or equal to" on the real line.

DEFINITION A.1 A relation R in a nonempty set S is called a *partial order* in S if the following three conditions are satisfied:

1. aRa (reflexive property).
2. If aRb and bRa, then $a = b$ (antisymmetric property).
3. If aRb and bRc, then aRc (transitive property),

where a, b, c denote arbitrary elements of S.

From now on, we shall follow custom and adopt the symbol $\leqslant$ to represent a partial-order relation, writing $a \leqslant b$ in place of aRb; the foregoing axioms then read (1) $a \leqslant a$; (2) if $a \leqslant b$ and $b \leqslant a$, then $a = b$; and (3) if $a \leqslant b$ and $b \leqslant c$, then $a \leqslant c$. As a linguistic convention, let us also agree to say (depending on the circumstance) that "a precedes b" or "a is a predecessor of b," or "b succeeds a," or "b is a successor of a" if $a \leqslant b$ and $a \neq b$. By a *partially ordered set* is meant a pair $(S, \leqslant)$ consisting of a set S and a partial order relation $\leqslant$ in S. In practice, one tends to ignore the second component and simply speak of the partially ordered set S or, when more precision is required, say that S is partially ordered by $\leqslant$.

If A is a subset of a partially ordered set S, then the ordering of S restricted to A is a partial ordering of A, called the *induced partial order*; in this sense, any subset of a partially ordered set becomes a partially ordered set in its own right. When considering subsets of a partially ordered set as partially ordered sets, it is always the induced order that we have in mind.

Let S be a set partially ordered by the relation $\leqslant$. Two elements a and b of S such that either $a \leqslant b$ or $b \leqslant a$ are said to be *comparable*. In view of the reflexivity of a partial order, each element of S is comparable to itself. There is nothing, however, in Definition A.1 that ensures the comparability of every two elements of S. Indeed, the qualifying adverb "partially" in the phrase "partially ordered set" is intended to emphasize that there may exist pairs of elements in the set that are not comparable.

DEFINITION A.2 A partial order $\leqslant$ in a set S is termed *total* (sometimes simple, or linear) if any two elements of S are comparable; that is, $a \leqslant b$ or $b \leqslant a$ for any two elements a and b of S.

A partially ordered set $(S, \leqslant)$ whose relation $\leqslant$ constitutes a total order in S is called a *totally ordered set* or, for short, a *chain*.

Let us pause to illustrate some of the preceding remarks.

Example A.1

In the set $R^{\#}$ of real numbers, the relation $\leqslant$ (taken with the usual meaning) is the most natural example of a total ordering.

Example A.2

Given the set Z_+ of positive integers, define $a \leqslant b$ if and only if a divides b. This affords a partial ordering of Z_+, which is not total; for instance, the integers 4 and 6 are not comparable, since neither divides the other.

Example A.3

Let S be the collection of all real-valued functions defined on a nonempty set X. If $f \leqslant g$ is interpreted to mean $f(x) \leqslant g(x)$ for all $x \in X$, then $\leqslant$ partially, but not totally, orders S.

Example A.4

For a final illustration, consider the set $P(X)$ of all subsets of a set X. The relation $A \leqslant B$ if and only if $A \subseteq B$ is a partial ordering of $P(X)$, but not a total ordering provided that X has at least two elements. For example, if $X = \{1, 2, 3\}$, and $A = \{1, 2\}$, $B = \{2, 3\}$, then neither $A \leqslant B$ nor $B \leqslant A$ holds. As regards terminology, any family of sets ordered in this manner will be said to be *ordered by inclusion*.

Let $(A, \leqslant)$ and $(B, \leqslant)$ be two partially ordered sets (when there is no danger of confusion, we write $\leqslant$ for the partial orders in both A and B). A mapping $f: A \to B$ is said to be *order-preserving* or an *order-homomorphism* if for all $a, b \in A$, $a \leqslant b$ implies that $f(a) \leqslant f(b)$ in B. A one-to-one order-homomorphism f of the set A onto B whose inverse is also an order-homomorphism from B onto A is an *order-isomorphism*. If such a function exists, we say that the two partially ordered sets $(A, \leqslant)$ and $(B, \leqslant)$ are *order-isomorphic*. When the partial order is the primary object of interest and the nature of the elements plays no essential role, order-isomorphic sets can be regarded as identical.

The next theorem emphasizes the fundamental role of our last example on partially ordered sets (Example A.4); it allows us to represent any partially ordered set by a family of sets.

THEOREM A.1 *Let A be a set partially ordered by the relation $\leqslant$. Then A is order-isomorphic to a family of subsets of A, partially ordered by inclusion.*

Proof. For each $a \in A$, let $I_a = \{x \in A | x \leqslant a\}$. It is not hard to verify that the mapping $f: A \to P(A)$ defined by $f(a) = I_a$ is an order-homomorphism of A into $P(A)$. Indeed, if $a \leqslant b$, then the condition $x \leqslant a$ implies $x \leqslant b$, and therefore $I_a \subseteq I_b$, or, equivalently, $f(a) \subseteq f(b)$. To see that f is one-to-one, suppose that $a, b \in A$ are such that $f(a) = f(b)$. Then the element $a \in I_a = I_b$, and hence $a \leqslant b$ by definition of I_b; likewise, $b \leqslant a$, from which it follows that $a = b$. Finally, the inverse f^{-1} is also order-preserving. If the inclusion $I_a \subseteq I_b$ holds, then $a \in I_b$ and so $a \leqslant b$. These calculations make it clear that A is order-isomorphic to a certain set of subsets of $P(A)$.

COROLLARY *For no set A is A order-isomorphic to $P(A)$.*

Proof. We argue that if $f: A \to P(A)$ is any order-homomorphism from A into $P(A)$, then f cannot map onto $P(A)$. For purposes of contradiction, assume that f does carry A onto $P(A)$. Define $B = \{a \in A | a \notin f(a)\}$ and $B^* = \{c \in A | c \leqslant a \text{ for some } a \in B\}$.

By supposition the set $B^* = f(b)$ for some element $b \in A$. If $b \notin B^*$, then, according to the definition of B, $b \in B \subseteq B^*$, a contradiction. Hence $b \in B^*$, and so $b \leqslant a$ for some a in B. From the order-preserving character of f we must have $B^* = f(b) \subseteq f(a)$. But then $a \in B \in B^* \subseteq f(a)$. The implication is that $a \notin B$, which is again a contradiction.

In an ordered set there are sometimes elements with special properties that are worth mentioning.

DEFINITION A.3 Let S be a set partially ordered by the relation $\leqslant$. An element $x \in S$ is said to be a *minimal* (*maximal*) *element* of S if $a \in S$ and $a \leqslant x$ ($x \leqslant a$) imply that $a = x$.

In other words, x is a minimal (maximal) element of S if no element of S precedes (exceeds) x. It is not always the case that a partially ordered set possesses a minimal (maximal) element, and when such an element exists, there is no guarantee that it will be unique.

Example A.5
The simplest illustration of a partially ordered set without minimal or maximal elements is furnished by the set $R^{\#}$ with the ordering $\leqslant$ in the usual sense.

Example A.6
In the collection $P(X) - \{\varnothing\}$ of all nonempty subsets of a nonempty set X (partially ordered by set-theoretic inclusion), the minimal elements are those subsets consisting of a single element.

Example A.7
Consider the set S of all integers greater than 1 and the partial order $\leqslant$ defined by $a \leqslant b$ if and only if a divides b. In this setting, the prime numbers serve as minimal elements.

It is technically convenient to distinguish between the notion of a minimal (maximal) element and that of a first (last) element.

DEFINITION A.4 Let S be a set partially ordered by the relation $\leqslant$. An element $x \in S$ is called the *first* (*last*) *element* of S if $x \leqslant a (a \leqslant x)$ for all $a \in S$.

Let us point out immediately the important distinction between first (last) elements and minimal (maximal) elements. Definition A.4 asserts that the first (last) element of a partially ordered set S must be comparable to every element of S. On the other hand, as Definition A.3 implies, it is not required that a minimal (maximal) element be comparable to every element of S, only that there be no element in S that precedes (exceeds) it. A minimal (maximal) element has no predecessors (successors), whereas a first (last) element precedes (succeeds) every element. Clearly, any first (last) element is a minimal (maximal) element, but not conversely.

First (last) elements of partially ordered sets are unique, if they exist at all. Indeed, suppose that the partially ordered set $(S, \leqslant)$ has two first elements, say x and y; then $x \leqslant y$ and $y \leqslant x$, so $x = y$ by the antisymmetric property. Thus x is unique, and we are justified in using the definite article when referring to *the* first element of S. A similar argument holds for last elements.

Let us introduce some additional terminology pertaining to partially ordered sets.

DEFINITION A.5 Let S be a set partially ordered by the relation $\leqslant$ and let A be a subset of S. An element $x \in S$ is said to be a *lower* (*upper*) *bound* for A if $x \leqslant a$ ($a \leqslant x$) for all $a \in A$.

We emphasize that a lower (upper) bound for a subset A of a partially ordered set is not required to belong to A itself. If A happens to have a first (last) element, then the same element is a lower (upper) bound for A; conversely, if a lower (upper) bound for A is contained in the set A, then it serves as the first (last) element for A. Note, too, that a lower (upper) bound for A is a lower (upper) bound for any subset of A.

A subset of a partially ordered set need not have upper or lower bounds (just consider $Z \subseteq R^{\#}$ with respect to $\leqslant$), or it may have many. For an example of this latter situation, one may turn to the family $P(X)$ of all subsets of a set X, the order being given by the inclusion relation; an upper bound for a subfamily $\mathscr{A} \subseteq P(X)$ is any set containing $\cup\mathscr{A}$, while a lower bound is any set contained in $\cap\mathscr{A}$.

As we know, a given partially ordered set need not have a first element, and if it does, some subset could very well fail to possess one. This prompts the following definition: A partially ordered set $(S, \leqslant)$ is said to be *well-ordered* if every nonempty subset $A \subseteq S$ has a first element ("with respect to $\leqslant$" being understood). The set Z_+ is well-ordered by the usual $\leqslant$; each nonempty subset has a first element, namely, the integer of smallest magnitude in the set.

Note that any well-ordered set $(S, \leqslant)$ is in fact totally ordered. Each subset $\{a, b\} \subseteq S$ must have a first element. According to whether the first element is a or b, we see that $a \leqslant b$ or $b \leqslant a$, whence the two elements a and b are comparable. Going in the other direction, any total ordering of a finite set is a well-ordering of that set. Let it also be remarked that a subset of a well-ordered set is again well-ordered by the restriction of the ordering.

Example A.8

Consider the Cartesian product $S = Z_+ \times Z_+$. We partially order S as follows: If (a, b) and (a', b') are ordered pairs of positive integers, $(a, b) \leqslant (a', b')$ means (1) that $a < a'$ (in the usual sense) or (2) that $a = a'$ and $b \leqslant b'$. This is called the *lexicographic order* of $Z_+ \times Z_+$ because of its resemblance to the way in which words are arranged in a dictionary. For instance, $(4, 8) \leqslant (5, 2)$, while $(3, 5) \leqslant (3, 9)$. To confirm that $\leqslant$ is a well-ordering of S, let $\varnothing \neq A \subseteq S$ and define $B = \{a \in Z_+ | (a, b) \in A\}$. Since A is a nonempty subset of Z_+, it has a first element, call it a_0. Now let $C = \{b \in Z_+ | (a_0, b) \in A\}$. Again the well-ordering of Z_+ under $\leqslant$ guarantees that C has a first element, say b_0. We leave it to the reader to become convinced that the pair (a_0, b_0) serves as the first element of A, thereby making S a well-ordered set relative to $\leqslant$.

A fundamental axiom of set theory, which has a surprising variety of logically equivalent formulations, is the so-called Well-Ordering Theorem of Zermelo (1904). The designation "theorem" notwithstanding, we take this to be an axiom—assumed and unproved—of our system. We state:

ZERMELO'S THEOREM *Any set S can be well-ordered; that is, there is a partial ordering $\leqslant$ for S such that $(S, \leqslant)$ is a well-ordered set.*

Accepting the existence of such orderings, we do not pretend at all to be able to specify them. Indeed, nobody has ever "constructed" an explicit function that well-orders an uncountable set. Moreover, the promised well-ordering may bear no relation to any other ordering that the given set may already possess; the well-ordering of $R^{\#}$, for instance, cannot coincide with its customary ordering.

Zermelo based the "proof" of his classical Well-Ordering Theorem on a seemingly innocent property whose validity had never been questioned and that has since become known as the axiom of choice. To state this axiom, we first need the definition of a choice function.

DEFINITION A.6 Let $\mathscr{C}$ be a (nonempty) collection of nonempty sets. A function $f : \mathscr{C} \to \cup\mathscr{C}$ is called a *choice function* for $\mathscr{C}$ if $f(A) \in A$ for every set A in $\mathscr{C}$.

Informally, a choice function f can be thought of as "selecting" from each set $A \in \mathscr{C}$ a certain representative element $f(A)$ of that set. As a simple illustration, there are two distinct choice functions f_1 and f_2 for the family of nonempty subsets of $\{1, 2\}$:

$$f_1(\{1, 2\}) = 1, \qquad f_1(\{1\}) = 1, \qquad f_1(\{2\}) = 2,$$
$$f_2(\{1, 2\}) = 2, \qquad f_2(\{1\}) = 1, \qquad f_2(\{2\}) = 2.$$

The question arises as to whether this selection process can actually be carried out when $\mathscr{C}$ has infinitely many members. The possibility of making such choices is handled by the axiom mentioned above:

AXIOM OF CHOICE *Every collection $\mathscr{C}$ of nonempty sets has at least one choice function.*

Since this general principle of choice has a way of slipping into proofs unnoticed, the reader should become familiar with its disguised forms.

For instance, one often encounters the following wording: If $\{X_i\}$ is a family of nonempty sets indexed by the nonempty set $\mathscr{I}$, then the Cartesian product $\times_{i \in \mathscr{I}} X_i$ is nonempty (it should be clear that the elements of $\times X_i$ are precisely the choice functions for $\{X_i\}$). For another common phrasing, which again expresses the idea of selection, let $\mathscr{C}$ be a collection of disjoint, nonempty sets. The axiom of choice, as we have stated it, is equivalent to asserting the existence of a set S with the property that $A \cap S$ contains exactly one element, for each A in $\mathscr{C}$.

Granted Zermelo's Theorem, we can see that a choice function f can be defined for any collection $\mathscr{C}$ of nonempty sets: Having well-ordered $\cup\mathscr{C}$, simply take f to be the function which assigns to each set A in $\mathscr{C}$ its first element. As was indicated earlier, Zermelo's Theorem was originally derived from the axiom of choice, so these are in reality two equivalent principles (although they are seemingly quite different).

Although the axiom of choice may strike the reader as being intuitively obvious, the soundness of this principle has aroused more philosophical discussion than any other single question in the foundations of mathematics. At the heart of the controversy is the ancient problem of existence. Some mathematicians believe that a set exists only if each of its elements can be designated specifically, or at the very least if there is a rule by which each of its members can be constructed. A more liberal school of thought is that an axiom about existence of sets may be used if it does not lead to a contradiction. In 1938, Godel demonstrated that the axiom of choice is not in contradiction with the other generally accepted axioms of set theory, assuming that the latter are consistent with one another. It was subsequently established by Cohen (1963) that the denial of this axiom is also consistent with the rest of set theory. Thus the axiom of choice is in fact an independent axiom whose use or rejection is a matter of personal inclination. The feeling among most mathematicians today is that the axiom of choice is harmless in principle and indispensable in practice—provided that one calls attention to the occasions of its use. It is also valuable as an heuristic tool, since every proof by means of this assumption represents a result for which we can then seek proofs along other lines.

A nonconstructive criterion for the existence of maximal elements is given by the so-called "maximality principle," which is generally cited in the literature under the name Zorn's Lemma. From the point of view of priority this principle goes back to Hausdorff and Kuratowski, but Zorn gave a formulation of it that is particularly suitable to algebra; he was also the first to state, without proof, that a maximality principle implies the axiom of choice.

ZORN'S LEMMA *Let S be a nonempty set partially ordered by $\leqslant$. Suppose that every subset $A \subseteq S$ that is totally ordered by $\leqslant$ has an upper bound in S. Then S possesses at least one maximal element.*

Zorn's Lemma is a particularly handy tool when the underlying set is partially ordered and the required object of interest is characterized by maximality. To demonstrate how it is used in practice, let us prove what is sometimes known as Hausdorff's Theorem (recall that by a chain is meant a totally ordered set):

THEOREM A.2 *Every partially ordered set contains a maximal chain, that is, a chain that is not a proper subset of any other chain.*

Proof. Consider the collection $\mathscr{C}$ of all chains of a partially ordered set $(S, \leqslant)$; $\mathscr{C}$ is nonempty, since it contains the chains consisting of single elements of S. Partially order $\mathscr{C}$ by inclusion and let $\mathscr{A}$ be any chain of $\mathscr{C}$ for the ordering $\subseteq$. We maintain that the union $\cup\mathscr{A}$ belongs to $\mathscr{C}$. Given elements $a, b \in \cup\mathscr{A}$, we have $a \in A \in \mathscr{A}$ and $b \in B \in \mathscr{A}$, for some A, B. Since $\mathscr{A}$ is a chain, either $A \subseteq B$ or $B \subseteq A$; suppose, for convenience, that $A \subseteq B$. Then a, b both lie in B, and since B is itself a chain (in S), it follows that $a \leqslant b$ or $b \leqslant a$. In consequence, any two elements of $\cup\mathscr{A}$ are comparable, making $\cup\mathscr{A}$ a chain in S. Since $\cup\mathscr{A}$ is clearly an upper bound for $\mathscr{A}$ in $\mathscr{C}$, Zorn's Lemma implies that $(\mathscr{C}, \subseteq)$ has a maximal member.

As another brief application of Zorn's Lemma, consider the following assertion: If $(S, \leqslant)$ is a partially ordered set, every chain of which has an upper bound, then for each $a \in S$ there exists a maximal element $x \in S$ with the property that $a \leqslant x$; in other words, a maximal element exists that is larger than the given element. For a proof of this, first observe that the set $J_a = \{y \in S | a \leqslant y\}$ satisfies the hypotheses of Zorn's Lemma (under the restriction of $\leqslant$); hence it has a maximal element x. But x is maximal in S, not merely in J_a. Suppose that $s \in S$ with $x \leqslant s$. Then $a \leqslant s$ (since both $a \leqslant x$ and $x \leqslant s$), and so $s \in J_a$. From the maximality of x in J_a it then follows that $s = x$, completing the argument.

Needless to say, we could just as well have phrased Zorn's Lemma in terms of lower bounds and minimal elements. The assertion in this case is that at least one minimal element in S exists.

Before concluding, let us state the following theorem.

THEOREM A.3 *Zermelo's Well-Ordering Theorem, the Axiom of Choice, and Zorn's Lemma are all equivalent.*

The deduction of these equivalences is somewhat involved, and the argument is not presented here; the interested reader can find the proofs in any number of texts on set theory.

Selected References

Our purpose here is to provide a list of suggestions for collateral reading and further study. Those works that are classified as general references roughly parallel the content of this book. The specialized sources more fully develop topics that are mentioned in the text and will carry the reader considerably beyond his or her present knowledge.

ABSTRACT ALGEBRA IN GENERAL

1. Allenby, R. B., *Rings, Fields and Groups: An Introduction to Abstract Algebra.* London: Edward Arnold, 1983.
2. Ames, D., *An Introduction to Abstract Algebra.* Scranton, Pa.: International, 1969.
3. Birkhoff, G., and MacLane, S., *A Survey of Modern Algebra* (4th ed.). New York: Macmillan, 1977.
4. Block, N., *Abstract Algebra with Applications.* Englewood Cliffs, N.J.: Prentice-Hall, 1987.
5. Burton, D., *Abstract and Linear Algebra.* Reading, Mass.: Addison-Wesley, 1972.
6. Clark, A., *Elements of Abstract Algebra.* Belmont, Calif.: Wadsworth, 1970.
7. Cohn, P. M., *Algebra* (2nd ed.), Vols. 1 and 2. New York: Wiley, 1982.
8. Crown, G., Fenrick, M., and Valenza, R., *Abstract Algebra.* New York: Marcel Dekker, 1977.
9. Dean, R., *Elements of Abstract Algebra.* New York: Wiley, 1967.

10. Fraleigh, J., *A First Course in Abstract Algebra* (3rd ed.). Reading, Mass.: Addison-Wesley, 1982.

11. Gallian, J., *Contemporary Abstract Algebra.* Lexington, Mass.: Heath, 1986.

12. Gilbert, J., and Gilbert, L., *Elements of Modern Algebra.* Boston: Prindle, Weber & Schmidt, 1984.

13. Gilbert, W., *Modern Algebra with Applications.* New York: Wiley, 1976.

14. Goldstein, L., *Abstract Algebra: A First Course.* Englewood Cliffs, N.J.: Prentice-Hall, 1973.

15. Hall, F. M., *An Introduction to Abstract Algebra*, Vols. 1 and 2. Cambridge: Cambridge University Press, 1969.

16. Herstein, I. N., *Abstract Algebra.* New York: Macmillan, 1986.

17. ———, *Topics in Algebra* (2nd ed.). Lexington, Mass.: Xerox Publishing, 1975.

18. Hillman, A., and Alexanderson, G., *A First Undergraduate Course in Abstract Algebra* (3rd ed.). Belmont, Calif.: Wadsworth, 1983.

19. Hungerford, T., *Algebra.* New York: Springer-Verlag, 1974.

20. Jacobson, N., *Basic Algebra I* (2nd ed.). San Francisco: Freeman, 1985.

21. ———, *Lectures in Abstract Algebra*, Vols. 1, 2, and 3. New York: Van Nostrand, 1951, 1953, 1964.

22. Jones, B., *An Introduction to Modern Algebra.* New York: Macmillan, 1975.

23. Kuczkowski, J., and Gersting, J., *Abstract Algebra: A First Look.* New York: Marcel Dekker, 1977.

24. MacLane, S., and Birkhoff, G., *Algebra* (2nd ed.). New York: Macmillan, 1979.

25. McCoy, N., *Introduction to Modern Algebra* (3rd ed.). Boston: Allyn and Bacon, 1975.

26. ———, and Berger, T., *Algebra: Groups, Rings and Other Topics.* Boston: Allyn and Bacon, 1977.

27. Marcus, M., *Introduction to Modern Algebra.* New York: Marcel Dekker, 1978.

28. Moore, J. T., *Introduction to Abstract Algebra.* New York: Academic Press, 1975.

29. Paley, H., and Weichsel, P., *A First Course in Abstract Algebra.* New York: Holt, Rinehart and Winston, 1966.

30. Pinter, C., *A Book of Abstract Algebra.* New York: McGraw-Hill, 1982.

31. Saracino, D., *Abstract Algebra: A First Course.* Reading, Mass.: Addison-Wesley, 1979.

32. Schilling, O., and Piper, W. S., *Basic Abstract Algebra.* Boston: Allyn and Bacon, 1975.

GROUP THEORY

33. Aschbacker, M., *Finite Group Theory*. Cambridge: Cambridge University Press, 1986.
34. Budden, F. J., *The Fascination of Groups*. Cambridge: Cambridge University Press, 1971.
35. Dixon, J. D., *Problems in Group Theory*. Waltham, Mass.: Blaisdell, 1967.
36. Gorenstein, D., *Finite Groups*. New York: Harper and Row, 1968.
37. ———, *Finite Simple Groups*. New York: Plenum, 1982.
38. Hall, M., Jr., *The Theory of Groups* (rev. ed.). New York: Macmillan, 1959.
39. Kargapolov, M. I., and Merzljakov, Ju. I., *Fundamentals of Group Theory*. New York: Springer-Verlag, 1979.
40. Macdonald, I., *The Theory of Groups*. Oxford: Oxford University Press, 1968.
41. Miller, G. A., Blichfeldt, H. F., and Dickson, L. E., *Theory and Applications of Finite Groups*. New York: Dover, 1961.
42. Polites, G., *An Introduction to the Theory of Groups*. Scranton, Pa.: International, 1968.
43. Robinson, D., *A Course in the Theory of Groups*. New York: Springer-Verlag, 1982.
44. Rotman, J., *An Introduction to the Theory of Groups* (3rd ed.). Boston: Allyn and Bacon, 1984.
45. Scott, W., *Group Theory*. Englewood Cliffs, N.J.: Prentice-Hall, 1964.
46. Suzuki, M., *Group Theory*, Vols. 1 and 2. New York: Springer-Verlag, 1982, 1986.

RING THEORY

47. Atiyah, M., and Macdonald, I., *Introduction to Commutative Algebra*. Reading, Mass.: Addison-Wesley, 1969.
48. Behrens, E.-A., *Ring Theory*. New York: Academic Press, 1972.
49. Bhattacharya, P., and Jain, S., *First Course in Rings, Fields and Vector Spaces*. New York: Wiley, 1977.
50. Burton, D., *A First Course in Rings and Ideals*. Reading, Mass.: Addison-Wesley, 1970.
51. Gray, M., *A Radical Approach to Algebra*. Reading, Mass.: Addison-Wesley, 1970.
52. Herstein, I. N., *Noncommutative Rings*. Carus Monograph 15. Washington, D.C.: Mathematical Association of America, 1968.

53. Kaplansky, I., *Commutative Rings.* Boston: Allyn and Bacon, 1970.
54. Lambek, J., *Lectures on Rings and Modules.* Waltham, Mass.: Blaisdell, 1966.
55. Larsen, M., and McCarthy, P., *Multiplicative Theory of Ideals.* New York: Academic Press, 1971.
56. McCoy, N., *Rings and Ideals.* Carus Monograph 8. Washington, D.C.: Mathematical Association of America, 1948.
57. ———, *The Theory of Rings.* New York: Chelsea, 1973.
58. Northcott, D., *Ideal Theory.* Cambridge, England: Cambridge University Press, 1953.
59. ———, *Lessons on Rings, Modules and Multiplicities.* Cambridge, England: Cambridge University Press, 1968.
60. Ribenboim, P., *Rings and Modules.* New York: Interscience, 1969.
61. Zariski, O. and Samuel, P., *Commutative Algebra*, Vols. 1 and 2. New York: Van Nostrand, 1958.

FIELD THEORY

62. Adamson, I., *Introduction to Field Theory* (2nd ed.). Cambridge, England: Cambridge University Press, 1982.
63. Artin, E., *Galois Theory* (2nd ed.). South Bend, Ind.: University of Notre Dame Press, 1944.
64. Bastida, J., *Field Extensions and Galois Theory.* Reading, Mass.: Addison-Wesley, 1984.
65. Edwards, H., *Galois Theory.* New York: Springer-Verlag, 1984.
66. Hadlock, C., *Field Theory and Its Classical Problems.* Carus Monograph 19. Washington, D.C.: Mathematical Association of America, 1978.
67. Lidl, R. and Niederreiter, H., *Finite Fields.* Reading, Mass.: Addison-Wesley, 1983.
68. ———, *Introduction to Finite Fields and Their Applications.* Cambridge, England: Cambridge University Press, 1986.
69. McCarthy, P., *Algebraic Extensions of Fields.* Waltham, Mass.: Blaisdell, 1966.
70. Nagata, M., *Field Theory.* New York: Marcel Dekker, 1977.
71. Stewart, I., *Galois Theory.* London: Chapman and Hall, 1973.
72. Winter, D., *The Structure of Fields.* New York: Springer-Verlag, 1974.

HISTORY

73. Bell, E. T., *Development of Mathematics* (2nd ed.). New York: McGraw-Hill, 1945.
74. Boyer, C., *A History of Mathematics.* Princeton, N.J.: Princeton University Press, 1985.
75. Burton, D., *The History of Mathematics: An Introduction.* Boston: Allyn and Bacon, 1985.
76. Crowe, M., *A History of Vector Analysis.* New York: Dover, 1985.
77. Eves, H., *An Introduction to the History of Mathematics* (5th ed.). Philadelphia: Saunders, 1983.
78. Kline, M., *Mathematical Thought from Ancient to Modern Times.* New York: Oxford University Press, 1972.
79. Novy, L., *Origins of Modern Algebra.* Leyden, The Netherlands: Noordhoff, 1973.
80. Struik, D., *A Concise History of Mathematics.* New York: Dover, 1948.
81. Temple, G., *100 Years of Mathematics.* New York: Springer-Verlag, 1981.
82. Van der Waerden, B. L., *A History of Algebra: From al-Khowarizmi to Emmy Noether.* New York: Springer-Verlag, 1985.
83. Wussing, H., *The Genesis of the Abstract Group Concept.* Cambridge, Mass.: M.I.T. Press, 1984.

Index of Terms

Index of Notation

The following is by no means a complete index of all the notations used in the text, but is rather a listing of certain symbols which occur frequently. Numbers refer to the page on which the symbol in question is first found.

$a \equiv b \pmod{n}$ — Integer a is congruent to the integer b modulo n, 85
$a|b$, $a \nmid b$ — a divides (does not divide) b, 32
(a, b) — Ordered pair of a and b, 10
$\gcd(a, b)$ — Greatest common divisor of a and b, 32
$\operatorname{lcm}(a, b)$ — Least common multiple of a and b, 40
$f: A \to B$ — f is a function from A into B, 15
$\operatorname{dom} f$, $\operatorname{rng} f$ — Domain and range of the function f, 14
$f(A)$, $f^{-1}(A)$ — Direct and inverse image of the set A under the function f, 20
f^{-1} — Inverse of the function f, 18
$f|A$ — Restriction of the function f to the set A, 20
$f \circ g$ — Composition of the functions f and g, 16
i_A — Identity function on the set A, 19
$P(A)$ — Power set of the set A, 9
$\varnothing$ — Empty set, 4
$\phi(n)$ — Euler phi-function, 123
C — Set of complex numbers, 286
Q — Set of rational numbers, 3
$R^{\#}$, $R_{+}^{\#}$ — Set of real (positive real) numbers, 3
U — Universal set, 5
Z — Set of integers, 3
Z_e, Z_o, Z_+ — Set of even, odd, and positive integers, 52
Z_n — Set of integers modulo n, 89
$+_n$, $\cdot_n$ — Addition (multiplication) modulo n, 90, 247
$\simeq$ — Is isomorphic to, 158

NOTATIONS FROM GROUP THEORY

$[a, b]$ — Commutator of the elements a and b, 139
aH, $a + H$ — Coset of the subgroup H, 125
(a) — Cyclic subgroup generated by the element a, 114
a^{-1}, $-a$ — Inverse of the element a, 58
A_n — Alternating group on n symbols, 100
$\operatorname{Aut} G$ — Set of automorphisms of the group G, 152
$C(a)$ — Centralizer of the element a, 121
$\operatorname{cent} G$ — Center of the group G, 100
e — Identity element of a group, 56
f_a — Left-multiplication function induced by the element a, 162
G_s — Group of symmetries of the square, 63
G_p — Set of elements in the group G of p-power order, 202
G/H — Quotient group of G by the subgroup H, 136
$G \times G'$ — External direct product of groups G and G', 196
$[G, G]$ — Commutator subgroup of the group G, 139
$[G:H]$ — Index of the subgroup H in G, 129
HK — Product of sets H and K, 118
$H \otimes K$ — Internal direct product of subgroups H and K, 192
$\operatorname{Hom} G$ — Set of homomorphisms from G into itself, 152
ι — Identity permutation, 94
$\operatorname{Inn} G$ — Set of inner automorphisms of the group G, 175
K_4 — Klein 4-group, 132
$\ker f$ — Kernel of the homomorphism f, 155
nat_H — Natural mapping from G onto G/H, 156
$N_K(H)$ — Normalizer of the subgroup H in the subgroup K, 210
$N(H)$ — Normalizer of the subgroup H, 147
$o(G)$ — Order of the group G, 108
$o(a)$ — Order of the element a, 116
σ_a — Inner automorphism induced by the element a, 174
sgn — Signature (sign) function, 98
$\operatorname{Sym} G$ — Symmetric group of the set G, 101
S_n — Symmetric group on n symbols, 92

NOTATIONS FROM RING AND FIELD THEORY

na	Integral multiple of the element a, 255
(a)	Principal ideal generated by the element a, 271
char R	Characteristic of the ring R, 255
cont $f(x)$	Content of the polynomial $f(x)$, 359
deg $f(x)$	Degree of the polynomial $f(x)$, 346
$F(a)$	Subfield generated by the element a and field F, 403
$F(a_1, \ldots, a_r)$	Subfield generated by the elements $a_1, \ldots, a_r$ and field F, 403
$G(K/F)$	Galois group of the extension K over F, 437
$GF(p^n)$	Galois field with p^n elements, 429
$H^{\blacktriangle}$	Fixed field of the group H, 438
IJ	Product of the ideals I and J, 273
$I + J$	Sum of the ideals I and J, 272
$\sqrt{I}$	Nil radical of the ideal I, 315
$[K:F]$	Degree of the extension K over F, 419
$N(\alpha)$	Norm of the element α, 335
$M_n(R)$	Set of $n \times n$ matrices with entries from R, 248
$\Phi_n(x)$	nth cyclotomic polynomial, 364
map (X, R)	Set of mappings from X into R, 249
ϕ_r	Substitution homomorphism induced by the element r, 348
$Q(\sqrt{n})$	Quadratic number field, 334
$Q_{cl}(R)$	Classical field of quotients of R, 295
$R[a]$	Set of polynomials in the element a, 348
R^*	Set of invertible elements of R, 246
$R[x]$	Polynomial ring in one indeterminant over R, 345
rad R	Jacobson radical of R, 317
$Z1$	Set of multiples of the identity element, 257
$Z(i)$	Domain of Gaussian integers, 325
$Z(\sqrt{n})$	Domain of quadratic integers, 336
ω	nth root of unity, 454

NOTATIONS FROM LINEAR ALGEBRA

$(a_1, a_2, \ldots, a_n)$	Vector with components a_k, 378
(a_{ij})	Matrix with entries a_{ij}, 248
det A	Determinant of the matrix A, 67
dim V	Dimension of the vector space V, 396
δ_{ij}	The Kronecker delta, 249
e_i	Unit vectors, 388
E_{ij}	Matrix units, 395
$F_n[x]$	Set of polynomials over F of degree $\leqslant n$, 382
I_n	Identity matrix of order n, 249
$T_2(Z)$	Set of upper triangular matrices of order 2 over Z, 268
$[S]$	Subspace spanned by the set S, 383
$U + W$	Sum of subspaces U and W, 382
$V_n(F)$	Set of n-component vectors over F, 378
$V_\infty(F)$	Set of infinite sequences over F, 379
$[x_1, \ldots, x_n]$	Subspace spanned by the vectors $x_1, \ldots, x_n$, 385
$V(F)$	Vector space over F, 377

Index of Mathematicians